FORGING THE SWORD

Forging the Sword

Selecting, Educating, and Training Cadets and Junior Officers in the Modern World

Edited by

Elliott V. Converse III

Imprint Publications

Chicago

1998

Cover design and illustration by Chris Hureau

Library of Congress Catalog Card Number 98-75809
ISBN 1-879176-28-9 (Paper)

Military History Symposium Series of the United States Air Force Academy, Vol. 5
Carl W. Reddel, Series Editor

Printed in the United States of America

To the Association of Graduates
a vital link
with the past, the present, and the future
for
United States Air Force Academy Cadets

Contents

Preface

Without question the most demanding profession of the late twentieth century is public military service. No other professional role incorporates such a wide range of decision-making challenges posing profound implications for nations and their societies—nationally, regionally, and globally. In the twentieth century, military decisions and policies possessed potential for ethical, economic, social, and political ramifications far beyond those of the nineteenth century, when military officers were more narrowly focused and when weapons were far less lethal. Moreover, the trend of increasing complexity in decision-making seems likely to continue, as military forces must deal with advancing technology, evolving biological and chemical threats, and rapidly changing and, sometimes, disintegrating societies and nations.

What education and training can possibly prepare young men and women to meet these challenges successfully and to serve the security needs of their public sponsors? No single, formulaic answer to this question exists. However, the diverse essays of this unique volume survey the varied international experience in twentieth-century officer preparation and suggest implicitly, and in some cases explicitly, a wide range of possible principles and practices to guide policy- and decision-makers.

Comparative study of this diverse experience—both in its international dimension and nationally among different services—can perhaps enlighten all of us. Military officers from different cultures and nations have long understood the commonality of their primary shared concerns because risking life and safety for patriotic reasons reduces the significance of cultural differences. Courage is recognized and valued by military professionals no matter the color and cut of the uniform. It is also a universal characteristic of modern professions, including the military, that modernization has resulted in a need for increasingly refined expertise and training. Civilians in general, and some military leaders in particular, have lagged in their appreciation of this development's pertinence to the military profession. This is sometimes due to arguments about the relative virtues of brain and brawn, about the value of education over training, and about the need for independent decision-makers versus disciplined subordinates. These and other issues are debated publicly and privately with special heat and emotion because the stakes are high—survival—both for the individual and society. As a result, reasoned discourse sometimes falls by the wayside.

However, the historical record is clear. The process of developing the habits of critical thought and continuous learning in senior military leaders must begin with cadets and junior officers, the primary subjects of the essays in this volume. It is unlikely that the characteristics of the great leaders that military academies have traditionally produced, such as those possessed by George C. Marshall of the Virginia Military Institute and Dwight D. Eisenhower of the United States Military Academy, will be developed by later professional military education programs if the seeds have not already been planted and nurtured from the beginning of an officer's career.

All decisions about organizing and directing development in young people are complex, none more so than those involving the inherent contradictions of life-and-death decision-making. Especially for those in leadership positions, the military professional lives and works on the knife-edge of these paradoxical elements. Ethical training and character development programs, notably those which give primacy to process over performance, are by themselves insufficient to meet this challenge of decision-making on the knife-edge. Moreover, two historical impulses push military education and training programs in the direction of discipline and destructive capability. One is the requirement that the warrior be willing to sacrifice life itself for a higher cause—the *sine qua non* of the profession. Neither civilian nor military leaders, in either democracies or dictatorships, will compromise on this point. All civilian and military leaders require this destructive capability and for this reason training historically has received the nod over education. The other is that history has over time favored the growth of destructive capabilities, including weapons capable of near planetary destruction. For this reason, the statement usually attributed to Georges Clemenceau—"War is too important to be left to the generals"—received currency and resonance. Nonetheless, the generals and the admirals, for better and for worse, look after the conduct of warfare, and the success of their education, training, and performance is vital for the short- and long-term survival of the civilians of all nations.

Today, the resolution of these issues is complicated for the military professional by the transitional stresses of the end of the Cold War. Redefinition—professionally, institutionally, and personally—is difficult. At the end of World War II, redefinition for the American military resulted in the creation of a new service, the United States Air Force, and shortly thereafter, in a new experiment in officer and cadet education, the United States Air Force Academy. Both were formed in the face of a continentally based adversary, and the clear delineation of an enemy disciplined the minds and energies of their creators. Also, the rich base of success in modern global warfare informed the founders of both institutions with the authority of experience, if not of intellectual application. At the Air Force Academy, Brig. Gen. Robert F. McDermott's brilliant curricular synthesis stimulated positive and far-reaching changes in United States military education. In this regard the military profession and the military services in every country have their own stories, certainly worthy of examination if not emulation.

Redefinition of military forces now occupies not only leaders of the two major adversaries of the Cold War, but also the political and military leaders of other nations. Because of the essential ties of the military profession to the industrial and information revolutions, the historical tensions between human and technological resources in the military services have been accentuated with the general reduction of military forces throughout the world. By definition, education and training are focused on the human resource, always a critical, central dimension in human conflict. However, military forces focused for generations on the leading edge of technological development, as epitomized by nuclear weapons, may find it difficult to redefine the military professional as a public military servant in the new domestic and international context and to develop education and training programs to that end. Nonetheless, it must be done.

In the vortex of the post–Cold War period, it was fitting that the Air Force Academy hosted the international meeting leading to this volume. Embodying an extraordinary experiment in military professionalism after World War II, as well as in higher education focused on public service, the Air Force Academy was flexibly shaped. The historical experience of creating a new, innovative institution was a positive stimulus to holding the symposium which produced most of these essays. Indeed, conscious no doubt of the academy's recent founding, the efforts of its Department of History to hold such a meeting in the late 1970s, as the academy completed its first quarter-century, were still-born.

This symposium benefitted from the support of many individuals and diverse organizations. Lacking a single sponsor, the symposium organizers drew upon the support of its greatest long-term sponsor, the Air Force Academy's Association of Graduates, to whom this volume is dedicated. Significant financial assistance was also provided by the Falcon Foundation, the Dean of the Faculty, the Humanities division, and the Major Donald R. Backlund Memorial Fund.

Moreover, this symposium volume would not have been possible without the direct contributions of the faculty of the Air Force Academy's Department of History and other individuals from various staff agencies. Under the department's leadership, human, material, and financial resources were bought to bear on the largest military history symposium held at the academy since the inception of the series in 1967. Lieutenant Colonel Tony Kern provided superb managerial leadership as the symposium's executive director, and he was adroitly assisted by Capt. John Abbatiello, as the deputy director. Captain Derek Varble and Lt. Ed Petka completed the management team. Heather Emory of the Superintendent's office was particularly helpful on protocol matters. Debbie Tome provided sound financial guidance and LaWanda Fisher provided outstanding secretarial assistance.

Carl W. Reddel
Series Editor

Introduction

Elliott V. Converse III

After warfighting itself, military education and training, especially the selection and preparation of officers, are among the most important activities engaged in by armed forces around the world.[1] Since the seventeenth and eighteenth centuries, when these practices became formal in Europe, nations have gradually erected education and training systems designed to provide officers with the values, knowledge, and skills deemed essential for their profession. There is and has been much variety in these arrangements, but today they generally follow a common pattern. Precommissioning programs, most familiarly in military academies, but also through other sources such as officer-candidate schools or courses of instruction based in civilian secondary or tertiary educational institutions, begin the process. For newly commissioned officers, these programs are almost immediately followed by training courses, originally known as schools of application, that offer instruction in the great diversity of military specialties—infantry, aviation, missiles, submarines, maintenance, intelligence, and the like. At intermediate points in their careers, many officers then attend a staff college and, when they are more senior, a war college.

The purposes, structure, content, and adequacy of these programs have attracted wide interest from both within and outside the military. Traditionally, officers have observed and reported on the military education systems of other nations.[2] Many countries, seeking to modernize their armed forces, have not only copied other systems—initially Western European, later U.S. and Soviet models—but have also imported foreign officers to help organize and instruct in their programs. Late nineteenth-century China and Japan are perhaps the best known examples. Worldwide, articles on education and training by serving and retired officers, and occasionally by civilians, appear regularly in professional military journals. The Russian journal *Military Thought,* for example, dedicates a section to these subjects and, since Communism's demise, has published several articles arguing for radical new directions in Russian officer preparation, including the infusion of humanistic studies.[3] Military services also periodically assess their officer education and training. The U.S. Air Force alone, in the four decades after its birth as an independent service in 1947, has conducted 120 such evaluations.[4] Additionally, high-level government commissions, often following a nation's defeat or poor performance in war, have been charged with investigating the state of officer education. In Great Britain, in the latter half of the nineteenth century, there were four in less than fifty years—1855, 1856 (Crimean War), 1869, and 1902 (Boer War).[5] In the United States since the 1970s, there have been two national-level, comprehensive inquiries, as well as numerous assessments by the government's General Accounting Office.[6] Most recently, in

1

1997, the Congressional Research Service of the Library of Congress issued a report on the U.S. service academies, and the Center for Strategic and International Studies, a private Washington, D.C. "think tank," published one on the overall system of American professional military education.[7]

Relatively little scholarship on officer selection, education, and training came from within the academic community until after World War II. Especially since the 1960s, however, social scientists, historians, and scholars from the discipline of education have produced an extensive body of literature. For social scientists, analyses of officer recruitment (particularly concerning social origins) and of the subsequent education and training of officers have been central to their study of the military in the contemporary world—its characteristics as an institution and a profession, its relation to the state and to society, and its role in social and political change. Such seminal works as Samuel Huntington's *The Soldier and the State* (1957) and Morris Janowitz's *The Professional Soldier* (1960); and the scholarship of Jacques van Doorn, Charles Moskos, and many others, often appearing in the interdisciplinary journal *Armed Forces and Society* (founded in 1974), all reflect the significance social scientists have attached to officer selection, education, and training.

Before World War II, a few military historians (such as Hans Delbrück in Germany) treated military history in its larger institutional and sociopolitical contexts, but most emphasized wars, campaigns, battles, and the great military and naval commanders. With respect to officer education and training, there were many "histories" of cadet academies. These, however, were rarely written by professional historians and were largely celebratory rather than analytical, praising the schools' customs and traditions, rigor of routine, and the exploits of the cadets and their later achievements as officers. Two-thirds of A. F. Mockler-Ferryman's turn-of-the-century history of Sandhurst in Great Britain, for example, details the results of the military academy's athletic contests.[8]

Some scholars writing before World War II did examine the history of officer selection and preparation more deeply. They were an integral part of Michael Lewis's *England's Sea Officers* (1939), the first of his several books on the social history of the British Navy, and also of Karl Demeter's *The German Officer-Corps in Society and State* (first published in 1930).[9] Demeter showed how middle-class penetration during the nineteenth and early twentieth centuries of an officer corps dominated by the nobility resulted in a struggle in the realm of officer selection and education. Throughout this period, he argued, traditional aristocratic qualities of "character" competed with, and usually prevailed over, bourgeois qualities of "mind" or "intellect." (In Section III of this volume, Horst Boog analyzes this conflict from its nineteenth-century origins through the Nazi era to its resolution in the postwar *Bundeswehr*.)

Beginning in the 1960s, large numbers of military historians around the world turned away from the so-called "drum and trumpet" history and concentrated instead on "the interaction of war with society, economics, politics, and culture."[10] In the United States, the resulting scholarship has been called the "new military history." It was, as others have pointed out, not really new.[11] Moreover, traditional military history was hardly abandoned. Nonetheless, like their social science colleagues, many historians (sometimes applying analytical methods and techniques from the social sciences) began to write about the broader aspects of war and the military profession.

Studies exemplifying the new military history have greatly expanded our knowledge of officer social origins, selection, education, and training. The bibliography on these subjects prepared for the U.S. Air Force Academy's 17th Military History Symposium in 1996 lists more than 1,600 English-language titles for over fifty nations. Some represent the scholarship of other disciplines, but most are histories, primarily secondary works published since 1945.[12]

Despite the increase in scholarship on the history of officer education and training, very little is comparative history. Social scientists have been much more likely than historians to adopt a cross-national, comparative perspective. For both groups, such an approach requires overcoming significant obstacles. Problems presented by language, by the lack of contemporary data or trouble in gaining access to archival materials, and by the conceptual difficulties of comparing time and place are forbidding. Even though military technology or the emergence of the military as a profession may be unifying threads, social and political change has been nowhere quite the same. Cultural differences are distinct and persistent. National education systems have evolved differently in both nature and purpose. So have national armed forces. Thus, there has been and continues to be enormous diversity in the ways nations identify and prepare their officers.

Contrasting the systems used by the United States and Israel illustrates the point. In the United States an officer normally must have a college degree and is selected and trained separately from enlisted personnel. The U.S. officer also usually specializes during his or her career in one of a wide range of professional pursuits (many nearly indistinguishable from comparable civilian occupations) necessary for the armed services to fulfill global, yet increasingly amorphous, national military commitments. In Israel no formal educational qualification for commissioning exists. Potential officers are selected on the basis of ability from the conscript pool during basic military training and, appropriate to Israel's well-defined and limited military tasks, there are few specialists. (In Section III of this volume, Gunther Rothenberg describes the origins and development of the Israeli system.)

However problematic the effort, cross-national studies of officer education and training may be very useful. In a recent comparison of the value orientation of cadets in the military academies of thirteen nations, Joseph Soeters, a professor of sociology at the Royal Netherlands Military Academy, notes an increasing "internationalization in military life." The number of United Nations military operations is on the rise, and formation of the French/German brigade and the 1st German/Dutch Army Corps indicates a trend in Europe toward combined military forces. In such circumstances, he argues, commanders of multinational forces are likely to profit from knowing not only national differences in values among their officers, but also similarities transcending national borders that constitute an "international military culture."[13]

The only comparative, book-length history published since 1945, with officer preparation as its exclusive focus, is Martin van Creveld's *The Training of Officers* (1990), an analysis of the origins and development of advanced officer education in France, Germany, Great Britain, the Soviet Union, and the United States.[14] Three other comparative volumes give the history of officer education and training substantial treatment: Frederick M. Nunn's *Yesterday's Soldiers* (1983), and *The Time of the Generals* (1992); and David B. Ralston's *Importing the European Army* (1991), an examination of the introduction of European military techniques and institutions into Russia, the Ottoman Empire, Egypt, China, and Japan from 1600 to 1914.[15] In addition to these books, comparative articles by Correlli Barnett, Richard Preston, Theodore Ropp, and John Shy survey the evolution of cadet and junior officer education in Europe and the United States.[16] In the first chapter of *The Victorian Army and the Staff College, 1854–1914* (1972), Brian Bond also provides a comparative overview in the context of the emergence of general staffs and staff colleges in Europe.[17]

Nearly all of the extensive work on the history of officer selection, education, and training published in the last three decades has been national history. Although a few article-length surveys have been written, there is as yet no comprehensive, single-volume account in English of the history of officer preparation in any nation's armed forces, or even in an individual service.[18] To follow this history, one must turn to the many articles covering particular aspects of the subject; to books incorporating the topic in larger frameworks or themes such as general military and single-service histories or studies of the development of the military profession, officer corps, and civil-military relations; and finally, of course, to histories of specific officer education and training institutions.[19]

Neither military academy, as noted, nor staff and war college histories have been known for depth or sophistication of analysis. But, in keeping with the orientation of the new military history, quite a few high-quality studies have been written in the last quarter-century. John P. Lovell's comparative examination of the postwar U.S. service academies, *Neither Athens Nor Sparta?* (1979), is a good example. One reviewer commented that Lovell's blending of history and organization theory put the book "in a class above the normal drivel written about the academies."[20] Accounts of the antebellum U.S. Naval and Military academies by Charles Todorich and James L. Morrison, Jr., respectively, are also excellent.[21] So are Richard A. Preston's two volumes on Canada's Royal Military College; C. D. Coulthard-Clark's history of Duntroon, Australia's Royal Military College; J. E. O. Screen's study of the mid-nineteenth-century Helsinki Yunker School that readied both Russian and Finnish officer aspirants for commissions; B. P. N. Sinha and Sunil Chandra's volume on the Indian Military Academy; and John Moncure's book on the Royal Prussian Cadet Corps and colleges from 1871 to 1918.[22]

Measured by the percentage of officers supplied in both peace and war, military academies have received lopsided attention. Commissions obtained after service in military units or on ships at sea, and through other paths to commissions, originating in the nineteenth century and expanding significantly in the twentieth, have provided more officers. Yet we know far less about the preparation of prospective officers trained "on-the-job," in officer-candidate schools (especially during wartime), or in militias and reserve programs. (Several essays in this volume contribute to our knowledge of these alternate commissioning sources.) Still, permanently established military academies have been around for a long time—about three hundred years; they have a well-defined institutional status, making them easier to study; and their graduates have often been disproportionately influential (again in terms of relative numbers) in the upper ranks of armies, navies, and air forces. Moreover, the appearance of military academies in Europe reflected acceptance of the concept that officers should receive formal education and training in order to carry out their responsibilities.

During the sixteenth and seventeenth centuries, as armies equipped with firearms and mobile cannon grew in size, evolved more complex tactics, and began to apply scientific principles to fortification and siege, more officers with more skills were needed. Proposals that future officers be trained in separate academies, rather than as pages at court or in a warrior's household, appeared in the late sixteenth century, and several opened across Europe in the first quarter of the seventeenth century.[23] All, however, were short-lived, their development interrupted by the Thirty Years' War (1618–48). Over the next hundred years, as European monarchs fashioned "absolute" states, they maintained large standing armies, and sponsored schools (sometimes in the form of separate cadet bodies attached to regular military units) as a way to create loyal and uniformly trained officer corps drawn from the nobility (while allowing wealthy and educated bourgeoisie into the technical arms— the artillery and engineers).[24]

Although the vast majority of army officers would not receive any systematic training, precommissioning academies and schools to train officers became an established feature of eighteenth-century military life. Artillery and engineering schools came first. France, Prussia, and Russia had artillery schools, and the latter two nations engineering schools, by the end of the century's first decade. The British Royal Military Academy at Woolwich (the "Shop") was founded in 1741 to train artillery and engineering officers and France opened an engineering academy at Mézières in 1749 and an advanced artillery school at La Fère in 1756. At the same time, other institutions were founded to train officer candidates destined mainly for the cavalry and the infantry: the Royal Prussian Cadet Corps in 1717; the Noble Cadet Corps in Russia in 1732; the *École Royale Militaire* in France in 1753 (which closed in 1787, but was succeeded after the Revolution by the *École Spéciale Militaire* in 1803 [from 1808 at St.-Cyr]); the Austrian military academy at Wiener Neustadt in 1752; and the British Royal Military College in 1802 (from 1812 at Sandhurst), the same year as West Point in the United States.

Formal training for naval officers began at about the same time as that for army officers. In 1670, the French organized companies of naval cadets, the *gardes de la marine,* and schools located in port cities in which to train them.[25] Under Peter the Great, the Russians started a School of Navigation in 1698 and established a Naval Academy in 1715.[26] In some countries, however, shore-based instruction for naval officer-candidates took hold more slowly. The British Admiralty opened a naval academy at Portsmouth in 1729, but it turned out only a tiny fraction of the Royal Navy's officers and closed in 1837 (in part, because the navy hierarchy believed that future officers should be prepared through practical experience at sea). To provide some general education for naval cadets, who usually went to sea between the ages of twelve and fourteen, Royal Navy ships often had a schoolmaster on board beginning in the late eighteenth century.[27] The U.S. Navy also employed schoolmasters (normally the chaplain) aboard its ships, and schools on shore to train future naval officers existed as early as 1802. Nonetheless the American Navy, like many others, considered active service on a warship at sea as the way to learn the profession; a naval academy comparable to West Point was not established until 1845.[28]

Because artillerists and engineers required knowledge of science and mathematics, most historians have linked the emergence of military academies to those aspects of warfare's advancing technology.[29] Yet John Moncure, author of a history of Prussia's cadet corps, points out that academies for cadets, however brief their tenure, came into existence well before artillery and engineering schools. In his view, political rather than technological imperatives were more important in initiating formal education and training for officers. The catalyst, writes Moncure, was the monarch's desire to transfer the officer nobility's loyalty from the locality to the crown and state.[30] Additionally, social and cultural factors rather than the technology of war sometimes drove the kind of education given to academy cadets. The curriculum at the *École Royale Militaire,* for example, was heavily mathematical, even though the sons of the lesser nobility enrolled there were being groomed for commissions in the cavalry and infantry. The curriculum, argues the historian David Bien, reflected a conviction by army reformers that eighteenth-century French society was corrupt—that the prevailing literary strain of Enlightenment culture produced officers who were amateurs and dilettantes. The reformers held that studying mathematics disciplined the mind and would develop in cadets—shielded at the *École Royale Militaire* from the surrounding culture in Paris—qualities of logical and precise thinking and a seriousness of purpose essential for regenerating the French officer corps.[31] Similarly, James Pritchard, an expert on the eighteenth-century French Navy, maintains that the system of *gardes de la marine* was designed "to recruit officers, seal them off from the civilian world of dangerous ideas, and provide a renewed basis for aristocratic culture, not create scientifically minded officers."[32]

Political, social, cultural, technological, and other forces have influenced officer formation from the time that systematic preparation originated in seventeenth and eighteenth-century Europe to the present. Since these beginnings, the key issues involved in selecting, educating, and training officers have remained much the same:

Who should be an officer? Where should officer candidates be trained—on-the-job or in a separate environment? Should they come from one program or several? What should the content of education and training be? How much theory and how much practice? How much general education is necessary and from what disciplines? Where should general education be obtained—in a military or in a civilian institution? How much professional instruction is essential prior to commissioning? Finally, probably the most consequential—what values should officers have and how should they be imparted?

The twenty-seven essays in this volume show how nations around the world have answered these questions in the twentieth century. The volume follows both a chronological and topical arrangement. The essays in Sections I and II cover aspects of officer education and training in six different nations prior to World Wars I and II, while those in Section V focus on the United States after 1945. The essays in Sections III, IV, and VI address key topics: in Section III, the relationship between officership and formal education; in Section IV, the impact of radical political change on officer education and training; and in Section VI, the connection between officer formation and civil-military relations in several developing nations. The authors of the four essays in Section VII present personal views of how future generations of officers should be prepared in order to meet the demands of their profession. Many of the essays demonstrate how closely officer formation and society have been associated, and the adverse, even devastating, consequences that have resulted when the two become separated, as occurred at times in Germany, Japan, the Philippines, and Russia. Articles on Germany, Japan, and interwar France and Italy also illustrate what happens when balance has not been maintained in officer preparation between theory and practice or between narrowly professional and broader dimensions of military education. In drawing on historical or personal experience, all of the volume's essays indicate directions for cadet and junior officer education and training in the century to come.

I owe thanks to many fine people who helped me with this book—to Colonel Carl Reddel, Permanent Professor and Head of the U.S. Air Force Academy's history department and general editor of the Academy's Military History Symposium Series, who gave me the opportunity to be its editor and who collaborated as a friend and colleague throughout the publication process; to Anthony Cheung, head of Imprint Publications, whose considerable experience made my part of the job much easier; and to all of the authors whose outstanding scholarship makes up this book and who cheerfully and promptly responded to my numerous questions. Two longtime friends—Dr. Thomas Keaney of the U.S. National War College, and Dr. David Spires of the University of Colorado at Boulder—helped with the editing, as did Lt. Col. Lorry Fenner, U.S. Air Force, also of the National War College, and Lt. Col. Elliott Gruner, U.S. Army, of the Air Force Academy's English department. Kimberly A. Emerson Scott skillfully and rapidly did most of the word processing. When she returned to full-time teaching, Liz Jones, Maria Mobley, and Charmagne Moss just as ably completed the task. The history department's administrative staff—Susan Smith, Sharon Mitterer, and Anna Perry also did some of the word processing and assisted me in countless other ways. The talented Christopher Hureau, of the 10th Communications Squadron, designed the book's striking cover from our rather hazy ideas. Finally, I am grateful for the superb support given to me by the Air Force Academy Library—Dr. Edward Scott, its director; Duane Reed, head of Special Collections; and many members of the Library's staff, especially John Beardsley, Trudy Pollok, and Elwood White. For over two years, my office has been in Special Collections— there could not have been a more pleasant place to work.

Notes

1. I. B. Holley, in his essay in Section I of this volume, notes the different meanings of the terms "training" and "education," and provides working definitions, as do Philip Caine and Michael Neiberg in Section V. James Toner summed up the distinction in terms of skills and values: "Here's how one fires an M-16 rifle; don't shoot it at friends." See James H. Toner, *True Faith and Allegiance: The Burden of Military Ethics* (Lexington: University Press of Kentucky, 1995), 39–40. For an extended discussion, see Kenneth Lawson, "Introduction: The Concepts of 'Training' and 'Education' in a Military Context," in Michael D. Stephens, ed., *The Educating of Armies* (New York: St. Martin's Press, 1989).

2. Several of the late nineteenth- and early twentieth-century reports have been published. See, for example, Henry Barnard, *Military Schools and Courses of Instruction in the Science and Art of War, in France, Prussia, Austria, Russia, Sweden, Switzerland, Sardinia, England, and the United States*, rev. ed. (1872; New York: Greenwood Press, 1969); Charles P. Echols, *Report Upon Foreign Military Schools* (West Point, N.Y.: United States Military Academy Press, 1907); William B. Hazen, *The School and the Army in Germany and France* (New York: Harper, 1872); Military Education Commission, Great Britain, *Accounts of the Systems of Military Education in France, Prussia, Austria, Bavaria, and the United States* (London: George E. Eyre and William Spottiswoode, 1870); James R. Soley, *Report on Foreign Systems of Naval Education* (Washington, D.C.: Government Printing Office, 1880); and Emory Upton, *The Armies of Asia and Europe: Embracing Official Reports on the Armies of Japan, China, India, Persia, Italy, Russia, Austria, Germany, France, and England* (New York: Appleton, 1878).

3. See, for example, A. I. Karmanov, "Greater Role for Humanities in Higher Educational Establishments of Russia's Defense Ministry," *Military Thought* (English version) 6 (No. 1, 1997); and I. I. Barsukov, "Putting Humanities into Military Education: Experience and Problems," ibid. 3 (No. 7, 1993).

4. Richard L. Davis and Frank P. Donnini, eds., *Professional Military Education for Air Force Officers: Comments and Criticisms* (Maxwell Air Force Base, Ala.: Air University Press, 1991), 1.

5. Richard A. Preston, *Perspectives in the History of Military Education and Professionalism*, Harmon Memorial Lectures in Military History No. 22 (USAF Academy, Colo.: U.S. Air Force Academy, 1980), 17–19.

6. U.S. Department of Defense Committee on Excellence in Education (Clements Committee, 1975–76), and U.S. House Armed Services Committee Panel on Military Education (Skelton Committee, 1989).

7. Robert L. Goldfinch, *The DOD Service Academies: Issues for Congress* (Washington, D.C.: Congressional Research Service, Library of Congress, 1997); and Report of the CSIS Study Group on Professional Military Education, *Professional Military Education: An Asset for Peace and Progress* (Washington, D.C.: Center for Strategic & International Studies, 1997).

8. A. F. Mockler-Ferryman, *Annals of Sandhurst: A Chronicle of the Royal Military College* (London: Heinemann, 1900).

9. Michael Lewis, *England's Sea Officers: The Story of the Naval Profession* (London: Allen & Unwin, 1939); and Karl Demeter, *The German Officer-Corps in Society and State, 1630–1945*, trans. Angus Malcolm (New York: Praeger, 1965).

10. Peter Paret, "The New Military History," *Parameters* 21 (Autumn 1991): 10.

11. For discussions of the new military history, see ibid.; John Whiteclay Chambers II, "The New Military History: Myth and Reality," *Journal of Military History* 55 (July 1991); Edward M. Coffman, "The New American Military History," *Military Affairs* 48 (January 1984); Peter Karsten, "The 'New' American Military History: A Map of Territory Explored and Unexplored," *American Quarterly* 36 (1984); and Richard H. Kohn, "The Social History of the American Soldier: A Review and Prospectus for Research," *American Historical Review* 86 (June 1981).

12. Elliott V. Converse III and Elwood L. White, *The History of Officer Social Origins, Selection, Education, and Training Since the Eighteenth Century: An Introductory Bibliography*, United States Air Force Academy Library Special Bibliography Series No. 90, September 1996.

13. Joseph L. Soeters, "Value Orientations in Military Academies: A Thirteen Country Study," *Armed Forces and Society* 24 (Fall 1997).

14. Martin van Creveld, *The Training of Officers: From Military Professionalism to Irrelevance* (New York: Free Press, 1990).

15. Frederick M. Nunn, *Yesterday's Soldiers: European Military Professionalism in South America, 1890–1940* (Lincoln: University of Nebraska Press, 1983); idem, *The Time of the Generals: Latin American Professional Militarism in World Perspective* (Lincoln and London: University of Nebraska Press, 1992); and David B. Ralston, *Importing the European Army* (Chicago: University of Chicago Press, 1991).

16. Correlli Barnett, "The Education of Military Elites," *Journal of Contemporary History* 2 (July 1967); Preston, *Perspectives in the History of Military Education and Professionalism;* John Shy, "Western Military Education, 1700–1850," in Monte D. Wright and Lawrence J. Paszek, eds., *Science, Technology, and Warfare: Proceedings of the Third Military History Symposium, U.S. Air Force Academy, 8–9 May 1969* (Washington, D.C.: Government Printing Office, 1971); and Theodore Ropp, "The Persistence of Imperial Ideas and Institutions in the Selection, Education, and Training of Officers," in Béla K. Király and Walter Scott Dillard, eds., *The East Central European Officer Corps 1740–1920s: Social Origins, Selection, Education, and Training* (Highland Lakes, N.J.: Atlantic Research and Publications, 1988).

17. Brian Bond, *The Victorian Army and the Staff College, 1854–1914* (London: Eyre Methuen, 1972).

18. For article-length surveys on individual nations, see Robin Higham, "The Selection, Education, and Training of British Officers, 1740–1920," *East Central European Officer Corps;* Harold D. Langley, "Military Education and Training," in John E. Jessup, ed., *Encyclopedia of the American Military: Studies of the History, Traditions, Policies, Institutions, and Roles of the Armed Forces in War and Peace*, 3 vols. (New York: Scribner's, 1994), 3:1525–73; and Richard Woff, *The Armed Forces of the Former Soviet Union: Evolution, Structure and Personalities*, 2d ed., 3 vols. (London: Brassey's, 1996), 1:chaps. B21–23.

19. See Select Bibliography in this volume.

20. John P. Lovell, *Neither Athens Nor Sparta?: The American Service Academies in Transition* (Bloomington and London: Indiana University Press, 1979); and Allan R. Millett, review of ibid., *American Historical Review* 85 (February 1980): 235–36. An older, but also exceptional comparative history of the academies is William E. Simons, *Liberal Education in the Service Academies* (New York: Teachers College, Columbia University, 1965).

21. Charles Todorich, *The Spirited Years: A History of the Antebellum Naval Academy* (Annapolis, Md.: Naval Institute Press, 1984); and James L. Morrison, Jr., *"The Best School in the World": West Point, the Pre-Civil War Years, 1833–1866* (Kent, Ohio: Kent State University Press, 1986).

22. Richard A. Preston, *Canada's RMC: A History of the Royal Military College* (Toronto: University of Toronto Press, 1969); idem, *In the Service of Canada: History of the Royal Military College Since the Second World War* (Ottawa: University of Toronto Press, 1992); C. D. Coulthard-Clark, *Duntroon: The Royal Military College of Australia, 1911–1986* (Sydney: Allen & Unwin, 1986); J. E. O. Screen, *The Helsinki Yunker School, 1846–1879: A Case Study of Officer Training in the Russian Army* (Helsinki: Suomen Hist. Seura, 1986); B. P. N. Sinha and Sunil Chandra, *Valour and Wisdom: Genesis and Growth of the Indian Military Academy* (New Delhi: Oxford and I.B.H. Publishing, 1992); and John Moncure, *Forging the King's Sword: Military Education Between Tradition and Modernization—The Case of the Royal Prussian Cadet Corps, 1871–1918* (New York: Peter Lang, 1993).

23. See chaps. 8 and 10 of J. R. Hale, *Renaissance War Studies* (London: Hambledon Press, 1983); and idem, *War and Society in Renaissance Europe, 1450–1620* (Baltimore, Md.: Johns Hopkins University Press, 1985).

24. Three excellent surveys that include substantial material on officer selection, education, and training are: John Childs, *Armies and Warfare in Europe, 1648–1789* (New York: Holmes & Meier, 1982); André Corvisier, *Armies and Societies in Europe, 1494–1789*, trans. Abigail T. Siddall (Bloomington and London: Indiana University Press, 1979); and Christopher Duffy, *The Military Experience in the Age of Reason* (New York: Atheneum, 1988).

25. Frederick B. Artz, *The Development of Technical Education in France, 1500–1850* (Cambridge, Mass.: Society for the History of Technology and the M.I.T. Press, 1966), 52–55.

26. Woff, 1:chap. B21–1.

27. See F. B. Sullivan, "The Royal Academy at Portsmouth, 1729–1806," *Mariner's Mirror* 63 (November 1977); and idem, "The Naval Schoolmaster during the Eighteenth and the Early Nineteenth Century," *Mariner's Mirror* 62 (November 1976).

28. Christopher McKee, *A Gentlemanly and Honorable Profession: The Creation of the U.S. Naval Officer Corps, 1794–1815* (Annapolis, Md.: Naval Institute Press, 1991), 155–209.

29. See, for example, Barnett, 19; Shy, 60–68; and Van Creveld, 14.

30. Moncure, 26–27.

31. David D. Bien, "Military Education in 18th Century France: Technical and Non-Technical Determinants," in Wright and Paszek, 51–59.

32. James Pritchard, "The French Naval Officer Corps during the Seven Years' War," in Department of History, U.S. Naval Academy, ed., *New Aspects of Naval History: Selected Papers from the 5th Naval History Symposium* (Baltimore, Md.: Nautical and Aviation Publishing Company of America, 1985), 64.

Section I
Junior Officer Preparation and the Great War

Despite the large standing armies created by Europe's major powers and the plans to mobilize reserves, no country was or could be prepared for the staggering number of officers that World War I would demand. Great Britain alone, with a regular and reserve army officer corps of under 20,000 before the war started, granted nearly 230,000 commissions after it began.[1] In the combatant armies, subalterns constituted all but a small fraction of the officer casualties. In *The Guns of August,* Barbara Tuchman, with an eye for the telling detail, observed that one entry on the plaque memorializing the graduates of the French military academy, St.-Cyr, who had fallen in battle, said simply: To the Class of 1914.[2]

Essays in this section by Tim Travers, I. B. Holley, Jr., and Holger Herwig examine the preparation junior officers in Great Britain, the United States, and Germany had received on the eve of the conflict. Their assessments identify both strengths and weaknesses that have meaning for officer education and training today.

Travers, a professor of history at the University of Calgary, outlines the evolution of British Army officer formation from the eighteenth century to the beginning of the war, a pattern of development not unlike that in most other European nations. Appropriately for the first essay in this volume, he squarely confronts the central question of officer education and training then and now: What difference does it make on the battlefield? Travers's answer for the British Army's regular junior officers is that the formal education obtained at Sandhurst or elsewhere, although increasingly important by 1914, was less significant than the leadership, traditions, and quality of training in the officer's regiment. The best junior officers, he concludes, were those who possessed the "whole 'package' stemming from family background, school, Sandhurst, and a high *esprit de corps* and a well-commanded regiment."

In 1914, according to Travers, newly commissioned British officers lacked training in practical, professional subjects, notably the influence of firepower on tactics. I. B. Holley, a professor emeritus of history at Duke University, contends that the U.S. Military Academy was also deficient in imparting professional skills. This was left to the Artillery School, the Engineer School, or other professional schools that graduates attended after they left the academy. At West Point, the academic curriculum stressed the study of mathematics to develop the "mental discipline" thought to be the basis of military discipline. West Point officers performed well, however, not because of this doubtful educational philosophy, but because the academy succeeded in producing officers "who displayed character, a strict code of ethics, physical and moral courage, habits of obedience and punctuality."

In the German Army and Navy, writes Holger Herwig (also a professor of history at Calgary), an officer's character was defined in terms of the supposed attributes of a particular social class and consistently outweighed formal education. This concept of character also included mastery of professional skills, the skills of battle. The German Army's failure in 1914, he argues, was not the fault of the junior officers who were technically proficient and well trained, but of commanders at the highest strategic and operational levels, who had not received the broad education in politics, government, and economics that might have helped them to perceive the political, diplomatic, and social dimensions of the war. Like the junior officers, they were products of a narrowly technical system of professional military education and sought only tactical solutions to the bloody stalemate on the Western Front. Herwig shows that the lack of political-social education had another effect. When German soldiers and sailors deserted or openly rebelled in the fall of 1918, neither senior nor junior officers were able to restore discipline. The German revolution soon followed.

Notes

1. Keith Simpson, "The Officers," in Ian F. W. Beckett and Keith Simpson, eds., *A Nation in Arms: A Social Study of the British Army in the First World War* (Manchester, England: Manchester University Press, 1985), 64.

2. Barbara Tuchman, *The Guns of August* (New York: Dell, 1963), 488.

Learning the Art of War: Junior Officer Training in the British Army from the Eighteenth Century to 1914

Tim Travers

At the battle of Albuera, in 1811, the colors of the Buffs was carried by young Lt. Matthew Latham:

> He was attacked by several French hussars, one of whom . . . aimed a stroke at Latham's head, which . . . sadly mutilated him, severing one side of his face and nose; he still struggled with the hussar, and exclaimed, "I will surrender it only with my life." A second stroke severed his left arm and hand, in which he held the staff, from his body. He then seized the staff in his right hand, throwing away his sword, and continued to struggle with his opponents . . . when ultimately thrown down, trampled upon and pierced by the spears of the Polish lancers, his last effort was to tear the flag from the staff . . . and thrust it into the breast of his jacket.

Surprisingly enough, Latham recovered from his injuries, but this story illustrates one important theme that will reoccur, namely, a powerful regimental loyalty by junior officers. Another theme is the remarkable devotion to duty by most junior officers when battle was joined in the eighteenth century, frequently resulting in brutal and painful wounds, and equally brutal medical treatment, or of course, death.[1]

Such attitudes are surprising, given the fact that two-thirds of commissions in the eighteenth century were purchased, and that there was no training of junior officers before they joined their regiments. But were these attitudes enough for military efficiency? There were only a few private military academies in Britain and on the Continent in the eighteenth century, although the Royal Military Academy, Woolwich, for artillery and engineer officers, had been founded in 1741. The vast majority of young infantry officers learned the art of war from two sources. The first was a large literature of guide books published during the eighteenth century, for example, Capt. Samuel Bever's *The Cadet* (1756), James Wolfe's *Instructions for Young Officers* (1768), and Lewis Lochee's *Essay on Military Education* (1773). There were also the various official military regulations published by the Board of General Officers through the century. Most of this literature concerned itself with the prevailing emphasis on drill and tactics. The second source of training for the new officer was simply learning on the job. This depended on the attitude of each regiment, and took the form of learning how to perform the drill, movements, commands and posts of the day, and the platoon or alternate firings. The new officer would be trained by a field officer or captain, or by a senior noncommissioned officer (NCO) of the regiment, so that, as the 1755 general orders said, all "newly appointed" officers were to attend parade every morning, ask questions of their superiors, who would teach them their duties "and tell them their faults and omissions . . . and let them know they are not to have their Pay to be Idle."[2]

The great drawback of this form of training and education was that junior officers learned their trade mainly in their troops and companies, and only occasionally with the regiment as a whole, which was often dispersed. No one trained with larger brigade units, and certainly never with the entire army. The result was poor initial performance in war, and maybe two or three campaigns before the army and the young officers became thoroughly competent. It also appears that there were some problems of junior officer quality in the 1790s. The adjutant general, for example, wrote of the disastrous Flanders campaign that:

> there is not a young man in the Army that cares one farthing whether his commanding officer, the brigadier or the commander in chief approves his conduct or not. . . . . His friends [family] can give him a thousand pounds with which to go to the auctions in Charles Street and in a fortnight he becomes a captain. Out of the fifteen regiments of cavalry and twenty-six of infantry which we have here, twenty-one are commanded literally by boys or idiots. . . .

In addition, officers in the 1790s were apparently often absent without leave, and frequently drunk.[3] These problems refer particularly to the 1790s and after, when the Napoleonic wars required a large increase in junior officers, but are otherwise exaggerated, because the government could dismiss the incompetent (thus the officer would lose his considerable investment in his rank), and the great majority of purchased commissions produced career officers who took the army seriously.

Nevertheless, the 1790s do appear to be a low point of junior officer efficiency, just at the time when large numbers of new officers were needed. During the Napoleonic wars the officer corps increased from around 3,000 in 1793 to over 10,500 in 1814. In this situation many new officers were appointed, officially not younger than sixteen, with the only requirement being literacy, plus the recommendation of someone who held the rank of major or above. Although 20 percent of these officers came via the militia, founded in 1803, and a few from the Royal Military College, Sandhurst, founded in 1802, the relatively small numbers graduated from the college (four hundred in 1803) meant that 65 percent of junior officers in the infantry and cavalry were still educated through reading manuals, and then drilled on joining the regiment, as in the eighteenth century.[4]

During the Napoleonic wars, as before, much depended on the spirit and efficiency of the particular regiment. In this regard, experiences differed considerably. Ralph Heathcote, a new cornet in the Royals, described that regiment's routine:

> About nine o'clock the trumpets sound for foot parade, when the different troops being formed before the stable doors march toward the centre of the barrack yard, and after being formed in line are examined by the major (viz. their dress and arms are inspected); then the sergeant major exercises the regiment, with which we have nothing to do. At ten o'clock I breakfast with some others in the mess-room. . . . At eleven o'clock all the subalterns are to go to the riding school, but if you don't go,

no notice is taken of it, excepting you were perhaps to stay away for weeks altogether, and at twelve the same subalterns have to attend foot drill, and then your business is done for the day.

On the other hand, William Bell, an ensign in the Eighty-ninth Foot, wrote: "I have been out at six o'clock in the morning for some time past—since I joined the regiment. We are drilled with the men exactly the same as the private soldiers. We began with the facings and went through all the steps and evolutions of the marching squads. We were then exercised with the firelock." Similarly, in the Forty-third Regiment, a young officer was not "considered clear of the adjutant until he could put a company through all the evolutions by a word of command, which he had already learned in the ranks. It generally took him six months in summer, at four times a day (an hour to each period) to perfect him in all he had to learn." The net result of all this was that some young officers and regiments were efficient, and some were not. But, as before, the experience of war against Napoleon's armies was really the best training ground for young officers prior to the army's gradual professionalization in the nineteenth century. Young officers therefore started out with a great deal to learn—in the words of the historian Michael Glover: "The young officers of the army [of the Duke of Wellington in the Peninsula] were a mixed lot. Almost the only thing they had in common was . . . an almost total ignorance of their profession. When a general in the Portuguese service asked permission to have an ensign as his aide-de-camp, Wellington replied, 'An ensign can be of little use to him, or to anybody else.'"[5]

In the eighteenth century, therefore, the young officer taught himself via manuals, and was trained on the job, with varying results according to the nature of the regiment he joined and the experience of campaign. By the mid-nineteenth century, however, there were basically three attitudes to junior officer education: first, that officers were born in the middle and upper classes as natural leaders; second, that junior officers should be molded by early military college instruction; and third, that officers should be trained in their regiments after early military education. Because of the growing Victorian emphasis on education for professional occupations, the focus tended to shift toward the second perspective—that officers should have a military education, although this did not exclude further training in the regiment. The military education option was also particularly strengthened by the abolition of the purchase system in 1871. Before this, as long as the purchase system continued, there was no obvious need, except inability to purchase a commission, for the young officer to obtain training and credentials at the Royal Military College, Sandhurst (although Woolwich was a different matter). Thus, between 1834 and 1838, of 397 commissions given in the line, only 86 went to Sandhurst cadets, while between 1860 and 1867, 3,167 commissions were by purchase, and only 836 without. There was also considerable opposition to military education by those who favored the first view, that officers were born and not made. Hence the Duke of Cambridge echoed the general military sentiment by saying "the British officer should be a gentleman first and an officer second." Similarly, Lt. Col. W. H. Adams, an instructor at Sandhurst, told an 1855 committee that "military education is but little valued by the greater part of the high military authorities. They consider . . . whether a man is professionally educated or not, it will make not the slightest difference with regard to his qualities as an officer. . . ."[6]

Nevertheless, the problems of the Crimean War, the demands of the empire, the example of other armies on the Continent, and the previously mentioned climate of middle-class professionalism, gradually moved formal military education forward. Here, of importance was the introduction in 1849 of an examination for first commissions, and then, another examination for promotion to captain. Furthermore, Edward Cardwell, secretary of state for war, announced that beginning in 1873 all candidates for a commission, except militia officers, had to study for a year at Sandhurst. Thus by the end of the century, the army settled on formal schooling (Sandhurst or Woolwich) and regimental training for the military education of new officers, and this remained true until 1914. However, limiting the officer corps to young gentlemen of the middle and upper classes, and thus "natural" leaders, also remained true.[7]

Yet there was some way to go before junior officers were thoroughly trained at Sandhurst or elsewhere, even after the Cardwell reforms. Between 1876 and 1914, Sandhurst still trained only 55 percent of new officers, mainly in the infantry and cavalry—most of the rest coming through the back door of the militia—usually because they had failed to enter Sandhurst. As an indication of the low value of Sandhurst education, between 1873 and 1878, the six highest scoring candidates at the entrance exam were allowed to skip the Royal Military College altogether, and joined their regiments immediately. Similarly, only 13 out of 180 officers in the 1891 class at Sandhurst reached brigadier general, and only 3 of these gained that rank during World War I. However, a well-known example of an officer who did gain high rank and fame, but who criticized Sandhurst, was J. F. C. Fuller, the future tank theorist. He entered Sandhurst in 1897, after the almost obligatory period of cramming at Captain James's Academy in London. At Sandhurst, Fuller recalled:

> We studied Philip's *Manual of Field Fortification* and Clery's *Minor Tactics* which was nothing like so instructive as *Hume's Tactics*, published years before. . . . Military Law was a jest. Once a week for two hours at a stretch we sat in a classroom and read the Manual, and when we had exhausted those sections dealing with murder, rape and indecency, we either destroyed Her Majesty's property with our pen-knives or twiddled our thumbs. Fortunately, our instructor was as deaf as a post, for this enabled us to keep up a running conversation. . . .

On the other hand, another famous cadet, Winston Churchill, the future prime minister, having entered Sandhurst with some difficulty, found the college useful, and "a hard but happy experience."[8]

Yet the experience of the Boer War (1899–1902) showed there was much wrong with Sandhurst and the army, although the junior officers generally had done better than their seniors. Consequently a committee was appointed to consider the education and training of officers, which reported in 1902. The report generally approved of Woolwich, but found serious problems at Sandhurst. Among the deficiencies at the Royal Military College were the following: instructors saw their careers there as a "step down hill," and so had no inducement to teach well; staff sergeants drilled the cadets and ran the riding school and gym rather than officers; there was no musketry or revolver shooting, in fact cadets had to join a club at a pound a term to shoot, and although they pipeclayed their belts, the cadets had their rifles cleaned for them; tactics were studied in isolation from military topography and engineering; and exams emphasized detail rather than principles, resulting in cramming, while originality and practicality were neglected—instructors regarding "with horror any deviation from a sealed pattern." In general, the committee kept on coming back to the idea that study at Sandhurst should be *practical* rather than theoretical, as was then the case. The committee concluded that military education was in a very poor state, and that cadets avoided keenness, so "the idea is . . . to do as little as they possibly can." Cadets graduated with few technical or practical skills, and had no interest in studying their profession. Nor did matters improve after they joined their regiments because the promotion of the indolent was as rapid as the industrious due to seniority rules and the interest or favor of senior officers and other social connections. Moreover, subsequent regimental training for the junior officer did not generally happen because captains did not train the subalterns; instead the junior officer had to rely on NCOs, and this resulted in a lack of initiative on campaign. Finally, junior officers did not see the men they should train, as these were generally on fatigues or acting as orderlies and other duties. Consequently, the committee stated prophetically that in a future campaign "British troops must incur grave risks of disaster." The committee ended by making various recommendations such as less theory at Sandhurst, practical promotion exams, promotion by merit, less expense for junior officers (especially in the elite cavalry regiments), and the appointment of an inspector general of military education.[9]

It does not appear that many of these reforms actually occurred at Sandhurst. NCOs continued to drill the cadets (and still do), and the army remained expensive and socially exclusive. But Sandhurst did expand to a two-year course in 1905 and then back to eighteen months in 1906. In 1907 a future field marshal, Bernard Montgomery, entered Sandhurst. His syllabus for the first term appears the same as before the committee recommendations, while the second term did not even include musketry, drill, or gymnastics. By his own account, Montgomery did well at his studies, was promoted to lance corporal in the Sandhurst hierarchy, and played rugby for the college. In the second term, however, Montgomery recalls that his Company B began a war with Company A, and with other companies:

> Our company became known as "Bloody B." . . . Fierce battles were fought in the passages after dark; pokers and similar weapons were used and cadets often retired to hospital for repairs. . . . The climax came when during the ragging of an unpopular cadet I set fire to the tail of his shirt as he was undressing; he got badly burnt behind, retired to hospital, and was unable to sit down with any comfort for some time.

Montgomery's sin was discovered, and he was reduced in rank, and nearly rusticated. It is significant, too, that Montgomery's professional interest in military studies did not occur at Sandhurst, but took place later through discussions with a brother officer in 1913.[10]

The impression remains that Sandhurst as the preferred route to a regular commission in the infantry and cavalry before World War I was rather more a rite of passage than a really efficient preparation for war. However, as Montgomery's escapade illustrates, *esprit de corps* was high, and the cadets were tough and adventurous. It is also worth noting that leadership was not taught at Sandhurst; it was simply assumed that social background and public (private) school education before military college already provided the necessary leadership qualities. Then, following Sandhurst, there was theoretically the same effort at regimental training as in the two previous centuries. For some six months there was basic training, drilling on the square, musketry, and tactics. After this period, the new officer was passed off the square by the adjutant. As in the eighteenth and nineteenth centuries this obviously varied by regiment. According to the historian of the Second Scottish Rifles, based in Malta before 1914, attention was paid to marksmanship and fitness, but "In any theatre there was a tendency for tactical training to be sketchy; in Malta it was almost completely ignored." Junior officers read *Field Service Regulations, Part 1 (Operations),* but "tactics were something of an academic study for officers rather than a vital part of the military knowledge of all ranks." Then, as a counterpart to basic training with the regiment, in the officers' mess the new officer was unofficially trained as a "gentleman and an officer," and existed as the lowest form of human life. Consequently, the probationary period often proved an unpleasant shock. For example, in the Second Scottish Rifles, a new subaltern could not speak to a senior officer without being spoken to first for six months, and then had to wait three years before he could stand on the hearth rug in front of the mess fire. Robert Graves, when joining the Second Battalion of the Royal Welch Fusiliers in France, where the peacetime habit of ignoring the new officer for six months still prevailed, found himself verbally attacked by the second in command at his first meal in the mess: "You there, wart! Why the hell are your wearing your stars on your shoulder instead of your sleeve?" Graves explained this was the practice in the Welsh regiment with which he had been serving. The colonel replied: "As soon as you have finished your lunch you will visit the master-tailor. Report at the orderly room when you're properly dressed." After another error, the colonel ordered him to parade every morning before breakfast for a month for an hour's saluting drill. Graves noted: "This was not a particular act of spite against me, but an incident in the general game of 'chasing the warts' [new officers] at which all conscientious officers played, and honestly intended to make us better soldiers." On the other hand, J. F. C. Fuller seems to have received a friendlier reception from his regiment, but found the mess conversation limited to "foxes, duck and trout."[11]

The image of the immediate pre-1914 Sandhurst-trained junior officers is that of a traditional emphasis on social correctness, sports (especially hunting), *esprit de corps,* and loyalty to the regiment. In some regiments it appears to have been "bad form" to work and show keenness, and "good form" to indulge in sports. Thus

Second Lieutenant Macleod, in the Royal Regiment of Artillery, remarked that with enough horses he could, and occasionally did, hunt six days a week. Perhaps as a result of this kind of officer training, Brig. Gen. John Gough led-off the annual conference of staff officers at Camberley in January 1914 with the highly relevant topic: "Can more be done to render officers competent commanders in war?" Gough believed much more could be done, and in regard to junior officers, he felt that they should be given far more responsibility, especially in company training. Along the same lines, the historian Keith Simpson notes that, "In the first few months of the war the regular officer class generally displayed traditional leadership qualities, although they frequently found it difficult to adapt to the requirements of modern warfare."[12]

How then did the junior officers perform in 1914–15, and can any link be found with their previous education and training? It will only be possible to survey some individual experiences, and so the conclusions drawn must be tentative. As a preliminary, it should be noted that there were various routes to a commission just before the war, and so there were differing backgrounds and educational experiences. Sixty-seven percent of these officers came via Sandhurst, 15 percent came from the universities via the Officer Training Corps, formed in 1908, another 15 percent came from the Special Reserve and Territorial Force, and 2 percent were commissioned from the ranks.[13] Because the majority of regular officers did come from Sandhurst, this will be the focus of attention. The 1914 campaign was unusual for the British Expeditionary Force (BEF) because much of it involved a continuous retreat, which on several occasions proved too difficult for the high command to control. Therefore, much of the responsibility for saving the BEF often depended on officers at the brigade and battalion level, and below these, at the company level, the junior officers.

One interesting document is the report of the surrender of the Gordon Highlanders on 27 August 1914. The report was written in a prisoner-of-war camp in Germany in September 1914 by the colonel of the regiment, W. E. Gordon. According to this officer, brigade headquarters gave no order to this regiment between 26 and 27 August 1914, and they and the neighboring Royal Scouts became isolated. At 2:30 P.M. on 27 August, Colonel Gordon saw B Company under Captain Fawke, Lieutenant Sandeman, and Second Lieutenant Turnbull, leave their trenches in disorder, but then the officers steadied the men. The whole regiment then retreated, but Colonel Gordon specifically reported that the junior officers did well, and one, Lieutenant Lyon, was shot dead upon entering a German occupied house, but the other junior officers then killed the German soldiers. During the retreat, the regiment and Colonel Gordon eventually surrendered. In his report, Gordon singled out for special praise Second Lieutenant Stewart, Lieutenant the Hon. A. Fraser, three other second lieutenants, and Lieutenant Lyon, who had been killed. No other Gordon's officer senior to these was mentioned. In this case, therefore, it seems that the junior officers, who probably would all have come through Sandhurst in this regular army regiment in the early days of the war, did as well or better than could be expected.[14]

The slight evidence so far supports the common sense assumption that poor senior officers produced a poor regiment, although junior officers could perform well. As Keith Simpson notes: "It is difficult for someone without military experience to realize just how important the personality, character, physical courage, professionalism and sheer leadership of a Commanding Officer is for an infantry battalion in war." In this context, Robert Graves gives an opposite example of a good regiment. The Royal Welch regiment only accepted junior officers who had distinguished themselves in the passing out examination at Sandhurst, who had been recommended by two officers of the regiment, and who possessed a guaranteed private income that enabled the officer to hunt, play polo, and keep up a social reputation. Because of this regimental attitude, writes Graves:

> The regimental spirit persistently survived all catastrophes. Our First Battalion, for instance, was practically annihilated within two months of joining the BEF. Young Orme, who joined straight from Sandhurst, at the crisis of the first battle of Ypres, found himself commanding a battalion reduced to only about forty rifles. With these, and another small force, the remnants of the Queen's Regiment, reduced to thirty men and two officers, he helped to recapture three lines of lost trenches and was himself killed.

It might be deduced that regimental attitudes and good regimental leadership helped young Orme very much in his heroics, but it is difficult to tell how much of this was due to his Sandhurst education.[15]

The influence of regimental leadership and efficiency on the actions of both junior and senior officers seems to have been the reason for the poor showing of the Bedford and Cheshire Regiments on 29 October 1914, as related by the staff officer Capt. Billy Congreve. According to Congreve, the Cheshires that day consisted of "a most hopeless CO [commanding officer] and an equally poor adjutant, and only half a battalion, the rest being 'lost.'" Congreve put the Cheshires and Bedfords into the line, but soon found them retiring: "The men said that they had had orders to retire. There were very few officers about and what there were seemed useless. The officer commanding—a captain—saluted me and called me 'Sir' which showed he was pretty far gone." Earlier in October, a report by Gen. H. L. Smith-Dorrien, commanding II Corps, also shows the Cheshires to have been a disheartened regiment, as well as the Dorsets, so that: "The Regiments concerned in that fight [22 October 1914] did not put up resolute resistance. . . ." Smith-Dorrien concluded: "I recognised how little reliance can now be placed on several Battalions. . . ." Presumably there were junior officers involved in these incidents, and they probably behaved according to regimental values and leadership, whatever their military education. On the other hand, Smith-Dorrien had considerable praise for the junior officers of the Royal West Kent Regiment in late August 1914, who continued to hold their trenches despite the loss of all senior officers: "This splendid battalion, with only two officers left—Lieutenant H. B. H. White, commanding, and 2nd Lt. J. R. Russell, maintained their position, fighting hard every day, until withdrawn on the 8th of November [1914]." Regimental spirit and attitudes seem the key to these events.[16]

Another factor, however, is very hard to evaluate, namely, the influence of personality, family, and school background. For example, to take two obviously very good prewar junior officers: Billy Congreve and Johnny Gough. Both came from distinguished military families, and both went to the best known public (private) school in England, Eton, and then to Sandhurst. Both did very well at Sandhurst, and then excelled in the early months of the war; both became staff officers (understandable for Gough, who attended the Staff College in 1904–5); both won the Victoria Cross, Gough in Somaliland in 1904, Congreve at the Somme in 1916; and both were killed, Gough in 1915, Congreve in 1916. How to separate out the impact of Sandhurst from the influence of personality, family, school, and regiment? All that can be said at this stage is that Sandhurst was part of the "package" that could produce these outstanding young officers.

On the other hand, the case of Bernard Montgomery is instructive. Montgomery came from a religious family, attended the well-known public school St. Paul's, but as noted above, left Sandhurst in disgrace. Despite this, he clearly became an outstanding young officer, even though he equally obviously belonged to a very poorly led regiment. He describes how on 26 August 1914, it was left to Montgomery's regiment, the Royal Warwickshires, to restore a difficult situation with an immediate attack. The commanding officer, Lieutenant Colonel Elkington, ordered no reconnaissance, prepared no plan, and had no covering fire. Montgomery recalled: "Waving my sword I ran forward in front of my platoon, but unfortunately I had only gone six paces when I tripped over my scabbard, the sword fell from my hand . . . and I fell flat on my face on very hard ground. By the time I had picked myself up and rushed after my men I found that most of them had been killed." Montgomery continued: "My Company commander was wounded and there were many casualties. Nobody knew what to do, so we returned to the original position from which we had begun the attack." Then bad staff work isolated the regiment, which for the next forty-eight hours chased after the retreating BEF; meanwhile Lieutenant Colonel Elkington retired to St. Quentin, where he and another commanding officer surrendered their respective regiments to the mayor of the town. The two officers were later cashiered. However, Montgomery's talents were soon recognized, and he was shortly made acting captain, and put in charge of a company of 350 men. This example shows that despite a poorly commanded regiment, and despite a checkered career at Sandhurst, it was possible for natural ability to emerge and be rewarded. Here it was the individual and not Sandhurst or the regiment that mattered.[17]

By and large, the junior officers of the regular army, well prepared by the self-confidence produced by family, school, and Sandhurst, did well and fought bravely, providing the particular regiment they joined reinforced their military training and education with strong *esprit de corps,* traditions, and leadership. In contrast, it may be useful to look at one or two of Kitchener's volunteer New Army officers, who did not attend Sandhurst, and compare their performances. One was Lt. C. A. Elliott, who joined the Ninth Battalion of the West Yorkshire Regiment in Gallipoli in 1915. In a remarkably frank letter to his mother, Elliott, having landed at Suvla in August 1915, related that his company was in reserve behind the East Yorkshire Regiment, who were firing at the Ottoman Turks: "Well, that was their funeral. We had water and cover and were quite happy." Ordered up to support the firing line, Elliott recalled that "We cursed the East Yorks . . . the Turks and the Kaiser for 10 solid minutes while the men put out their fires and got into their equipment." The next day, 9 August, the Turks attacked, and Elliott was shot in the leg. His first thought was "one of elation—I'm wounded I can clear out, and the second of thankfulness, I was not knocked down and I can still walk." Elliott took the bolt out of his rifle and threw it away, then found a casualty clearing station. Elliott's New Army attitude does seem rather different from the regular army officers of 1914 on the Western Front.[18]

Another New Army officer at Gallipoli, Second Lieutenant Drury, joined the Sixth Battalion, the Royal Dublin Fusiliers. He came direct from the Trinity College officer training corps, with a New Army commission, but remarked that many of his senior officers were regulars. Drury commented on what seems to have been a heavy training program with the regiment in Ireland and England, where at least two officers were sent home as unsuitable. Drury and his regiment then also landed at Suvla in August 1915, and appeared to spend much time trying to prevent neighboring regiments from leaving the line. Thus, on 9 August, Drury found parties of the Lincoln and Borderer Regiments "all over the place, lying up in funk holes. Finally . . . a whole lot of these two regiments started running away like mad, shouting out that they were cut to ribbons etc." Drury and a fellow junior officer had to threaten them with revolvers, "even to the officers." Another junior Dublin's officer accompanying Drury, hit one officer over the head with his telescope to prevent him leaving. "The Queen's West Surreys [also] bolted the minute the heavy shelling started. . . ." Drury reflected "that if we had had our own 10th Division here complete, we could have smashed the Turk defence and got to our objectives." Even allowing for obvious bias toward his own regiment and his own actions, it does seem that Drury and his fellow junior officers of the Sixth Royal Dublins performed quite well at Suvla, while other New Army junior officers with other nonregular army regiments did not. The difference was probably the regular officers, leadership, and *esprit de corps* of the Sixth Royal Dublins, as he wrote in 1914 "we seemed like a regular battalion and not a 'Kitchener Crowd.'"[19]

Clearly, it is dangerous to extrapolate from just a few examples, but a tentative conclusion is that it was the regiment, with its commanding officer, traditions, and training habits, that was a key factor in molding junior officers, as much as prior military education at Sandhurst or elsewhere. Yet probably the very best junior officers were those with the whole "package" stemming from family background, school, Sandhurst, and a high *esprit de corps* and a well-commanded regiment. No doubt this was particularly the case with elite regiments such as the Grenadier Guards.[20] Another factor influencing junior officer performance in 1914 was the simple statistic that regardless of the benefits of such a "package," far too many officers got killed in 1914. The British official historian, Sir James Edmonds, noted that by the end of 1914, the battalions that had landed in Europe in August comprised on average only one officer and thirty other ranks. Losses of officers for the whole campaign in 1914 were 3,627. Most of these were junior officers, and according to Keith Simpson, the average length of time a subaltern served in France before becoming a casualty or leaving, was six months. One reason for this loss of junior officers was addressed by a subaltern:

> Here was the secret of the British Army's weakness vis-à-vis the Germans. The latter could use NCOs where we had to use officers, and thousands of promising young officers were killed doing lance-corporal's work, simply because section commanders [NCOs] had not it in them to exercise sufficient authority to carry out their task.

This resulted from the over-officering of companies in the infantry battalions, where a company commander had five junior officers, when probably two good officers would have been sufficient. All first-hand accounts remark on the loss of officers (particularly the junior ranks), so that Sir John French, the commander in chief of the BEF, was compelled to write to his corps commanders in September 1914: "Army Corps commanders will, I am sure, have noticed and deplored, as I have, the terribly high proportion of casualties amongst officers as opposed to the rank and file. . . ." On the other hand, the junior officer casualties reflected the high *esprit de corps* and loyalty to regiment and duty of these officers, as in the days of Lieutenant Latham in 1811. These attitudes were probably produced in 1914 by that "package" of self-confident social background, school, military education, and good regiment. Reflecting this, Robert Graves, in an early uncritical mood, wrote to a friend in 1914 when he joined up: "France is the only place for a gentleman now. . . ."[21]

In conclusion, then, first, there was a mixture of continuity and change from regimental training in the eighteenth century to a mixture of military college, militia, and regimental training in the nineteenth and early twentieth centuries, with a growing number of officers eventually attending Sandhurst by 1914. Second, the continuity aspect relates to the class origins of the officer corps (officers are born, not made), together with regimental on-the-job training. Both these aspects continued from the eighteenth century to 1914. On the other hand, the change aspect relates to the increasing importance of military education by the late nineteenth century, but judging by the results of the 1902 commission on officer training, this military education requirement came too late to bite very deep into the army before 1914. Third, junior officer military education appears to have been barely adequate to the needs of those two hundred-odd years, and not terribly efficient, as the examples of the American War of Independence, the Crimean War, and the Boer War, to name a few campaigns, indicate. However, by 1914, there was some improvement. Fourth, for junior officers, it is difficult to separate out the relative influence of Sandhurst from the other elements of the "package" of family background, public school, and regimental training. Forced to choose, one might argue that the regiment was still the most influential element on the junior officer in 1914, and junior officers in 1914 generally performed as well as their regiments allowed.

Last, there is a general impression that the great majority of new officers entering the deadly year of 1914 did so with a rather narrow, traditional educational ethos or culture, which stressed moral and sporting qualities such as courage, loyalty, duty, character, and a very large emphasis on sports, rather than practical training—in short, the need to be both a "gentleman and an officer." These moral and sporting qualities undoubtedly helped the BEF survive in 1914 at very great sacrifice and cost to junior officers but they were perhaps rather narrow qualities, and were more useful for imperial and colonial warfare than for the conditions of 1914. As the journalist Philip Gibbs remarked of British junior officers, having observed the BEF in 1914: "They had the manners of caste, the touch of arrogance which belongs to a caste, in power. Every idea they had was a caste idea, contemptuous in a civil way of poor devils who had other ideas, and who were therefore guilty . . . of shocking bad form." But Gibbs also wrote of them: "There was something superb in those simple, self-confident, normal men, who made no fuss, but obeyed orders, or gave them, with a spirit of discipline which belonged to their own souls and was not imposed. . . ."[22]

A final question: what would a useful practical education have been for the junior officer just before 1914? Probably the largest single omission in their education was a proper understanding of firepower, and the influence of that vital element on tactics. But that critical omission was also shared with their senior officers, who for the most part gave the junior officer very little help in this regard, and so condemned many to a painful awakening to the realities of the 1914 battlefield.

Notes

1. Michael Glover, *Wellington's Army in the Peninsula, 1808–1814* (Newton Abbott, London and Vancouver: David & Charles, 1977), 166.

2. J. A. Houlding, *Fit for Service: The Training of the British Army, 1715–1795* (Oxford, England: Clarendon Press, 1981), 273 and passim.

3. Michael Yardley, *Sandhurst: A Documentary* (London: Harrap, 1987), 17; Richard Glover, *Peninsular Preparation: The Reform of the British Army, 1795–1809* (Cambridge, England: Cambridge University Press, 1963), 163–64.

4. Glover, *Wellington's Army in the Peninsula*, 40–44.

5. Ibid.

6. Brian Bond, *The Victorian Army and the Staff College, 1854–1914* (London: Eyre Methuen, 1972), 17, 19, 56; Hew Strachan, *Wellington's Legacy: The Reform of the British Army, 1830–1854* (Manchester, England: Manchester University Press, 1984), 126; similarly, Lt. Gen. Sir Frederick Hamilton, *Origins and History of the First Grenadier Guards* (1874), cited in J. M. Craster, *"Fifteen Rounds a Minute": The Grenadiers At War, August to December 1914* (London: Macmillan, 1976), 5.

7. Yardley, 31–32, 38; Strachan, 128, 136, 140; Bond, 29–30; and E. M. Spiers, *The Late Victorian Army, 1868–1902* (Manchester, England: Manchester University Press, 1992), 107ff. Sandhurst was not outstanding in the early years of the nineteenth century for those who did graduate, since between 1813 and 1837, of 420 cadets, only 3 reached lieutenant colonel. As late as the 1860s it was extremely rare to find a Sandhurst or Woolwich graduate among the well-known generals of the day.

8. Keith Simpson, "The Officers," in Ian F. W. Beckett and Keith Simpson, eds., *A Nation in Arms: A Social Study of the British Army in the First World War* (Manchester, England: Manchester University Press, 1985), 64; Yardley, 38; Ian F. W. Beckett, *Johnnie Gough, V.C.: A Biography of Brigadier-General Sir John Edmond Gough, V.C., K.C.B.* (London: Tom Donovan, 1989), 8; Maj. Gen. J. F. C. Fuller, *Memoirs Of An Unconventional Soldier* (London: Nicholson & Watson, 1936), 3–4, 5–6; Winston Churchill, *My Early Life* (London: Butterworth, 1930), 73.

9. *Report of the Committee appointed to consider the Education and Training of Officers of the Army*, Command Paper 982, 1902, vol. 10, 15–39. Similar comments appear in L.S. Amery, *The Problem Of The Army* (London: Edward Arnold, 1903), 192–210, who particularly attacked the system of militia officers getting regular commissions through the back door (196).

10. Yardley, 54; B. L. Montgomery, *The Memoirs of Field Marshal the Viscount Montgomery of Alamein* (London: Collins, 1958), 23–25; Nigel Hamilton, *Monty: The Making of a General, 1887–1942* (London: Hamlyn, 1982), 48, 57.

11. Simpson, "Officers," 67–68; John Baynes, *Morale: A Study of Men and Courage, The Second Scottish Rifles at the Battle of Neuve Chapelle, 1915* (New York and Washington, D.C.: Praeger, 1967), 48, 28; Robert Graves, *Goodbye to All That* (Harmondsworth, England: Penguin, 1960), 106, 109, 117–18; Fuller, 6.

12. Baynes, 114–16, Lyn Macdonald, *1914* (Harmondsworth, England: Penguin, 1989), 157–58; Beckett, 149; Simpson, "Officers," 83.

13. Simpson, "Officers," 64.

14. Col. W. E. Gordon, "Report in Connection with 26 and 27 August 1914. Actions before Audencourt and Bertry, France," 16 Sept. 1914, Torgau, Germany, *Gordon Highlanders Surrender, 27 August 1914*, War Office (hereafter WO) 141/37, Public Record Office, Kew Gardens, London (hereafter PRO). This is a newly released document.

15. Keith Simpson, "Introduction," in [J. C. Dunn], *The War the Infantry Knew 1914–1919* (1938; London: Jane's, 1987), xlii; Graves, 78.

16. Maj. Billy Congreve, Diary, 29 Oct. 1914, in Terry Norman, ed., *Armageddon Road, A VC's Diary, Billy Congreve* (London: William Kimber, 1982), 67–68; Smith-Dorrien to GHQ, 22 Oct. 1914, II Corps War Diary, WO 95/629; idem, "Report on Operations of II Corps, 11th October to 18th November 1914," WO 95/630; idem, War Diary, 25 Sept. 1914, Cabinet 45/206; all in PRO. I am grateful to Nikolas Gardner, a doctoral history student at the University of Calgary, for these Smith-Dorrien references.

17. Nigel Hamilton, 78–81.

18. Elliott to his mother, 8403/63, National Army Museum, London.

19. Captain Drury, Diary, Sixth Royal Dublin Fusiliers, vol. 1, 1914, 7–8, 14; and 9 and 10 Aug. 1915, 7607/69, ibid.

20. Witness the bravery and leadership of three young lieutenants of the Grenadier Guards at Villers-Cotterets on 1 September 1914. See Craster, 54.

21. Brig. Gen. Sir James Edmonds, *Military Operations, France and Belgium, 1914*, 14 vols. (London: Macmillan, 1925–48), 2:465, 467; Simpson, "Officers," 86–87, and A. A. Hanbury-Sparrow, *The Land-locked Lake* (London: Arthur Barker, 1932), 130, 87; Baynes, 177–79; French to corps commanders, 16 Sept. 1914, I Corps War Diary, WO 95/588, PRO; Graves to Hartman, 25 Oct. 1914, Special Miscellaneous Collection, M4, Imperial War Museum, London.

22. For ethos and culture, see Albert Palazzo, "Tradition, Innovation, and the Pursuit of the Decisive Battle: Poison Gas and the British Army on the Western Front, 1915–1918" (Ph.D. diss., Ohio State University, 1996), chap. 2; Philip Gibbs, *The Soul of War* (Toronto: McClelland, Goodchild & Stewart, 1916), 340–41; see also idem, *Now It Can Be Told* (New York and London: Harper, 1920), 15 and 67, for a story of a clearly very inexperienced junior officer in 1914.

Training and Educating Pre–World War I United States Army Officers

I. B. Holley, Jr.

This essay is devoted to the training and education of junior United States Army officers in the decades prior to World War I. It is of the utmost importance that we define these terms, for they are not, as often carelessly employed, interchangeable and synonymous. Training is undertaken to develop proficiency. Those who train impart what they believe to be the best way to perform a task. It can be tested objectively by trial or performance. Education is quite different. It aims to help one learn how to develop perspective, values, and the ability to cope with change, novelty, and uncertainty. Those who seek to educate cannot claim to have the "answers." They can challenge and stimulate thought, but if they attempt to give answers they indoctrinate but do not educate.

Between the Civil War and World War I—more than fifty years, more than a generation without a major war—training and education had to cope with shifting objectives. Although I shall touch upon the full range of junior officer preparation, my emphasis will be on the development of future officers at the United States Military Academy. To what ends should West Point direct its efforts? To prepare for Indian warfare? To prepare for the burdens of empire thrust up by the Spanish-American War and the Philippines or the Panama Canal? Or should the major effort be devoted to preparing for a major conflict on the scale of World War I?

As every officer knows, training and education are not confined to a cadet's days at the academy. They continue throughout one's career. Newly commissioned lieutenants soon realize that they have only just begun their apprenticeships. As the grizzled old Indian fighter Gen. Jim Parker put it, "everyday I was learning some new thing about commanding troops, . . . especially after I had attained high rank."[1] Successful officers understood that self-education was the road to high command. Probably the outstanding example of this is Gen. George S. Patton, Jr. Like many another achiever he read widely in military history. But he went still further. After completing the Staff College at Leavenworth, he wrote back annually for the map problems, the tactical problems of the course. These he worked out for himself, year after year. He did not ask for the school solutions.

The Staff College, the War College, and the various branch schools were certainly vital ingredients in preparing officers of the pre–World War I generation. So too, the appearance of professional journals played a significant part. Starting with the *Cavalry Journal* in 1888, the *Artillery Journal* in 1892, down to the belated appearance of the *Infantry Journal* in 1904, these did much to sustain the pace of self-education.[2] But so did the mentoring of officers who took their subordinates in hand as did Lt. John M. Palmer's company commander Capt. George A. Cornish who loaned his copy of Carl von Clausewitz's *On War*, opening Palmer's eyes to that great work. Or again, consider Maj. Gen. John F. Morrison in the Philippines requiring subordinates to memorize his detailed verbal orders precisely and promptly, inculcating an excellent mental discipline. Who is to say this was any less important in the forming of junior officers than was mastering conic sections in math at West Point?

An assessment of education at West Point before World War I must begin with the institution's recruitment practices. As we say of computers, GIGO—garbage in, garbage out. If the selection process is flawed, undoubtedly this will be reflected in the officers commissioned four years later. The system of congressional appointments was designed to insure that no one political party or faction dominated the officer corps. Cadets were to represent every corner of the nation. This was certainly a democratic safeguard. Unfortunately the qualifications of appointees were not all of an acceptable standard. Some members of Congress held competitive exams and their appointees tended to do well at the academy. But the overall statistics suggest a rather discouraging record. From the end of the Civil War to the turn of the century, only about one-fourth of those appointed to West Point eventually graduated. Of the 200 appointees to the Class of 1885, for example, only 85 were finally admitted even though the exam they took after arriving at the Point could scarcely be called rigorous. Only 39 of the original 200 appointees were graduated as second lieutenants.[3] One might conclude from this high attrition rate that the academy was intellectually rigorous, but manifestly there were many factors other than intellectual standards involved.

Hazing by upper classmen, the sheer physical demands of cadet life, and the constraints of the rigid discipline imposed by regulations all conspired to weed out those who felt personal freedom was more desirable than a free education. Cadet John McAuley Palmer received demerits for failing to take a bath on one occasion and more demerits on another occasion for "lingering too long in the bath." Even those who stayed the course and won their commissions tended to be scathing in their comments on cadet life. Looking back years later, Col. T. B. Mott compared the academy to a combination of Frederick the Great's military schools, a Jesuit seminary, and a Tibetan lamasery.[4]

If the cadet's life at the academy was hard, it was meant to be. How else could the army inculcate the habit of absolute obedience in the most minute particulars? The occasional excess in the practice of hazing or the absurdity of a regulation here or there should not obscure the important role played by constraints and pressures in acclimating high-spirited youngsters to the meaning of discipline. As Plutarch said, "The wildest colts make the best horses—if properly broken." General Parker, who taught at West Point at one stage in his career, put it succinctly: "The two things which I regard as of greatest value in the course at West Point are the code of honor and discipline. . . . It is this discipline which they bring to the army, the leaven which creates a martial army in a democratic country."[5]

We are all familiar with the slogan, the by-word, of the academy: Duty, Honor, Country. One does not lie, cheat, steal, or tolerate those who do.[6] These are the attributes of a gentleman. Some cadets were gentlemen when they arrived. They had been reared in the substance of the code. For those who reached the academy less favored, the institution faced the difficult task of acculturating by instruction, example, and by condign public punishment for those who violated the code, usually by ignominious expulsion.[7] As they say in the Royal Navy, it is necessary now and then to hang an admiral *pour encourager les autres!*

Those who know the academies today are aware that organized athletic programs play an important role in developing future officers. When I taught at West Point, I was impressed that the academy had about a dozen football fields and *everybody,* not just the varsity, engaged in the rough and tumble of contact sports. But it was not always so. Organized football did not appear until the end of the nineteenth century. Cadets did engage earlier in various types of ball but this was informal and played in their regular uniforms.[8]

The Board of Visitors in 1884 declared that the objective of the United States Military Academy was to develop the "mental, moral, and physical qualities"[9] required by officers. At the time, prescribed exercise was largely confined to instruction in horseback riding and fencing. The instructor in fencing was known as the Master of the Sword. Not until 1947 was the post redesignated as director of physical education.[10] As late as 1910 the form used to evaluate officers still had a section for rating "swordsmanship." This, in spite of the fact that Chief of Ordnance Steven Vincent Benet, on the basis of medical records showing negligible numbers of sword wounds in the entire Union Army during the Civil War, had recommended abolishing the cavalry saber.[11]

Having touched on the moral and physical aspects of the academy scene, let me turn now to the classroom. The system of instruction which persisted well down into the twentieth century was the pattern laid down by Superintendent Sylvanus Thayer at the beginning of the nineteenth century. Wisely, he insisted on small sections seldom numbering more than eight or ten cadets. Such small numbers held the promise of fruitful interplay between instructor and student. It approximated Mark Hopkins's ideal of the teacher on one end of the log and the student on the other.

Unfortunately this opportunity for interaction was seldom exploited. In the typical class, cadets were handed slips of paper with a question. They stood and recited an answer or worked the problem on the board. All subjects, with the exception of languages and drawing were taught this way. Recitations, virtually verbatim from the text, were "correct." The instructor listened to the recitation and graded it. This could hardly be called teaching. Instructors had little or no contact with students outside of the classroom; that was left to the tactical officers.[12]

For those who maintain that the only real education is self-education, the academy system could be said to work because it forced cadets to get an understanding of the assigned work on their own, perhaps aided by an abler fellow student in the barracks. But parroting a textbook could scarcely be called education. There were many complaints that the system was stultifying.[13]

Although the senior officers who served as department heads had tenure, most of the faculty consisted of junior officers on three- or four-year assignments. While a few of the senior faculty in engineering managed to take short courses at the Massachusetts Institute of Technology, the junior faculty members were usually woefully unprepared to teach. They had little or no special preparation beyond what they had learned about their subject as cadets. As Capt. John M. Palmer realized when he reported for faculty duty at the academy in 1901, nine years after his commissioning there, "I was deeply conscious of my limitations as an instructor. I had stood fairly well in chemistry while a cadet, but with no postgraduate training I was ill-prepared to teach that interesting science."[14]

The rising tide of criticism about the academy curriculum coupled with the army's sorry performance in the Spanish-American War led to a ferment for reform after 1900. In 1901 the Board of Visitors sought a broadening of the curriculum with less emphasis on math and engineering and more work in history and English. Other critics charged with considerable justice that the academy failed to develop officers who understood how to lead citizen soldiers who make up the mass of the army in wartime.[15] An example of an officer who failed to learn how to elicit the best from enlisted men, recalled by Col. T. B. Mott, was the one who said, "I don't give a damn what you think. I'm paid . . . to do your thinking for you."[16] When Mott wrote back to the academy suggesting a way to improve the teaching, one of the professors replied accusing him of being "a young upstart" for his audacity in daring to criticize the lessons on fortification. These lessons, one might note, were introduced in 1820 and were still being taught in 1885 in Mott's cadet days![17]

The reform movement did manage to bring many changes. Secretary of the Army Elihu Root secured appropriations from Congress for a building program that gave the academy its neo-Gothic buildings which we know today. After the turn of the century modern weapons were introduced, the athletic program was vastly enhanced, and cadets engaged in more realistic maneuvers. But the curriculum and the classroom teaching methods remained substantially untouched. The main source of resistance to change in this arena was the Academic Board made up of the department heads, all senior, tenured professors. They out-voted and out-maneuvered even the superintendent who customarily retained his office only for a single tour.[18]

One bright spot in the otherwise mediocre setting was the library. In 1902 E. H. Holden, an academy graduate, who had been president of the University of California and later director of the Lick Observatory, accepted the post of academy librarian. He instituted an aggressive program of annual book buying and set the faculty to work indexing professional journals.[19] By 1915 he had built the library collection to 90,000 volumes. All this was promising, but it is noteworthy that as late as 1911 cadets had restricted access to the library which was closed at night. Librarian Holden finally managed to secure permission to remain open until 9:30 on week nights. Even then the cadets checked out few books because the faculty assigned no research problems.[20]

After 1900, with officers confronting Spanish-speaking populations in Cuba and the Philippines, the Academic Board finally agreed to increase the amount of Spanish taught. But how? They reduced what little history was then being taught. Adding to this folly, they ordered the displaced history instructors to assist the language instructor. Clearly they seemed to regard instructors as interchangeable parts without reference to particular qualifications. Under pressure from the advocates of history, in 1907 they created a Department of History and English. When Professor J. C. Adams was recruited from Yale to conduct this two-headed monstrosity, he urged that the two utterly different disciplines be separated. But the Academic Board resisted this move.[21]

Before we leap to the conclusion that the academy curriculum before World War I was hopelessly out-of-date and the Academic Board comprised of arch-conservatives, we should try to understand their pedagogical philosophy. To begin with, they argued that the objective sought was "mental discipline," not "utility."[22] Practical professional subjects were best acquired in the various professional schools attended after leaving the academy—the Engineer School, the Artillery School, and the like. The Staff College at Leavenworth, and the War College were intended to top-off an officer's professional qualifications. The function of the academy was to lay a foundation of "mental discipline" which was essential to "military discipline." And the study of mathematics, they believed, was the approved route to "mental discipline."[23]

Of course there were many critics of the whole "mental discipline" school of thought. As one faculty member rather shrewdly observed, "Mathematics does not teach how to observe, how to generalize, how to classify." When the Academic Board argued that math was essential for engineering and ordnance, the quick response was that only 10 percent of the academy graduates went into these fields. The critics pointed to the writings of William James and John Dewey who denied the validity of the whole "mental discipline" approach, an education which tended to produce graduates who literally hated intellectual effort. They alleged that the "mental discipline" school failed to realize that the best teaching involved cultivating enthusiasm for learning.[24]

There was much cogency to the arguments offered by the critics. However, the ultimate test lay with the performance of the graduates. They may not have been intellectually prepared as well as they might have been. But with trifling exceptions the academy turned out officers who displayed character, a strict code of ethics, physical and moral courage, habits of obedience and punctuality. The remarkable success of many academy graduates who left the army to enter the world of business or the professions suggests that their education for all its shortfalls had considerable efficacy. But we are more interested in the officers it produced to command the army.

The performance of West Point graduates in World War I really did support the claim that character could be inculcated.[25] Roger Nye, in his study of the academy during the era of educational reform in the first quarter of the twentieth century, spells out what was accomplished by West Point as a character building institution. The academy imparted, he says, the attributes of a gentleman: honesty, courage, integrity, responsibility, and the obligation to serve worthy causes along with the more military qualities of obedience and loyalty, accuracy in official reporting, a refusal to equivocate even to the point of harm to self, and courage to death in battle. In short, the training and education offered by the academy asked more of officers than was expected of civilian society. This was no small accomplishment.[26] But the greatest accomplishment of the academy was probably what tough old Gen. James Parker saw: it took in independent, individualistic, typically willful youngsters from all walks of American democratic life and imparted those martial qualities without which an effective army is impossible.

Notes

1. James Parker, *The Old Army: Memories, 1872–1918* (Philadelphia, Pa.: Dorrance, 1929), 10.

2. Stephen Ambrose, *Duty, Honor, Country: A History of West Point* (Baltimore, Md.: Johns Hopkins University Press, 1966), 210.

3. William B. Skelton, "Samuel P. Huntington and the Roots of the American Military Tradition," *Journal of Military History* 60 (April 1996): 335; Allan R. Millett, *The General: Robert L. Bullard and Officership in the United States Army* (Westport, Conn.: Greenwood Press, 1975), 33.

4. I. B. Holley, Jr., *General John M. Palmer, Citizen Soldiers, and the Army of a Democracy* (Westport, Conn.: Greenwood Press, 1982), 705; Millett, 34–37; and David R. Mets, *Master of Airpower: General Carl A. Spaatz* (Novato, Calif.: Presidio Press, 1988), 6–9.

5. Parker, 202.

6. Thomas J. Fleming, *West Point* (New York: William Morrow, 1969), 281; Roger Nye, "The United States Military Academy in an Era of Military Reform, 1900–1925" (Ph.D. diss., Columbia University, 1968), 140, 160–83.

7. Nye, 164–65.

8. Fleming, 257; Nye, 120.

9. Millett, 40.

10. Marty Maher, *Bringing Up the Brass: My 55 Years at West Point* (New York: McKay, 1951), 72ff.

11. Brig. Gen. S. V. Benet, ed., *Annual Reports and Other Important Papers Relating to the Ordnance Department* (Washington, D.C.: Government Printing Office, 1878–90).

12. Fleming, 233–34.

13. Millett, 39.

14. Holley, 129.

15. Ambrose, 215.

16. Fleming, 256.

17. Ibid., 223–24, 237.

18. Ibid., 280ff.

19. Robert K. Johnson, *The Library System of the United States Military Academy* (Rochester, N.Y.: University of Rochester Press, 1956), 30; Superintendent, U.S. Military Academy to Adjutant General, "The Library, U.S. Military Academy," reprint in *Journal of the U.S. Cavalry Association* (October 1902): 199.

20. Nye, 193–94.

21. Ibid., 198; Fleming, 281.

22. Ambrose, 208.

23. Nye, 104ff., 131.

24. Ibid., 235–37.

25. Ibid., 18.

26. Ibid., 139.

"You Are Here to Learn How to Die": German Subaltern Officer Education on the Eve of the Great War

Holger H. Herwig

> Majesty, as long as you have this officer corps, you easily can do whatever you want; should that no longer be the case, things would be much different.
> —Bismarck to Wilhelm II, 1897

The historiography of officer education has run the proverbial gamut from analyses of curricula to statistical breakdowns of classes and crews by geographic origin, religion, and social class. In 1976, Ian Dixon broke new ground with his highly controversial book, *On the Psychology of Military Incompetence*. Chapter *bon mots* such as "Bullshit," "Socialization and Anal Character," "Anti-Effeminacy," "Mothers of Incompetence," and "Education and the Cult of Muscular Christianity," while undoubtedly not pleasing most academy administrators and cadets, nevertheless enlivened the debate over officer education. No such work has yet been attempted for the German military. Nor shall I undertake it. Instead, let me begin this essay by outlining the kind of force that I examine in this paper.

On the eve of World War I, the peacetime strength of the German Army was 29,000 officers, 106,000 noncommissioned officers, and 616,000 men. Put differently, the officer corps comprised 3.8 percent of armed strength. Roughly 1,200 officers entered the army every year. There were two paths to officer candidacy: 20 to 30 percent of applicants came through the venerable cadet schools (*Kadettenschule*)— nine in Prussia and one each in Bavaria and Saxony; the rest through direct appointment by regimental commanders as "aspirant officer" (or *Fahnenjunker*). By either route, social origins and "character" were as important as formal education. In each case, the prospective candidate had to be a decent Christian, Catholic or Protestant; Jews were tolerated neither in the active nor in the reserve officer corps of the Prussian and Saxon Armies before 1914. Final appointment to the grade of officer had two components: unanimous election by the officer corps of a regiment, followed by "commissioning" by the king of Prussia, Bavaria, Saxony, or Württemberg (there was no unified "German Army" until the outbreak of the Great War in August 1914).

The Prussian Army was first and foremost a *corps royal*, a praetorian guard— that is, the guarantor of internal stability and monarchical order. As Kaiser Wilhelm II poignantly reminded Chancellor Bernhard von Bülow in a celebrated New Year's epistle in 1906: "First gun down the socialists, then behead them and render them harmless—if need be by a blood bath—and then war outside our borders. But not the other way around and not too soon."[1] The army's secondary mission was to defend the territorial integrity of the Reich and, if need be, to conduct war against hostile foreign powers.

While history books abound with the sagas of such great captains as Helmuth von Moltke (the Elder), Erich Ludendorff, Paul von Hindenburg, Erich von Manstein, and Heinz Guderian, the backbone of the Prussian officer corps consisted of the relatively junior ranks of second lieutenant to captain (and perhaps major). Advancement was slow and pay was poor. On average, *Fahnenjunker* entered the officer corps at age nineteen or twenty. It then took fifteen years to pass through the grades of second and first lieutenant (*Leutnant* and *Oberleutnant*); thereafter came ten years as captain (*Hauptmann* or *Rittmeister*). For the vast majority of officers, careers came to an involuntary end in their mid-forties as they failed to pass what the Germans called the "major's corner" (*Majorsecke*). The lucky few who were promoted to major remained in that grade for six to eight years; successful lieutenant colonels required only twelve to eighteen months to reach colonel.

For the vast majority of officers, promotion proceeded strictly according to seniority (*Anciennität* or *Dienstalter*). Special promotions were reserved for the few leading lights of the General Staff. In this way, the army hoped to eliminate unhealthy rivalry within each grade. Every year, every officer was formally evaluated by way of a brutally blunt efficiency report (*Qualifikationsbericht*) prepared by his regimental commander. Apart from character and military performance, the officer was rated on suitability for promotion and for command position. The Military Cabinet based its annual promotion decisions on these reports.

On the eve of the Great War, second lieutenants were paid 125 Reichsmark (RM) per month (about the same as a highly skilled industrial worker), lieutenants 200, and captains between 182 and 425. From this pay, the army kept one-third for uniform outlays for infantry officers. Parental subsidies to subalterns—ranging from 45 RM per month in the infantry, to 75 in the artillery, and to 150 in the cavalry—were essential for years to pay for uniforms, horses, dancing lessons, fencing instruction, and "proper" social activities. A good steed could cost up to 2,000 RM. Still, the total cost of education of around 6,000 RM in the Cadet Corps compared favorably to 15,000 for a career in the church, 18,000 in medicine, and 25,000 in the law. And whereas a second lieutenant began to draw pay at the latest by the age of twenty, his compatriots in the church only at twenty-six and in the civil service at twenty-eight.[2]

The German Army of 1914 was symmetrical and organized by twos: its twenty-five corps consisted each of two divisions, each of two brigades, each of two regiments. Of interest especially for this study of subalterns, each regiment—usually commanded by a colonel—consisted of three battalions, each of four companies, each of three platoons. The peacetime battalion numbered 20 officers and 600 men. Since the Franco-Prussian War of 1870–71, infantry was heralded as the "queen of battle." The Field Regulations of 1906—with which Prussia-Germany entered the Great War—were clear on the subject: "Infantry is the premier arm. In unison with artillery, it destroys the enemy. It alone breaks his last resistance. It bears the main burden of combat and makes the greatest sacrifices. But for this, it also gathers the greatest glory."[3] Less charitably, some General Staff officers privately referred to infantry as little more than moving targets for the artillery.

As stated earlier, the military careers of nearly one out of every three or four Prussian officer candidates started at a *Kadettenschule*. And since half of all German generals in 1914 and half of all Prussian-born field marshals of World War II came from the Cadet Corps, this institution deserves a closer look.

Under the Second Empire, Prussia maintained eight preparatory cadet schools (*Voranstalten*) with 1,690 billets as well as the main cadet school at Gross-Lichterfelde with 1,000. Cadets entered the *Voranstalten* at age eleven; annually, 240 joined the army as officers between the ages of eighteen and twenty. Most remembered a Spartan existence of drill, unquestioned discipline, cold, and hunger. At the preparatory cadet school at Köslin, for example, the day started at 6 A.M. and ended at 9:15 P.M.; in between were five class periods of fifty minutes each as well as gymnastics, Matins, Vespers, and inspections. Over his eight or nine years of formal education, a cadet was taught 42 hours of mathematics, 40 of Latin, 34 of French, 28 of German, 18 of earth sciences, 17 of history, and 13 of physics. The curriculum was hardly demanding: Gen. Emory Upton noted that West Point cadets learned in one year the entire corpus of mathematics that Prussian cadets had to master in four. Class size was kept to about twenty cadets.[4]

But the "real" education took place outside the classroom. First and foremost, it consisted of forging "character," that ill-defined, politically incorrect, and elusive concept. For the German officer-to-be, it revolved around a sense of honor and duty, obedience and self-reliance, cleanliness and orderliness, camaraderie and honesty. Hazing of cadets, as long as it did not cross the boundary into cruelty, was accepted. Homosexuality and masturbation were ignored ("don't ask; don't tell").

Additionally, "character" meant learning the rudimentary skills needed for a line officer: close-order drill, rifle marksmanship, athletics, field exercises, riding, and fencing. And third, there was the ultimate skill to be learned: dying. The future novelist Ernst von Salomon vividly remembered as a first-year cadet at Karlsruhe being introduced to close-order drill by a Captain von Flotow with the words: "You are here to learn how to die." When Ernst queried his brother, three years senior in the Cadet Corps, about this, he received the uplifting response: "The most wonderful thing in life would be to perish as a 20-year-old lieutenant in some roadside ditch outside of Paris."[5]

Cadets matriculated by way of examinations that had not changed since 1844. The written tests lasted seven hours a day for four days and were followed by an hour-long oral examination. It required what many cadets called "fatal skill" to fail. In other words, very few cadets failed their examinations. Successful candidates were commissioned as second lieutenants; the very few unsuccessful ones were commissioned as *Fahnenjunker*, placed one year behind on the seniority list, and forced to retake the officer's examination at a later date.

The choice of a regiment was critical since a junior officer would spend the greater part of his career there. About ninety regiments—including all guards and cuirassiers units, most grenadiers, and many other cavalry formations—were considered exclusive preserves of the nobility and hence off-limits to commoners. Bonds forged during these formative years were permanent. In May 1918, the commanders of the three attacking armies of Operation Michael in France—Georg von der Marwitz, Otto von Below, and Oskar von Hutier—delighted in that they had been meritorious cadets (*Selektaner*) in the class of 1875 at Gross-Lichterfelde.[6]

Most fresh-baked second lieutenants were posted to the east and west frontiers to regiments with high numbers and low prestige. They were first and foremost "doers," not thinkers, having been taught the skills of battle rather than the nuances of strategy or even operations. A German officer led a rather dreary and monotonous existence for the first two dozen years of service. This centered on drill and maneuver. A mountain of manuals and regulations occupied his full attention and determined his every step. The daily roster consisted of drill in the field, on the parade square, and on the rifle range; schooling of recruits; field maneuvers; and more of the same all over again. For many of the rank and file recruits, the army was their first contact with personal hygiene, soap, prostitutes, and venereal diseases. The lieutenant or captain was the natural contact to deal with this.

In terms of command, a lieutenant of infantry was expected to take charge of a platoon of eighty men, whose squads or sections each of eight to nine men were under a noncommissioned officer. The company, consisting of 250 rifles, was usually headed by a captain; the battalion of 1,000 by a major or lieutenant colonel. The regiment, commanded by a colonel, likely was the highest echelon of the army with which subalterns had contact. About half the rank and file were reservists; three out of every four met government standards for literacy.[7]

In short, the subaltern's tasks were physical and routine, not cerebral and challenging. He was expected to drill his men to the point where they not only were willing to march into fire, but to know how and why they were doing so, and where they were going. When the shooting started, lieutenants and captains were not expected to enjoy long careers: by the time of the great 1916 offensives at Verdun and the Somme, the life expectancy of a lieutenant of infantry attacking enemy positions was about 120 seconds.

During peacetime, nonaristocratic subalterns serving in high-numbered regiments were scattered along the Empire's borderlands. Mostly, these were overgrown hamlets of about 10,000 inhabitants—peasant farmers, tradesmen, merchants, and bureaucrats. Joseph Roth graphically captured this life in his novel of the Austro-Hungarian Army in Galicia, *The Radetzky March*. Each hamlet consisted of a village square, a town hall, a church, a district court, a hotel, and a railroad station. The streets had no names, the huts no numbers. Year in, year out, subalterns drilled their men. Each day the rifle battalion marched out of its barracks over crunching gravel— and then returned, tunics and boots splattered with gray mud. Gradual decay was the lot for most: despair, gambling, and debts; sharp *Schnaps*, fast women, and sinister loan sharks.[8] Special Courts of Military Honor, established during the Napoleonic reform era to handle the flood of nonnoble officers, guided their conduct as officers and gentlemen.

For those who survived service in the borderlands, there were two avenues of relief: marriage and the War Academy (*Kriegsakademie*). The former was probably the most difficult because the state openly discouraged marriage among its junior officers for fear that this would reduce their dash and daring, their willingness to take risks. A prospective groom had first to find a woman socially acceptable to the king, and then to produce evidence that he could support her in the style befitting an officer. The royal "marriage consent" was usually given (or denied) by the regimental commander, but only after an intensive background check into the prospective wife's origins, upbringing, education, and social class. This hurdle passed, lieutenants (by 1900) had to show evidence of 2,500 RM private annual income beyond their salary; captains of 1,000 RM. Not surprisingly, officers placed marriage advertisements in newspapers to attract rich partners. Many a noble guards officer was not beneath "fertilizing the tines of his baronial crown" through marriage with a rich middle-class (and even Jewish) heiress.[9] A good marriage obviously enhanced one's chances of turning the "major's corner."

The second route came down to merit—and a supportive regimental commander. The venerable Prussian *Kriegsakademie*, founded by Carl von Clausewitz, offered good prospects for promotion. But the road was hard. Each year about 1,000 subalterns applied for the War Academy's 160 billets. They were nominated by their colonels and took grueling written exams anonymously at corps headquarters. Applicants had to have been officers for five years and on tours of active duty for three—and still five years away from promotion to captain. If admitted, a first lieutenant was appointed to the three-year course only for one year at a time; admission to the next year depended on performance.

Attendance at every lecture was mandatory. Required courses included tactics, ancient and modern military history, military and general geography, hygiene, military and international law, weapons, fortifications, General Staff work, administration, and communications. Courses were weighted, with tactics, General Staff work, and military history counting most. Students could choose their science option from among mathematics, physics, chemistry, physical geography, and surveying; their language option from among French, English, Russian, and Japanese.[10] Above all, the academy stressed the skills required for mission-oriented tactics (*Auftragstaktik*). During the winter term, officers received tactical instruction in the classroom; in the summer term, they were evaluated for their leadership abilities in the famous staff rides (*Schlussreise*). Careers were decided at this critical point. About one in ten graduates was deemed fit for service with the operational branches of the General Staff and the field divisions.[11] The demand for advanced military education was so great that the Kingdom of Prussia also maintained ten so-called "war schools" (*Kriegsschulen*) that by 1913 offered seven-month-long "crammer" courses in tactics, weapons, terrain, fortresses, drill regulations, and French or Russian.

Germany's other armed service, the Imperial Navy, by and large, was a mirror image of the army in terms of its officer education and code of behavior. And why not? The Prussian military model in general and its *Kriegsakademie* in particular were adopted by China, Japan, Turkey, and most South American countries (save Peru).

Thus, the navy copied the *Kriegsschule* with the *Marineschule* and the *Kriegsakademie* with the *Marineakademie*. It adopted the Prussian Army's courts of military honor, officer elections (by squadron and flotillas), annual fitness reports (*Qualifikationsberichte*), royal marriage consent, dueling code, and even inspectorates for education and artillery. Like the army, it required relatively high private family subsidies—about 7,500 RM by 1914—as an indirect financial barrier to keep out undesirable elements such as Social Democrats. Furthermore, it adopted the Prussian Army's disdain for formal education in favor of building "character," and its bias against technical specialties such as engineering. Like the army, it did not allow officers to vote in local, state, or federal elections, and it eschewed instruction in politics and economics in favor of dancing and riding. The *Kaiserliche Marine* shared the army's hard-line stance against admitting nonbaptized Jews. That its Admiralty Staff remained but a pale copy of the General Staff was due simply to Adm. Alfred von Tirpitz's fear that this institution could turn into a rival of his Navy Office. But while the army swore a personal oath of allegiance to the King of Prussia, the navy as a federal institution did so to the German Emperor.[12]

To be sure, there were differences between the two services. Since Tirpitz created the High Sea Fleet almost from scratch, officer advancement was much more rapid: on average, ensign at twenty-one years of age, second lieutenant at twenty-four, first lieutenant at twenty-nine, and lieutenant commander (major) at thirty-six.[13] Second, the Naval Academy had to struggle for legitimacy. For many naval officers, it was mainly an opportunity to sail the Kiel Fjord—much like golf, in the words of a recent chief of the U.S. Army's historical program, became the main attraction of Carlisle Barracks. Not until 1904 did the *Marineakademie* introduce a core curriculum that featured mathematics, naval architecture, and electrical as well as mechanical engineering. But four years later, the academy's director suggested dropping mathematics as it simply was too difficult for what across the North Sea the British naval historian, Sir Julian Corbett, once maliciously termed, the "unused organs of Naval officers."[14] Little wonder that the navy never developed a "general staff" analogous to that of the Prussian Army, or that 80 to 85 percent of *Seeoffiziere* (executive officers) never advanced their education beyond that of their cadet years.

The former naval officer Alexander Rost offered an anecdote that for him summed up the navy's subservience to the august Prussian Army in these matters. Shortly before World War I, a number of naval officers were bowling in Wilhelmshaven. The alley got hot. A dispute broke out whether officers could shed their jackets when bowling. The conundrum was solved by way of a telegram to that ultimate arbiter of military decorum, the Potsdam Guards. The reply was expectedly condescending: "If you *must* bowl, then by all means do it in shirt-sleeves."[15]

In summary, neither the army nor the navy managed to reconcile the basic conflict between building "character" and formal education in an age of rapid industrial and technological advancement. Both agreed that formal education for junior officers tended only to confuse the mind—much like Ludendorff thought about reading Clausewitz! Classroom learning clouded an officer's vision, slowed his reflex reactions, and caused him to think. Cadets went to Lichterfelde or to Karlsruhe "to learn how to die." For the next two decades, as subalterns, they drilled recruits on how to follow them into and to conduct battle. For that, very few, if any, needed theoretical physics or advanced calculus. A 1913 primer on how to become a naval officer reflects that view: "The training officer observes the midshipman at sailing and at handling the steam pinnace, knows who has a clear nautical outlook and can make quick decisions, and who panics. He also gets to know [the midshipman] on his horse, knows who can take the hurdle even without formal equestrian training. . . . He gets to know [him] at dancing lessons, and knows who can cut a figure on the dance floor and who suffers from a lack of social graces . . . ."[16] Little wonder that Vice Adm. Eberhard von Mantey, a former inspector of education, in the 1920s as official historian of the war at sea lamented: "We were a Prussian army-corps transplanted on to iron barracks."[17]

The outbreak of war in 1914 found junior officers every bit as prepared for battle as their grandfathers had been in 1866 or 1870—and probably better prepared than their counterparts in the Austro-Hungarian, British, French, and Russian Armies. They had learned well what the historian Gerhard Ritter called their *Kriegshandwerk*, the mechanics of warfare. They knew how to march their charges to the marshaling points of the German railway net, how to bring them up to the frontiers, how to deploy them once there, and how to lead them into battle. That the German recipe for victory, the Schlieffen plan, broke down at the Marne by mid-September was not their fault.

The German Army's failings in 1914 rested with its senior commanders. In their dogged pursuit of building "character" and their persistent disdain of formal education, they had prepared the army for a mid-nineteenth-century "cabinet war." In the process, they had ignored the warning signs of the American Civil War and the Russo-Japanese War that the future would bring a "peoples' war" of the industrialized masses. While many General Staff officers deep in their hearts feared that the rapid technological advancements in small-arms fire, artillery, howitzers, machine guns, and wire might break even the grandest advance, apparently few were willing to draw the necessary conclusions. Almost to a man, they insisted that a war of movement was possible, that the "cult of the offensive" remained supreme, that any conflict would be short, that million-men armies still could live off the land, and that war remained a legitimate way to solve the nation's ills, real or perceived.

Specifically, Gen. Alfred von Schlieffen gambled the nation's existence on a single high-risk campaign born of despair and devoid of political-diplomatic input. The plan was coordinated neither with the chancellor nor with the Habsburg ally.

Apparently lost was the realization that Otto von Bismarck through diplomacy had prepared the way for each of the three wars of unification in 1864, 1866, and 1870. Schlieffen stipulated thirteen corps for the decisive right wing alone—at a time when he had but five available. He foresaw a siege of Paris with seven or eight corps— which existed neither in reality nor yet even on paper.[18] He sidestepped the lethal firepower of modern infantry: each minute of combat, a modern brigade of 3,000 men with artillery expended a volume of fire equal to the volley and salvo firing of the Duke of Wellington's entire army of 60,000 men at the Battle of Waterloo.[19] And he demanded that more than 2 million men advance 400 miles in forty days, while simultaneously defeating the numerically superior combined armies of Belgium, France, and Great Britain.

These were dire failings at the grand strategic and highest operational levels— not at company, or battalion, or regimental levels. There, all unfolded according to plan in August 1914. The Military Telegraph Section of the Great General Staff mobilized 3.8 million men and 119,764 officers without a hitch. This gigantic force was transported to the front in 312 hours by 11,000 trains. For sixteen days, 560 fifty-four-car trains thundered across the various Rhine bridges. Thereafter, the 950 infantry battalions of the west army and the 158 of the east army marshaled, resupplied, and advanced into battle.[20] Morale and confidence were high; food and ammunition were low.

The campaign of August–September 1914 revealed all the glaring errors and omissions in German planning. Communications were abysmal. Gen. Helmuth von Moltke (the Younger), chief of the General Staff, for the critical days as his armies approached the Marne River, sat virtually alone in a former girls' school in Luxembourg, out of touch with his field commanders. The latter operated "blind," without military intelligence or operational instructions from Moltke.

Schlieffen's "ostrich-like refusal" to address logistics also dogged the grand design in the west. Artillery, whose firing tables were forty years old, quickly expended its allotment of 1,000 rounds per tube. The 84,000 horses of the First Army alone ate up 2 million pounds of fodder each day, and soon fell by the wayside due to starvation and exhaustion. Each corps consumed 130 tons of food per day, requiring 1,168 railway wagons for resupply. When the Belgians destroyed their rail system, the armies went without. During the critical days of September, most German railheads were 80 to 100 miles behind the front.[21] Neither improvised narrow-gauge field railways nor horse-drawn wagons could haul the vast supplies required by thirty-three corps from these railheads to the front. Motor supply was out of the question: 60 percent of Moltke's 4,000 trucks broke down before the armies ever reached the Marne. In any case, it would have taken 18,000 trucks to move just the right wing. Subalterns and rankers paid the price for these failings.

The year 1914 ended with stalemate on the Western Front. Moltke was replaced by Erich von Falkenhayn. The latter informed the government that the war could not be won, and counseled a negotiated peace in the east. Chancellor Theobald von Bethmann Hollweg would have none of it, and instead offered the nation a program of vast territorial expansion. In four months of fighting, the Reich had suffered 800,000 casualties, including 18,000 mostly junior officers; 116,000 men had been killed, almost three times the total for the Franco-Prussian War. No contingency plan existed. No reassessment of the war took place. That crusty curmudgeon, Field Marshal Gottlieb von Haeseler, a member of the Cadet Corps of 1852, at the tender age of seventy-eight almost alone looked reality in the eye: "It seems to me that the moment has come in which we must try to end the war."[22]

Instead, the German Army refined its *Kriegshandwerk*, hoping that enhanced technical efficiency would break the stalemate. From 1915 to 1918, this work was conceived of and carried out almost exclusively by officers in the grades of captain to lieutenant colonel. After all, they were the ones with the most immediate combat experience, and whose ranks had sustained the greatest losses in the first half-year of the war. In Austria-Hungary, for example, three out of every four officers killed, wounded, or missing were lieutenants and captains. Those "missing or lost" made up 48 percent of total officer strength (compared to 25 percent for Russia and 16 for Germany), while those "killed" formed but 8.7 percent (compared to 25 percent for Russia and 16 to 18 for Germany).[23]

In the realm of artillery, the greatest reform came with the "creeping" or "rolling" barrage first introduced on the Eastern Front in 1915. German troops were retrained to advance at a distance of only fifty meters behind the so-called *Feuerwalze*—a murderous wall of fire that moved against the enemy according to preplanned firing tables. These were initially set at 100 meters per minute; later at 200 per ten minutes. In May, Gen. Hermann von François noted its effect at the Battle of Gorlice-Tarnów: "Trees break like matches, huge trunks are hurled through the air, the stone walls of houses cave in, fountains of earth rise from the ground. Ten o'clock! The mortar fire subsides, the artillery advances its barrage. Shrill whistles. The first assault wave breaks over the trenches against the enemy."[24] It is testimony to the expert training and trust developed by German junior officers commanding infantry that their men followed the wall of fire, confident that the artillery would advance its fire as preplanned. The man who pioneered the revolutionary "creeping barrage," Georg Bruchmüller, was but a retired lieutenant colonel on temporary active duty, a rank in which he languished for four years of the war.[25]

With regard to tactical innovation, the Germans, with the help of captured French documents, in mid-1915 devised a novel, elastic defense-in-depth. It no longer consisted of the rigid linear trench system roughly one mile deep, but rather of an outpost zone, a battle zone, and a rearward zone that ranged between six and eight miles deep. The new defense forced the enemy to expend its forces against several echelons of troops arranged in depth; counterattacking infantry supplied the resiliency, or elasticity, of the system. German units deployed on reverse slopes wherever possible; artillery was commanded at the divisional level; and all combat arms were fully integrated into the defensive units. Allied soldiers now charged into a deadly killing ground: entrenchments and barbed wire, protected by enfilading concrete machine-gun posts and artillery echeloned in depth.[26] Hermann Geyer, the officer who in

December 1915 devised Germany's elastic defense-in-depth in the "Principles of Command for the Defensive Battle in Position Warfare," spent the entire war as a captain. In fact, the same Captain Geyer two years later would develop the new assault tactics with which Ludendorff hoped to win the war in France.

Third, the German Army pioneered the development of "storm battalions" (*Sturmbataillonen*). By the summer of 1916, General von Falkenhayn ordered that every division train and equip a battalion of storm troops. Each was to consist of 20 officers and 1,052 noncommissioned officers and men organized into a staff, four storm companies, one machine-gun company, one artillery battery, one section of flame-throwers, and one section of trench mortars. The soldiers, mostly volunteers under the age of twenty-five, were outfitted with steel helmets, spades, axes, hatchets, picks, wire cutters, and carbines. Training courses were established at Seboncourt for ten officers at a time.[27] General Ludendorff later added hand-grenades, gas masks, daggers, and light machine guns to the troopers' armament, and by 1917 ran four- to six-week courses to train them. During the great Michael offensive in France in 1918, a typical storm battalion, led by a captain and four lieutenants, consisted of twenty-four light machine guns, eight trench mortars, eight light mortars, eight flame-throwers, four light artillery pieces, and heavy machine guns.[28] Willy Rohr, the creator of the *Sturmbataillonen*, like Geyer spent the entire war as a captain.

Why? Given both the destruction of General Staff records by Allied air forces in 1944 and 1945 and the refusal of the trustees of the Ludendorff and Hindenburg papers to make them available to scholars, we can only guess that there existed at general headquarters a bias, overt or covert, against promoting wartime innovators. General Staff officers close to Ludendorff and Hindenburg (Max Bauer, Wilhelm von Leeb, Gerhard Tappen, Albrecht von Thaer, Georg Wetzell) on average moved up three rungs on the promotion ladder during the war.

Germany's insistence that officers be apolitical and that they receive no instruction in politics or government also came home to roost by the middle of the war. Without an economic "general staff" and with no prewar plans extant on how to feed sixty million people undergoing maritime blockade, civilians and soldiers alike became acquainted with cold, hunger, and disease. The so-called "turnip winter" of 1916–17 constituted a caesura in wartime suffering.

The food crisis first rocked the Imperial Navy. In August 1917, the roughly 100,000 sailors of the High Sea Fleet, radicalized by close contact with socialist shipyard workers and enervated by boredom, lack of action, and no prospect of peace, rebelled against their officers. Subsequent investigations showed that food was a major cause of the revolt. Sailors were reduced to eating what they called "wire entanglement" (*Drahtverhau*)—a nauseous concoction of "75% water, 10% sausage, 3% potatoes, 2% peas, 1% yellow turnips, and small amounts of beef, fat, and vinegar."[29] Officer messes, in contrast, were well stocked with food, liquor, and tobacco. Iron Crosses had been liberally distributed to officers, but seldom to ratings. Daily drill had become rigid and stultifying, with every advantage for officers, real or imagined, magnified into class warfare.

Since many of the navy's junior officers—second and first lieutenants as well as a few lieutenant commanders—had volunteered for the U-boat service, the men more and more came under the command of reserve officers and hastily commissioned sea cadets. Many of these lacked the tact and the patience necessary to deal with battle-experienced enlisted men twice their age. The result was rebellion at Wilhelmshaven. Several thousand sailors from the battleships *Prinzregent Luitpold, Friedrich der Grosse, Kaiser, Kaiserin,* and *Westfalen* as well as the cruiser *Pillau* refused to obey orders and demanded more equitable treatment. Military police and shore patrols quickly quelled the revolt, but not before five death sentences and numerous prison terms had been handed down. Interestingly, the rebellion gripped only those units that lay idle in port, the capital ships, and bypassed those constantly in action and commanded by subalterns, torpedo-boats and U-boats. Much the same malady afflicted the Austro-Hungarian fleet in its Adriatic ports early in 1918.

It is hardly surprising that the Prussian Army in the last year of World War I sought victory first and foremost by falling back on what it had always done best—military training and small-unit tactics. Put differently, Ludendorff placed his faith entirely in Captain Geyer's refined tactics to break the strategic stalemate in the west. A vast host of captains and majors from the roughly seventy assault divisions chosen to spearhead Operation Michael underwent extensive weeks-long training in eighty-men cohorts at Sedan and Valenciennes.[30] Still, Ludendorff recognized that the army lacked seasoned officers, and hastily advanced young and inexperienced subalterns, mainly from reserve formations, to command companies. Training was especially critical under such conditions.

Ludendorff once more called on Geyer to draft the assault manual for Michael. In January 1918, Geyer produced "The Attack in Position Warfare," featuring infiltration tactics by small units commanded by lieutenants and captains.[31] Tactical virtuosity had replaced strategic calculus at German headquarters. Ludendorff refused even to use the word strategy. And when queried by his staff about Michael's operational objective, he curtly responded: "I object to the word 'operation.' We will punch a hole into [their line]. For the rest, we shall see."[32] In the verdict of the historian Dennis Showalter, Ludendorff never rose above the intellectual level of a regimental colonel commanding infantry.

German forces scored spectacular initial battlefield gains, but the overall campaign quickly lost focus. By the end of March 1918, it ground to a halt after a series of unfocused small attacks with limited objectives against first the British, then the French, and then again the British. Diary after diary of lieutenants commanding infantry companies in the Bavarian Army attests to the fact that the assault, while tactically brilliant, had been devoid of any operational (let alone strategic) purpose.[33] After the Second Battle of the Marne in July 1918, the German Army was spent, physically exhausted and morally defeated.

In late 1918, both army and navy suffered internal collapse. The officers of the land forces especially faced a severe crisis. Inexperienced in politics and in dealing with rebellious soldiers, they proved unable to master a rapidly disintegrating army.

Gen. Wilhelm Groener, Ludendorff's successor as first quartermaster-general, in the first week of October estimated the number of "missing" or deserters at anywhere between 200,000 and 1.5 million![34] Four weeks later, barely a dozen fully-ready combat divisions guarded the Belgian Channel and the Upper Rhine.

In desperation, Groener called a meeting of fifty of his most senior commanders. Of the thirty-nine officers who made it to the council on 9 November, only one guaranteed that the men stood squarely behind their Supreme War Lord; only one believed that the army would still fight for "King and Fatherland"; and twenty-seven bluntly stated that the army would not obey orders to march against Bolsheviks and socialists at home.[35] Reserves moving up to the front had been met with shouts of "strike breakers" and "war prolongers" even by the elite Potsdam Guards. In short, the officer corps of Frederick William I, Frederick the Great, Gebhard Blücher, and the Elder Moltke had suffered a near-complete breakdown. The historian Wilhelm Deist described this disintegration as "a veiled military strike."[36]

With regard to the navy, the debacle of August 1917 was repeated. Too many sea cadets and junior officers from the reserves had been hastily promoted into command positions. A major organizational shuffle between August and October 1918 resulted in 48 percent of fleet captains and 45 percent of first officers being relieved of their posts; on many ships, both captain and first officer had been reassigned. The resulting breakdown in authority and discipline did not go unnoticed. Lt. Cdr. Bogislav von Selchow of the Admiralty Staff warned the fleet that it took two to execute a command, "namely the one who commands—he is very important—and the one who obeys; he is also important."[37]

Fleet commanders discussed Selchow's warning—and then ignored it at their own risk. On 29–30 October 1918, sailors on the battleships *König, Kronprinz Wilhelm,* and *Markgraf* doused the fires in the boilers, hoisted the red flag of revolt, and, joined by thousands of rebellious sailors from other ships, headed ashore to foster revolt. Senior naval officers in an act of utter political naiveté dispersed the rebellious sailors, once back aboard their ships, among Germany's North Sea and Baltic Sea ports. There they stormed city halls and police barracks, and then headed inland to spread the revolt. That his darling "boys in blue" started the German revolution was probably the Kaiser's most bitter disappointment in the fall of 1918.

Did the German armed forces study the causes of their collapse? And did they draw "lessons" from it? The answer to both is "yes." With regard to its future officer corps, the *Reichsmarine* of the Weimar Republic paid off the vast majority of captains and admirals and retained the best subalterns. In March 1919, Vice Adm. Adolf von Trotha, head of the newly created Admiralty, candidly stated: "I want to preserve the smallest seed so that when the time comes, a useful tree will grow from it."[38] The "seeds" collected and nurtured were junior naval officers, many of them lieutenant commanders such as Wilhelm Canaris, Rolf Carls, Otto Ciliax, Karl Dönitz, Günther Lütjens, and Erich Raeder. The "tree" that grew from them twenty years later was the Third Reich's *Kriegsmarine.*

Above all, the navy was determined never again to suffer revolution. What the historian Michael Salewski has called the "taboo" of rebellion in 1917 and revolution in 1918 drove its steadfast adherence to National Socialism down to 1945.[39] With a pathos that is still difficult to fathom, Admiral Raeder on 3 September 1939 penned a special situation report, vowing that this time 'round the navy would not remain idle in port, but rather would "die gallantly."[40] True to his word, Raeder sent every available warship out on the open seas—until all were either sunk or severely damaged. His successor, Grand Admiral Dönitz, dispatched 739 U-boats and 30,003 sailors to their watery graves.

The *Reichswehr* under Gen. Hans von Seeckt retired its line officers en masse, and almost exclusively retained junior General Staff officers. It continued the prewar myth of an apolitical officer corps, shielding that corps from the vicissitudes of German politics in the 1920s and burdening it with an education and promotion system that was almost brutal in its emphasis on performance and learning. As the historian David Spires has shown, the *Reichswehr* strove for a youthful officer corps of superior quality, eliminated ineffective units and programs, and placed mastery of technical knowledge above social origins as the key to success. Seeckt encouraged his charges to foster the narrow professionalism of the tactical-technical specialist rather than the broad outlook of the generalist. This overriding emphasis on obedience and duty encouraged neither individual initiative nor independent thought—not to speak of civil courage and moral fortitude.[41]

With regard to "lessons" learned, the brightest of the Weimar Republic's disguised general staff, the *Truppenamt,* spent the 1920s wrestling with the problem of how to conduct a war of the industrialized masses in the wake of collapse and revolution. These officers, such as Heinz Guderian, Wilhelm Heye, and Carl-Heinrich as well as Joachim von Stülpnagel, like Bruchmüller, Geyer, and Rohr before them, were primarily captains, majors, and lieutenant colonels. Later, they became the generals and field marshals of the Third Reich.

Interestingly, the major "lessons" they drew were political.[42] The armed forces, so the argument ran, should never again allow themselves to be sucked into the daily grind of party politics as they had under Ludendorff and Hindenburg. Instead, officers would concentrate on their narrow military specialties—*Kriegshandwerk*—and agree to go to war again only when a government appeared that promised to spare them negotiating with trade unions, managing the nation's labor and food supplies, defining war aims, and maintaining domestic order. They were overjoyed when such a possibility presented itself in 1933.

Notes

1. Bernhard von Bülow, *Denkwürdigkeiten,* 4 vols. (Berlin: Ullstein, 1930–31), 3:197-98.
2. Daniel J. Hughes, *The King's Finest: A Social and Bureaucratic Profile of Prussia's General Officers, 1871–1914* (Westport, Conn.: Praeger, 1987), 67–68.
3. See *Handbuch zur deutschen Militärgeschichte 1648–1939,* 5 vols. (Munich: Bernard & Graefe, 1979), 3:159.

4. John Moncure, *Forging the King's Sword: Military Education between Tradition and Modernization—The Case of the Royal Prussian Cadet Corps, 1871–1918* (New York: Peter Lang, 1993), 143, 153.

5. Ernst von Salomon, *Die Kadetten* (Reinbek bei Hamburg: Rowohlt, 1968), 28, 36.

6. Tagebücher v. Hutier, entry for 20 May 1918. Bundesarchiv-Militärarchiv, Freiburg, Kriegsgeschichtliche Forschungsanstalt des Heeres, W-10/50640.

7. See the superb chapter "The Virgin Soldiers" in Dennis E. Showalter, *Tannenberg: Clash of Empires* (Hamden, Conn.: Archon Books, 1991), 105ff.

8. See chapter 9 of Joseph Roth, *The Radetzky March* (Woodstock, N.Y.: Overlook Press, 1995), 125ff.

9. Hughes, 95–102.

10. See *Handbuch zur deutschen Militärgeschichte,* 4/2: 120, for a detailed program listing. Most officers took either French or Russian.

11. Emil Obermann, *Soldaten, Bürger, Militaristen, Militär und Demokratie in Deutschland* (Stuttgart: Cotta, 1958), 93. See also Manfred Messerschmidt, "German Military Effectiveness between 1919 and 1939," in Allan R. Millett and Williamson Murray, eds., *Military Effectiveness,* 3 vols. (Boston: Allen & Unwin, 1987–88), 2:243–44. By the 1930s, tactics and military history dominated the curriculum.

12. For an example of the oath, see Holger H. Herwig, *The German Naval Officer Corps: A Social and Political History 1890–1918* (Oxford, England: Clarendon Press, 1973), 67.

13. Hermann Lützow, *Die Seeoffizier=Laufbahn* (Berlin: Eisenschmidt, 1913), 10.

14. Donald M. Schurman, *Julian S. Corbett, 1854–1922: Historian of British Maritime Policy from Drake to Jellicoe* (London: Royal Historical Society, 1981), 44; Friedrich Forstmeier, "Probleme der Erziehung und Ausbildung in der Kaiserlichen Marine in Abhängigkeit von geistiger Situation und sozialer Struktur," *Marine-Rundschau* 63 (1966): 195.

15. Alexander Rost in *Die Welt am Sonntag,* 47 (1979), 46.

16. Lützow, 25.

17. Cited in Herwig, *German Naval Officer Corps,* 100.

18. See Gerhard Ritter, *The Schlieffen Plan; Critique of a Myth* (London: Wolff, 1958).

19. John Keegan, *The Mask of Command* (New York: Viking Penguin, 1987), 248.

20. Reichsarchiv, *Der Weltkrieg 1914 bis 1918,* 14 vols. I: *Die Grenzschlachten im Westen* (Berlin: Mittler & Sohn, 1925), 69.

21. Martin van Creveld, *Supplying War: Logistics from Wallenstein to Patton* (Cambridge, England: Cambridge University Press, 1977), 140; Jehuda L. Wallach, *The Dogma of the Battle of Annihilation: The Theories of Clausewitz and Schlieffen and Their Impact on the German Conduct of Two World Wars* (Westport, Conn. and London: Greenwood Press, 1986), 110.

22. Cited in Arnold Rechberg, *Reichsniedergang. Ein Beitrag zu dessen Ursachen aus meinen persönlichen Erinnerungen* (Munich: Musarion Verlag, 1919), 21.

23. War Minister von Krobatin's calculations of 31 March 1915. Österreichisches Staatsarchiv-Kriegsarchiv, MKSM, 69-9/9.

24. Hermann von François, *Gorlice 1915. Der Karpethendurchbruch und die Befreiung von Galizien* (Leipzig: Koehler, 1922), 47–48.

25. David T. Zabecki, *Steel Wind: Colonel Georg Bruchmüller and the Birth of Modern Artillery* (Westport, Conn. and London: Praeger, 1994), 30, 78.

26. Holger H. Herwig, "The Dynamics of Necessity: German Military Policy during the Great War," in Millett and Murray, 1:95; Reichsarchiv, *Der Weltkrieg 1914 bis 1918.* XII: *Die Kriegführung im Frühjahr 1917* (Berlin: Mittler & Sohn, 1939), 40–51.

27. Falkenhayn's memorandum of 27 June 1916. Bundesarchiv-Militärarchiv, Freiburg, PG 3/ 305 Sturmabteilung.

28. Ibid.

29. Holger H. Herwig, *"Luxury" Fleet: The Imperial German Navy 1888–1918* (London and Atlantic Highlands, N.J.: Ashfield Press, 1987), 230.

30. Übungskurs Sedan, September 1917. Bundesarchiv-Militärarchiv, Freiburg.

31. Erich Ludendorff, *Urkunden der Obersten Heeresleitung über ihre Tätigkeit 1916/18* (Berlin: Mittler & Sohn, 1920), 641; and Bruce I. Gudmundsson, *Stormtroop Tactics: Innovation in the German Army, 1914–1918* (Westport, Conn. and London: Praeger, 1989), 149–52.

32. Cited in Crown Prince Rupprecht of Bavaria, *Mein Kriegstagebuch,* 3 vols. (Berlin: Deutscher Nationalverlag, 1929), 2:372.

33. See chapter 10 of Holger H. Herwig, *The First World War: Germany and Austria-Hungary 1914–1918* (London: Arnold, 1997).

34. *Der Weltkrieg 1914 bis 1918.* XIV: *Die Kriegführung an der Westfront im Jahre 1918* (Berlin: Mittler & Sohn, 1944), 666, 697, 760.

35. Ibid., 716.

36. Wilhelm Deist, "Der militärische Zusammenbruch des Kaiserreichs. Zur Realität der 'Dolchstosslegende,'" in Ursula Büttner, ed., *Das Unrechtsregime. Internationale Forschung über den Nationalsozialismus,* 2 vols. (Hamburg: Christians, 1986), 1:101–29.

37. Bundesarchiv-Koblenz, "Logbuch Bogislav von Selchow," 37:7368.

38. Cited in Herwig, *German Naval Officer Corps,* 265.

39. See Michael Salewski, *Die deutsche Seekriegsleitung 1935 bis 1945,* 3 vols. (Frankfurt: Bernard & Graefe, 1970–75), I.

40. Gerhard Wagner, ed., *Lagevorträge des Oberbefehlshabers der Kriegsmarine vor Hitler 1939–1945* (Munich: Lehmanns, 1972), 19–21.

41. David N. Spires, *Image and Reality: The Making of the German Officer, 1921–1933* (Westport, Conn. and London: Greenwood Press, 1984), 127–29.

42. See Wilhelm Deist, "Die Reichswehr und der Krieg der Zukunft," *Militärgeschichtliche Mitteilungen* 45 (1989): 81–92.

Section II
Interwar Junior Officer Education and Training

Along with armed forces of millions, World War I had brought the mechanization of the battlefield and, with the airplane, the skies overhead. During the two decades before the next total war, national systems of officer education and training reflected efforts to assess the implications of these changes. Results were mixed. German education and training produced officers as technically proficient in the new style of warfare as in the old, but with the same general ignorance of war's political and strategic dimensions as their predecessors. In Great Britain, officers returned to their prewar ethos (gentlemen naturally possessing the requisites of leadership). With some exceptions, they focused on their regiments and not on professional studies, apparently preferring mastery of equitation to considering the internal combustion engine's significance for future war.[1] At Sandhurst, army cadets graduated from the eighteen-month course with inadequate instruction in motor vehicles, but, as late as 1937, with 125 curricular hours in horsemanship.[2] At the same time, however, the British were the first to establish an academy for air officer cadets—Royal Air Force College, Cranwell, in 1919.

After the Great War, the continued reliance on mass-mobilization armies meant that large numbers of reserve junior officers and improved training programs to prepare them for combat would be required. The first two essays in this section, by Jennie Kiesling on France and Brian Sullivan on Italy, show just how badly those two nations failed in the latter respect between the wars. Kiesling, who teaches history at West Point, finds the source of France's defeat in 1940 in the "failure of leadership at all levels." In the army, four-fifths of the officers were reservists who either had little opportunity or little inclination to develop professional skills. In part, this stemmed from a cost-conscious legislature often unwilling to provide funds for training, but more fundamentally from an educational philosophy that gave primacy to general education over professional training—"learning how to think over what to think about." In contrast to reservists who had little active service, regular officers could acquire professional expertise on top of their general educations. Kiesling, however, cautions against drawing the wrong lesson from the French experience. Professional education and training, she argues, should not come at the expense of the ability to think critically or the sense of "moral and philosophical sophistication" cultivated by a general education. Those attributes of mind enable officers to understand what they are fighting for and why force may be justly and rationally applied.

Like their French counterparts, Italian reserve junior officers had little practical professional training. Brian Sullivan, a faculty member at the U.S. National Defense University, points, however, to different reasons for the Italian shortfall. To keep himself in power, Benito Mussolini bought the support of the army's upper ranks, enhancing their professional and financial security by maintaining numbers of field-grade and general officers far in excess of needs throughout most of the 1920s and 1930s. The cost of this political arrangement was borne by the junior officers of the army and the Fascist militia, overwhelmingly reservists after the mid-1920s. Resources that should have gone to field training for reserve junior officers (especially in the doctrine and employment concepts for mobile and mechanized warfare developed by the Army General Staff after 1935) went instead to paying for the army's bloated upper ranks and for Italian adventures in Ethiopia and Spain, and pacification operations in East Africa. The result, Sullivan concludes, was that reserve junior officers performed ineffectively—an important factor in the defeats suffered by Italian arms in 1940–41.

Political consequences of a much different kind—those resulting from fundamental political revolution—influenced officer selection and determined the nature and quality of junior officer education and training in the Red Army. Mark von Hagen, director of Columbia University's Harriman Institute, describes how the demands of revolutionary ideology clashed with the need to create a professionally competent officer corps. Compounding this conflict was the multinational and multiethnic character of the Soviet state with its diverse languages, uneven levels of literacy and economic development, and medley of military traditions and training. All of these factors, along with military expansion, reorganization, and purges of the officer corps in the 1930s, held back achieving standardization, let alone incorporating the principles and techniques of mechanized warfare in officer education and training. "Only during World War II," writes von Hagen, "did the army learn its craft on the job and at a frightening cost in manpower."

Critics of officer education and training methods often claim that the process stresses conformity to accepted theory and techniques and fails to encourage creative thought and innovative solutions. In the section's last essay, Charles Kirkpatrick, a U.S. Army historian, details the concentration on the "orthodoxy of approved doctrine and methods of staff and command" that permeated all levels (except at the War College) of the U.S. Army's school system during the interwar period. This emphasis, contends Kirkpatrick, was the principal reason that American officers, many of whom had served only in company grades for the greater part of two decades, emerged rapidly in World War II to "become generals who could organize, train, and wield divisions in battle." Kirkpatrick's analysis is a useful reminder that the exceptional tactician or strategist, a George Patton or a George Marshall, is likely to appear no matter the nature of his military schooling, but mastery by the "undistinguished middle" of professional fundamentals—usually not an imaginative undertaking, can be just as critical.

Notes

1. Brian Bond, *British Military Policy between the Two World Wars* (Oxford, England: Clarendon Press, 1980), chap. 2.

2. John Smyth, *Sandhurst* (London: Weidenfeld & Nicholson, 1961), 199–200.

Educated But Not Trained: Junior Officers in Interwar France

E. C. Kiesling

The French Army's defeat in 1940 stemmed not from material or doctrinal inferiority to the *Wehrmacht* but from failures of leadership at every level. Senior commanders committed the army to an inflexible strategy, field-grade officers declined to exploit the "Sitzkrieg" to train their troops, and subalterns failed to display the effective combat leadership necessary to redeem a desperate situation. Behind such a comprehensive failure of leadership, one expects to find a defective system for selecting, educating, and training leaders. In this respect, it would be useful to isolate the causes of French inadequacy, whether to deduce prescriptions for contemporary use or to assure ourselves of the soundness of our own institutions. In fact, deficiencies in the preparation of French company-grade officers between the two world wars suggest complex, even dismaying, lessons about future junior officer education and training.

French officer education evokes images of the prestigious academies of Saint-Cyr and the *École Polytechnique,* which together graduated perhaps 650 sublieutenants a year during the 1930s. A few hundred more new officers came annually from programs for noncommissioned officers in the infantry, cavalry, artillery, and armor at Saint-Maixent, Saumur, Poitiers, and Versailles respectively. About 300 other noncommissioned officers received direct commissions in the technical branches. The processes shaping the approximately 1,200 new regular officers each year, however, are peripheral to the thesis of this paper, just as the 32,000 regular officers whose ranks they replenished had less impact on the army's battlefield performance than did its 133,000 reserve officers. Too many discussions of French leadership have saluted respectfully in the direction of Saint-Cyr and the *École Supérieure de Guerre* without acknowledging that regular officers comprised only about one-fifth of the corps as a whole and that graduates of the esteemed staff college were a mere eight percent of that one-fifth. This paper will examine the education and training of regular junior officers in some detail, but only to identify French thinking about junior officer preparation generally. It will then argue that, however sound the French Army's mechanisms for producing regular officers, France failed to meet the more difficult challenge of training the huge numbers of reserve officers required by a mass mobilization army.

Regular officers came from three sources. The most prestigious route was the two-year course at either Saint-Cyr or *l'École Polytechnique* followed by a year (two for artillery and engineers) at the appropriate *école d'application.*[1] Of 32,000 officers in service in 1932, 7,522 were products of these premier academies.[2]

Saint-Cyr was purely a military school, founded in 1808 as the *École Impériale Militaire*, but the curriculum reflected the French notion that military subjects were only a small part of an officer's professional education. Required courses covered history, geography, topography, drawing, applied science, law, administration, and foreign languages. On the military side, cadets devoted the first year to infantry skills and the second to those of their eventual arm of service—usually infantry or cavalry.[3] The intervening summer was spent in an infantry battalion, a familiarization period unnecessary before 1914, when future cadets had to serve as privates even before entering the academy.[4] In sharp distinction to the Royal Military College at Sandhurst and the United States Military Academy at West Point, Saint-Cyr shared the French Army's general disinterest in organized sport. The daily predawn ride, followed by a drill session, was alleged to have had no purpose other than to inure future officers to suffering. The only other required exercise was fencing. Only because some cadets volunteered did the academy field a decent rugby side and a crew to row against the rival *Polytechniciens*.[5] The Cavalry School at Saumur required horseback riding to instill courage and encouraged automobile racing to nourish the cavalry spirit.[6]

Engineer and artillery officers came from the *École Polytechnique*, France's foremost school of science and engineering. Created in 1794 and transformed into a military school under Napoleon in 1804,[7] the "Poly" was a state-supported engineering school whose students wore military uniforms, studied military subjects, and had as their motto *"pour la patrie . . . et pour la gloire."* Graduates repaid the cost of their education with a period of military service.[8] Unlike the similar American academy at West Point, the *Poly* was only a preparation for further education at an advanced school of civil, military, or naval engineering or artillery. In the meantime, the *Polytechnique* offered an academic and technical curriculum intended explicitly to produce leaders whose fitness to command rested on their trained ratiocination. It produced "an intellectual aristocracy, in which the stress [was] on mental self-confidence—perhaps essential qualities in any power élites."[9] One graduate bolsters his claim that *Polytechniciens* were better trained even than the graduates of the staff college with the observation that two phrases recur in the graduates' posthumous citations: "clarity and speed of decision" and "precision of orders."[10]

Admission to Saint-Cyr and the *École Polytechnique*, as to any of the *grandes écoles* at the pinnacle of the French educational hierarchy, was highly competitive even though the pool of qualified candidates was, in a day of undemocratic secondary education, relatively small.[11] Applicants to Saint-Cyr had to be between eighteen and twenty-two years of age and to have passed part one, and preferably part two, of the baccalaureate examination. The *Polytechnique* accepted only men between twenty and twenty-three who had passed both parts of the baccalaureate. To take the baccalaureate required seven years of demanding secondary education at a *lycée*.[12] After achieving the baccalaureate, prospective applicants for a *grande école* normally spent between one and three years in additional *lycée* courses designed to prepare them for the admissions examinations. That one earned the baccalaureate at eighteen but could not enter the *Polytechnique* until the age of twenty confirms that two years of additional instruction were deemed essential, and the minimum age for admission was raised to twenty-one in 1934.[13] None of this education was free, and scholarships were rare; the *lycée* and the baccalaureate, and the *grandes écoles*, remained middle-class prerogatives.[14] Because Saint-Cyr, alone among the *grandes écoles*, accepted applicants with the "classical" as well as the more technical "scientific" *bac*, it was possible, as the age standards indicate, to be admitted without postbaccalaureate study. Even so, a 1902 critic—admittedly writing from a cavalryman's point of view—complained that admission to Saint-Cyr depended on a "purely intellectual examination" and a doctor's attestation that the candidate was neither crippled nor blind. Twenty-four years later, the academy's commandant complained that its educational entrance requirements led young men to sacrifice their physical fitness.[15]

The second route to a regular army commission was for noncommissioned and reserve officers and involved a two-year course at one of the branch schools: Saint-Maixent for infantry and tanks, Saumur for cavalry, Poitiers for artillery, or Versailles for engineers and mechanics.[16] Admission to these programs was by comprehensive written examinations covering French, literature, mathematics, and physics (with additional subjects for engineer and artillery candidates). Candidates also took oral examinations in history, geography, and trigonometry.[17] In principle, graduates of these programs were eligible for any assignment, but their greater age and slighter education put them at a disadvantage relative to the self-perpetuating elite produced by Saint-Cyr and the *École Polytechnique*.

Finally, some experienced noncommissioned officers received direct commissions to fill the army's need for instructors, administrators, and supervisors of technicians. These men, however, were not on the track for command.

The process by which the regular army selected and prepared its junior lieutenants reveals much about the French attitude toward education. Most striking is the emphasis on what the French called "general culture" rather than professional training. General culture was acquired in the long years at the *lycée;* the baccalaureate degree, virtually obligatory for a successful career, was proof that one had attained it. General culture included specific bodies of knowledge—especially French language and literature, mathematics, a foreign language, and either science or classical languages—but it also represented a way of thinking. The goal of French secondary education was to teach the art of *bien penser,* thinking well. Behind French education was the theory that, "By the right use of the power of reason every problem can be solved, every intellectual difficulty surmounted."[18]

The army fully acknowledged the primacy of learning how to think over learning what to think about. A post–World War I General Staff proposal to merge the two great military schools specified that the order of priority at the unified academy would be general culture, professional development, and, finally, military education.[19] At the French war college, the *Centre des Hautes Études Militaires,* forty-one of the lectures in the annual course fell under the rubric of general culture while only twenty-six covered military topics. A 1917 treatise by a Lieutenant Caillet, *"Le nouvel officier d'infanterie en guerre, ce qu'il doit savoir"* (the new infantry officer in war, what he needs to know) insists that the three attributes required of an officer are general culture, conscience, and energy (defined as willpower combined with courage). Caillet describes general culture as "the ability to understand or assimilate everything that could be useful . . . including a critical spirit which allows one to go beyond the narrow constraints of military regulations." He argues that general culture gives the officer the intellectual resources to understand that discipline must include initiative and intelligence. Moreover, it prepares him to appraise situations, to explain them to his soldiers, and to prepare contingency plans. Above all, general culture allows one "to rise above immediate situations and not to be stopped by factors that paralyze simpler souls."[20] Ten years after Lieutenant Caillet described the importance of general culture, a Colonel Lucas in an article in the official *Revue militaire française* also elevated above professional knowledge the intellectual qualities of the commander, which he identified as imagination, judgment, prudence, and audacity.[21]

A preference for general over professional education also marked the four paths to a reserve commission. The only reserve officers with a generally high level of professional preparation were a relatively small group composed of officers and noncommissioned officers retired from the active army. Much less well prepared were the 10 percent of reserve sublieutenants who had by law to be promoted from the pool of reserve noncommissioned officers.[22] The third path to a reserve commission was through one of six *grandes écoles,* the elite institutions at the top of the French educational hierarchy. Dedicated to preparing men for state service, these schools provided three years of obligatory military courses and, at least in the early 1930s, bestowed the officer's commission immediately upon graduation. Veterans not of military service but of the *lycée,* preparatory courses, and three years of advanced technical study, these new officers reported as sublieutenants to special "officer student" platoons at the branch training centers. Later in the 1930s, the government determined to save money by delaying the award of the commission. Thereafter, *grande école* graduates spent six months at Saint-Cyr, Saint-Maixent, Saumur, or Versailles in the rank of "aspirant." In either case, they finally joined their units as sublieutenants for the remaining six months of their active duty obligation.[23]

The élite *grandes écoles* graduated relatively few potential officers, and the most common route to a reserve commission, accounting for about three-quarters of the total, was through a two-year course at a faculty of letters, science, theology, law, or medicine.[24] These institutions offered an optional two-year military program of 240 hours and 12 half-day exercises leading to a certificate of *préparation militaire supérieure* (PMS).[25] The top 5 percent graduating with the PMS certificate joined the *grande école* products in the "officer student" or "aspirant" platoons at Saint-Cyr, Saint-Maixent, Saumer, or Versailles.[26] The remainder of the top half of each graduating class also went to the branch schools, but as privates in "student officer" (as opposed to "officer student") platoons.[27] After four months and a successful examination, they became aspirant officers. Six months later they were commissioned as reserve sublieutenants.

If their obligation had not already been met, the new sublieutenants would then report to a unit for their remaining active duty commitment. This was a "particularly delicate" stage in the green officers' training. Unaccustomed to troop service and unready for command, they required guidance, advice, and circumspect direction.[28] What they received, however, was all too often the disdain of the regular officers for men who were, from their point of view, merely elevated members of the year's conscript class. The Ministry of War considered the fledgling sublieutenant's "delicate" transitional period not an opportunity for training but a needless expense. Rather than paying supernumerary sublieutenants, the ministry proposed in 1935 to defer commissioning until the end of the active service period.[29] The month or so in grade may have served little purpose (and often included a period of leave), but the proposal to send new officers home without any troop service at all was another manifestation of the French Army's faith that education could take the place of training.

Because of the intense competition to avoid spending twelve months of conscript service as a private, the quality of the officer candidates was high. In 1933, for example, the army inducted 14,000 holders of the PMS certificate and could afford to employ the bottom half of the group as reserve noncommissioned officers.[30] Although the number of university graduates far exceeded the reserve officer vacancies, the Third Republic insisted that commissions be open to the vast majority of Frenchmen without a university education. Thus, conscripts lacking the PMS certificate could attain their colonel's permission to sit an examination for admission first to a preparatory platoon and then to a student officer platoon.

Details of the actual training are hard to come by, but Robert Felsenhardt's published account is revealing (unless its author had a strong sense of self-deprecation). Felsenhardt, a university graduate with the PMS certificate who applied to enter a reserve student officer platoon in 1922, claims to have received eighteen marks out of a possible twenty on the entrance examination simply for applauding his examiner's demonstration of bayonet drill. He reports that his four months at Saint-Maixent taught him only to install barbed wire, handle wire cutters, and slither through mud. As a sublieutenant in the 107th Infantry Regiment, Felsenhardt's manifest inability to lead troops on parade or to instruct them in weapons handling relegated him to preparing the regiment's Bastille Day fête.[31]

Reserve officer selection clearly demonstrates the priority the French Army gave to education over professional experience. Armed with a degree from a top-flight school, a young Frenchman could become a reserve officer without any active military service beyond his six months in a training platoon composed of fellow aspirants. Under the one-year service regime of 1930–35, those holders of the higher military preparation certificate who spent four months in the student reserve officer platoons and then six as aspirants were commissioned before they met any troops and were then immediately released from active service.

Before leaping to conclusions about the quality of reserve officers commissioned in this fashion one ought to reflect upon the confidence that the French had in this program, a confidence based largely on a conviction that their nation's educational institutions offered an incomparable background for future leaders. These young men possessed not only a solid technical education but disciplined minds and a deep appreciation for French culture. They believed that culture to be founded on a uniquely French understanding of the power of the human mind. As soldiers, they were motivated to fight both for France and for the commitment to knowledge and reason that France embodied. But French junior officers had to build professional competence on top of their intellectual foundations. The prospects of success differed strikingly for regular and reserve officers, and the difference is one with significant consequences for understanding French performance in 1940.

The first few years of the regular officer's career were his initiation to troop service, and he probably had little inclination for study. If, however, he had career ambitions, he had only a brief respite from academics before turning his attention to the entrance examination for the staff college, the famous *École Supérieure de Guerre*. Applicants to the college had to be between twenty-eight and thirty-eight and generally held the rank of captain. Most enrolled in a two-year correspondence course to which they devoted a minimum of twenty hours a week. Even so, two-thirds failed the twenty-seven hours of written examinations covering a divisional map exercise, general culture, military history, geography, topography, and English or German. The 120 or so who succeeded travelled to Paris for a week of oral examinations and an equitation test. About eighty were finally accepted into each class, most of them succeeding only at the second attempt. Although few French officers were admitted to the staff college, that many more took the exams or, at least, studied for them gave the school an intellectual leavening effect on the army as a whole.[32]

Before actually entering the *École Supérieure de Guerre*, each successful candidate spent six months doing a brief familiarization tour in each branch other than his own. In particular, officers were encouraged to use this period to qualify as air observers.[33]

The staff college course was notoriously difficult. As at other French schools from the *lycée* on up, the faculty strove less to impart information than to teach a method of thought and action. The course was taught largely through *cas concrets,* case studies, in which the students had to decide how to handle a given set of circumstances and then to issue the appropriate orders. The cases were designed to be difficult, in one American participant's view, artificially so. Major R. K. Sutherland felt that instructors outdid themselves in creating unconventional problems so as to demonstrate the validity of the approved methods of operation under the most extreme conditions. Unimpressed with this pedagogical style, the American acknowledged that it "teaches mental agility and inculcates the ability to adapt oneself to extreme situations," but suggested that "a student should have as a base of departure the ability to solve rapidly and accurately an ordinary problem, acquired through considerable practice, before being given the extremely unusual situations calling for equally unusual solutions."[34] General Marie-Eugène Debeney, the postwar director of the *École Supérieure de Guerre*, insisted, however, that the method not only honed decision, judgment, and good sense, but drew upon the imagination as well.[35]

Debeney did not mention the most notorious feature of the college, the intensity of the time pressure imposed on the students. Afternoon exercises deliberately compressed five hours of work into three, and the ten-day homework exercises distributed on Saturday mornings consumed twenty to fifty hours of each student's evenings and weekends. The purpose of these daunting assignments appeared to an English participant "not to be so much to attain the impossible as to train officers to think and work quickly and in a somewhat rattled state of mind."[36] To make sure that the physical tiredness of the officers approximated wartime conditions, the day began with a mandatory horseback ride or fencing session at 7 A.M. The only people in Paris who worked longer were hospital interns.[37]

The interwar course at the *École Supérieure de Guerre* contained three elements: general culture, military education, and languages.[38] The military subjects covered in the course included the roles of the various arms and services, military history (i.e., the Great War), intelligence, and the role of the navy. The general culture lectures, once again the primary element of the course, focused on geography and political science.

The French had no doubts about the quality of the education offered at the *École Supérieure de Guerre*. As a member of the 1927–28 class claimed, graduates had "confidence in their ability to think. . . ."[39] Addressing the French parliament on 25 September 1920, President Alexandre Millerand announced that the *École Supérieure de Guerre* would be known in the future as *l'école de la victoire.*[40] Foreign armies took French claims seriously and eagerly sent officers to study in Paris.[41] Although in the thirty years before 1914 only fifty-two foreigners from twelve countries had attended the French staff college, between the two world wars thirty-eight countries sent more than 500 officers. A course for colonels and generals in 1926 had a ratio of forty foreigners to only twenty-nine Frenchmen. Poland, moreover, called upon French guidance to establish a Polish military academy.[42] (The Francophobes and Germanophiles who dominate military history written in the English language will point out that the rival *Kriegsakademie* had been shut down by the Versailles Treaty. There is no reason to believe, however, that contemporary armies had forgotten why the Germans had signed that treaty.)

The staff college was but the most famous of a series of educational opportunities offered to regular officers in the French Army. Some of its graduates could look forward to attending the *"École des Maréchals"*—the *Centre des Hautes Études Militaires.* From 1936 onward, a select handful reached the apex of French military education, the *Collège des Hautes Études de la Défense Nationale.*[43] Colonels and generals who missed out on the *Centre des Hautes Études Militaires* could attend either a combined arms course at the *École Militaire* in Paris or a branch course for cavalry, artillery, or aviation.[44] Each branch also offered a three-week command course for newly promoted majors as well as a variety of specialized courses.[45] Unlike the *Centre des Hautes Études Militaires* and the *Collège des Hautes Études de la Défense Nationale,* the combined arms course and the branch courses were devoted to professional knowledge rather than general culture.

The French high command encouraged officers to continue their professional development outside of structured courses. At the end of the Great War, the General Staff's historical section received the grander title of historical service and the mission to disseminate the lessons of the recent conflict throughout the army. Officers found another source of information in the official branch journals: *Revue d'infanterie, Revue de cavalerie, Revue d'artillerie, Revue du génie, Revue des forces aériennes, Revue de l'intendance,* and *Revue des troupes coloniales.* Especially important was the General Staff's own *Revue militaire française.* In addition to articles on domestic and foreign military subjects, *Revue militaire française* surveyed the contents of recent French journals and offered a bibliographical section to guide officers in their professional reading and garrison libraries in their acquisitions.[46]

Although all of the professional journals aimed at French officers were published under official auspices (and implicit censorship), the range of opinions expressed was wide. For example, the publication in 1934 of the Austrian Ludwig Ritter von Eimannsberger's *Der Kampfwagenkrieg* evoked conflicting responses. In two different numbers of the 1937 *Revue d'infanterie* a Captain Soury denied the validity of Eimannsberger's conclusions, while a Major Gervaise argued that the Austrian's ideas paralleled those of the French high command. In *Revue militare française,* a Captain Glain praised Eimannsberger's logic and concluded that his ideas, however contrary to French premises, deserved further study. Cautiously, Glain signed the piece with the letter "X," but his name appeared, nonetheless, in the journal's table of contents.[47]

So unorthodox were some of the articles in the professional military journals that Gen. L. A. Colson complained that the *Revue militaire française* ought to eliminate "speculative" writings in favor of "constructive studies" designed to reinforce official doctrine. He proposed that the General Staff ought to suggest topics to trustworthy officers and to guide them "towards the ideas and lessons to be emphasized."[48] To Colson's argument for closer editorial control over the *Revue militaire française,* the General Staff replied that the review was already too much an official instrument and should be replaced by a free-thinking journal independent of the military hierarchy.[49] Indeed, in 1937, the General Staff abolished the *Revue militaire française* in favor of the new *Revue militaire générale.* In introducing the new publication, Marshal Franchet d'Esperey heralded the end of "the era of intellectual dictatorship."[50] The dispute between Colson and his fellow generals over the role of the official professional journals suggests both that French soldiers took education seriously and that the high command did not stifle independent thought.

Further evidence of the high command's commitment to education was the official effort, of which the bibliographical surveys in *Revue militaire française* were only one facet, to encourage officers to read. As of 1 April 1932, French garrison libraries (not including those of the Ministry of War, the service schools, or individual regiments) contained 1,573,000 volumes distributed among 320 garrisons in France, North Africa, and the Levant. A contemporary study of library use suggests that the books were not only available but were read. Dividing the number of books checked out in each of seventeen sample garrisons by the number of officers stationed there, the study concluded that the number of books read by the average officer in 1931 ranged from sixty-six at Besançon and fifty-eight at Verdun to thirty at La Rochelle.[51] Of course, one may wonder what they read. It is dismaying to learn from Anthony Adamthwaite that, although a French translation of Gen. Heinz Guderian's *Achtung Panzer* had been distributed to garrison libraries, as of the winter of 1937–38 not one copy's pages had been cut.[52]

If the French regular officer between the wars did not lack for educational opportunities, his chances to exercise command were limited. The ratio of generals to divisions was five to one. The army had 100 regiments for 236 colonels, 300 battalions for 1,247 majors, and 1,200 companies for 3,689 captains.[53] Still, there is no reason to believe that the French officer was worse off in this regard than were his contemporaries in other armies or are today's soldiers. For staff college graduates, at least, regulations insured a period of command at every grade.

Officer education looks rather different when the focus shifts from the regular officers, and especially from the favored graduates of the military academies and the staff college, to the reserve officers who made up 75 percent of a regiment's wartime establishment.[54] At the end of the commissioning process, the new reserve sublieutenant was usually a man with a first-rate general education who now needed professional training and experience to match his intellectual preparation. What he got, however, was the freedom to return to civilian life. His further preparation to lead men in battle would come through reserve duties and participation in the *écoles de perfectionnement des officiers de réserve*, both of which were of limited utility.

The reserve training obligation varied somewhat over the course of the interwar period (reserve exercises were not held from 1919 to 1927), but the eventual standard was that every French officer was liable for 120 days of reserve duty, normally in ten-day increments every two years. During that period, officers were, ideally, to command an active or a reserve unit. In the absence of a unit needing a commander, reserve officers were supposed to take turns commanding a formation provided for them by a regular army unit.[55] This scheme was a fantasy far divorced from the real condition of the French Army.

For one thing, reservists could only fulfill their obligations when the legislature funded training. So many exercises were cancelled for financial reasons that, in 1934, the Ministry of War took special steps to provide training for the 5,000 men who had done no training during their five years as reserve lieutenants.[56] When large numbers of reserve officers were called up, such as the 17,000 in 1933, 18,000 in 1934, and 20,000 in 1935, there were no troops for them to command.[57] Three hundred reserve officers called up for the Haute Marne maneuvers of 1936 merely watched as the 15th Motorized Infantry Division practiced an attack against the 21st Infantry Regiment and 107th Artillery Regiment.[58] Observing the cadre exercises at Valenciennes at the end of May 1937 were 500 otherwise unemployed reserve officers.[59] Whenever reserve troops were present for major maneuvers, so too were plenty of regular officers also in need of command experience and less likely to trip up in front of journalists, parliamentary observers, and foreign attachés.

Most reserve officer training took the form of cadre convocations, the call-up of a division's full complement of active and reserve officers. Getting acquainted was a primary objective, and units were encouraged to provide opportunities for *la vie en popote*—the comradeship of the officers' mess.[60] Of course, the utility of these gatherings depended on the degree of participation, and reserve officers could be less than diligent. Fifty-nine officers out of 230 called up for the 22nd Infantry Division's exercise of 20–26 September 1936 failed to appear, most of them without having been excused. Because of poor attendance, the convocations proved an unreliable way for higher commanders to meet their officers and to relieve the unsuitable. The price was paid upon mobilization in September 1939, when a typical division discovered that its reserve officers "came from the four corners of France and did not know one another. Those who possessed some practical notion of their role on campaign could be counted on one's fingers."[61] The junior officers' portion of a cadre convocation lasted for three days, during which they studied their unit's *journaux de mobilisation,* verified personnel assignments, inspected stored supplies and equipment, attended demonstrations of new weapons, and participated in terrain reconnaissances. Half a day was allocated to a demonstration of new equipment—if the mobilization centers had the required materiel. Under the circumstances, one is not surprised to discover an October 1940 assessment characterizing reserve officers by their "apathy, their lack of initiative, and their unconcern for their men." Criticized for failing to exercise their authority, they responded resentfully, "Command? . . . But we aren't used to it!"[62]

The defects in reserve officer training become more glaring when put in the context of the army reorganization laws of 1927 and 1928. The General Staff had accepted the twelve-month service law only with the stipulation that the reserve system be strengthened by bonding soldiers together into permanent units. Reservists were to be assigned to units derived from their active service regiments and to perform their training exercises within their complete wartime organizations.[63] Thus, "Men who will fight together in wartime must already have had the opportunity to get to know one another in time of peace. They must already know their leaders and their leaders must know them."[64] For reasons beyond the scope of this paper, permanent reserve units proved chimerical, and there proved to be no institutional means to ensure contact between the officers and soldiers of reserve units. To the extent that French units were organized geographically, men may have known their officers from civilian life. Given the weak professional skills of many reserve officers, familiarity may have contributed less to unit cohesion than to contempt for the officers' pretensions to command.[65]

The poor preparation of reserve officers was no secret within the French Army and received repeated emphasis in unit evaluations. The high command attempted to improve reserve leadership by organizing a *Union nationale des officiers de réserve,* disseminating the journal *L'officier de réserve,* and, above all, encouraging regular participation in courses offered at regional *écoles de perfectionnement des officiers de réserve.*[66] These voluntary courses consisted of twelve half-day sessions, which the officer was expected to repeat each year. The incentive to enroll was a railway pass allowing the bearer first-class travel on a third-class ticket and promises of future promotion in the *Légion d'honneur*. Of 34,000 officers enrolled in the inaugural course of 1926, only 3,500 attended with sufficient regularity to earn the rail pass, numbers that improved in 1927 to 42,000 and 12,000 respectively. In 1928, 22,000 were deemed to have been "assiduous," 26,700 in 1929, and 33,000 in 1930—still only a quarter of the eligible officers.[67] Frustrated, Minister of National Defense Edouard Daladier ordered a six-day training period in 1938 for all officers who had not attended the reserve schools, and in the following year sponsored a law making participation compulsory on penalty of extra reserve training.[68] Whether or not in response to Daladier's efforts, attendance reached 50,012 in 1938.[69]

If the lecture given by French armor expert Lt. Col. Jean Perré on "The Modern Tank" is a fair sample of the course, poor attendance was not the only defect of French reserve officer education.[70] Perré spent fourteen pages on definitions (What is a tank? What are its essential elements—armor, armament, speed, strategic mobility, weight?) before offering a six-page discussion (including further definitions) of employing tanks in the attack. Perré's purpose appears to have been to convince his audience that French armor doctrine derived irrefutably from the essential nature of the tank, that, given the definitions, "the deductions are obvious."[71] There was little in Perré's rationalist discourse likely to have been of practical use to a green lieutenant in the field.

In 1936, in another effort to improve the overall quality of the reserve officers, the army established special field training periods to prepare 80 reserve captains and 200 reserve lieutenants annually for promotion. Another special program sent reserve artillery officers to courses on new mortars and antitank guns at the *École de Perfectionnement de Sous-Officiers de Carrière*.[72] By 1938, about 800 of the army's 133,000 reserve officers participated each year in the three programs.

For the reserve officer who desired to increase his professional knowledge, there were a few other avenues. He could volunteer for extra (and probably unpaid) reserve duty, but he would not be with his own unit. He was entitled to check out books from the local garrison library. Like the seemingly hapless Robert Felsenhardt, he could learn by teaching courses in the *écoles de perfectionnnement* for reserve noncommissioned officers. Lieutenants and captains over thirty years old could apply for service with the General Staff. Those selected attended a two-week regional course followed by a three-week course at the *École Supérieure de Guerre* and were expected to keep up-to-date by attending the *école de perfectionnement* in one of nine major cities.[73]

Most reserve officers probably shared the attitude of Capt. Marc Bloch. A brilliant historian and energetic officer in two world wars, Bloch died courageously under *Gestapo* torture but could not be bothered to attend reserve officer courses.[74] One is not surprised to hear of a reserve machine gun officer's assurance to a British officer that he would learn after mobilization how to lay out a fire plan. These officers were more characteristic than those reserve lieutenants from a Maginot Line "crew" who were ready in 1939 because they had spent their peacetime Sundays measuring fields of fire.[75]

We have no definitive way to evaluate the performance of French junior officers in 1940 or to distinguish between the products of the regular and reserve training programs. The brief successes of units blessed with manageable assignments show that many French officers knew how to fight and how to lead. The casualty lists suggest that they understood the second lieutenant's primary role as well.[76] Still, the story as a whole begs for the conclusion that the reserve officers upon which France depended went to war with a solid educational background but without the training they needed.

It would be easy to jump from the reserve officers' lack of professional skills to the facile conclusion that the French Army overemphasized educational preparation in selecting officers and put too little weight on further training. The next seductive leap is to infer that the lesson to be learned from the French defeat is that today's army can afford to sacrifice education in "general culture" for technical and professional knowledge. That, however, would be a serious misreading of the French experience. French failure stemmed not from a mistaken belief in the importance of general education but from structural problems that limited the amount and utility of the training received by reserve officers after commissioning. Whatever the intentions of the high command, it found no way to provide reservists with opportunities to serve with troops, to attend professional schools, and to command. This essential professional experience would have been possible only in a country willing to spend huge sums on national defense and to commit its population to frequent periods of military training. France, like most countries, had to live not with the best military arrangements imaginable but with what was possible.

One lesson to learn from the French experience is that, because of the limited time and money available for professional military education, armies will be forced to choose between general education and professional skills. As it is easier to see how more battles are lost by officers who cannot handle a machine gun or read a map than by those who have not read Shakespeare or the *Gettysburg Address,* the temptation will be to focus on training and to relegate general studies to the realm of commendable spare-time activities.

But if that fine historian Marc Bloch's assessment of the fall of France is correct, then the truer lesson of the French defeat suggests a different course of action. Bloch attributed France's collapse to national educational institutions that had failed to imbue French citizens with a sense of common cause overriding differences of class and politics. Too many right-wing officers believed abstractly in France but were not committed to die for the Third Republic. For too many left-wing soldiers, the officers were the real enemy and the war a distraction from the struggle against capitalism. If Bloch is correct, then one cannot say of even the educated French officer that, like Oliver Cromwell's soldier, "he knew why he fought, and he loved what he knew." How much worse off will be the modern army that allows professional military training to push out the study of history, literature, politics, language, and philosophy—subjects that French cadets were expected already to have mastered. The first casualty of such a policy would be the critical thinking skills now, for example, so little in evidence in the products of American high schools. Perhaps more dangerous still would be a loss of moral and philosophical sophistication. Two questions, in particular, that may face any officer on the battlefield: "What am I fighting for?" and "what do the laws of war have to do with life and death?" can best be answered by someone whose patriotism and professionalism are anchored in a sound general education.

Notes

1. See Gen. A. Tanant, "Nos grandes écoles II: Saint-Cyr," *Revue des deux mondes* 7 (March–April 1926), and Maurice d'Ocagne, "Nos grandes écoles IV: L'École Polytechnique," *Revue des deux mondes* 7 (May–June 1926).

2. Joseph Monteilhet, *Les institutions militaires de la France 1814–1932* (Paris: Felix Alcan, 1932), 432.

3. For Saint-Cyr's contribution to the various arms, including the air force, see Rene Desmazes, *Saint-Cyr: Son histoire, ses gloires, ses leçons* (Paris: La Saint-Cyrienne, 1948), 154–57.

4. The year of enlisted service prescribed for officer candidates by the law of 21 March 1905 was not exactly like that for ordinary conscripts. Pvt. Charles de Gaulle of the 33rd Infantry Regiment had the opportunity to address his entire battalion and was promoted to corporal after only six months of service. See Jean Lacouture, *De Gaulle, The Rebel, 1890–1944*, trans. Patrick O'Brian (New York and London: Norton, 1990), 16–17.

5. Lt. A. G. Salisbury-Jones, "The Sandhurst of France: Some Impressions of the École Spéciale Militaire de Saint-Cyr," *Army Quarterly* 6 (April 1923). Salisbury-Jones was the only Englishman to graduate from Saint-Cyr before World War II. See Georges Merz, *Saint-Cyr: Livre d'or* (Paris: Office français de diffusion artistique et littéraire, 1962), 88.

6. Marcel Dupont, "Nos grandes écoles XIII: Saumur," *Revue des deux mondes* 7 (November–December 1930): 58.

7. d'Ocagne, 515–16.

8. Reduction of the military obligation from six years to one in 1924 increased the number willing to enter military service from 51 in 1924 to 111 in 1925. See d'Ocagne, 529. The service obligation was increased again in 1930. See Col. François-André Paoli, *L'Armée française de 1919 à 1939*, 4 vols. (Paris: Ministère des armées. État-major de l'armée de terre. Service historique, 1969–71), 4:193.

9. W. D. Halls, *Education, Culture, and Politics in Modern France* (Oxford, England: Halls-Pergamon, 1976), 227.

10. "Clarté et rapidité de la décision" and "précision des ordres" in Jean Pierre Callot, *Histoire de École Polytechnique* (Paris: Les Presses modernes, 1958). More tendentiously, Callot suggests that the absence of *Polytechniciens* from the leadership of the French Army may explain the debacle of 1940.

11. French tertiary education was essentially vocational. The highly competitive *grandes écoles* trained men for service to the state in administration, engineering, applied science, and war. Only a tiny percentage of the population could aspire to attend an institution whose diploma has been described as "a long-range rocket which, barring accidents, can propel you all the way to retirement." See Alain Peyrefitte, *The Trouble With France* (New York: Knopf, 1981), 201.

12. Details of the *lycée* curriculum can be found in John E. Talbott, *The Politics of Educational Reform in France, 1918–1940* (Princeton, N.J.: Princeton University Press, 1969), 17.

13. Paoli, 4:193.

14. Halls, *Education*, 221; Talbott, 18, 21. At a rough estimate, 10 percent of French males attended a *lycée*, 5 percent went beyond the baccalaureate to tertiary education, and one-tenth of 1 percent entered a *grande école*.

15. *Saumur. Son rôle et son avenir. Réflexions d'un officer de cavalerie* (Paris and Nancy,

France: Librairie Militaire Berger-Levrault, 1902), 24; Tanant, 56. Although Charles de Gaulle's *lycée* record was excellent, he required nine months of special study at the *Collège Stanislas* before entering Saint-Cyr. See Lacouture, 10.

16. The officer courses at Saint-Maixent and Poitiers were extended from one year to two in November 1927 with promotion to sublieutenant after the first year. Saumur adopted the two-year program in 1928. See Paoli, 3:183. For more on these courses, see Gen. L. Borie, "Nos grandes écoles XIV: Saint-Maixent," *Revue des deux mondes* 7 (January–February 1928) and J. P[erré] and R. S., "Les cadres de l'infanterie et des chars de combat: Leur recrutement et leur instruction," *Revue d'infanterie* 94 (June 1939).

17. These are the standards set by the instruction of 27 November 1933 described in Paoli, 4:193.

18. W. D. Halls, *Society, Schools and Progress in France* (Oxford, England: Pergamon, 1965), 6.

19. Paoli, 1:91–92.

20. Lieutenant Caillet, *Le nouvel officier d'infanterie en guerre, ce qu'il doit savoir* (Paris: Berger-Levrault, 1916), 10–12.

21. Lucas, "Des qualités du chef," *Revue militaire française* 26 (October–December 1927).

22. General Baratier, a military correspondent with close contacts within the General Staff, disparages this source of officers in *Le Temps,* 24 Aug. 1934, 1–2.

23. The high command believed that this period offered valuable opportunities for fruitful intercourse between candidates for regular and reserve commissions (Paoli, 3:182), but a contemporary British observer described the relations between active and reserve officers as "not quite as cordial as might have been expected." See Great Britain, War Office, General Staff, *Monthly Intelligence Summary* [hereafter War Office, *MIS*] (September 1936): 192. For that matter, a fracas in 1931 between officer candidates from Saint-Cyr and Saint-Maixent ended the practice of joint summer field exercises. See General Thoumin, "Les écoles d'infanterie de 1919 à 1939," *Revue historique de l'armée* 10 (1954): 93.

24. The word "university" does not do justice to the French institution. General education ended with the baccalaureate, and tertiary education occurred either at "faculties" devoted to a single professional subject or at the more technical *grandes écoles*. The faculties offered lectures but no required course work and prepared students for the *licence*, a degree achieved by examination. See R. D. Anderson, *Education in France 1848–1870* (Oxford, England: Clarendon Press, 1975), 27–28, 224.

25. Contemporary accounts of officer training are found in J. P[erré] and R. S., "Les cadres," and Charles R. Alley, "Recruiting and Training of Infantry Reserve Officers" (trans. of an article by Maj. Henri Mayerhoeffer, *Revue d'infanterie* 73 (September 1928) (G-2 14,150-W) United States National Archives, Record Group 165, 2015-928/6. All material from the United States National Archives (hereafter USNA) is from Record Group (hereafter RG) 165 (Military Intelligence) and is cited by the original document number (e.g., G-2 14,150-W) and carton/file/document number (e.g., 2015-928/6). Interestingly, the PMS programs were run not by the army but by private organizations. See Wm. W. Harts, "Selection and Training of Reserve General Staff Officers," (G-2 13,180-W) USNA, RG 165, 2015-928/3. Below the PMS certificate was the *brevet de préparation militaire* available through public schools and entitling a conscript to his choice of regiment.

26. Thoumin, 95.

27. This dual system, in which graduates of elite schools were "officer students" while those from the faculties were "student officers," led some critics to oppose the accelerated commissioning of the *grandes écoles* graduates and others to advocate extending the privilege to every school that sought it.

28. Alley, 7.

29. Minister of War, "Note sur les retards apportés à la nomination des officiers de réserve provenant du contingent," 26 Jan. 1935, Service Historique de l'Armée de Terre (hereafter SHAT), carton 6N428/5.

30. Ibid. The 1933 figure is from General Baratier, *Le Temps,* 5 Oct. 1934, 1. See also

Thoumin, 95, and J. P[erré] and R. S., "Les cadres." Fourteen thousand was a very large percentage of a total university population of about 60,000 in 1939. See Peyrefitte, 70.

31. Lt. Robert Felsenhardt, *1939–1940 avec le 18ᵉ Corps d'Armée* (Paris: Editions La Tête de Feuilles, 1973), 26–27.

32. The point is from Gen. Marie-Eugène Debeney, "Nos grandes écoles VIII: L'École supérieure de guerre," *Revue des deux mondes* 7 (January–February 1927): 103.

33. Maj. Ralph G. Smith, "Report on Methods of Instruction at the ESG, Paris," 13 July 1937 (G-2 23,635-W) USNA, RG 165, 2277-C-193/1, 2.

34. Maj. R. K. Sutherland, "Report on the Course at the E.S.G. 1928–29 and 1929–30," 30 June 1930 (G-2 16,437-W) USNA, RG 165, 2277-C-151, 4.

35. Debeney, 90–91.

36. G. V. Y. Waterhouse, "Some Notes on the E.S.G., Paris," *Army Quarterly* 8 (1924): 329.

37. Lt. Col. Jean Delmas, "Aperçu historique des méthodes," Association des Amis de l'École supérieure de guerre, *Bulletin Trimestriel* 57, 1st trimester, 1973, 24, and Smith, "Report," 9.

38. Delmas, 22. A two-year syllabus appears in Smith, "Report." The courses are listed in "The E.S.G., Paris," *Journal of the Royal United Services Institute* 70 (February 1925).

39. Delmas, 37.

40. Debeney, 87.

41. " . . . au lendemain de la victoire de 1918—dont le monde entier n'a pas encore oublié qu'elle est essentiellement l'oeuvre de notre pays—l'armée française jouit d'un prestige inégalé; l'accés de nos écoles militaires est pour recherché." See Merz, 88.

42. Delmas, 38. One domestic constituency was less impressed by the staff college. According to Gen. Gaston Galliffet, "The cavalryman has three enemies: marriage, obesity, and the École de Guerre," *Saumur,* 10.

43. For the highest college, see Eugenia C. Kiesling, "A Staff College for the Nation in Arms: Le Collége des Hautes Études de la Défense Nationale" (Ph.D. diss., Stanford University, 1988).

44. For a sample syllabus, see "Le cycle d'information des officiers généraux et colonels en 1926," *Revue militaire française* 23 (January–March 1927). For a first-hand account of one such course, see "The Tactical Tendencies of the French Army," War Office, *MIS*, June 1928.

45. For a list of courses, see Paoli, 4:202–6.

46. Commandant Delbé, "Les bibliothéques d'officiers," *Revue militaire française* 47 (January–March 1933): 98–99.

47. Captain Soury, "Le combat contre les engins cuirassés par le major von Schell de l'état-major général de la Wehrmacht," *Revue d'infanterie* 91 (July–December 1937); Major Gervaise, "Tactique des engins blindés," *Revue d'infanterie* 91 (July–December 1937); Captain Glain, "Analyse d'ouvrage du général autrichien von Eimannsberger. La guerre des char," *Revue militaire française* 56 (April–June 1935).

48. Colson, 3/EMA No.6585, 11 Oct. 1935, SHAT, 7N4013/1.

49. 3/EMA, "Note pour le Général Chef de l'E.M.A. au sujet de la crise de la littérature militaire et des moyens d'y remédier," 5 Dec. 1935, SHAT, 7N4013. I discuss this incident in Eugenia C. Kiesling, *Arming against Hitler* (Lawrence: University Press of Kansas, 1996), 126–27.

50. Franchet d'Esperey, "Indépendance et imagination," *Revue militaire générale* 1 (1937): 11. For editorial restrictions on the *Revue militaire générale*, see Gen. Paul Azan, "But et programme," *Revue militaire générale* 1 (1937): 7, and Gamelin, 3/EMA No.7282, 17 Nov. 1936, SHAT, 7N4013. Faris R. Kirkland, who argues that the *Revue militaire française* was "suppressed" in 1936, ignores the replacement journal. See Kirkland, "The French Officer Corps and the Fall of France—1920–1940" (Ph.D. diss., University of Pennsylvania, 1982), 164, 250.

51. Delbé, 102.

52. Anthony Adamthwaite, *France and the Coming of the Second World War, 1936–1939* (London: Frank Cass, 1977), 166.

53. Monteilhet, 430.

54. For the active-reserve breakdown, see Harts, "Selection and Training," (G-2 23,635-W) USNA, RG 165, 2015-928/3.

55. H. H. Fuller, "Individual Training. Training Periods for Reserve Officers," 3 May 1937

(G-2 23,396-W) USNA, RG 165, 2015-735/27, 1, 4.

56. General Baratier, *Le Temps*, 24 Aug. 1934.

57. The figures are from War Office, *MIS*, November 1933, December 1934. This discussion of reserve officer training comes from Kiesling, *Arming*, 108–12.

58. *Le Temps*, 29 Sept. 1936, 3. See also the report of British observers of 36th Infantry Division exercises at Souge Camp, War Office, *MIS*, August 1934, 64, and Thoumin, 102.

59. *Le Temps*, 30 May 1937, 4.

60. Gamelin, "Convocation des Cadres des divisions et unités de formation," 13 Jan. 1933, SHAT, 7N4008/4. For the schedule of a representative convocation, see General Champon, "Rapport," 3 Nov. 1936, SHAT, 7N4019.

61. Col. F. Soulet, *La 36ᵉ Division d'infanterie à l'honneur 1939–1940* (Toulouse, France: Imprimerie Fournie, 1945), 13.

62. Robert Bourget-Pailleron, "Le soldat de la dernière guerre," *Revue des deux mondes* 8 (September–October 1940): 160–61.

63. 1/EMA, "Note au sujet de l'organisation générale de l'Armée," 6 Nov. 1925, quoted in Paoli, 3:59; "Instruction générale pour la convocation des réservistes en 1927," SHAT, 7N2333/11; Brig. Gen. Stanley H. Ford, "Active duty training periods for reserve enlisted men, noncommissioned officers and officers," 26 Nov. 1930 (G-2 16,893-W) USNA, RG 165, 2015-928/15; Maj. James B. Ord, "Training Instruction for Reservists French Army (Effective 1931)" (G-2 17,436-W) USNA, RG 165, 2015-928/19.

64. 1/EMA, "Note au sujet de l'organisation générale de l'Armée."

65. For the reservist's contempt for his officers, see Commandant Altairac, "Rôle de l'officer de réserve appelé à accomplir une période dans un bataillon de réservistes," *Revue d'infanterie* 81 (1932): 75, and other sources cited in Kiesling, *Arming*, 108.

66. Sample lectures and issues of *L'officier de réserve* can be found in SHAT, 7N4006. There were 1,108 reserve officer schools in 1929–30. See Paoli, 3:149.

67. A. Niessel, "Les écoles d'officiers de réserve," *Revue des deux mondes* 7 (January–February 1931): 407, SHAT, 7N4008/3.

68. "Projet de loi No. 5043 (Deputé Daladier), presented 17 January 1939," 27 May 1938, SHAT, 7N2322.

69. Horace H. Fuller, "Changes in Articles of the Recruiting Law Dealing with Individual Training," 28 Mar. 1939 (G-2 24,907-W) USNA, RG 165, 2015-901/104.

70. Published as J. Perré, "Le char moderne. Ses possibilités et ses servitudes. Son emploi dans l'attaque," *Revue d'infanterie* 92 (January–June 1938).

71. Ibid., 634.

72. Thoumin, 102–3.

73. Harts, "Selection and Training."

74. Marc Bloch, *Strange Defeat: A Statement of Evidence Written in 1940*, trans. Gerard Hopkins (New York: Norton, 1968), 5. Bloch earned his reserve commission in the Great War but had a superb educational pedigree—the *Lycée Louis-le-Grand* and the *École Normale Supérieure*.

75. André Maurois, *The Fall of France*, trans. Denver Lindley (London: John Lane The Bodley Head, 1941), 47–48.

76. "Les saint-cyriens de 40, en tout cas, n'ont été inférieurs en rien à leurs anciens de 14," Merz, 93. Three hundred forty-five *Polytechniciens* and about 800 *Saint Cyriens* died during World War II. See Callot, *Histoire*, and Desmazes, 157. Two works rectifying the notion that the French Army simply collapsed in 1940 are Robert A. Doughty, *The Breaking Point: Sedan and the Fall of France, 1940* (Hamden, Conn.: Archon Books, 1990), and Jeffrey A. Gunsberg, "The Battle of the Belgian Plain, 12–14 May 1940: The First Great Tank Battle," *Journal of Military History* 56 (1992).

The Primacy of Politics: Civil-Military Relations and Italian Junior Officers, 1918–1940

Brian R. Sullivan

The Politics of Retention

Bad training in peacetime results in excessive casualties or lost battles in wartime. This holds true particularly when junior officers are not well prepared for combat. If those who lead an army's small units against the enemy prove incompetent, the negative consequences can expand from the tactical to the operational and even to the strategic level. The disasters suffered by Italian ground forces in 1940–41 provide notorious examples of this cause-and-effect relationship.

Even in militarist states armies remain at peace longer than they make war. Furthermore, throughout the nineteenth and twentieth centuries, most European armies, including the Italian, did more than project power and defend national boundaries. These other duties, some self-imposed, included maintaining internal security, inculcating nationalism and literacy in conscripts, providing advisers to the government, and protecting certain social values, usually of a conservative and patriotic nature. Often such activities, reinforced during prolonged periods of peace, can detract from a critical mission for any army in peacetime—training for war.

An army's leaders can become more interested in political and institutional issues than with readying themselves and their service for armed conflict. If concerns about maintaining power, remolding society, promulgating an ideology, or crushing political opposition are the central goals of a government, an already-politicized army in such a state can be distracted even more from its proper obligations: deterring enemies and, if necessary, defeating or conquering them. These conditions afflicted the Italian Army, as well as both the Liberal and Fascist governments that ruled Italy between the two world wars, and they help explain why the army, especially its junior officers, proved so unfit for combat in 1940. The origins of this debacle lay some twenty years in the past.[1]

Benito Mussolini bullied his way into becoming Italian prime minister in October 1922. He had threatened use of his black-shirted Fascist party militia to cause a civil war if denied control of the government, but the support of the Italian Army officer corps was more instrumental to his seizure of power. Pro-Fascist sympathy was especially strong in the Army Council, the group of the highest ranking generals who advised the minister of war on military affairs. The Council effectively outflanked the postwar army chiefs of staff, depriving the position of much of its prewar authority. Gen. Armando Diaz, senior member of the Council and the victorious army chief of staff in 1917–18, is reported to have warned King Vittorio Emanuele III that the army would do its duty if commanded to fire on the *Camicie Nere* (Black Shirts) but it would be better that it not be put to such a test. By indicating that they were not willing to stop Mussolini, the generals had left the king and his ministers virtually impotent.[2]

As a result, the Black Shirts' March on Rome was unopposed, and the king appointed Mussolini to head the government. Once in office, Mussolini reassured the generals that he would not use the *Camicie Nere* to supplant the army or to overthrow the monarchy. To give substance to this promise, Mussolini legally transformed his armed followers into an official paramilitary internal security force, the Voluntary Militia for National Security, in February 1923. This measure led to a degree of discipline being imposed on the Black Shirts, who began receiving some military training from pro-Fascist retired army officers. Over the next six months, some militia units acquired a modicum of military skills. In September 1923, the first combat units of the militia were organized: three regiment-size formations of one thousand Black Shirts each. These units deployed to Libya for operations against Arab guerrillas.[3]

Although the purpose was disguised, the Fascist militia remained a force for political intimidation. Black Shirts bullied voters during the parliamentary elections of April 1924. Even though the Fascists had captured the majority of seats in the chamber of deputies, *Camicie Nere* spread terror in those districts where constituents had elected other parties' candidates. That June, Mussolini's bodyguards kidnapped and murdered Giacomo Matteotti, a prominent Socialist deputy who had protested these crimes. The discovery of Matteotti's corpse provoked outrage. In an attempt to reassure the public and the army, Mussolini ordered the militia to display obedience to the law by swearing allegiance to the king. This October 1924 act made the Black Shirts formal members of the armed forces, supposedly subservient to the crown rather than to Mussolini.

Nonetheless, anti-Fascist sentiment mounted throughout the fall of 1924. The king and the Army Council realized that Mussolini could remain in power only by abolishing the constitutional order. That seemed preferable to the likely alternative: a Socialist government certain to reduce the ability of the army and the king to interfere in politics. Simultaneously, Mussolini's subordinates realized that either they had to break the anti-Fascist opposition or be swept from power.

In late December, the leaders of the militia stormed into Mussolini's office. They demanded that he choose between using the militia to neutralize their political enemies or being replaced by another Fascist leader. Backed tacitly by the monarchy and the army, Mussolini instituted a dictatorship in January 1925 and expanded his powers over the next several years. Yet for all his boasts of having established a totalitarian state, Mussolini remained aware that the king's army could remove him at any time. The *Regio Esercito* (Royal Army) remained far more powerful than the Fascist militia. In consolidating his regime, Mussolini made special efforts to please and pamper the upper ranks of the army officer corps. In particular, he enhanced the professional and financial security of senior officers.[4]

Since the end of World War I, Italian field-grade and general officers had faced a prolonged crisis involving promotion, tenure, and pay. In August 1914, the regular officer corps had numbered approximately 15,900—including 8,200 second and first lieutenants, and 5,300 captains. Thus, junior officers totalled 13,500 and field-grade officers 2,200. In the *Regio Esercito's* uppermost ranks were 354 colonels and 178 generals. Between mid-1914 and mid-1917, the Italian Army had expanded from its peacetime strength of twenty-five infantry and three cavalry divisions to its wartime peak of sixty-five infantry and four cavalry divisions, along with large increases in the number of alpine and independent artillery units, as well as the creation of new formations such as aircraft squadrons and assault battalions. To command and staff these units, thousands of noncommissioned officers (NCOs) and reserve officers had received regular commissions, and thousands of regular officers had been promoted two or three grades. By November 1918, 13,500 men had been added to the regular officer corps. Even with the deaths of 3,400 regular officers and the retirement for wounds or incompetence of another 4,000, the professional Italian officer corps had swollen to approximately 22,000. This included 6,700 lieutenants and 8,200 captains, for a total of 14,900 junior officers. The total of field-grade officers nearly had tripled to 6,400 (the number of colonels increasing almost four-fold to 1,308). The number of generals had risen to 513.[5]

The excess of officers to needs was exacerbated by Italian Army regulations. Unlike the American and British Armies, the *Regio Esercito* did not promote officers to temporary ranks, with reversion to a lower rank at war's end. Nor did the Italian Army during war—in contrast to the German and French Armies—place outstanding officers in positions ordinarily held by those of higher rank. In World War I, for example, the *Regio Esercito* had no junior officer equivalents of Germany's Erwin Rommel, who as a first lieutenant had commanded a battalion-size force of mountain troops on the Italian Front. Instead, when Italian officers were promoted, they held their rank permanently. An Italian officer only commanded a unit commensurate with his rank. But in 1919, the nation could not support an army of the size to justify such a large regular officer corps. Italy had emerged from the conflict burdened with the immense cost of World War I, and it labored under the economic depression which followed. In November 1919, Italy's parliament fixed the army's peacetime strength at thirty infantry divisions. It was expected that the number would double in wartime. In the meantime, parliament suspended officer promotions indefinitely.[6]

In September 1919, as a partial solution to the problem, parliament had authorized a new military status: *posizione ausiliaria speciale* (PAS). Regular officers with at least ten years' service could request assignment to PAS but they also could be transferred to it against their will. Officers in PAS received roughly 80 percent of the pay they had gotten on active duty. For pension purposes, each year in PAS status counted as a year of active duty. Once determined, however, PAS pay rates remained fixed and were not affected by new pay scales subsequently introduced for active-duty officers. Officers in PAS status were not eligible for promotion under the seniority system that in ordinary times automatically raised regular officers step by step to the rank of colonel. However, when an officer in PAS became eligible to retire, he would be given a pension equal to that he would have received had he remained on active duty and been promoted by seniority. Officers in PAS would be transferred to full pension status only four years after they reached the maximum age for active duty at the rank they held when entering PAS. Thus, captains in PAS would begin receiving pensions at age fifty-four, not fifty, and colonels at age sixty-two, not fifty-eight. PAS status did not preclude an officer being recalled to active duty if needed. Given the international situation and the virtual bankruptcy of the Italian state following the Great War, however, such recalls seemed unlikely. By February 1920, 250 generals, 1,250 field-grade officers, and 100 captains had been assigned to PAS. These measures, along with voluntary retirements, dropped the officer corps to 18,900.[7]

Faced with continued economic pressure, the Italian parliament approved a further reduction in army strength in April 1920. Although the number of infantry divisions remained at thirty, each infantry regiment was reduced from three to two battalions and most cavalry regiments from five to three or four squadrons. The reduction required a professional officer corps even smaller than that of 1914—just 15,000. The number of generals was reduced to 156 and junior officers to 12,300, but field-grade officer slots were set at 2,600, including 478 colonels. Although this meant a significant decrease from the 6,400 majors, lieutenant colonels, and colonels of December 1918, it was still an increase of nearly 20 percent over the figures of mid-1914. Moreover, the number of regular colonels had increased by 35 percent. Clearly parliament was pampering field-grade officers at the expense of their juniors. Nonetheless, by the end of 1921, another 2,100 generals and field-grade officers and 300 more captains had been placed in PAS. As economy measures, officers' pay was reduced by about 4 percent and, more serious, PAS status was limited to ten years. At the end of that term, an officer in PAS would be forced into permanent retirement. With rampant inflation eating into their frozen salaries, the standard of living of PAS officers began to drop. Even though officers in PAS could hold other jobs, the national depression had led to high unemployment rates and few could find positions in 1919–21. A sense of deep resentment spread throughout the entire Italian officer corps. To a large degree, these grievances and Fascist promises to rectify matters if they came to power, explain the favorable attitude of army officers to the March on Rome.[8]

After Mussolini became prime minister, he appeared to keep his promises to the officer corps. In January 1923, acting on the advice of General Diaz, whom he had appointed minister of war, Mussolini used his special decree legislative powers to augment the number of officers on active duty by some 19 percent. More significant, the percentage of field-grade officers within the total officer corps rose to 18 percent, compared to a prewar 13 percent. Another 8,000 army officers, mostly recalled from PAS status, were assigned to the Fascist militia as military instructors. Although true Fascists retained the highest positions in the militia, the army officers recalled to active duty as Black Shirts were grateful and could be expected to be loyal to their new service. Both the expansion of the regular officer corps and the creation of thousands of new militia officers provided positions for almost all who wished to return to active duty from PAS. By 1924, the army had nearly 19,000 officers on active duty. Together with the militia officers the total came to about 27,000—some 5,000 more than the regular officers on active duty in late 1918 and nearly double the number of mid-1914.[9]

Yet Diaz had agreed to help Mussolini balance the budget by reducing army expenditures. The war minister could do this only by reducing training, suspending purchase of new equipment, and greatly decreasing the number of conscripts called to duty. The result was an army of grossly undermanned units with glacially slow officer promotion rates. Although Diaz ended the 1919 freeze on promotions in 1924, only 170 officers were elevated in rank that year—less than 1 percent of the total officer corps. Pay raises were granted but only to the highest ranking officers. As a result, although a colonel received somewhat more than double the pay of a second lieutenant, a major general received two-and-a-half and a full general four-and-a-half times the salary of a colonel. More scandalous, the total cost of pay, quarters, food, and other allowances for the officer corps amounted to 28 percent of the army budget. At the same time, equipping, housing, feeding, and training each year's conscripts consumed just 30 percent.[10]

This system did little but preserve a large regular officer corps. Diaz realized this and tried to persuade Mussolini to increase spending on the army. Much to the general's dismay, Mussolini refused. Yet, so long as Diaz held office, the aura surrounding him protected his military arrangements from effective criticism inside or outside the army. By early 1924, however, Diaz had decided to leave office in the hope that his successor could obtain larger army budgets.[11]

Following Diaz's departure, Mussolini pursued a program of drastic military reform. To that end he appointed the energetic and imaginative Lt. Gen. Antonino Di Giorgio to succeed Diaz. Di Giorgio agreed to the small budgets Diaz had found unacceptable. Nonetheless, the new war minister believed that he could still enact major reforms, and developed his plans for the army's renovation over the second half of 1924. During those months the Matteotti Crisis nearly toppled Mussolini from power, but the Duce was still able to establish the Fascist dictatorship in January 1925. Then his health broke down under the strain of the previous six months. While Mussolini convalesced, the war minister perfected his reform plans.

Di Giorgio proposed to reduce the number of peacetime army units drastically. He would devote the resulting savings to replacing the army's outdated artillery; to motorizing, first the artillery, and then the infantry; to the intensive and realistic training of combat units; and to reeducating the officer corps in the new mobile warfare that had emerged in 1917–18. The war minister also planned to expand the regular officer corps as the base for a large wartime army. In a number of ways, Di Giorgio was inspired by the German *Reichswehr,* as organized by Gen. Hans von Seeckt. Italian military intelligence and Fascist party agents, following Mussolini's interest in forming an alliance with the German radical right and its military supporters, had gathered a good deal of information on the *Reichswehr.* Based on that knowledge, Di Giorgio hoped to transform the *Regio Esercito* into a rough copy of von Seeckt's army in terms of intellectual dynamism, high standards of combat proficiency, and imaginative experimentation with new doctrinal and technological concepts. Furthermore, Di Giorgio envisioned the militia serving as the equivalent of the so-called "Black *Reichswehr,*" the secret body of reserves created by von Seeckt to circumvent the restrictions of the Versailles Treaty. Of course, such a reorganization of the militia would make it far more an army reserve organization but far less "the armed guard of the Fascist revolution."[12]

Di Giorgio also proposed to expand the powers of the army chief of staff at the expense of the Army Council. Since the chief of staff was subordinate to the war minister, Di Giorgio was attempting to marginalize the conservative senior generals who had guided the army to victory in 1918 and had dominated it since. Naturally, the Army Council opposed Di Giorgio's reforms. Nonetheless, the Fascist-dominated chamber of deputies approved them in early December 1924. Presentation of the proposals to the senate was postponed until April, due to the political crisis and Mussolini's need to recover his health.[13]

Meanwhile, an unintended aspect of Di Giorgio's reforms had created serious controversy. By reducing the size of the army so sharply, the reforms would make the militia larger than the peacetime *Regio Esercito.* Despite the army's acceptance of the Fascist dictatorship, its leading generals were not prepared to lose their implicit control over Mussolini. Furthermore, the regular army officers who had been transferred from PAS to the *Camicie Nere* feared that the militia rank and file and Fascist party leaders might take advantage of the army's reduced strength. For all their gratitude over employment at full pay, the last thing the ex-PAS officers in the militia wanted was a real Fascist revolution. It might lead to the overthrow of the monarchy and the imposition of a radical republican regime in Rome.

In a strange alliance with the Army Council to block Di Giorgio's reform proposals, the new secretary of the Fascist party, Roberto Farinacci, announced his own opposition in February 1925. Farinacci resented Di Giorgio for his adamantine refusal to allow Fascist propaganda inside the army. But the party secretary also saw Di Giorgio as a significant threat to the Fascist regime for his antipathy to converting the militia into an independent military force and his desire to subordinate it to the army as an apolitical military reserve. Farinacci felt assured enough of support from within the Fascist party to ignore Mussolini's wishes in the matter. Moreover, Di Giorgio had shown himself to be politically inept. Mussolini decided that he could not afford to support his war minister in opposition to both the party hierarchy and the Army Council. Furthermore, Mussolini had discovered that Di Giorgio had been using army intelligence to spy not only on the Army Council but on the Duce himself. In early April, Mussolini announced in the senate his decision to withdraw Di Giorgio's proposals. A few days later, the war minister resigned.[14]

There would be other attempts to reform and renovate the Italian Army over the next fifteen years. Some would enjoy partial success, particularly in the mid-1930s.

But Di Giorgio's failure to modernize and improve the effectiveness of the *Regio Esercito* represented the loss of the most promising opportunity. In many ways, the repulse of Di Giorgio's efforts in 1925 can be linked directly to the defeats of 1940–41. This holds particularly true for the wretched state of the army's junior officer corps when Italy entered World War II.[15]

In May 1925, Mussolini enlarged his control over the armed forces. He appointed himself Di Giorgio's successor and simultaneously took over the navy and air force ministerships. He also created a new post, chief of the Supreme General Staff; its occupant would advise him on all military matters. The Army Council lost its influence, and Mussolini refused to augment the military budget. Still, he had ended any threats to the selfish interests of the army's senior officers. Furthermore, Mussolini forced Farinacci to subdue any rebelliousness and independence in the militia. Thereafter, the *Camicie Nere* would serve the state, not the party. In early 1926, acting in his capacity as war minister, Mussolini reorganized the army once again. He gained some savings by modestly reducing the number of active army regiments from 125 to 111 and by converting each of the army's 30 divisions from 4 infantry regiments to 3. Thereafter, each division contained one, three-regiment infantry brigade, rather than two, two-infantry brigades as previously. At the same time, he authorized the army to increase conscription.

Yet the army could not afford to keep so many recruits on active duty for the full eighteen months of their military obligation. As a result, only about three-quarters of each year's inductees served a year and a half; the other 25 percent were returned to civilian life after six months' training. When even these measures left the army more conscripts than it could afford to maintain, the period of service was reduced unofficially by three to six months, depending on the category of conscripts. Consequently, for two-thirds of the year, each of the army's regiments consisted of only one full-strength battalion.[16]

Mussolini instituted other changes affecting the officer corps. To compensate for the loss of a brigade command from each division, he increased the number of corps from ten to eleven, created thirty new mobilization inspectorates in the zones from which each division was formed, and appointed a general officer as inspector for each combat arm. These new posts required twenty-five new general-officer positions. To placate field-grade officers who had lost fifteen regimental commands and staffs, Mussolini increased their numbers by 800 to allow for many promotions. Furthermore, he established 450 new posts for captains. But to pay for this wave of promotions, the Duce slashed the number of active-duty lieutenants from 7,150 to 3,730. As a result, the regular officer corps dropped from 18,370 to 15,800.

Under Diaz, generals had formed 1.2 percent of the officer corps; Mussolini's reorganization raised that percentage to 1.5. Similarly, the percentage of field-grade officers rose from 17.7 to 27.5 and that of captains from 30.7 to 39.8. But the percentage of lieutenants in the regular officer corps dropped from 50.4 to 31.1. With Fascist censorship of the press, no newspaper stories appeared on the fate of the 3,420 regular lieutenants summarily added to the ranks of Italy's unemployed; no editorials protested or deplored their fate. Instead, the War Ministry simply announced the dismissals as an economic necessity. It noted that, in future, especially in war, the army's subalterns would be drawn overwhelmingly from reserve officers.[17]

Within a few years, the relatively large number of captains and majors retained on active duty presented new personnel problems. As these officers were promoted to major or lieutenant colonel, they were more than the army needed. Since for eight months of the year each of the army's regiments consisted of only one active-duty battalion, there was little or nothing during that time to occupy the majority of majors and lieutenant colonels. Mussolini and his military advisers considered retiring all field-grade officers passed over for promotion, but rejected the idea for fear of creating anti-Fascist discontent within the officer corps. Instead, by a decree of April 1930, 1,220 lieutenant colonels and 1,475 majors were shifted from the army's combat arms to make-work positions in the thirty mobilization centers. In late 1932, another law raised the retirement age for officers passed over for promotion, and also shifted captains considered too old for combat and unfit for promotion to "command" arsenals or offices. Finally, a series of laws enacted between mid-1934 and mid-1937 created a number of new military organizations that offered other posts to the army's excess captains and field-grade officers. This legislation assigned the militia to Italy's antiaircraft and coastal defense artillery units, which were expanded in number; organized a new command structure to train, inspect, and supervise these batteries; and posted hundreds of active-duty army officers to these command and staff positions. Finally, a frontier guard command was instituted and placed in charge of border defenses. The frontier commands provided posts for some forty brigadier generals and colonels, and a correspondingly larger number of other field-grade officers and captains.[18]

Such reorganization helped maintain the warm relationship between the Fascist dictatorship and the regular officer corps. Although statistics are lacking, it seems that the regular lieutenants retained on active duty were mostly the sons or nephews of higher ranking officers. Some of the dismissed lieutenants did find positions in the Fascist militia. But many were forced to find nonmilitary employment. These young men, however, enjoyed little influence within Italian society and their fate did not overly trouble the Fascist regime. As for the potentially more dangerous but superfluous captains, majors, and lieutenant colonels, they certainly preferred to remain on active duty guarding magazines, running offices, or supervising air, coastal, or border defenses, than to be forced into retirement. Their gratitude to Mussolini grew stronger after the onset of the world depression in Europe in 1930.

The army's size and budgets expanded from 1934 on, at first in response to Nazi German pressure on Austria, later to the demands of the Ethiopian War, of Italian intervention in the Spanish Civil War, and of the growing Axis hostility toward France and Britain. Nonetheless, the grave disparity between the numbers of officers and men in the *Regio Esercito,* and the resources available for their training and equipment, would remain. These deficiencies particularly affected the junior officer corps, which, from 1926 was composed overwhelmingly of reservists. Their interests had been neglected in favor of maintaining a large and satisfied senior officer corps loyal to the Fascist regime. In effect, Mussolini looked to the army to sustain internal security and to that end his military policies succeeded. Even more so, the militia had been organized and trained to keep Mussolini in power. But when his foreign policy led to Italian involvement in a new European conflict, the Duce would discover that he had formed neither an army nor a Fascist militia ready for war.

The Realities of Training

The regular junior officers of the Italian Army received a good professional education in the 1920s and 1930s. In general, it equaled that provided to the subalterns in the French and British Armies at St.-Cyr and Sandhurst during the same period. Given the surplus of lieutenants in the immediate postwar years, the Italian military academies were closed from 1918 to 1920. They reopened not to educate new second lieutenants but to instruct the thousands of first lieutenants and captains who had received regular commissions during World War I. Finally, in March 1923, the normal military education system resumed. The standards, however, had been raised from those of the pre-1915 period. Previously, candidates for admission had needed only two years of secondary education. From 1923, the requirements for admission were a four-year academic or technical secondary school diploma, a commission as a reserve second lieutenant, and the passing of a rigorous entrance examination.

Students spent two years at the Modena or Turin academies. Those who were second lieutenants went through an intellectually demanding, physically rigorous course of instruction. They lived within a strictly disciplined barracks-like military environment, maintained unchanged since the nineteenth century. The one serious deficiency in the curriculum of both schools was a lack of sufficient, realistic training for combat. Too much time was spent studying tactics and operations in the classroom and too little on exercises in the field approximating wartime conditions. Upon completion of their education, graduates were promoted to first lieutenant. Between 1926 and the mid-1930s, the academies graduated 400–500 first lieutenants each year. In the years 1936–40, these numbers rose to 1,100–1,200 annually.

Depending on the branch from which they had entered and to which they would return, most new first lieutenants received additional training. Administrative and supply officers were posted to active-duty assignments after graduation. Infantry, cavalry, and commissariate officers, however, were sent to a school of application for an additional year. Artillery and engineer officers were given two more years at their branch school of application. As the history and traditions of the Italian Army indicated, the latter were the most carefully selected subalterns and were given the best professional education. Nonetheless, even these elite of the *Regio Esercito*'s junior officers were taught too much theory and too few practical combat and other field skills.[19]

As noted earlier, even during the lean years of the 1920s and the early 1930s, the

Regio Esercito depended on reserve officers to command its platoons—and often its companies, troops, and batteries—and to hold other subordinate positions. As the army expanded from the mid-1930s onward, the need for reserve subalterns rose considerably. Furthermore, in the months before the Ethiopian War, the Fascist militia began forming its own divisions and, by the spring of 1936, there were seven. While many of the officers who served in these Black Shirt units came from the army, the formation of divisions and smaller militia combat units still required thousands more *Camicie Nere* junior officers.[20]

As early as 1928, the then undersecretary of war, Gen. Pietro Gazzera, reaffirmed previous calculations that Italy would need to form sixty divisions in case of war. Over the next decade, however, Mussolini moved from his public stance as defender of the status quo in Europe to Hitler's ally in the quest to bring the continent under Nazi-Fascist domination. By the fall of 1939, Gen. Alberto Pariani, serving as both undersecretary of war and army chief of staff, was preparing to expand the *Regio Esercito* and the militia to 126 divisions. While these divisions would be only two-thirds the size of those planned by Gazzera in 1928 and would not be ready before late 1942, they would still need enough lieutenants and captains to officer the equivalent of eighty-four divisions with three infantry regiments each. This meant a requirement for roughly 275,000–325,000 reserve junior officers capable of serving in combat or combat support units. Even Gazzera's more modest projections had called for as many as 225,000–250,000 such officers.[21]

From 1925, the Fascist regime, the army, and the militia had attempted to create such huge numbers of reserve officers. That year, Italian universities and engineering schools were required to offer electives in a range of military subjects—ballistics, military communications, cartography, fortifications, explosives and poison gas, mechanical traction, and military history—all well taught by knowledgeable officer instructors the army supplied at no cost to the schools.

Starting in 1935 such courses were made mandatory for all male students, both in secondary and higher level schools, and a three-year curriculum of fifteen hours per semester was specified. To fill the greatly expanded need, both active *Camicie Nere* and retired *Regio Esercito* officers were assigned to teach. Most were hardly gifted and the quality of instruction declined accordingly. When administration of the enlarged program passed from the army to the Ministry of Education, corruption entered the system due to the lucrative contracts for the tens of thousands of textbooks used in the military courses. The books, authored by men with connections at the top of the Fascist party or the Ministry of Education, were not necessarily the best available. As a result of these changes, subject matter treatment quickly degenerated from that of 1925–35 to far more superficial instruction mixed with much jingoistic propaganda in favor of fascism, war, xenophobia, and imperial expansion. Despite the bogus nature of the new required courses, their introduction gave the army the excuse to drop these subjects from the officer-candidate curriculum. The quality of reserve junior officer training deteriorated.[22]

In 1927, the Fascist regime ordered all physically fit male university graduates to enter the armed forces as officer candidates. Most enrolled in the *Regio Esercito* at the rate of some 6,000 per year. Since these numbers exceeded the army's need for second lieutenants, 1,000–2,000 were sent annually to NCO schools and promoted to sergeant on completion of training. Both sets of university graduates spent five months in the classroom, then one month in the field on tactical exercises at one of the fifteen instruction centers attached to each army corps. At the approach of the Ethiopian War, the courses were accelerated, as well as simplified, as mentioned above. Intake of officer candidates rose to about 8,000 a year. As international tensions grew in Europe from 1936 onward, the army began inducting even more officer candidates, averaging 14,000–15,000 annually. But the resources were not provided to train properly such growing numbers. In particular, not enough well-trained regular officers were available as instructors, while sufficient quantities of modern arms, ammunition, and equipment were lacking for live-fire and other field exercises.

Meanwhile, the militia had begun to instruct both its own officer candidates and some destined for the army. In 1925, the Fascist party had instituted *Gruppi universitari fascisti* (GUF) at all institutions of higher learning. These organizations provided ardent Fascist students with military training and ideological formation. The success of GUF organizations inspired the militia high command to propose a more ambitious program. In 1929, the militia gained permission to institute and run officer-candidate courses for university students in their last two years of school. These consisted of three, twenty-day instruction periods at the end of each semester, followed by three-and-a-half months of training after graduation. Most students who took these courses did their postgraduate training at army centers. But members of the GUF, as well as some other graduates, went on to militia officer-candidate schools to be commissioned as Black Shirts. Before 1935, the militia was training 3,000–4,000 officer candidates each year. By the late 1930s, however, even though the army had doubled the number of its own officer-candidate centers, the militia was training and commissioning about 10,000 of its own officers each year.

The differentiation of officer candidates roughly followed social and political lines. Aristocratic and upper middle-class young men with conservative political views and often Catholic inclinations tended to join the army. Pro-Fascist middle and lower middle-class graduates, often with anticlerical attitudes, frequently entered the militia. The previous uniformity in the ideological formation, education, and training of junior officers was replaced by an ever-widening gap among regulars, reservists, and *Camicie Nere*. The old sense of special caste and honor enjoyed by the monarchist officer corps greatly diminished. As a result, the prestige and morale of the *Regio Esercito*'s regular junior officers began to drop from the mid-1930s onward.[23]

After they had completed their initial period on active duty and returned to civilian life, reserve junior officers were legally required to attend annual refresher courses. These lasted fifteen days for second lieutenants, twenty days for first lieutenants, and thirty days for captains. Militia reserve officers of all ranks were obliged to participate in ten to twelve days of instruction yearly. Select army reserve captains were assigned to intensive courses in arms employment. These consisted of fifteen

days of classroom instruction during the winter months, then a fifteen-day application course in the summer. Only 450 captains were admitted to such courses each
year, far too few considering that reserve captains in the combat arms would number
some 60,000 in mid-1940. According to regulations, no reserve first lieutenants could
be promoted without taking a two-week preparatory course for company, troop, or
battery commanders. All of these courses, however, became increasingly superficial
in the second half of the 1930s as the demand for more junior officers placed the
army's resources and militia training establishments under ever-increasing strain.
Greater numbers of reserve officers were not called up for training periods, and when
junior officer reservists did train, insufficient ammunition for fire exercises with machine guns and other crew-served weapons left them increasingly ignorant of combined arms tactics. Only a relatively small minority of army reserve officers fought in
Ethiopia or Spain, since regular junior officers received priority in assignment to
command small combat units. As a result, few reserve junior officers had combat
experience at the outbreak of World War II and most could not be considered well
trained. Of the 79,900 reserve junior officers put through training courses by the
army and militia in 1935–39, only 30,000 were considered fit to be mobilized in September 1939.

These 30,000 were only the best available. This did not mean they were ready for
war. Reserve officers were limited by age for combat service. No lieutenants over
forty, no captains over forty-five, no majors and lieutenant colonels over fifty, were
eligible for such demanding duty. Junior officers up to the age of fifty-six, however,
could be assigned to combat service units or to garrison duty. In practice, this meant
that when Mussolini ordered mobilization in September 1939, reserve officers who
had last served on extended active duty in the 1920s or early 1930s were recalled as
captains; reservists who had seen active service in the mid- or late 1930s were reacti-
vated as first lieutenants. Still, the need for captains was so great that a good many
mobilized reserve first lieutenants who had not taken the prescribed courses were
promoted anyway in September–October 1939.[24]

Mussolini himself recognized the severity of these problems. At a meeting of the
Army Council in May 1939, the Duce had noted: "Reserve officers as a body are poor.
. . . The majority of reserve officers are not equal to their responsibilities."[25]

Intense training courses for recalled reserve officers were instituted in September 1939. But after deciding not to enter the war immediately, Mussolini demobilized
most reservists for economic reasons in October–November. The reserve officer
training courses were disbanded. Men recently appointed to captain's rank and
assigned as company, battery, and troop commanders were sent home without proper
training and even the chance for practical knowledge gained from day-to-day experience with their units. When the army was mobilized again in May 1940, the courses
were not resumed. Instead, the best reserve junior officers were pulled out of their
combat units and used as instructors in the regimental basic training camps for new
enlisted recruits. These measures disrupted unit combat training, left many reserve
junior officers insufficiently prepared for war, and rendered many divisions unfit for

the fast-approaching Alpine Campaign. Another 5,000 army reserve junior officers and several hundred militia subalterns had been commissioned and trained over the winter and spring. Only about one-third of the reserve junior officers and the enlisted men they led into battle against the French in June 1940, however, could be considered properly trained.

By summer 1940, the army had seventy-one divisions in the southern European–Mediterranean region and one in East Africa; the militia had three divisions in Libya and one in Ethiopia.[26] As a rule, battalions and squadrons contained only two or three regular officers. This meant that almost all platoons, companies, batteries, and troops were commanded by recently recalled and poorly trained reserve junior officers. The small numbers of regular field-grade officers and the inadequate training of reserve junior officers resulted in an armywide system of rigid supervision imposed from above. This stifled any initiative at the battalion level or below. The problem was even worse in the militia divisions where battalions and even legions were often commanded by reserve officers. The age of many recalled reserve officers worsened these shortcomings. Many captains and even a good number of lieutenants were in their late thirties; reserve majors generally were in their mid-to-late forties. The same held true for their *Camicie Nere* counterparts. These men frequently lacked the vigor to lead combat units effectively in battle. As a result of these deficiencies, below the regimental/legion level, commanders ordinarily waited for directions and rarely exercised independent judgment even in emergency situations.[27]

These problems could have been addressed and might have been solved had intense and realistic reserve officer training continued over the fall, winter, and spring of 1939–40. But Mussolini had not expected to enter the war in the immediate future, nor had he wished to create discontent by keeping large numbers of reserve officers on active duty. In retrospect, both decisions proved to be major blunders. Despite the intricate system for training reserve junior officers erected since the late 1920s, the content of instruction had ranged from fair to very poor. Additionally, army budgets had been too small until the mid-1930s for the arms and equipment necessary to train reservists for modern war. Moreover, the intake of conscripts had been too low to provide reserve officers with sufficient command experience. Perhaps worst of all, the regular officer corps was always too small throughout the 1920s and 1930s to educate reserve junior officers in the new army doctrine adopted after 1933.[28]

The Duce could not be faulted for spending too little money on the army. Indeed, he had nearly driven his country to bankruptcy with his lavish military expenditures. But the majority of the tens of billions of lire the army received after 1933 had been expended on the conquest of Ethiopia, intervention in the Spanish Civil War, and the ongoing pacification of the East African empire. Other monies had gone for the salaries and comforts of the regular officer class. Consequently, when Italy faced the prospect of a major war in 1939–40, the large army budgets of the previous six years had been used in ways that did little to prepare the *Regio Esercito*. Italian soldiers and Black Shirts were poorly trained, equipped, and clothed. The army's artillery was outmoded, its tanks pitifully weak in armor and firepower, its vehicles few in number, its medical services woefully inadequate, and its ammunition supplies far too low. Worst of all, Italian military doctrine and the training of Italian junior officers to carry it out were hopelessly inadequate to the realities of modern war in 1940.[29]

For all of its deficiencies, the Army General Staff had developed the basic concepts of motorized and mechanized warfare by 1938–39. But while the tactical and operational concepts of the Army General Staff were far more advanced than is usually recognized, the arms and equipment to carry them out did not exist in anything like sufficient quantities. Worse, the army's junior officers—especially its reservists—had not been adequately instructed in the doctrine developed in the second half of the 1930s. In fact, the lack of equipment and instructors had left the great majority of reserve junior officers inadequately trained even in the older approaches to warfare.[30]

Finally, both the army and militia suffered from discipline problems, although of a very different nature. Army junior officers were taught to form a rigid and distant relationship with the men under their command. Contact between officers and men was maintained by generally brutal and poorly educated corporals and sergeants. In many ways, up to 1940, the *Regio Esercito* preserved social attitudes reminiscent of the nineteenth-century British Army. This resulted in a brittle army in which command and control tended to collapse under severe battlefield conditions.

In contrast to their army counterparts, *Camicie Nere* were encouraged to promote camaraderie among officers, noncommissioned officers, and men. Although the officers obtained through the GUF were generally dedicated, a good many Black Shirt officers had been commissioned purely for political "merit." Some of the latter insisted on service in militia combat units to enhance their careers. Furthermore, the Black Shirts did not select their rank and file with particular care. Nor were prevailing notions of Fascist comradeship compatible with strict discipline imposed from above. As a result, militia combat units suffered frequently from slack and often negligent attitudes about cleanliness of persons, arms, and equipment; the necessity for hard training and physical fitness; and attention to details and professional standards. *Camicie Nere* units that served in the Ethiopian War fought fairly well (at the time, they were heavily staffed with army officers). A later indication of deep-seated problems within the Black Shirts was the disintegration of a number of militia units at the Battle of Guadalajara in March 1937. Nonetheless, while the discipline of *Camicie Nere* serving in Spain was tightened considerably thereafter, the militia as a whole continued in its often ill-disciplined ways. Needless to say, relations between the two military organizations—especially between their junior officers—varied from cold to hostile.[31]

The repulse of the Italians in the French Alpine Campaign of June 1940; the heavy losses inflicted by the small garrison of British Somaliland on invading Italian forces in August 1940; the botched Italian offensive in northern Greece in October–November 1940; the Italian failure to halt the Greek conquest of southern Albania during November 1940–January 1941; the rout of the Italian Tenth Army in Egypt and Cyrenaica by Commonwealth forces in December 1940–February 1941; the British conquest of Italian East Africa from February to November 1941; the bloody yet fruitless Italian counteroffensive in Albania in March 1941: all revealed the short-comings of the *Regio Esercito* and the *Camicie Nere* in a stark and bloody manner. These defeats marked the nadir of Italian military performance. They cost some 400,000 Italian casualties and still sully the reputation of Italian arms.[32]

While these humiliations sprang from a number of factors, one stands out promi-nently. The majority of the lieutenants, captains, and their militia counterparts who led Italian soldiers and Black Shirts into battle in 1940–41 knew how to fight and die bravely. But they did not know how to fight well. Most of the blame for this defi-ciency in training can be ascribed to the Fascist regime and to the high-ranking generals who gave it political support. For reasons that had little to do with military effectiveness, quantity had been stressed over quality. The political reliability of the regular officer corps had been considered far more important than preparing the reserve officer corps for war. Yet the funds and perquisites used to guarantee peace-ful civil-military relations in the Fascist regime made it impossible to maintain a large, regular junior officer corps, ensuring that the army's combat performance would rely on its reserve lieutenants and captains.

Mussolini preached war as the noblest human activity and as the ultimate test of the worth of the society that waged it. Yet despite his bombast about Fascist military power, the Duce devoted far more attention to the political uses of the army and the militia than in preparing either to wage war effectively. One outstanding example is Mussolini's abuse of the spearhead of his ground forces—the junior officer corps. His policies ensured that when Italy entered World War II, it would stumble into military defeat. Perhaps nothing better illustrates the corruption, incompetence, and stupidity of Fascism and all its works.

Notes

1. These issues are explored in Brian R. Sullivan, "A Thirst for Glory: Mussolini, the Italian Military and the Fascist Regime, 1922–1936" (Ph.D. diss., Columbia University, 1984).

2. Adrian Lyttelton, *The Seizure of Power: Fascism in Italy 1919–1929* (New York: Scribner's, 1973), 91–93; Antonio Repaci, *La marcia su Roma. Mito e realtà*, 2 vols. (Rome: Canesi, 1963), 2:386; Giorgio Rochat, *L'esercito italiano da Vittorio Veneto a Mussolini* (Bari, Italy: Laterza, 1967), 398–408; and Oreste Bovio, *Storia dell'esercito italiano* (Rome: Ufficio Storico, Stato Maggiore dell'Esercito [hereafter USSME], 1996), 245.

3. Virgilio Ilari and Antonio Sema, *Marte in orbace. Guerra, esercito e milizia nella concezione fascista della nazione* (Ancona, Italy: Nuove Ricerche, 1988), 283–94.

4. Lyttelton, 135–48, 237–68; Giorgio Rochat, "The Fascist Militia and the Army, 1922–1924," in Roland Sarti, ed., *The Ax Within: Italian Fascism in Action* (New York: New Viewpoints, 1974), 43–56.

5. Lucio Ceva, *Le forze armate* (Turin, Italy: UTET, 1981), 127; Bovio, 180, 203–7, 213–14, 219–20; Rochat, *L'esercito italiano*, 137.

6. Bovio, 241, 244–45; David Fraser, *Knight's Cross: A Life of Erwin Rommel* (London: HarperCollins, 1993), 63, 65; Ceva, 197.

7. Bovio, 241–42, 246; Rochat, *L'esercito italiano*, 147–50.

8. Rochat, *L'esercito italiano*, 164 67.

9. Virgilio Ilari, *Storia del servizio militare in Italia*, 4 vols. (Rome: Centro Militare di studi strategici, 1989–91), vol. 3, *"Nazione Militare" e "Fronte del Lavoro" (1919–1943)*, 59–60, 73; Ceva, 192.

10. Ceva, 197–98; Carlo Jean, "La relazione Belluzzo," *Rivista Militare* 10 (June 1978).

11. Ilari, *"Nazione Militare,"* 63–69.

12. Ibid., 69–70; Alan Cassels, *Mussolini's Early Diplomacy* (Princeton, N.J.: Princeton University Press, 1970), 133, 161–65; James S. Corum, *The Roots of Blitzkrieg: Hans von Seeckt and German Military Reform* (Lawrence: University Press of Kansas, 1992), 29–143, 177–82.

13. Emilio Canavari, *La guerra italiana. Retroscena della disfatta*, 2 vols. (Rome: Tosi, 1948), 1:125–28; Rochat, *L'esercito italiano*, 525–46; Filippo Stefani, *La storia della dottrina e degli ordinamenti dell'esercito italiano*, 3 vols. (Rome: USSME, 1984–87), 2/1:63–65; Ilari, *"Nazione Militare,"* 70–82.

14. Ilari, *"Nazione Militare,"* 82–88; Aldo Giambartolomei, "I servizi segreti militari italiani," *Rivista Militare* 106 (May–June 1983): 62; Ambrogio Viviani, *Servizi segreti italiani 1815–1985*, 2 vols. (Rome: Adn Kronos, 1985), 1:186.

15. Stefani, 2/1:64–66; Rochat, *L'esercito italiano*, 551–61.

16. Lyttelton, 277–99; Ilari, *"Nazione Militare,"* 97–99.

17. Rochat, *L'esercito italiano*, 581–84; Ilari, *"Nazione Militare,"* 92–97, 101.

18. Bovio, 265–66, 271–72; Mario Montanari, *L'esercito italiano alla vigilia della 2a guerra mondiale* (Rome: USSME, 1982), 25–26; Ilari, *"Nazione Militare,"* 140–46.

19. *Enciclopedia militare*, 6 vols. (Milan, Italy: Il Popolo d'Italia, 1927–33), 1:51; Ceva, 262; Emilio De Bono, *Nell'esercito nostro prima della guerra* (Milan, Italy: Mondadori, 1931), 73–79; Brian R. Sullivan, "The Italian Armed Forces, 1918–40," in Allan R. Millett and Williamson Murray, eds., *Military Effectiveness*, 3 vols. (Boston: Allen & Unwin, 1987–88), vol. 2, *The Interwar Period*, 202.

20. The Fascist militia employed rank and unit designations drawn from those of the Roman Army. A *sottocapomanipolo* corresponded to a second lieutenant, a *capo manipolo* to a first lieutenant, and a *centurione* to a captain. A *manipolo* (maniple), *centuria* (century), *coorto* (cohort) and *legione* (legion) corresponded, in order, to a platoon, company, battalion, and regiment. However, battalion-size *Camicie Nere* combat units were called battalions, not cohorts. See Ilari and Sema, 356, 371–76.

21. Sergio Pelagalli, "Il generale Pietro Gàzzera al ministero della guerra (1928–1933)," *Storia contemporanea* 20 (December 1989):1012–15; Brian R. Sullivan, "The Impatient Cat: Assessments of Military Power in Fascist Italy, 1936–1940," in Williamson Murray and Allan R. Millett, eds., *Calculations: Net Assessment and the Coming of World War II* (New York: The Free Press, 1992), 107, 117.

22. *Enciclopedia militare*, 3:338; Ferruccio Botti and Virgilio Ilari, *Il pensiero militare italiano dal primo al secondo dopoguerra (1919–1949)* (Rome: USSME, 1985), 322–33; Giordano Bruno Guerri, *Giuseppe Bottai, un fascista critico* (Milan, Italy: Feltrinelli, 1976), 174–75; Alberto Baldini, *Elementi di cultura per il cittadino italiano* (Rome: Nazione Militare, 1935), passim; Curio Barbasetti di Prun, "L'Istituto superiore di Guerra," in Pietro Maravigna, ed., *Un secolo di progresso italiano nelle scienze militari (1839–1939)* (Rome: Società italiana per il progresso delle scienze, 1939), 77–82; Edoardo Scala, "Per la preparazione dei nostri ufficiali," *Rivista di Fanteria* 3 (September 1936).

23. Umberto Spigo, *Premesse tecniche della disfatta* (Rome: Faro, 1946), 49–50; Pelagalli, 1018, 1032; Montanari, *L'esercito italiano alla vigilia*, 218, 225–28; Ilari, *"Nazione Militare,"* 370–77.

24. Montanari, *L'esercito italiano alla vigilia*, 221–23; Enrico Serra, *Tempi duri: guerra e resistenza* (Bologna, Italy: Il Mulino, 1996), 8–11; Ugo De Lorenzis, *Dal primo all'ultimo giorno. Ricordi di guerra (1939–1945)* (Milan, Italy: Longanesi, 1971), 24; Ilari, *"Nazione Militare,"* 101, 377–81; Spigo, 50–51.

25. Montanari, *L'esercito italiano alla vigilia*, 26.

26. When Italy entered World War II in 1940, the army numbered 56,500 Italian officers (21,500 regular and 35,000 reserve) and 1,255,800 Italian NCOs and men. The combat units of the Black Shirts numbered 2,200 officers and 66,300 NCOs and men. In addition, the army included 400 Albanian officers and 11,000 NCOs and men, 24,000 Libyan troops, and 181,900 East African soldiers. In Italian East Africa, 5,100 army officers and 900 militia officers commanded the 67,100 soldiers and Black Shirts, and East African troops. There was a smaller ratio of Italian officers to colonial troops in Italian East Africa, as opposed to the officer-enlisted ratio in the so-called "metropolitan army." This reflected the higher quality of Italian colonial officers and their generally superb Eritrean and Somali NCOs. *L'esercito italiano tra la 1a e la 2a guerra mondiale, Novembre 1918—Giugno 1940* (Rome: USSME, 1954), 331–32; Antonio Giachi, *Truppe coloniali italiane tradizioni, colori, medaglie* (Florence, Italy: Grafica-Lito, 1977), 16–19; Ilari, *"Nazione Militare,"* 415–16, 423, 427.

27. Ugo Marchini, *La battaglia delle alpi occidentali* (Rome: USSME, 1947), 113–15; Rex Trye, *Mussolini's Soldiers* (Osceola, Wisc.: Motor Books, 1995), 30–33; Brian R. Sullivan, "The Italian Soldier in Combat, June 1940–September 1943: Myths, Realities and Explanations," in Paul Addison and Angus Calder, eds., *Time to Kill: The Soldier's Experience of War in the West 1939–1945* (London: Pimlico, 1997), 183.

28. Sullivan, "Impatient Cat," 116–24.

29. Sullivan, "Italian Armed Forces," 171, 190–93.

30. Stefani, 2/1:292–309, 453–89, 543–58, 636–54; Montanari, *L'esercito italiano alla vigilia*, 9–29; Botti and Ilari, 215–30, 250–53; Brian R. Sullivan, "Fascist Italy's Involvement in the Spanish Civil War," *Journal of Military History* 59 (October 1995):708–13.

31. Sullivan, "Italian Soldier in Combat," 179–80; John F. Coverdale, *Italian Intervention in the Spanish Civil War* (Princeton, N.J.: Princeton University Press, 1975), 181–86, 225–48, 275–76; Sullivan, "Fascist Italy's Involvement," 706–8; Ilari and Sema, 298–301, 304–13, 318–30.

32. For a description of these campaigns, see MacGregor Knox, *Mussolini Unleashed 1939–1941: Politics and Strategy in Fascist Italy's Last War* (New York: Cambridge University Press, 1982), 128–33, 150–65, 232–38, 249–60; Mario Cervi, *The Hollow Legions: Mussolini's Blunder in Greece 1940–41* (London: Chatto & Windus, 1972), 124–257; Anthony Mockler, *Haile Selassie's War: The Italian-Ethiopian Campaign, 1935–1941* (New York: Random House, 1984), 241–382. The debacles suffered by Italian forces in North Africa and East Africa from December 1940 to April 1941 resulted in the loss of many records, making it impossible to know the extent of casualties with precision. Statistics concerning the Albanians, Libyans, and East Africans in Italian service are particularly uncertain. Purely Italian losses suffered in the campaigns specified, however, can be approximated at 48,000 dead, 54,000 wounded, 67,000 severely sick or frostbitten, and 232,000 prisoners (including additional wounded and sick). These were heavy casualties for a country of some 42 million, amounting to nearly 1 percent of the entire population. See Virgilio Ilari, *Storia del servizio militare in Italia*, vol. 4, *Soldati e Partigiani (1943–1945)*, 28, 33; Mario Montanari, *Le operazioni in Africa Settentrionale*, 4 vols. (Rome: USSME, 1990–93), vol. 1, *Sidi el Barrani (Giugno 1940–Febbraio 1941)*, 216, 237, 279, 314, 385; and *L'esercito italiano tra la 1a e la 2a guerra mondiale*, 332.

Confronting Backwardness: Dilemmas of Soviet Officer Education in the Interwar Years, 1918–1941

Mark von Hagen

For most of the interwar years, the primary challenges concerning Soviet military education revolved not so much around the content of that education as it did around the character of the army itself, the place and status of officers in that army, and the social and ethnic composition of the officer corps. Reaching a consensus on these matters occupied much of the period between the Revolutions of 1917 and the outbreak of World War II. Those contextual aspects form the bulk of this essay. As that consensus was being fought out, officer education at the very highest levels of the Red Army was conducted by former officers of the Imperial Army until new graduating classes were able to replace them. The other fundamental shift in the interwar period was from an army that was just beginning at the end of the Old Regime to consider the challenges of new technologies in artillery, aviation, and communications to a renewed interest in those developments after the Red Army began to be outfitted with modern weaponry as a consequence of the industrialization plans of the 1930s; in the interim, a technologically underdeveloped Red Army in the 1920s focused its educational programs on morale and the inculcation of discipline rather than on technical skills. In a significant sense, the Red Army's educational programs did not have the chance to become routinized nor tested much in actual combat until World War II, partly due to the army's rapid expansion, but also to the deadly purges of the officer corps. Following the war, the Soviet Army carefully studied the lessons of modern warfare and only then was able to consolidate the initiatives of the 1930s.

The Contexts of Officer Education

The leadership of the Red Army, from the very first days of its existence, faced a virtually intractable set of dilemmas that they understood in part as the legacy of backwardness from the autocracy and in part as the demands placed on them by their own claims to build an army of a new type, one that was appropriate for a regime calling itself socialist. The officer corps of the Old Regime had been closely identified with the social privileges and status of the gentry estate, despite a considerable degree of democratization forced by the attrition of the Great War. Both the reformist and revolutionary wings of the opposition in Russia viewed the officer caste as a bulwark of despotism and a preserve of reaction. To a very large degree, the revolutions of 1917 were driven by soldiers' movements to assert their rights against the officers; soldiers' deputies to the Petrograd Soviet pushed through Order No. 1, which democratized the army by granting soldiers full civil and political rights and subjecting major army decisions to the deliberations of elected soldiers' committees.[1] As a consequence, both the Provisional Government's liberal-moderate socialist coalition parties sanctioned measures and the Bolshevik party during 1917 pursued a politics of demoralization and disorganization that undermined the authority and succeeded in earning the enmity of that officer corps. The Provisional Government appointed commissars to represent "revolutionary power" in the army's units; among their responsibilities were monitoring the behavior of the officers and acting as a conciliatory force between increasingly discredited and unpopular officers and their subordinates. Conservative and right-wing politicians looked to the army's elite to overthrow the revolutionary experiment and to restore order in the ranks; not only did they pin their hopes on Gen. Lavr Kornilov's march on Petrograd in August 1917, but later during the year and in later years hoped that military officers would overthrow the Bolshevik regime and restore Russia, one and indivisible.[2]

When the Bolsheviks seized power in Petrograd in November 1917, they entertained the utopian hope that the European proletariat would follow their revolutionary example and turn their weapons against their own domestic bourgeoisies; V. I. Lenin's Decree on Peace, which appealed to the belligerent powers to come to a negotiated end to the Great War, was drafted in that euphoric spirit of the first postcoup days. The Bolshevik leadership hoped to preside over the demobilization of the Imperial Army and begin the new era of socialist society without much thought to a future army. When those initial illusions foundered on the realities of a renewed German offensive and the outbreak of armed domestic resistance to the Soviet regime, the Bolshevik leadership painfully reconsidered its earlier opposition to organizing a military force and resolved, at least provisionally, not only to begin conscripting soldiers, but to recruit officers from the former Imperial Army as well. All this was defended, however, as a temporary compromise with the socialist ideal of a workers' militia, an ideal now postponed until the triumph of world revolution and the ensuing peace. The socialist writings on militias generally agreed on the pernicious character of standing armies and professional officer corps. This antiprofessional sentiment shaped the wide range of quasi- and para-military experiments that emerged during 1917 and 1918, especially the Red Guards or factory-based militia units.[3]

The consequences of these compromises with revolutionary doctrine included an overwhelmingly peasant army led by largely Tsarist-era officers and noncommissioned officers (NCOs), two developments which flew in the face of Bolshevik class prejudices; Bolsheviks suspected peasants and even many workers, not to mention white-collar workers and former officers, of lacking the requisite proletarian consciousness they felt necessary to safeguard the gains of the revolution. A powerful opposition within the party to these policies forced Army Commissar Leon Trotsky to recognize the institution of the political commissar, adapting the practice of the

Provisional Government, to serve as watchdogs over the behavior of the Tsarist officers, now dubbed military specialists to distinguish them from the Red Commanders who were being trained in crash courses (often by other Tsarist officers) and promoted from within the Red Army's enlisted ranks.[4] For all their profound hesitations about the loyalties of the former Imperial officers, the Bolsheviks reluctantly agreed that they needed some measure of professional military expertise to command their new army. By the end of 1918, more than 22,000 former officers and 128,000 former NCOs were in the Red Army. They served out of a variety of motivations, some coerced by the holding hostage of their families but others genuinely won over to the new regime; among their numbers were those who betrayed the Soviet cause at the first opportunity and thereby fueled the opposition. But without them the Soviet regime would not have survived the Civil War, let alone emerged victorious over the array of anti-Bolshevik domestic and foreign armies.[5]

By the end of the Civil War, the Bolsheviks had not resolved the fundamental ambiguities in their attitudes toward military professionals. Many in the party insisted that the end of hostilities provided the opportunity to revive the ideal of the socialist militia, based on an armed working class and urban population, trained in universal military service for defensive purposes in programs that only minimally disrupted the working life of the nation. The victory in the Civil War did not eliminate the opposition to a professional officer corps among influential segments of the party, who continued to warn of a Bonapartist threat to the Revolution. These party spokesmen demanded a complete demobilization of the Red Army and a thorough-going purge of the still suspect officers of the Old Regime. Even those Bolshevik leaders who had come to a grudging respect for the military specialists agreed that the new regime needed to train its own, loyal corps of Red Commanders in a spirit of "proletarian consciousness" and internationalism. The future of the armed forces in the Soviet state was one of several focal points in the struggles that dominated the party congresses of 1920, 1921, and 1922. As the regime felt more confident in its own survival and as the Red Army itself gained in legitimacy as a postrevolutionary institution enjoying widespread support among the new elites, the notion that military professionalism was a long-term priority also found increasing numbers of adherents in the leadership.[6]

Class Origins

Lenin and his immediate entourage in the party's Central Committee became disillusioned in the capacity of the workers' spontaneous energies and skills to organize the new society, but many others in the party, whether inspired by Waclaw Machajskii's critique of intellectuals' usurpation of the socialist parties or one of the other available contemporary antielitist platforms, found it more difficult to part with their often anarchist-syndicalist convictions.[7] One of the tradeoffs the party leadership made in acquiescing to what was now acknowledged as the need for elites in socialist society was to insist that those elites be drawn from the putatively more loyal democratic classes, the workers and peasants. It was assumed that by promoting workers and peasants into management positions, the socialist character of the new state would be assured, whatever other compromises were deemed necessary to provide for the smooth and effective functioning of a state organism.

Class criteria thereby were enshrined in Red Army conscription policies, but even more so in recruiting the commanders of the workers-peasants' Red Army.[8] In this respect, the new regime came up against one of the consequences of Russia's relative backwardness, the mass illiteracy of the workers and peasants who were the pool for the future officer corps. Already during the Civil War, the army's political enlightenment staff, an offshoot of the institution of the commissars, had devised a range of adult education programs to bring culture to the soldiers and junior officers during lulls in the fighting. It was clear, however, that such basic education would become an essential component of all junior and even senior officer training for the foreseeable future.[9]

Of course, class membership was never claimed to be wholly sufficient to guarantee the proper class consciousness in military men, however important it was. After all, the Bolsheviks had plenty of experience with putative proletarians who failed to "understand" their genuine class interests and sided at one time or another with enemy forces or rival socialist parties. Gradually and to some degree imperceptibly, proletarian class consciousness became conflated with loyalty to the Bolshevik-led state and willingness to lay down one's life for that cause. To keep secure that "class consciousness" it was necessary to subject socialism's new citizenry to large doses of what became known as "political education." Unlike the American tradition that insists on its armed forces remaining politically neutral and that has sought to exclude partisan politics from the military and to keep the military out of political frays, the Red Army was very consciously intended to be politicized virtually from the start. Officers were encouraged to join the Communist party and to take a leading role in party politics both in their own units and as delegates to the civilian party's congresses and conferences.[10] Both contemporaries during the Red Army's interwar history and Western observers since have assumed that political considerations often outweighed professional competence in the army; clearly, such tensions mitigated against developing a full-fledged meritocracy in the Red Army (and in many other organizations) and limited the degree to which the military controlled its own professional interests.[11]

In addition to the tensions that characterized relations between political officers and regular commanders, the Soviet security forces also kept a hand in military life, not only serving as a watchdog over soldiers' behavior, but also monitoring officers' performance and attitudes.[12] Local military commissariats not only processed recruits for the Red Army during the annual call-up, but were additionally responsible for recruitment into the security police (OGPU/NKVD). OGPU/NKVD officials insistently demanded that the best conscripts, defined by physical, educational, and political criteria, be sent to the security police. At least beginning in the early 1930s and no doubt more intensively following 1934 when OGPU/NKVD ranks expanded rapidly, army authorities frequently complained that they were regularly assigned the second-best recruits.[13] The struggle with the security police over the best recruits, when combined with already exclusionary class and political principles that were applied during annual recruitment drives, deprived the military of many of the best available cadres, at least by Soviet criteria. Clearly, then, the relationship of the Soviet state in the interwar period to its officer corps was ambivalent; they were showered with awards, status, and considerable privilege in the new order, but they were never entirely trusted by that same regime. The ambivalence was born of the conflicts that dated from the Civil War, the prejudices of the first generation of Bolshevik leaders against the officers of the Old Regime, and the continual disagreements within the peacetime army over the proper balance between professional training in military affairs and political loyalty.

The National Question in the Red Army

If all this were not enough to fill the plate of those assigned to train the Red Army's officer corps, another issue emerged by the end of the Civil War, namely, the charge that the Red Army was acting as an institution of Russification in the territories inhabited by large non-Russian populations. The Bolsheviks, as part of their approach to the "national question," had insisted that Great Russian chauvinism was an evil to be extirpated wherever it was detected; without substantial success in eliminating the legacy of discrimination against the non-Russian nations of the Russian Empire, the international socialist revolution could make little headway among those peoples who had felt themselves to be the victims of historical national oppression. The Imperial Army had been sensitive to the dilemmas of outfitting an army in a multinational state (ethnic Russians were less than one-half of the empire's population by the middle of the nineteenth century) and chose to maintain a largely Slavic conscript pool, while recruiting a far more cosmopolitan officer corps to command the soldiers.[14] During the last decades of the Old Regime, the Imperial officer corps was overwhelmingly Russian by ethnicity, but large numbers of non-Russians served in the highest ranks as well.[15]

During the Revolution, the Bolshevik party had made common cause with several non-Russian nationalist parties that demanded a renegotiation of the relations among nations and, particularly, special consideration for those previously oppressed in the empire. Although the Bolsheviks only hesitantly compromised their insistence on the primacy of class with a series of what they understood to be temporary concessions to the principles of nation, they found themselves at the end of the Civil War facing a political environment in which national identity and nationalist ideologies had come to play very important roles. All around the peripheries of Bolshevik Russia, several proto-states had come and gone, but all had claimed some measure of allegiance to the principles of national sovereignty, whether they were the states whose independence the Bolsheviks reluctantly acknowledged (Poland, Finland, the Baltic States) or those whose claims to sovereignty they ended with Red Army occupation (Ukraine, Belorussia, Georgia, Armenia, Azerbaijan, and the Central Asian territories). As part of the program of national sovereignty, those states had, with varying degrees of success, formed national armies.

Because of the special role that the party assigned the Red Army as the "school of socialism" and a model institution in the Soviet state, it was clear that charges from non-Russians that the Red Army was serving to Russify the non-Russians could not go unaddressed. As part of the reforms of the 1920s,[16] Army Commissar Mikhail Frunze insisted that "our army is not Great Russian nor Russian (*rossiiskaia*), but the army of the Union of S[oviet] S[ocialist] R[epublics]" and he devised a five-year plan to create national military formations from among the non-Russian peoples of the Soviet Union.[17] This accorded with the general party and state policy of *korenizatsiia* (translated variously as indigenization or nation-building), a package of administrative measures intended to increase the representation of non-Russian minority peoples in public offices and institutions, to encourage the use of non-Russian languages in local administration, and to promote non-Russian cultures. This radical experiment in "affirmative action" and "multiculturalism" was applied in the army to overcome the legacies of Great Russian chauvinism and to inculcate in soldiers and commanders ideas of "brotherhood and solidarity of the peoples of the USSR." Ukrainian, Polish, Latvian, Tatar and many other citizens now could choose whether to serve in units made up of soldiers from their own ethnic group.

M. Zakharov, the author of the official manual for commanders on the nationality issue, summarized the views of the leadership. One of the most tragic legacies of the "past centuries of oppression and slavery," he wrote, was the uneven economic and cultural development of the various peoples of the Soviet Union. Gradually all these peoples would be drawn into all aspects of state-building. But, of course, not everything could be accomplished at once; consequently, those who in the past had no military training or native skills would not be called to military service on an equal footing with the majority of the population. Some would be subjected to preinduction training, others would form volunteer units, others would gradually be called up but at first in limited numbers. Zakharov insisted, however, that in the Soviet Union there is to be "no place for any division of peoples into two groups—those obliged to serve and those exempt, as backward, or even worse, unreliable."[18]

If vocal party spokesmen had protested the dilution of class principles and revolutionary democracy that they saw in the mass call-up of peasants and the employment of former Imperial officers in the Red Army, there was even less consensus in the party ranks on the wisdom of this ethnophilia.[19] Military men of both prerevolutionary and Soviet vintage warned of the dangers these measures posed to the unity of the armed forces, especially if non-Russian languages were introduced into units, but the Twelfth Party Congress made only minor concessions to these critics in the resolutions adopted in 1923. At first it was not clear whether these formations would constitute a permanent part of the Red Army or only represent a transitional stage until all formerly backward nationalities had attained the same level of culture and military skills as the most advanced peoples of the multiethnic state.

Spokesmen for at least three positions on the formations could be found among both the Bolshevik leadership and the Red Army: those who advocated the national formations as a permanent feature of the Red Army and as the core of a future multinational army that would lay claim to the entire world, those who stood unalterably opposed to any such experiments (either on grounds of military efficiency or on grounds of their hostility to any encouragement of national sentiment in any form and in any place), and those who saw them as a temporary expedient until the new socialist state had attained more progress in bringing all its peoples to the same level. At the same time, whatever the range of positions on these experiments, the Bolshevik leadership maintained a persistent ambivalence toward these units—as they did on so many other features of the Red Army—that was rooted in the Civil War experience when anti-Bolshevik forces used the slogan of national armies as rallying points for their nationalist movements. This was especially true in Ukraine, Crimea, and Transcaucasia, but elsewhere as well.

This ambivalence was reflected within the structure of the reformed army itself. The national formations were not intended to be more than a small part of the Red Army's total numbers. The primary model remained units composed along regular "anational" principles of conscription, which usually meant a predominantly Russian or Ukrainian makeup. These units were deployed extraterritorially, their sole language of command was Russian. But even within some "regular" units, national (non-Russian) subunits were organized, complicating matters even further. Here, too, disagreements emerged over what should distinguish these subunits from the more full-fledged national formations and how these subunits should be deployed. A part of the officer corps felt that rather than isolating soldiers of one nationality from those of other nationalities, all should be encouraged to follow the example of Russian soldiers and thereby overcome their "backwardness." These officers felt that isolation would only prolong their backward condition. But the Red Army's top leadership argued a different line in the mid-1920s: soldiers in the relatively isolated national subunits would make better fighting material because they would not feel themselves so alienated from their native traditions, nor would they face the harassment and chauvinistic attitudes particularly from Russian soldiers and officers as readily as they might with no special provisions made for them.

Even if they might come to some measure of agreement on the desirability of the national experiments, the military leadership still faced a fundamental dilemma as it tried to expand the number of national formations and subunits, namely, a shortage of qualified personnel to train and command them. The simple fact was that most of the minority nationalities could not put forward enough cadres, whether in the civilian or military spheres, with either the requisite administrative experience or tested political loyalty to the new regime to staff the rapidly burgeoning bureaucracies. In the army as elsewhere, good intentions to involve all the nationalities in defending the nation and thereby give formerly disfranchised nations a stake and role in the political order came up against formidable cultural, political, and economic barriers.

There was little opportunity, for example, to create officer corps from those peoples who had little or no previous military traditions or training. Wherever possible, non-Russian soldiers were promoted to officer rank and placed in command of the national formations; when such were not to be had, attempts were made to insure that at least the commanding officer knew enough of the language to make his command effective. As a result, most of the officers continued to be Russians or Ukrainians, with significant numbers of Tatars, Jews, and others who had been promoted during or after the Civil War. But even in the case of nationalities whose soldiers had long traditions of military experience and relatively large numbers of command personnel, especially the Ukrainians (the second largest national group after the Russians themselves), there were not enough officers for both the national formations (in the case of Ukrainians they were actually referred to as territorial formations rather than Ukrainian national formations because of an ever greater legacy of hostility among the civilian and military leadership toward the Ukrainian national movement than toward others) and the national subunits within larger regular units. Constant transfers of personnel to achieve some measure of national coordination created logistical headaches and nightmares for administrative staff and proved very disruptive and costly.[20]

With the gradual emergence of the Union administrative structure, local republican and national authorities eagerly set up schools to train officers in their native languages; each government felt it a matter of pride to have one or more national military schools.[21] But here, too, the non-Russian authorities faced an exaggerated version of the general national dilemma, the weakly developed network of primary schools and the vast numbers of illiterate and semiliterate citizens. If the Russian population suffered from high illiteracy, then the non-Russians, because of the discriminatory educational policies of the autocracy, experienced this backwardness with even greater acuity. Moreover, the schools mushroomed with little or no attempt to coordinate their activities nationally with the actual needs of the military or with much of a real sense of the material and cultural resources of the existing local administrations.

As a rule, military manuals and political texts did not exist in languages other than Russian, even if the officer cadets were literate in their native languages. Several languages did not have the full range of equivalent military vocabularies to allow translation from modern Russian. Throughout the 1920s a debate raged in military circles about the "language of command" in the Red Army. On one side were old soldiers who insisted that whatever the benefits of the new regime's nationality policy outside the Red Army, the Red Army itself could not afford the divisive tendencies that would be introduced by a policy of several languages of command. How, they asked, in the event of a real national emergency, would the army command be able to use soldiers and officers who understood no Russian? A small number argued, in contrast, that national units respecting native traditions and training soldiers and officers in their native languages would be more loyal and combat effective. Eventually a compromise was reached, at least on paper, that all officers, even those of non-Russian nationality and who commanded national formations, would be expected to know Russian as well as their native language.[22]

The Revolutionary Military Council (*Revvoensovet*), the highest organ of military power, worked out detailed instructions on language policy that revealed the gradual reemergence of Russian as the language of administrative convenience. It was to be used in all correspondence with higher headquarters (from the divisional level upward), in all correspondence with other units (not of the same nationality), for all economic reporting from the regimental level upward, and for all operational work. What was left for the non-Russian languages included: most combat training, political work and other extracurricular cultural and schooling activities, daily correspondence within the units, correspondence with local organizations, and all administrative-financial correspondence within the units.[23] To further these complicated and contested agendas of military and national policy, the Red Army published seven newspapers in languages other than Russian by the end of 1925.[24]

As one consequence of the often drastic differences in historic military experience of the non-Russian nations, some national military schools for officers designed programs that were considerably longer than those for other nationalities. For many of the peoples who had some military traditions, these were nevertheless not the traditions of modern European armies, but rather those of nomadic or tribal warfare. For such peoples, the barracks regime of initial military training was even more foreign than it might be for Russians or Ukrainians; after all, many Russians and Ukrainians at least had some experience of factory life so that large buildings with their regimented existence were not altogether unfamiliar. But for many other peoples, regimentation and barracks life required great tact from the commanding officers and other supervisory personnel.[25]

In other words, future officers in the Red Army needed considerable exposure to ethnographic education in order not to offend their subordinates and thereby undermine the intended effects of the nationalities' policies to overcome decades and centuries of ethnic hostilities. Of course, organizational improvisation could not change the political culture of the army nor could it alter the negative perceptions many former subject peoples nurtured about the Russian Army and Russian (or Slavic) soldiers. Officers in the old army frequently derided the non-Russians, and soldiers imbibed many of these prejudices and attitudes. In delicate situations, any insult, whether intentional or otherwise, was perceived very acutely by the national minority soldiers. Any improvements in interethnic relations were complicated during the 1920s by continued fighting, especially in Ukraine and Turkestan, where the partisan bands and *basmachi* refused to give up their fight against the new regime. Commanders and political workers were encouraged to introduce courses for soldiers that treated themes of local history, especially the contributions made by the local population to "the struggle against tsarism."

The history of the Khar'kov School of Red Sergeants illustrates the politics of nationality in the Red Army. The school was located in a very symbolically important former theological school in Khar'kov. It opened in late June 1920 and counted three hundred students by August. Officers taught the military subjects, while for general education Ukrainian teachers were brought in from local civilian schools. At the beginning of October, the entire student population was sent to the front and the remaining officers and two teachers were arrested. Only in mid-October were classes resumed. At first there were no textbooks for the courses and GUVUZ (the educational and training administration) warned that it could not provide textbooks in Ukrainian, so the school itself began translating various statutes and other terminology. In its plan for general education, the school included three hours weekly for Ukrainian language, four for Russian, four for political literacy, and two for Ukrainian studies (*ukrainovedenie*) of the total thirty-one weekly hours of instruction.[26] Some of these battles had to be refought in later years. For example, in November 1922 the pedagogical council of the Khar'kov School of Red Sergeants voted unanimously to introduce Ukrainian language in its curriculum and in place of the two hours of Russian language instruction.[27] By June 1923, however, when the school began writing all its correspondence in Ukrainian, the central authorities protested and demanded a translation into Russian.[28]

The Frunze Reforms, "Militarization," and Officer Education

After Mikhail Frunze replaced Leon Trotsky as army commissar in 1923, Communist party and military authorities began elaborating and implementing a series of reforms intended to put the army on a stable footing and to instill a firmer sense of discipline and order than had been the practice since 1918.[29] The reform team's first move was to begin gradually phasing out the practice of dual command, and, commensurately, raising the responsibilities and status of the commanders and demoting the position of commissars. (Many commissars were offered higher military education so that they could make the shift to commander positions rather than accept the lower status of political education workers.) More officers were admitted to party membership with the attendant status and access to political influence. Military schools and academies were reorganized and expanded to speed up the production of loyal Red Army commanders. A new and more rigorous attestation system was instituted that set regular procedures to review and promote officers completing higher education and other terms of duty. To further bolster the authority and status of the senior officers, the Red Army resolved to expand the numbers and responsibilities of the junior officers (*mladshii komsostav*), whose position was envisioned as something closer to NCOs in most Western armies. Junior officers were freed from the requirement to attend all political classes with the troops, although they still were obliged to attend some. This change was justified on the grounds that junior officers needed more time to fulfill their strictly military functions, especially the coordination of training and drill with their superior officers.

Frunze evinced a particular interest in and attention to questions of discipline and morale, taking as one model the tradition of Catherine the Great's talented and victorious generalissimo, Field Marshal Alexander Suvorov.[30] In that spirit, the reform team introduced measures assigning more punitive powers to junior officers in a new disciplinary code. Frunze rehabilitated from the lexicon of the Imperial Army the word *mushtra* and advocated a return to a harsher form of drill-sergeant discipline than the Red Army had practiced since its formation. The chief spokesman of the new hard line on disciplinary matters was Aleksandr Sediakin, who was appointed director of the Training and Drill Administration of the Red Army in 1927.[31] The centerpiece of the new campaign was the first national military criminal code, promulgated in October 1924.

The party, however, did not significantly scale back its demands for the political indoctrination of its officers nor its expectations that officers (and to a lesser degree soldiers) would play an important role in any nationwide campaign, such as the collectivization drive that got underway at the end of the 1920s. It was the collectivization campaigns that most threatened the military's perception of itself as a professional fighting force. After the party and agricultural administrations decided to use the army in 1929–30 to train rural cadres in large numbers, the military hierarchy rebelled and warned of dire consequences for officers' authority and soldiers' discipline if such campaigns continued. Although never again was the army called up in such wide-ranging programs that diverted energies and manpower away from military training, still the peculiar political economy of Stalinism allowed for considerable intervention in military training in the interwar period that did not sit well with professional officers.[32]

The costs that this political dimension exacted on the hopes some in the military's high command had for professionalization were clear in the maneuvers conducted during the second half of the 1920s. In evaluating Soviet maneuvers, German observers were very critical. They noted especially that lower rank officers had "a complete aversion to assuming responsibility and a great lack of initiative. The result was that in the absence of a formal order nothing was done."[33] During the second half of May 1927, Kliment Voroshilov (who had succeeded Frunze as army commissar) conducted an inspection tour of the Ukrainian District—possibly to prepare for the late summer maneuvers—and found serious deficiencies. Among other shortcomings, Voroshilov criticized the district command for failing to put the new regulations and directives into practice, for falling down in combat training and the level of marksmanship, and for poor administration and discipline. He warned that, unfortunately, the Ukrainian District was not exceptional in this regard, but more likely typical of the Red Army as a whole.[34]

The 1928 Kiev maneuvers reinforced the Germans' impressions that the Red Army was far from ready for any large-scale offensive operations and that officers remained deficient in tactical training. General Werner von Blomberg, for one, was impressed by the prominence of Civil War veterans in the Soviet high command. These men, Blomberg reported, had received their basic military experience in a war that had "little in common with war against a modern well-armed Power."[35] A Soviet report in autumn 1928 indicated a still high rate of turnover among the officer corps (53 percent in the past year); and despite a significant rise in the level of average education among officers, still too few officers had attended the central army staff academies.[36]

Motorization, Mechanization, and New Challenges for Officer Education

Until the first five-year plan began the development of a native industrial base with a defense orientation, the Red Army could not seriously contemplate introducing mechanized warfare in its doctrine, strategic plans, or training. At the beginning of the 1930s the first "mechanized brigade" was formed, but the army as a whole had only three hundred tanks and one hundred armored cars. The other arm that formed a crucial part of the new military thinking was tactical aviation, which was viewed as a kind of "air artillery" to be employed in conjunction with mechanized units, artillery, and armor. The first attempts at maneuvers with this combination were held from 1931 to 1933. These new formations coexisted with a mass army so that the Red Army would operate as a synthesis of two armies, one a shock echelon and the other, large infantry formations exploiting a decisive breakthrough. Soviet military doctrine came to rely on an assault by the motorized shock elements while the enemy was still mobilizing. Should the enemy not be defeated, then the mass infantry army would settle in for a defensive war of attrition.[37]

One of the consequences of even this limited mechanization of the Red Army was the realization that the old political-administrative structures which governed the life of the military needed reshaping. That reform, dubbed in Soviet sources as "organizational measures" (*organizatsionnye meropriatiia*), began in June 1934 with the transformation of the Army and Naval Commissariat into a unified People's Commissariat of Defense and the replacement of the former *Revvoensovet* by a much reduced Military Council attached to the commissar.[38] The reform entailed a considerable centralization of military decision-making in the Commissariat. As a result, the army too became burdened with the "super-bureaucratization" that infected all major sectors of the Soviet political economy in the 1930s, with the Politburo engaged in constant scrutiny of even the most insignificant details of administration and planning.[39] Another feature of the 1934 reforms was the launching of a considerable expansion of the armed forces with a commensurate increase in the army's budget.[40] These measures led almost inexorably to a reconsideration of the principle of a mixed cadre-territorial organization; the following year that compromise began to be resolved in favor of the cadre units.[41] The territorial militias had been responsible for preinduction training; those responsibilities were reassigned to an overhauled Osoaviakhim (Society for Defense, Aviation, and Chemistry).[42] Parallel to the abandonment of the territorial system, gradually all Ukrainian and other nationality formations were also disbanded. The Ukrainian officer schools were also closed. In 1935 the Ukrainian Military District was reorganized into three new districts, the Kiev, Khar'kov, and Odessa, thereby integrating the territory of the Ukrainian SSR into the all-union military commissariat.

Raymond Garthoff dates "stabilization of the new officer corps" to the period of Marshal Mikhail Tukhachevsky's rise within the Soviet defense establishment from 1931 to 1937. Tukhachevsky hoped to create a professional group of specialists in military art and technology, as was the intention of military reformers in the late nineteenth century. But with the professionalization came certain social changes that were incompatible with official Soviet ideology, namely, a striving by many officers to restore some version of the previous system of hierarchy and ranks, replete with insignia and expanded privileges for military men and their families. The higher officers benefited from new salary scales, which, however, widened the social gap between them and junior officers, as well as between both and rank-and-file conscripts. During the 1930s officers and their families also had access to an expanding network of special stores, theaters, and clubs. In 1937, the year of Tukhachevsky's execution, a modern rest sanatorium in Sochi was opened for Red Army officers.[43]

During the same period, a number of measures raised the prestige and welfare of the Red Army officer corps. On 22 September 1935, formal ranks and distinctions were introduced.[44] Two months later the rank of marshal was conferred upon Semen Budenny, Kliment Voroshilov, Aleksandr Egorov, Vasilii Bliukher, and Mikhail Tukhachevsky; Iona Iakir and three other military district commanders were named army commander 1st grade. The September decree furthermore provided immunity from arrest by civil authorities for all but junior officers, without prior authorization from the commissar of defense. And, to make an officer's career more attractive, special attention was paid to raising living standards and expanding privileges for the officer corps. Finally, in line with a campaign to provide commissars with more purely military skills, political workers were required to pass normal military school examinations. Senior officers occasionally demanded the removal of political commissars whom they found lacking.[45]

Conclusion

All these changes posed formidable challenges to those responsible for officer training. Not only was the Red Army in the constant throes of reorganization and simultaneous expansion, but the change in the status of the officer corps, plus the new demands of motorization and mechanization threw educational plans into turmoil.[46] The failures were evident in the maneuvers that were staged nearly every year during the 1930s. The level of tactical training among junior officers remained at a low level in 1936, according to the testimony of the British military mission, who attended the September maneuvers of the Belorussian Military District. Colonel Giffard Martel concluded that the Red Army resembled the Imperial Army in its "tactical clumsiness" and its advantage of great physical toughness. He observed very little radio communication, poor reconnaissance, and "little skill" in handling tank forces. Like its Imperial predecessor, the Red Army remained a bludgeon, albeit with "armored spikes" on its head.[47] Despite the winding down of the nationality experiment, which had never been allowed to prove its potential merits but instead introduced new conflicts and chaos into army life, and despite the gradual retreat from considerations of class background in favor of military training, the rise of Stalin and his clique to their murderous ascendancy in the 1930s prevented the military leadership, even if it had been so inclined, from raising the educational levels of the officer corps to the requisite level by the outbreak of World War II. The army had been seen as the ideal laboratory for too many social and cultural experiments before it had had a chance to form its own professional sense of identity and to set its own standards for professional evaluation. Political loyalty came to be seen as the ultimate criterion for success, and even that was often no guarantee that an officer would escape the ravages of the Terror and its aftershocks. Only during World War II did the army learn its craft on the job and at a frightening cost in manpower.

Notes

1. The best accounts of soldiers' behavior during 1917 are Allan K. Wildman, *The End of the Russian Imperial Army,* 2 vols. (Princeton, N.J.: Princeton University Press, 1980, 1988); and Mikhail Frenkin, *Russkaia armiia i revoliutsiia* (Munich: Logos, 1978).

2. See histories of the White movement: Anna Procyk, *Russian Nationalism and Ukraine: The Nationality Policy of the Volunteer Army during the Civil War* (Edmonton and Toronto: Canadian Institute of Ukrainian Studies Press, 1995); Peter Kenez, *Civil War in South Russia, 1918: The First Year of the Volunteer Army* (Berkeley: University of California Press, 1971); and idem, *Civil War in South Russia, 1919–1920: The Defeat of the Whites* (Berkeley: University of California Press, 1977).

3. On socialist ideas of military organization, see Martin Berger, *Engels, Armies and Revolution* (Hamden, Conn.: Archon Books, 1977), and Sigmund Neumann and Mark von Hagen, "Engels and Marx on Revolution, War and the Army in Society," in Peter Paret, ed., *Makers of Modern Strategy from Machiavelli to the Nuclear Age* (Princeton, N.J.: Princeton University Press, 1986), 262–80. On the Red Guards, see Rex Wade, *Red Guards and Workers' Militias in the Russian Revolution* (Stanford, Calif.: Stanford University Press, 1984).

4. On the origins of the commissar, see John Erickson, *The Soviet High Command: A Military-Political History, 1918–1941* (New York: St. Martin's Press, 1962), 41–45. On the party's wrenching debates on the issues of officers and commissars, see Francesco Benvenuti, *The Bolsheviks and the Red Army, 1918–1922* (Cambridge, England: Cambridge University Press, 1988).

5. See A. G. Kavtaradze, *Voennye spetsialisty na sluzhbe Respubliki Sovetov, 1917–1920 gg.* (Moscow: Nauka, 1988).

6. See Mark von Hagen, *Soldiers in the Proletarian Dictatorship: The Red Army and the Soviet Socialist State, 1917–1930* (Ithaca, N.Y.: Cornell University Press, 1990), chaps. 3–5.

7. Lenin parted with his scruples against elites relatively quickly upon coming to power. See his justification for a return to the principles of one-man management in industry, "The Immediate Tasks of the Soviet Government," April 1918; and his polemic with the antielitists in the party, among them the Left Communists, "'Left-Wing' Communism—An Infantile Disorder," April 1920.

8. When political workers identified trouble in any military unit or found friction between

officers and soldiers, their first instinct was to find a "social" explanation based on class. A typical explanation was that unruly soldiers were from the middle peasantry or, even worse, had managed to mask their true social identity as kulaks; officers who could not get along with the political workers or who dismissed political work as superfluous and a nuisance were discovered to have noble or bourgeois roots or, if they had come up from workers or peasants, spent too much time with officers of less democratic origins.

9. See von Hagen, *Soldiers in the Proletarian Dictatorship*, 89–114, 152–63.

10. Ongoing efforts to embrace as much of the officer corps and rank-and-file troops as possible resulted in approximately half the soldiers and 67.8 percent of officers listed as party members in 1933 and 1934. See K. E. Voroshilov, *Stat'i i rechi* (Moscow: Partizdat, 1937), 573, 575.

11. For a history of the relations between party and army, see Roman Kolkowicz, *The Soviet Military and the Communist Party* (Princeton, N.J.: Princeton University Press, 1967); Timothy J. Colton, *Commissars, Commanders, and Civilian Authority: The Structure of Soviet Military Politics* (Cambridge, Mass.: Harvard University Press, 1979); and Erickson, *Soviet High Command*. For contemporaries' concerns about the sacrifice of military professionalism to political loyalty, see S. Ivanovich (V. I. Talin), *Krasnaia armiia* (Paris: Sovremennye zapiski, 1931); and Mikhail Frunze's reassurances to the military specialists serving in the Red Army that there was "no possible way" that the army command could achieve 100 percent "communization" of its officer corps, even if it wanted to do so. The non-Communist officer will have "a secure place in the ranks of the Red Army," he promised in 1924. See his "Itogi plenuma RVS SSSR," 20 December 1924, in *Sobranie sochinenii*, 3 vols. (Moscow and Leningrad: Gosizdat, 1926–29) 2:182.

12. This control was effected by an extensive network of special section (*osobyi otdel*) employees who conveyed regular reports from their units to higher level agencies. Gradually these security officers made their presence felt throughout the entire system of promotions and transfers through the requirement that their reports be part of any officer's personnel records.

13. See the typical correspondence about annual recruitment drives in Rossiiskii gosudarstvennyi voennyi arkhiv (RGVA), f. 25899, op. 3, d. 1572, ll. 279–82, 354–61, 371–84.

14. The best English-language summaries of Imperial recruitment policies are John L. H. Keep, *Soldiers of the Tsar: Army and Society in Russia, 1462–1874* (Oxford, England: Clarendon Press, 1985); and Elise Kimerling Wirtschafter, *From Serf to Russian Soldier* (Princeton, N.J.: Princeton University Press, 1990), chap. 1; see also M. Zakharov, *Natsional'noe stroitel'stvo v Krasnoi Armii* (Moscow, 1927), chap. 1. On the officer corps before the Miliutin reforms of the 1870s see Hans-Peter Stein, "Der Offizier des russischen Heeres im Zeitabschnitt zwischen Reform und Revolution (1861–1905)," in *Forschungen zur osteuropaeischen Geschichte* (Berlin, 1967), 13:351–63.

15. Especially prominent were the Baltic Germans, Finns, Poles, and Caucasians, primarily Georgians, varying from 15 to 21 percent depending on rank. See John A. Armstrong, "Mobilized Diaspora in Tsarist Russia: The Case of the Baltic Germans," in Jeremy Azrael, ed., *Soviet Nationality Policies and Practices* (New York: Praeger, 1978), 63–104; J. E. O. Screen, "Finnish Officers in the Russian Army and Navy during the Autonomy Period (1809–1917)," *Siirtolaisuus Migration* 4 (1981): 1–7; P. A. Zaionchkovskii, *Samoderzhavie i russkaia armiia na rubezhe XIX–XX stoletii, 1881–1903* (Moscow: Mysl', 1973), 196–202; and Stein, 457–67.

16. For a general discussion of the reforms, see von Hagen, *Soldiers in the Proletarian Dictatorship*, pt. 2; I touch only peripherally on the reforms in the area of nationality policy, about which more below. See also I. B. Berkhin, *Voennaia reforma v SSSR (1924–1925)* (Moscow: Voenizdat, 1958), 124–42.

17. M. V. Frunze, "Itogi i perspektivy voennogo stroitel'stva," in *Na novykh putiakh* (Moscow: Voennyi vestnik, 1925), 56.

18. Zakharov, 27–28.

19. See Yuri Slezkine, "The USSR as Communal Apartment," *Slavic Review* 53 (Summer 1994).

20. For some of these difficulties, see Zakharov, chap. 5. For a good summary of other problems in the politics of *korenizatsiia*, see Gerhard Simon, *Nationalism and Policy toward the Nationalities in the Soviet Union*, trans. Karen Forster and Oswald Forster (Boulder, Colo.: Westview, 1991), chaps. 2–3.

21. In Ukraine, normal officer schools for training lower ranking officers (*uchylyshcha*)—of which three were Ukrainianized—were located in Kiev, Bila Tserkva, Poltava, Odessa, Khar'kov, Mykolaiv, Chuhuiv, and other cities. Higher ranking officers were sent to the military academies in Leningrad and Moscow.

22. This requirement was defended on several grounds. First, as long as most of the technical literature on military affairs was not translated into non-Russian languages, crucial information would be available in the foreseeable future only or primarily in Russian; hence, no officer could expect to take advantage of the available literature for his own advancement if he could not read Russian. Second, officers who wanted to advance their careers with higher military education in the central academies and service branch schools, would suffer disadvantages if they did not know Russian; thereby the discriminatory patterns inherited from the Imperial practices would be perpetuated in the new regime. Third, officers would have to know Russian to be able to coordinate their units' operations with those of other units in the event of war, which might not necessarily share the same ethnic composition. Finally, in the event of war, officers would be called from wherever they could be found to replace those fallen at the front. It was unrealistic and practically impossible to expect that the reserves would always be available from officers of the same nationality as those who had fallen.

23. A proposal to conduct basic training in two languages in parallel fashion whenever possible was further elaborated to reflect the diversity of interethnic relations in the Soviet Union. For example, Ukrainian and Belorussian were close enough to Russian that soldiers could learn basic commands in both languages and should be so trained. For Georgian, Armenian, and Azeri soldiers, whose languages were not close to Russian, but whose military vocabulary was developed enough to have corresponding words for the requisite Russian-language commands, parallel training was desirable in limited cases. For all other nationalities, Russian was to be the primary language of basic command.

24. Two were in Ukrainian and one each in Tatar, Georgian, Armenian, Turkic (*tiurkskii*), and Uzbek.

25. The Civil War experience had alerted military officials to the sensitivity of non-Russian soldiers to diet and sanitary habits. Zakharov, in his manual, listed the types of foods preferred by each people and suggested that field canteens adapt to local tastes. He also warned that public bathing was not customary for soldiers from Central Asia. Soldiers would be led to the bathhouse and not want to undress in front of other soldiers. Usually, if one of the more "experienced" older soldiers would come in and strip unceremoniously, the younger ones would follow suit. See Zakharov, 57–58, 78–79.

26. "Doklad o postanovke uchebnogo dela v Tsentral'noi shkole chervonnykh starshin," November 1920, RGVA f. 25084, op. 1, d. 4, ll. 33–34.

27. "Vypiska iz protokola No. 10 zasedaniia pedagogicheskogo soveta," 24 Nov. 1922, RGVA f. 25084, op. 1, d. 10, l. 34.

28. GUVUZ to head of School of Red Sergeants, 9 June 1923, RGVA f. 25084, op. 1, d. 10, l. 105.

29. For a more detailed account of these reforms, see von Hagen, *Soldiers in the Proletarian Dictatorship*, esp. chap. 5.

30. F. D. Khrustov, *Frunze o voinskom vospitanii* (Moscow: Voenizdat, 1946), 63. See also M. V. Frunze, "Lenin i Krasnaia Armiia," *Sputnik politrabotnika* 18 (1925): 1–5.

31. See Aleksandr Sediakin, "Puti stroitel'stva boesposobnoi armii," *Voennyi vestnik* 8 (1925): 214–15.

32. See von Hagen, *Soldiers in the Proletarian Dictatorship*, chaps. 6 and 8.

33. For the report of the maneuvers in the Ukrainian Military District during August 1927, see Militaer. Angelegenheiten, 9534H/671631-6, cited in Erickson, 260.

34. On Voroshilov's tour, see Berkhin, 376–77.

35. Blomberg's report is discussed in Erickson, 266–67.

36. "Protokol No 12," 23 Oct. 1928, TsGASA, f. 25899, op. 1, d. 76, l. 132.

37. V. D. Mostovenko, *Tanki: Ocherki iz istorii zarozhdeniia i razvitiia bronetankovoi tekhniki* (Moscow: Voenizdat, 1958), 85. Mostovenko dates the first mechanized brigade to May 1930. In 1932 a mechanized corps was formed from the brigades. For more on the history of Red Army tank doctrine, see Mary Habeck, "Reinventing War: The Development of Armored Doctrine in Germany and the Soviet Union, 1919–1939" (Ph.D. diss., Yale University, 1996).

38. For the text of this reform, see H. J. Berman and M. Kerner, *Documents on Soviet Military Law and Administration* (Cambridge, Mass.: Harvard University Press, 1955), no. 3.

39. Erickson, 372. For some sense of this stifling attention to detail and requirement for central approval of decisions, see the memoirs of Alexander Barmine, *One Who Survived* (New York: Putnam, 1945), 220.

40. The cadre forces increased to 940,000 (nearly twice the number reached in the Frunze reforms); the army's budget, which had stood at 1,420,700,000 rubles in 1933, jumped to 5,000,000,000 rubles in 1934. See *League of Nations Armaments Year-book*, 15 vols. (Geneva: League of Nations, 1924–38), 14:848.

41. See the speeches of military leaders at the Seventeenth Party Congress: Voroshilov, 224–35; Tukhachevsky, 464–66; and Bliukher, 629–31, in *XVII s"ezd VKP(b). Stenograficheskii otchet* (Moscow: Politizdat, 1934).

42. Stalin and Vlas Chubar (newly elevated to the Politburo) severely criticized the leadership of Osoaviakhim (headed by R. P. Eideman) in a secret decree of 8 August 1935. See Merle Fainsod, *Smolensk under Soviet Rule* (Boston: Unwin Hyman, 1958), 333–34.

43. Raymond Garthoff, "The Military as a Social Force," in Cyril E. Black, ed., *The Transformation of Russian Society* (Cambridge, Mass.: Harvard University Press, 1960), 331–33.

44. *O vvedenii personal'nykh voennykh zvanii nachal'stvuiushchego sostava RKKA i ob utverzhdenii polozhaniia o prokhozhdenii sluzhby komandnym i nachal'stvuiushchim sostavom RKKA, 22 September 1935* (Moscow: Voenizdat, 1935), 22 pp.

45. Erickson, 393. See also Trotsky's evaluation of what he termed a "revolution" in the Red Army in 1935, *The Revolution Betrayed* (New York: Pathfinder, 1972), chap. 8.

46. Roger R. Reese, *Stalin's Reluctant Soldiers: A Social History of the Red Army, 1925–1941* (Lawrence: University Press of Kansas, 1996), esp. chap. 5.

47. Lt. Gen. Sir Giffard Martel, *The Russian Outlook* (London: Michael Joseph, 1947), 21, 23–26. Capt. B. H. Liddell Hart evaluated the Red Army in 1937 as follows: new and old ideas were "strangely intermingled," with, for example, horsed cavalry racing into tank-infested areas; many foreign ideas were assimilated without being properly digested; the actual handling of the tanks by their crews was not equal to the higher tactical handling of armored forces; there was an undue disregard for modern fire methods; and it would be better to rely on the mechanized forces as such, for there was a risk of communications breaking down in the mass offensive. Liddell Hart, "The Armies of Europe," *Foreign Affairs* 15 (January 1937): 246.

Orthodox Soldiers: U.S. Army Formal Schools and Junior Officers between the Wars

Charles E. Kirkpatrick

On the evidence of the performance of its senior officers during World War II, it is customary to conclude that the United States Army was doing something right in the professionally arid years between the two world wars. This essay contends that officer education, and most particularly junior officer education, was the key factor. Educating junior officers, a crucial task for any army, was a central concern of the U.S. Army throughout the interwar period but one constrained by the special circumstances of small forces, restricted budgets, limited command opportunities, and painfully slow promotions. Chiefly because all of those circumstances conspired to make promotion so slow, it was not unusual for a captain to be forty years old or older, and thus still a "junior" officer. Whatever their ages, these junior officers faced circumstances that have not been repeated. This paper shows how, despite the limiting factors, the service created through its school system successful senior commanders from long-serving junior officers.

Between 1920 and 1940, the U.S. Army was deficient in virtually every criterion normally used to measure a nation's military power.[1] A penurious Congress kept the army to an average of 14,000 officers and 120,000 other ranks for almost twenty years, although the National Defense Act of 1920 had authorized a much larger force.[2] Well equipped in 1920 with surplus materiel from World War I, ten years later obsolescence and age had reduced weapons and equipment to superannuated junk. Small budgets, overwhelmingly devoted to maintaining existing installations, equipment, and other recurring expenses prevented replacement, and the service's technology consequently remained pegged to 1918 standards throughout the interwar years.[3]

During World War I, soldiers of an army that for years essentially had been a frontier constabulary had learned to think organizationally and tactically in terms of divisions and corps. After 1919, however, there was no money to build new posts that could house large units, even assuming that congressmen could be persuaded to abandon the many little frontier forts that were so dear to their constituents.[4] The staff thus had to forego plans for large tactical formations and parceled the army's few troops out in small packets to prewar installations. By 1931, the army dispersed twenty-four of its regular regiments among forty-five different posts, thirty-four housing one battalion or less. Seven more regiments and separate battalions were stationed at nineteen other small forts. In such circumstances, it was no wonder that training above the battalion level was generally impossible, even if such training could be funded. In any case, units were at a fraction of their wartime strength and soldiers spent much of their time in work details to maintain decrepit buildings, equipment, and installations, rather than in training.[5] By 1932, the army's actual combat forces within the continental United States amounted to no more than 60,000 men out of a total strength of 134,957.[6] Progressive budget cuts further restricted the amount and quality of training.[7]

These were serious problems, but Gen. Douglas MacArthur, chief of staff during the worst of the Depression years, was most concerned with the condition of what he regarded as the heart of the army—its officer corps. In his annual reports to the secretary of war, MacArthur stressed that a quality officer corps was the one vital element in a modern army. Undesirable as it might be, the service could manage on short rations and with poor weapons and equipment. Without an efficient corps of leaders, however, it was doomed to destruction when war came. MacArthur worried that ongoing budget reductions threatened to undermine the quality of the officer corps, a concern that had some merit, considering the circumstances of the day.

Officer promotions were very slow in the small army; most men served as lieu-tenants and captains for as much as two decades. Years of low pay and stultifying routine did little to hone the professional abilities of officers who did not attain their majorities until they passed the milestone of their fortieth birthdays.[8] MacArthur repeatedly stressed the problems caused by slow promotions, reminding the secre-tary of war that leaders required the increased responsibility of higher ranks to con-tinue their professional development. He stressed that skilled officers were the products of continuous training and experience, and that the process admitted of no short cuts. "Stagnation," he wrote in 1935, "destroys initiative, saps ambition, and encourages routine and perfunctory performance of duty."[9]

Yet, when war came to the United States in 1941, a series of anonymous junior officers of varying ages emerged from the limbo of MacArthur's fifth-rate force to become the colonels and generals who capably and, on the whole, successfully, commanded and administered an army of millions. How did the inauspicious circum-stances of the interwar army contrive to produce men of such ability? General J. Lawton Collins, a division and corps commander in World War II and subsequently chief of staff, speaking for many of his peers, said that the "thing that saved the American Army was the school system."[10]

Collins was referring to the army's formal school system, the one bright spot to which MacArthur could point in his annual reports and an edifice of which he was quite proud.[11] Essentially a product of Secretary of War Elihu Root's reforms of 1903, the army's hierarchy of schools tutored officers in a manner that MacArthur de-scribed as "progressive, practical, and comprehensive."[12] An officer might not serve with a well-equipped unit, or with one that had anything approaching its full comple-ment of soldiers, and he might not have many opportunities to command troops in field maneuvers. On the other hand, the service tried to compensate by training him in a sensible and functional way from the day he reported to his first duty station until he retired. Ideally, a line officer would, in the course of a long career, succes-sively attend garrison schools, a branch basic course, a branch advanced course, the renowned Command and General Staff School, and the Army War College. The peculiarity of the interwar years was that, because of the inordinately slow pace of promotion, company-grade officers attended all of those schools. Junior officer education thus came, in this unique instance, to encompass the entire formal school system.

Fonts of Orthodoxy

When answering the complaints of Command and General Staff School students that the curriculum at Fort Leavenworth was excessively rigid, Brig. Gen. Edward L. King counseled that "Before you can be heterodox, you must first learn to be orthodox."[13] That statement may safely be taken as an authoritative summary of the purpose of the army's schools of the era. They were to teach the details of the profession of arms but, most importantly, they were to teach approved doctrine.[14] The curricula of the various schools were progressive, both in complexity and in content, and their spacing at various career points at least theoretically allowed an officer to put his most recent training to practical use before going on to more complex matters. No one imagined that the small U.S. Army could go to war as it was then constituted, and the focus of service school instruction was explicitly on the mobilization mission. The assistant commandant of the Cavalry School spoke for all the branch schools when, while acknowledging that his school's formal task was to create successful and knowledgeable cavalry officers, confessed that its practical purpose was "to give such instruction to the graduates of this school as will enable them to train and produce, when needed in war, well-trained cavalry leaders and cavalry troops."[15] Thus, the emphasis on orthodoxy. Only officers who had mastered army doctrine for training, administration, and tactics could return to their regiments to "instruct that mythical being, the intelligent private."[16]

The process began in unit schools, which preceded branch schools. After commissioning, an officer typically went to a regiment, where he served with troops for as many as six to eight years. In that interval, he attended the garrison schools, which had existed in various forms since 1891. The garrison's troop school, a seven-month course for recently appointed officers of the line, came first. Taught by unit officers and sergeants, the troop school systematically introduced the lieutenant to the basic details of his daily duty. Thereafter, the novice joined his peers in the regiment's year-round garrison school that instructed him in basic tactics, logistics, staff work, troop leading, training methods, intelligence, supply, and the various details of maintenance and care of troops and public property.[17]

These schools, although spread throughout the army, shared a strong family resemblance. They appear universally to have concentrated on the practical. Future Chief of Staff George H. Decker's first tour of duty was with the 26th Infantry; its garrison school drilled him in essential tasks such as conduct of training and supply.[18] Paul L. Freeman's regimental school included Spanish because of the unit's mission along the Mexican border.[19] Language training was also typical of the garrison school of the 15th Infantry, stationed in China. Officers serving with that regiment had to study Chinese.[20] Lucian K. Truscott recalled that the 17th Cavalry used "Moss's Manual," a compendium of customs of the service and related material, as a basic text, although the course naturally stressed the practical aspects of mounted service.[21] James H. Polk, another cavalryman who served at Fort Bliss in 1933 and 1934, received certificates of training from the 8th Cavalry regimental school in such subjects as "Horseshoeing for Officers" and "Mess Management."[22] Very much the school of the soldier for officers, the garrison school prepared the ground for subsequent, and more sophisticated, training.

Although there were exceptions, the first resident school characteristically came when the officer was a senior lieutenant or junior captain. Somewhere around his fourth year of service, the officer attended the resident company, battery, or troop officer's course of his arm.[23] The nine-month course was roughly the same in all of the branches. Building upon the common foundation laid by the garrison schools, the courses stressed the basic tactics, logistics, and technical matters peculiar to the branch.[24] This instruction was overwhelmingly practical, and its constant attention to tactics and techniques occasionally frustrated the brighter officer. Bradford G. Chynoweth, for example, a man who was critical of the formal military establishment throughout his career, complained bitterly that the Engineer School before World War I was "intellectual torture." Of the instructors, he commented that "devoid of teaching talent, they substituted a Jove-like inflexibility" whose "only perceptible goal was numerical accuracy."[25]

Chynoweth's complaint, the cry of an intellectual snared in a den of pragmatism, overdraws the rigidity of instruction in the basic service schools, especially after their reorganization following World War I. Still, it points up the functional and technical nature of the courses. Other and more successful officers praised their branch schools, which taught the tactics and techniques of company and staff duties for the battalion and regiment. By 1925, officers accustomed to thinking of soldiering as a purely common sense business had begun to accept the novel idea that the academics of the company officers' course were both necessary and desirable. Lucian Truscott considered the Cavalry School's Troop Officer Course at Fort Riley a high point of his career. While the horse was still the centerpiece of instruction in his day, the school had acknowledged mechanization and was beginning to think in terms of how it might be used in the army. In any case, Truscott and others perceived graduation from the school as a cachet of success.[26] Officers commissioned later in the 1920s were even more convinced that the basic school was an essential. Typical of those taking much away from the school experience was Aubrey S. Newman, who commanded a regiment during World War II. Newman considered his nine-month company officers' course at the Infantry School invaluable and particularly praised the series of twenty-two graded problems in tactics and the field maneuvers that involved the school regiment.[27]

The branch advanced course was a more detailed and sophisticated nine-month program aimed at officers of longer experience and greater knowledge. Only field-grade officers and captains in the first thousand sequence numbers of the promotion list could attend and, testimony to the fact that the advanced course came somewhat later in the career, every student officer had to be under fifty years old when the course started.[28] It focused on the tactics and command of battalions through reinforced brigades, and was largely a technical curriculum that grounded officers in the details of operations of larger units of their branch. Many officers of the era commented favorably on their senior branch schooling, but the Infantry School at Fort Benning during the years when George C. Marshall was assistant commandant received the most lavish praise from its graduates. Matthew B. Ridgway summarized its value for his generation when he commented that the school taught him that "simplicity is the basic factor in any tactical plan."[29]

One utilitarian significance of the branch advanced course was that it was the usual prerequisite for attending the Command and General Staff School, the army's premier service school. With a Leavenworth diploma in hand, an officer was eligible for duty on the General Staff, the portal to wider professional horizons.[30] Student officers in the grades of captain through lieutenant colonel arrived at the Staff School somewhere around their sixteenth year of service. While curricula of the preceding schools insured expertise in a particular arm or service, the student officer at Leavenworth for the first time systematically studied the employment of various arms and services together in the division and corps.

Intensely detailed, Leavenworth instruction lasted for one to two years, the course varying in length according to the army's needs.[31] It was difficult and demanding in the sense that there was a great deal of information to be mastered. That is not to say that the subject matter itself was particularly abstruse or sophisticated. An almost universal student comment was that the Leavenworth course was the most difficult thing any officer had done in his career, and in terms of sustained study, that was probably true.[32] Expert about his own arm, the Leavenworth student was expected to assimilate details of arms and services with which he had no experience, and then to learn how to wield them as a unified weapon. The workload was considerable, but the material was not inherently difficult. Rumors of suicides among student officers, frequently cited in memoirs as proof of the Leavenworth course's difficulty, appear, however, to have no basis in fact.[33]

The Command and General Staff School, like the subordinate branch schools, concentrated on mastery and integration of factual information, and on the staff planning process, rather than on creative tactical thinking.[34] The point was often missed by the many critics of Leavenworth's infamous "school solution." Student officer doggerel might indicate that the school solution was both mindless and dangerous, as the epitaph on the tombstone of a gullible Coast Artillery Corps officer was alleged to have read:

> Here lie the bones of Lieutenant Jones
> A product of this Institution.
> In his very first fight
> He went out like a light
> By using the School Solution![35]

Valid as the complaint might have been, that sort of criticism missed the purpose of the exercise, for the correct answer was never the real goal.[36] While the set problem might convey useful tactical lessons, it was actually a vehicle to get across a common body of staff procedures, tactical language, and methodology for solving tactical problems. The most perceptive officers, many of whom simultaneously praised and condemned the Leavenworth teaching method, understood this. That most officers never grasped this subtlety was unimportant, for they nonetheless derived the intended benefits of the course.[37] The result was that Leavenworth graduates could function effectively on any U.S. Army staff, anywhere in the world. This was what Omar Bradley meant when he remarked that Leavenworth was "good mental discipline."[38] Maxwell Taylor explained Leavenworth's utility in imposing a common standard on two decades of graduates when he said that the school turned out "well-trained potential commanders and general staff officers, all speaking the same professional language, following the same staff procedures, schooled in the same military doctrine, and thus ready to work together smoothly in any theater of war."[39]

To this point, the curriculum of army schools was technical, rather than theoretical, consciously structured to assure a consistent, minimum level of professional competence. Neither the branch schools nor the Command and General Staff School were intended to create free-thinking strategic geniuses. Instead, the schools constituted the army's baccalaureate instruction. Those officers who did particularly well at the Command and General Staff School usually went on almost immediately to the Army War College, to become what Gen. Leslie J. McNair later called "the Army's Ph.Ds."[40]

Not every officer attended the Command and General Staff School, and fewer still were fortunate enough to be admitted to the War College, where almost half of the student body eventually became generals.[41] While the commandant at Leavenworth stressed orthodoxy, the assistant commandant at the War College took a contrary position, telling students in 1925:

> I believe I speak the truth when I say that no one helps his rating by blindly accepting the view of the Faculty on any subject. This is distinctly a college— where we learn from an exchange of ideas and not by accepting unquestioned, whether the views of the Faculty, or the views of the student.

"We reach our own conclusions, faculty and student," he assured them, "following a full and free discussion of the subject."[42]

The War College was thus, certainly by comparison with Leavenworth, a university in the broadest sense, where student officers ranging in rank from captain to colonel worked together in seminars modeled on those in civilian graduate schools. There, unlike at the branch schools and the Staff School, the faculty encouraged creative and innovative thinking. The subject matter fostered such an approach, for it dealt with higher staff duties, strategic planning, and politico-military issues. Students prepared committee reports, read broadly, and wrote war plans, a far cry from the graded problems in their previous army schools.[43]

Most officers deemed the year at the War College a period of thoughtful introspection. Instead of divisions and corps, they began to think in terms of armies and theaters, the logistical arrangements necessary to support overseas operations, and the strategic questions related to major war. Troy Middleton found his year there to have been very satisfactory from a personal point of view. In addition to his committee work, he had the time to read widely, and spent many hours in the Library of Congress.[44] Maxwell Taylor described his year at the War College as "deliberative," and the school itself as a place where he had the opportunity to educate his military judgment.[45] Thomas Handy similarly characterized the War College as "the one place you could sit down and think."[46] On balance, the best evidence that the War College was qualitatively different from Leavenworth and the other army schools is that its student committees during the 1930s produced sophisticated plans anticipating the RAINBOW War Plans of World War II.[47]

Some officers wound up in other "postgraduate" army schools, including the Army Industrial College, founded in 1924, and the senior schools of other services and other countries. Throughout the decade between 1925 and 1935, army officers studied at the Naval War College and the Air Corps Tactical School. Around forty officers a year were enrolled in French, Italian, British and, eventually, German military schools, or went abroad to study foreign languages, while others studied at civilian universities.[48]

Schools as Surrogates

By the time the successful officer had graduated from the War College, he had been trained in the junior schools and educated, to a degree, in the senior ones. He had spent an average of about four and one-half years as a student, and if he had not yet become, to use General King's phrase, a master of orthodoxy, that was not the fault of the schools. Of course, every officer did not attend every school because public law limited the number of officers in school at any one time to 2 percent of the total officer corps.[49] Thus, the percentage of any commissioning year group attending schools diminished over time. Postings to schools were generally spaced out, as well. Using the commissioning year groups of 1917 and 1918 as examples, it is possible to generalize that an officer typically attended his branch troop officers' course at about the fifth year of service, the branch advanced course at about the ninth, Command and General Staff School the following year (although this varied widely with different commissioning year groups), and the Army War College either the succeeding year or several years thereafter. For officers in year group 1918, at least, promotion to major typically came in the twentieth or twenty-first year of service, so that those officers who went through the entire school system did so with a rank that still marked them as junior officers.[50]

Ideally, the officer went on after graduation from any of the army's schools to apply the lessons he had learned in duty with troops. Unfortunately, the army did not have enough units to implement such a scheme, and many of the units that did exist were not up to strength, so that the value of the assignment was arguable. Peacetime troop command, in particular, was the great laboratory of experience for the aspiring officer. But commands were difficult to obtain, and battalion commands, as Thomas T. Handy explained, were even scarcer than company commands:

> You didn't have much duty with troops. Take majors in the Field Artillery; if you count up the number of majors and the number of battalions, about one in five maybe would ever have a chance to command a battalion. They just didn't exist and lots of times when you got one it wasn't much of a battalion.[51]

Such a situation obtained across the army, with the exception of units stationed overseas. The garrisons in Panama, the Philippines, and China retained units at least at full battalion strength, although a regiment might as often be organized with two battalions, instead of the authorized three.[52] Within the United States, only those organizations stationed at the schools such as the 29th Infantry at Fort Benning, remained at war strength. The resulting scarcity of real battalion and regimental assignments meant that officers had little opportunity to build their experience by working in the various echelons of tactical and administrative staffs.

The schools and the school regiments played a part in officer training that was certainly unforeseen when they were established. At various points in their careers, officers commonly returned to serve on the staffs and faculties of the various schools, and those assignments became the surrogates for the formative staff duties the officers could not obtain in the field army. More importantly, a fortunate minority also returned to command and staff duty in the school regiments.

Officers commissioned in 1917 and 1918 are again illustrative. Matthew Ridgway served in various capacities at West Point from 1918 through 1924. J. Lawton Collins was an instructor at West Point from 1921 through 1925, at the Infantry School from 1927 through 1931, and at the War College from 1938 through 1940. Lucian Truscott, after completing both the basic course and the Cavalry School's special Advanced Equitation Course, taught there between 1927 and 1931, as well as at Leavenworth from 1936 through 1940. There are many similar examples.[53] While at the schools, these men continued to study their profession, particularly benefiting from the stimulus of association with other officers. Again, one of the best examples is the Infantry School under Marshall, which turned out to be the cradle of many of the more noted American field commanders of World War II.[54] The practical benefit was that these officers, while serving as school secretaries, executive officers, quartermasters, and the like, had the challenge of dealing with real administrative and organizational problems on a relatively large scale. The experience proved invaluable.

The school regiment likewise served a dual function, the importance of which emerged only slowly. From the start, the intention had been to have a competent demonstration unit that could be used for instructional purposes in a purely practical way.[55] Certainly it enhanced tactical training to be able to carry out the various maneuvers, particularly since the schools generally allowed student officers to function briefly as commanders and staff officers during the exercises. The school regiment's more indirect value in command and staff training also eventually became evident.

In 1935, George Marshall left the Infantry School and assumed command of Fort Screven, Georgia. There he found, to his shock, that the young officers reporting to him did not have the same abilities as the men he had known while at Fort Benning, the home of the Infantry School. They appeared to "be from another litter," he complained:

> Each year I noted the splendid effect on the young graduates from West Point of a tour with the 29th Infantry. However, when at Fort Screven and Fort Moultrie, I was honestly horrified by the lack of development—if not the actual deterioration—in the young lieutenants who had reported directly from West Point.[56]

Those lieutenants who enjoyed the opportunity to command troops in the 29th Infantry had an inestimable advantage over those who had not, for they saw that great rarity of the interwar army, the war strength unit.

Aubrey S. Newman, reflecting on his training as an officer, considered his early troop duty to have been crucial. While he believed sophisticated schooling later in his career was important, he also noted of the schools that "you can learn about high level things when you get there . . . but nothing can replace a deep understanding of what goes on at company level."[57] Thus, he believed himself to have been extraordinarily fortunate that his first duty assignment was with the 29th Infantry. "*That*," he emphasized, "was the most valuable experience . . . with full strength units."[58] Many officers who attained high rank during and after World War II had served in the 29th Infantry or in one of the other school regiments, where they shared that rare experience Newman believed so important.[59] Just to observe such a unit was valuable. When Clyde D. Eddleman arrived at the Infantry School in 1932, he had served only in cadre units. "It was the first time," he said, that "any of us had ever seen a war strength outfit, as the 29th Infantry was."[60]

The law of unintended consequences applied to army schools during the interwar era. Efficient, if somewhat unimaginative, pedagogical institutions, they also helped to compensate for the field army's inability to give officers adequate opportunities to translate their academic preparation into worthwhile professional skills. Schools provided the theory, exercises, and appropriate drill in the business of becoming a soldier, but experience was essential, as Thomas T. Handy emphasized with the homely aphorism that "no guy every got to be an expert mechanic reading a book."[61]

Conclusions

It has become an accepted practice to deride the quality of American generalship in World War II by characterizing those officers as conservative, plodding, and unimaginative on the battlefield.[62] Certainly, very few led with the flashy brilliance of George Patton, himself recently criticized as a commander whose flair for publicity overshadowed his shortcomings as a general.[63] Fewer still displayed the gifts for coalition warfare that made Dwight Eisenhower such a successful supreme commander, although one about whom the majority British judgment has always been that he was not a good battlefield commander. But, setting aside such criticisms, it is also clear that the great majority of American generals performed capably, waging war in a style that accorded well with the American temperament. If not intuitive tacticians, these American commanders almost universally shared a superb managerial ability that appears to have been largely unmatched among other Allied officers. Inexperienced, unblooded, and relatively junior in the years immediately before the war, men who had, by and large, been junior officers rapidly became generals who could organize, train, and wield divisions in battle—units few of them had ever seen before. Reflecting on this, Williston Palmer later expressed his surprise that the number of his peers who had "stepped right up and hit the ball was very high."[64]

The remarkable thing, to paraphrase the expression, is not how badly American generals did the job. The remarkable thing, considering their backgrounds, is that they did it at all. There were few real geniuses in the U.S. Army of 1941—or, indeed, in any army of that period. Apply the definition of tactical genius rigorously, and few generals of any nation seem to measure up. Oddly, those commanders generally acknowledged as geniuses were those who fought on the defensive, or at such a disadvantage in materiel and manpower that tactical acumen was the only way to redress the balance. Thus, for example, one thinks of the brilliance of Stonewall Jackson's Valley Campaign in the Civil War in the face of overwhelming Union Army strength, or that of Erwin Rommel's desert battles in North Africa, which teetered on the brink of success even though the Allies had enormous superiority in every category of military power and were advised of his intentions through Ultra intercepts. The fact is that the Allies did not particularly need that kind of brilliance in World War II. The industrial might of the Western alliance, coupled with its overwhelming military strength, placed more of a premium on competence than on tactical genius. It is worth remembering, too, that the U.S. Army was itself largely an amateur force, having expanded from under a quarter of a million to over eight million men by 1945. It is possible to argue that such an army, in which whole battalions might have only a handful of professional soldiers, was not an instrument that could be used with much finesse. Commanding those troops capably was the key thing and, from the point of view of the service, the abilities of those average men of whom Williston Palmer spoke were consequently much more important than the spectacular accomplishments of the gifted few. That the army did well can be attributed more to the sum of their average talents than to the creative tactical and operational flair of the handful of intuitive battlefield commanders. The intriguing question is therefore not what influences shaped the George Pattons, but how the army educated what West Point once called the "undistinguished middle."[65]

This is the context in which Handy's judgment about army schools must be understood. In an army deficient in so many ways, the school system provided an escape from the perils MacArthur had drawn to the secretary of war's attention during the Depression. Army formal schools were a peculiar adaptation to the circumstances of the age. They assured that the officer corps assimilated the orthodoxy of approved doctrine and the methodology of staff and command. The United States was fortunate that the median level of native ability was so high, as Maxwell Taylor observed of his class at the Command and General Staff School, characterizing it as "knee-deep in talent."[66] Talent, though, was not universal, and the army had to make full use of all of its regulars when mobilization came in 1941 and 1942. Talented, gifted officers could and did see to their own professional educations. For them, the schools were merely polish. For the average officer, however, the schools were the essential experience. Orthodoxy, school solutions, grinding detail, and technique gave the average officer the essential knowledge ably to command a division or, occasionally, a corps.

Seen in this light, King was correct about the need to be orthodox and Handy was correct that the schools literally "saved" the army by assuring the quality of its officer corps. The successes of U.S. Army commanders in World War II, growing as they did from the service's focus on the rearing of junior officers through the interwar years was, finally, a belated testimony to the wisdom of Douglas MacArthur and the other chiefs of staff of the interwar army. MacArthur, in particular, was castigated at the time by airpower advocates and armor and mechanization enthusiasts because he did not spend money to develop those aspects of modern warfare, and castigated retrospectively by historians studying those narrow slices of the history of the military art. MacArthur, however, had a limited budget and had to spend it carefully. The eventual successes of the junior officers of the 1920s and 1930s make clear that MacArthur made the right decision. The "heart of the Army," a well educated officer corps, could—and eventually did—effectively wield whatever weapons the nation placed into their hands.

Notes

1. Former Chief of Staff Peyton C. March, in *The Nation at War* (Garden City, N.J.: Doubleday, 1932), 341ff., called the army impotent and characterized it as weaker than the force permitted Germany under the stringent provisions of the Treaty of Versailles. Chief of Staff George C. Marshall found little qualitative change by 1941, declaring in the *War Department Annual Report, 1941,* 48, that "as an army we were ineffective."

2. For strength figures, see *Annual Report of the Secretary of War to the President.* Figures are also tabulated in *Historical Statistics of the United States: Colonial Times to 1970,* 2 pts. (U.S. Department of Commerce, Bureau of the Census: Government Printing Office, 1976), 2:1141, and in Russell F. Weigley, *History of the United States Army* (New York: Macmillan, 1967).

3. In 1940, Congress appropriated funds to bring the army up to the strength authorized by the National Defense Act of 1920. Until that time, little money was appropriated for general improvements in military power. Through the preceding decade, only 13.9 percent of the funds could be spared for new equipment and force modernization. The records of the War Plans Division (WPD), War Department General Staff (WDGS), contain frequent references to such budgeting constraints. See especially file WPD 4302-1, 28 May 1940, U.S. National Archives and Records Administration, Record Group 165, WDGS. Depletion of World War I supplies and equipment meant harder times for the army, even when the budget was relatively constant. See Irving J. Phillipson, "War Department Participation in the President's Budget," *Lectures Deliv-*

ered at Army Industrial College 11 (1935), in the library of the Industrial College of the Armed Forces, Washington, D.C.

4. A single-minded drive for economy motivated both political parties after World War I. For the effect on the army, see John W. Killigrew, "The Impact of the Great Depression on the Army, 1929–1936" (Ph.D. diss., Indiana University, 1960); Harold W. Rood, "Strategy Out of Silence: American Military Policy and the Preparation for War, 1919–1940" (Ph.D. diss., University of California at Berkeley, 1961); and Robert A. Miller, "The United States Army During the 1930s" (Ph.D. diss., Princeton University, 1973). For the latter part of the 1930s, see Elias Huzar, *The Purse and the Sword: Control of the Army Through Military Appropriations, 1933–1950* (Ithaca, N.Y.: Cornell University Press, 1950), who points out the particular restrictions on construction, with Congress reserving the right to make specific appropriations for all projects in excess of $20,000.

5. Annual Report of the Chief of Staff for the Year Ending June 30, 1931, in *Report of the Secretary of War to the President, 1931* (Washington, D.C.: Government Printing Office, 1931), 40. The only real change in the stationing of units during the era came when units were removed from the active rolls. The interwar army was a service of small and generally isolated units, few at full strength. For a detailed picture, see the U.S. Army's *Station List* (Washington, D.C.: Government Printing Office, annually). Further details of how severely units were skeletonized may be found in the service's various professional journals. The *Coast Artillery Journal,* for example, carried a section devoted to unit news which indicated that many of the coast artillery regiments in the continental United States had as few as two or three firing batteries. General Charles L. Bolté, among others, noted that troop labor details "took precedence over any systematic training" between the wars. See Conversations between Gen. Charles L. Bolté and Arthur J. Zoebelin (Senior Officer Debriefing Program, U.S. Army Military History Institute, Carlisle Barracks, Pa., 1971), sec. I, 48–49.

6. Annual Report of the Chief of Staff for the Year Ending June 30, 1932, in *Report of the Secretary of War to the President, 1932* (Washington, D.C.: Government Printing Office, 1932), 59.

7. Army budgets throughout the period remained small. Through the first half of the 1930s, Congress appropriated roughly as much per year as in the preceding decade, but the money did not go nearly as far because of the need to replace arms and equipment. Army budgets also included funds for the Air Corps, which received a substantial share of the money available for new equipment. The Fiscal Year 1934 budget was roughly $26 million less than that for the preceding year, leading to all sorts of economies, including a 15 percent pay cut one month for soldiers. For War Department expenditures by year, see *Historical Statistics of the United States,* 2:1114.

8. For a statistical treatment of this topic, see Edward M. Coffman and Peter F. Herrly, "The American Regular Army Officer Corps Between the World Wars: A Collective Biography," *Armed Forces and Society* 4 (November 1977). In three sample years, Coffman and Herrly found that majors' mean ages were 42 years in 1925, 45.5 years in 1933, and 48.1 years in 1940. Perhaps more significantly, the mean age of a captain in the Depression year of 1933 was 43. The Historical Records Branch (HRC) of the U.S. Army Center of Military History in Washington, D.C. holds extensive, although abbreviated, personnel files (HRC 201) [hereafter CMH 201] on army general officers. A selective review of the files of one hundred men who became generals during or immediately after World War II confirms the data that Coffman and Herrly adduced in their sampling of the *Army Register.* General George Marshall referred to the problems caused by slow promotions in testimony before the House of Representatives, Committee on Military Affairs, in 1940. He noted that men commissioned in the years from 1920 through 1930 were still captains and averaged forty-five years old. See H. A. DeWeerd, *Selected Speeches and Statements of General of the Army George C. Marshall* (Washington, D.C.: The Infantry Journal, 1945), 45–48.

9. Report of the Chief of Staff, U.S. Army, 1935. Extract from the *Annual Report of the Secretary of War, 1935* (Washington, D.C.: Government Printing Office, 1935), 5–6.

10. J. Lawton Collins, *Lightning Joe: An Autobiography* (Baton Rouge: Louisiana State University Press, 1979), 3. Thomas T. Handy, wartime chief of the War Plans Division of the War Department General Staff and deputy chief of staff under George Marshall, used virtually the same words. The schools, he said, "saved the army." See Conversations between Gen. Thomas T. Handy and Lt. Col. Edward M. Knoff, Jr. (Senior Officer Debriefing Program, U.S. Army Military History Institute, Carlisle Barracks, Pa., 1976), sec. II, 8. Other officers of the interwar generation made similar assertions. For examples, see Bolté Interview, sec. I, 48–49; Conversations between Gen. Maxwell D. Taylor and Col. Richard A. Manion (Senior Officer Debriefing Program, U.S. Army Military History Institute, Carlisle Barracks, Pa., 1972), sec. II, 13; and Matthew B. Ridgway to author, 17 Dec. 1988.

11. The structure of the school system varied over time, and the changes are reflected in the annual general orders and circulars of the War Department. For the creation of the system, see *Annual Reports of the War Department for the Fiscal Year Ended June 30, 1903,* vol. 4, *Military Schools and Colleges; Record and Pension Office; Military Parks; and Soldiers' Homes* (Washington, D.C.: Government Printing Office, 1903); and the National Defense Act of 1920. Details about the schools, their administration, and assignment of officers to courses are to be found in various U.S. Army regulations. Regulations for 1935 illustrate the mature school system: AR 605-145, Officer Assignments; AR 350-460, Cavalry School; AR 350-600, Field Artillery School; AR 350-110, General Service Schools; AR 350-5 (General listing of schools—Infantry School, Command and General Staff School, Army War College, and Army Industrial College); and AR 350-700, Coast Artillery School. This discussion is based on the foregoing and upon app. F, "Chart of Army Educational System," in *Report of the Secretary of War to the President, 1929* (Washington, D.C.: Government Printing Office, 1929), which summarizes the detail contained in the relevant regulations.

12. Report of the Chief of Staff, U.S. Army, 1935. Extract from the *Annual Report of the Secretary of War, 1935,* 73.

13. Recalled by Bradford Grethen Chynoweth, then a student at Ft. Leavenworth, in *Bellamy Park: Memoirs by Bradford Grethen Chynoweth* (Hicksville, N.Y.: Exposition Press, 1975), 122.

14. The purposes of the schools evolved considerably. According to War Department General Order (WDGO) 107, 8 August 1901, professional knowledge was one of the basic principles of military training. Of it, the order specified that "the *art* of war is the application of its principles to actual practice and of more importance than the *science*" (emphasis in original). Army schools of the interwar decades, however, stressed skills and techniques, rather than the general principles referred to in WDGO 107. By the 1939 school year, the Cavalry School specified that its three major objectives were "to teach the duties of a cavalry officer and, in detail, the techniques and tactics of cavalry. To give a working familiarity with the techniques and tactics of associated arms. To teach the best methods of imparting instruction." See Academic Division Regulations, The Cavalry School, Fort Riley, Kans., 1939–1940, 1. The artillery school was more terse about its mission: "to teach artillerymen to shoot." Charles M. Hunter, "U.S. Artillery and Missile School 1911–1957" (Fort Sill, Okla.: Department of Publications and Nonresident Training, USAAMS, unpublished manuscript, 1957), 3.

15. R. J. Fleming, "Mission of the Cavalry School With Comments on Modern Cavalry and Cavalry Training," *Cavalry Journal* 38 (January 1929).

16. In the words of an anonymous humorist, writing in *The Rasp,* yearbook of the Cavalry School (Ft. Riley, Kans., 1923), 21.

17. Although under the control of local commanders, troop and garrison schools were regulated by War Department General Orders. See WDGO 70, 20 Apr. 1910, periodically amended and updated, which specified the conduct of garrison school instruction.

18. Conversations between Gen. George H. Decker and Lt. Col. Dan H. Ralls (Senior Officer Debriefing Program, U.S. Army Military History Institute, Carlisle Barracks, Pa., 1972), sec. I, 5.

19. Conversations between Gen. Paul L. Freeman and Col. James N. Ellis, ibid., 1974, sec. I, 13.

20. Study of Chinese in the 15th Infantry appears to have been an initiative of Brig. Gen. Fox Conner when he commanded the American garrison in China. For comments on this, see Freeman Interview, sec. I, 34; Albert C. Wedemeyer, *Wedemeyer Reports!* (New York: Holt, 1958), 47–48; Leslie Anders, *Gentle Knight: The Life and Times of Major General Edwin Forrest Harding* (Kent, Ohio: Kent State University Press, 1985), 85; Conversations between Gen. Matthew B. Ridgway and Col. John M. Blair (Senior Officer Debriefing Program, U.S. Army Military History Institute, Carlisle Barracks, Pa., 1972), vol. 1, sec. II, 9.

21. Lucian K. Truscott, Jr., *The Twilight of the U.S. Cavalry: Life in the Old Army, 1917–1942* (Lawrence: University Press of Kansas, 1989), 6–8. Other manuals Truscott studied included *Cavalry Drill Regulations, 1914; Field Service Regulations, 1914; Rules for Land Warfare; Manual of Interior Guard Duty; Engineer Field Manual; Manual for Stable Sergeants; Manual of Physical Training;* and the *Army Cook's Manual.* "Moss's Manual" was the basic guide for the junior officer: James A. Moss, *Officers' Manual* (Menasha, Wisc.: Banta, 1917).

22. James H. Polk Papers, U.S. Army Military History Institute, Carlisle Barracks, Pa., and Polk to author, 4 Jan. 1989.

23. As late as 1929, Chief of Staff Charles P. Summerall reported that the service could not yet implement its stated policy of having all officers attend the basic school of their branch within four years of being commissioned, although he expected that the army would achieve some uniformity on this sometime in the early 1930s. Schooling was linked with funding in curious ways. Summerall wrote that the Infantry School could not achieve the goal because of the shortage of quarters at Fort Benning and the scarcity and high cost of renting houses in adjacent Columbus, Georgia. Thus, he found it necessary to decrease the quota of students at the Infantry School. See *Report of the Secretary of War to the President, 1929,* enclosing Report of the Chief of Staff, 101–2.

24. Instruction was similar at all schools of the arms. Concerning both troop/battery/company officers' courses and advanced courses, see Robert Arthur, *The Coast Artillery Corps School, 1824–1929* (Ft. Monroe, Va.: Coast Artillery School, 1928); K. T. Blood, "The Coast Artillery School Courses for Officers of the Army, School Year 1934–1935," *Coast Artillery Journal* 77 (May–June 1934); "Coast Artillery School Moves to San Francisco," *Coast Artillery Journal* 88 (May–June 1946), which contains a summary of the school history; J. S. Switzer, "Department of Experiment—the Infantry School," *Infantry Journal* 18 (March 1921); "Infantry School Opens," *Army and Navy Journal* 77 (16 September 1939); Helen C. Phillips, "United States Army Signal School 1919–1967" (Ft. Monmouth, N.J.: U.S. Army Signal Center and School, MS, 1967); W. M. Grimes, "The Cavalry School, 1919, 1929," *The Cavalry Journal* 38 (April 1929); and CMH file 228.01, HRC 352, Schools-Army, which contains general orders for establishing and organizing the schools and summaries of their curricula through the 1920s and 1930s.

25. Chynoweth, 66.

26. Truscott, chap. 4.

27. Aubrey S. Newman to author, 30 Dec. 1988.

28. Marshall memorandum for the commandant, subj: Selection of Infantry Officers for the Advanced Course, Infantry School, and for the Command and General Staff School, 9 Jan. 1928, in Larry I. Bland and Sharon R. Ritenour, eds., *The Papers of George Catlett Marshall,* vol. 1, *"The Soldierly Spirit" December 1880–June 1939* (Baltimore, Md.: Johns Hopkins University Press, 1981), 324–26.

29. Matthew B. Ridgway, *Soldier: The Memoirs of Matthew B. Ridgway* (New York: Harper & Brothers, 1956), 41–42. Forrest C. Pogue, *George C. Marshall: Education of a General,*

1880–1939 (New York: Viking, 1963), discusses the Infantry School in chap. 15. For other comments on the Infantry School renaissance and the stimulating atmosphere of the school under Marshall, see Anders, 120–22; Conversations between Gen. William M. Hoge and Lt. Col. George R. Robertson (Senior Officer Debriefing Program, U.S. Army Military History Institute, Carlisle Barracks, Pa., 1976), sec. I, 46; and Collins, 56–57.

30. Amendments to the National Defense Act of 1920 required all appointees to the War Department General Staff to be graduates of the general staff college, a provision later interpreted to mean the Command and General Staff School. For a discussion of this point, see Stetson Conn, "The Army War College, 1899–1940: Mission, Purpose, Objectives," unpublished manuscript prepared for the commandant of the War College in 1964 and filed in CMH 228.01, HRC 352, Schools-Army. In general, appointment to the General Staff Eligible List was the decisive point in an officer's career, inasmuch as it determined whether he would ever serve beyond regimental level.

31. For a discussion of the Leavenworth curriculum and the decisions that underlay the changes from a one-year to a two-year course, see "A Military History of the U.S. Army Command and General Staff College, Fort Leavenworth, Kansas, 1861–1963" (n.p., n.d., but issued by the Command and General Staff College in 1963). On the length of the school, also see the report of the chief of staff in *Report of the Secretary of War to the President, 1928* (Washington, D.C.: Government Printing Office, 1928), 27, in which General Summerall discusses the decision to return to a two-year course; and the *Report of the Secretary of War to the President, 1929*, 101–2, in which Summerall deplores the fact that "all officers of satisfactory record cannot be given an opportunity to attend the Command and General Staff School." On the early development of the school, see Timothy K. Nenninger, *The Leavenworth Schools and the Old Army: Education, Professionalism, and the Officer Corps of the United States Army, 1881–1918* (Westport, Conn.: Greenwood, 1978).

32. Even Maxwell Taylor, a very bright man, termed it "a very tough course." See Taylor Interview, sec. I, 30–31. William M. Hoge agreed that he had to work hard. See Hoge Interview, sec. I, 45. E. N. Harmon, explaining that his years at Leavenworth were the most difficult of his training, wrote that he "studied upstairs and downstairs, often far past midnight," and that his "disposition at home became as mean as that of a starving prairie wolf." See E. N. Harmon, *Combat Commander: Autobiography of a Soldier* (Englewood Cliffs, N.J.: Prentice-Hall, 1970), 49–50.

33. I have found no evidence of any suicide among student officers enrolled in the Command and General Staff School, although memoirs of graduates almost always mention this "fact." The common feature in these recollections is that the suicide always took place in a preceding class. Virtually the same story was recounted by graduates before World War I and by those of the mid-1930s; I have heard Leavenworth students of my own era recount the story. It is probably apocryphal.

34. This point stands out in a survey of the curriculum focus at Leavenworth between 1919 and 1940. Curriculum evolution may be followed in the annual reports of the commandant of the Command and General Staff School, which are filed in the Combined Arms Research Library at Fort Leavenworth, Kans. For a concise summary of curriculum development, see "Evolution of Regular Course, CGSC," a Command and General Staff College unpublished briefing paper ca. 1987, copy in CMH 228.01, HRC 352, Schools-Command and General Staff College.

35. The school solution was a technique in general use, not just at Leavenworth. This specific verse is in the Coast Artillery School files, U.S. Army Air Defense Artillery School Museum archive, Fort Bliss, Texas.

36. On this point, see Decker Interview, sec. I, 32. General Decker said of Leavenworth that "it's not the specific tactics or the techniques that are really important here. It's how I learned and how the students learned to make sound decisions, given a group of basic facts." See also, Conversations between Gen. John E. Hull and Lt. Col. James W. Wurman (Senior Officer De-

briefing Program, U.S. Army Military History Institute, Carlisle Barracks, Pa., 1973), sec. III, 11, on the key instruction about "how to arrive at the appropriate decision."

37. For one analysis of this issue, see Charles E. Kirkpatrick, "Filling the Gaps: Reevaluating Officer Professional Education in the Interwar Army, 1920–1940," paper presented at the meeting of the American Military Institute (now the Society for Military History), 14 Apr. 1989, Virginia Military Institute, Lexington, Va.

38. Omar N. Bradley (with Clay Blair), *A General's Life* (New York: Simon & Schuster, 1983), 61. Nenninger addresses the question of the specific value of Command and General Staff School to officers who subsequently commanded troops in World War II in "Creating Officers: The Leavenworth Experience, 1920–1940," *Military Review* 69 (November 1989).

39. Maxwell D. Taylor, *Swords and Plowshares* (New York: Norton, 1972), 29–30. There are many concurrences. See, among others: Hull Interview, sec. III, 11; Conversations between Gen. Mark W. Clark and Lt. Col. Forest S. Rittgers, Jr. (Senior Officer Debriefing Program, U.S. Army Military History Institute, Carlisle Barracks, Pa., 1976), sec. I, 85; Taylor Interview, sec. II, 13; Ridgway Interview, vol. 1, sec. II, 24–25; Conversations between Lt. Gen. Edward M. Almond and Capt. Thomas G. Fergusson (Senior Officer Debriefing Program, U.S. Army Military History Institute, Carlisle Barracks, Pa., 1975), 31; Handy Interview, sec. II, 32–35; Bolté to author, 26 Dec. 1988; Ridgway to author, 17 Dec. 1988; Kinnard to author, 4 Jan. 1989; Stephen E. Ambrose, *Eisenhower,* vol. 1, *Soldier, General of the Army, President-Elect, 1890–1952* (New York: Simon & Schuster, 1983), 80; Collins, 56–57; and Harmon, 49–50.

40. Quoted in Handy Interview, sec. I, 51–52.

41. Roughly the top 10 percent of Leavenworth graduates went on to the Army War College. See Conn, "Army War College." The school averaged seventy-nine army graduates per year for a twenty-one-year period; three-fourths were combat arms officers. By 1946, 859 Army War College graduates had become general officers. For these and other statistics, see Army War College to Chief of Historical Division, Department of the Army, subj: Army War College, 20 June 1951, in CMH 228.01, HRC 352, Schools-Army War College.

42. Quoted in Conn, "Army War College."

43. On the curriculum and procedures of the War College, see George S. Pappas, *Prudens Futuri: The United States Army War College, 1901–1967* (Carlisle Barracks, Pa.: Alumni Association of the U.S. Army War College, 1967); and Harry P. Ball, *Of Responsible Command: A History of the U.S. Army War College* (Carlisle Barracks, Pa.: Alumni Association of the U.S. Army War College, 1984). Also see Conn, "Army War College." A contemporary view is Oswald H. Saunders, "The Army War College," *Military Engineer* 26 (March–April 1934).

44. Frank James Price, *Troy H. Middleton: A Biography* (Baton Rouge: Louisiana State University Press, 1974), 94.

45. Taylor, *Swords and Plowshares*, 37; and Taylor Interview, sec. II, 11–12.

46. Handy Interview, sec. I, 51–52.

47. Henry Gole, "War Planning at the War College in the Mid-1930s," *Parameters* 15 (Spring 1985).

48. See the *Report of the Secretary of War* for various years. Students enrolled in foreign schools amounted to 38 officers in 1924, 34 in 1925, and 36 in 1926. Totals varied but little into the next decade: 41 in 1931, 47 in 1932, and 42 in 1933. At the same time, about thirty foreign officers attended U.S. service schools. In his report to the secretary of war, General MacArthur commented that a large part of the value of having officers attend foreign schools, the schools of the Army Air Corps, Navy, and civilian universities, was that it tended to protect the army's pedagogical system against excessive inbreeding.

49. The National Defense Act of 1920, as amended, stipulated in Section 127a, Miscellaneous Provisions: "The Secretary of War is hereby authorized, in his discretion, to detail not to exceed 2 per centum of the commissioned officers . . . of the Regular Army in any fiscal year as students," with the added proviso that "The number of officers so detailed shall, as far as practicable, be distributed proportionately among the various branches." Chiefs of a branch

selected officers from their branches to attend all army schools, and there is evidence that some patronage was not unusual. A fixed number could attend schools in any given year, and their names were announced in orders and in lists occasionally published in the branch journals. For examples of the latter, see "Students at the Cavalry School," *Cavalry Journal* 38 (October 1930): 598, which names fourteen officers selected for the advanced course, eighteen officers for the advanced equitation class, and twenty-eight officers for the troop officer's course for academic year 1930–31. In 1933, the Economy Bill cut quotas of student officers and faculty by 51 percent and 33.3 percent, respectively, forcing consolidation of the basic and advanced courses in many schools in 1934.

50. CMH 201 summary for 1917 accessions. This group included many of the corps and division commanders of World War II. Among them: Norman D. Cota, Ernest N. Harmon, Robert W. Hasbrouck, Matthew B. Ridgway, Mark W. Clark, John E. Hull, Lucian K. Truscott, Jr., Holmes E. Dager, Charles L. Bolté, and J. Lawton Collins. It was not entirely a group of success stories, for year group 1917 also included Alan W. Jones, commander of the ill-fated 106th Infantry Division in the Battle of the Bulge.

51. Handy Interview, sec. II, 8, 10. In the interwar army, battalion command was customarily exercised by majors. Handy referred to the fact that most units operated at far below authorized strength, and many were frankly skeletonized.

52. *Station List* (Washington, D.C.: Government Printing Office, annually), details fluctuations in troop strength.

53. CMH 201 summary. Officers of the interwar generation had no formal "career pattern" in the modern sense, and their careers were diverse. Ridgway, for example, followed his patron Frank McCoy into a number of diplomatic missions. Albert C. Wedemeyer had extensive duty as an aide-de-camp. Others had one or more tours as professors of military science in Reserve Officers Training Corps detachments at civilian universities, or served as instructors to the state national guards. My survey, however, identified assignment one or more times as an instructor as a reasonable generalization about the careers of officers commissioned from 1909 through 1935.

54. Marshall's faculty included J. Lawton Collins, Charles L. Bolté, Omar Bradley, William M. Hoge, Harold R. Bull, Truman Smith, Philip B. Peyton, James A. Van Fleet, Edward M. Almond, Edwin F. Harding, Joseph W. Stilwell, Gilbert R. Cook, Clarence R. Huebner, and Bradford G. Chynoweth. Many officers mentioned the value of associating with other bright men at schools as being related to later success commanding troops during World War II. See Hull Interview, sec. III, 7–8; Anders, 120–21; Hoge Interview, sec. I, 45; Martin Blumenson, *Mark Clark* (New York: Congdon & Weed, 1984), 29; Taylor Interview, sec. I, 26; and Decker Interview, sec. I, 32.

55. The school regiments became increasingly important as time went on. "Hardly a day has passed in the present school year that there has not been some phase of instruction imparted by school troops to a class," according to the *Cavalry Journal,* which called the use of the Cavalry School's reinforced brigade "habitual." See Grimes, "Cavalry School."

56. Marshall to Croft, 13 Feb. 1935, in *Marshall Papers,* 1:455–58.

57. Newman to author, 30 Dec. 1988.

58. Ibid. Emphasis in original.

59. CMH 201 summary. Alumni of the 29th Infantry during Newman's assignment to that unit included: Maj. Gen. Asa L. Singleton, Maj. Gen. Holmes Dager, Lt. Gen. Andrew D. Bruce, Lt. Gen. Reuben E. Jenkins, Lt. Gen. Withers A. Burress, and Brig. Gen. LeGrande Diller. See Newman's manuscript article, "WWII Generals: Colonels Who Were Ready," in A. S. Newman Papers, U.S. Army Military History Institute, Carlisle Barracks, Pa. This article was published in amended form as "They Trained a Winning Army," *Army* (September 1968). Author's Interview with Gen. Albert C. Wedemeyer, 24 Apr. 1987, Friends' Advice Farm, Md.

60. Conversations between Gen. Clyde D. Eddleman and Lt. Col. L. G. Smith and Lt. Col. M. G. Swindler (Senior Officer Debriefing Program, U.S. Army Military History Institute, Carlisle Barracks, Pa., 1975), sec. I, 29.

61. Handy Interview, sec. I, 8.

62. One of the clearest statements of the thesis is Russell F. Weigley, *Eisenhower's Lieutenants* (Bloomington: Indiana University Press, 1981), 729. Disregarding the frankly polemical views of David Irving, for example, there are still thoughtful historians who state the thesis more forcefully, although some of those studied arguments are, unfortunately, partisan attempts merely to hold up one or another national group in the Allied coalition as having produced the better generals. The essays in Allan R. Millett and Williamson Murray, eds., *Military Effectiveness*, vol. 3, *The Second World War* (Boston: Allen & Unwin, 1988), recognize the challenges confronting senior American leaders in World War II.

63. John Ellis, *Brute Force: Allied Strategy and Tactics in the Second World War* (New York: Viking, 1990), 299, 386–87, 420, 436.

64. Conversations between Gen. Williston B. Palmer and Lt. Col. H. L. Hunter (Senior Officer Debriefing Program, U.S. Army Military History Institute, Carlisle Barracks, Pa., 1972), 31.

65. Pragmatic army schooling did not teach the art of war, nor is it correct to say that, the Army War College excepted, army schools were involved in education, as distinct from training. For the superior officer who could command armies and army groups, grasp the geopolitical complexities necessary to write war plans, and create and carry out high-level policy, schools were merely the springboard. Such men became intellectual soldiers on their own. I have offered an explanation of this process in "Filling the Gaps."

66. Taylor Interview, sec. I, 26.

Section III
Officership and Formal Education

In the course of the twentieth century, the level of formal, general education required for commissioned service has risen, although to varying degrees in different nations. In the United States a four-year college degree is, with few exceptions, mandatory; in other countries, a university degree is so desirable that military services provide opportunities for officers, or officer candidates, to obtain them—through civilian institutions, as in Great Britain, or as in Germany, through government-sponsored "armed forces universities."

Several factors account for the rise in the formal education standard. To some extent, more schooling has been necessary to keep pace with warfare's advancing technological complexity. In large part, the requirement for more formal education has also reflected social change. In nineteenth-century Europe, for example, an officer corps dominated by a nobility that did not value education opened more and more to increasingly powerful middle classes that did. Additionally, the general level of education has risen worldwide, demanding officers with more sophisticated leadership skills. Moreover, since World War II, access to higher education has expanded dramatically in nearly every part of the world.[1] With the growth in opportunities for higher education, armed forces have been pressed to attract society's best and brightest. To compete for these individuals, to retain officers once commissioned, or to ease their postretirement transition to civilian life, military establishments have made it possible for officer candidates, or officers, to obtain a basic university, or even an advanced degree after commissioning. Thus, although certainly related to a desire to improve the knowledge base of officer corps to meet warfare's changing demands, the rise in formal educational requirements has also been connected to external social pressures and to internal institutional concerns—an interaction explored in essays in this section on Germany, Great Britain, and Israel.

Horst Boog, senior director of research in the German armed forces' office of military history before his retirement, analyzes the role of the *Abitur* (the state-administered examination permitting university entry) as a qualifying factor in German officer selection from the nineteenth century to the present. Using this focus, he demonstrates that the conflicts created by political, economic, and social change, rather than the need for a well-educated officer corps, ultimately determined who would receive a commission in Germany until after World War II. Although German Army and Navy leaders recognized education's importance for modern armed forces and sought to raise the officer corps' educational level, a prospective officer's "character" (essentially meaning correct social origins under the Wilhelmine Empire and the Weimar Republic, and heroic ideological fighter under the Nazis) took precedence over educational background in officer selection. Furthermore, argues Boog, these irrational criteria (reinforced by Germany's geographic location) formed the basis for the German belief that "the conduct of war and military leadership were arts that could be mastered more by practice than by knowledge. . . ." Emphasized in German military education and training, this attitude produced officers unparalleled in technical competence, but mostly unaware of war's connection to society, economy, and politics.

In her essay on the development of British systems of officer entry, education, and training in the four decades after 1945, Cathy Downes leaves little doubt that factors other than a concept designed to produce effective junior officers often influenced the direction taken by those systems. Thus, from the end of World War II through the late 1960s and early 1970s (when the effort was abandoned), British military academies attempted to offer a program of general, university-level education, not primarily as the best way to prepare officer candidates, but to attract top-quality people to the military services in a society where opportunities for higher education were expanding rapidly. Similarly, the trend for each service to concentrate officer entry and training programs at a single location, according to Downes, reflected less an overall scheme to improve the systems, but "reactive responses to unanticipated events, or anticipated pressures (particularly to save money) either internal to the armed forces or in the broader society." For Downes, the New Zealand Defence Force's senior researcher, these and other examples characterizing postwar British officer entry and preparation, suggest the wisdom of a long-term strategy or plan for these systems based on desired outcomes.

In sharp contrast to the trend elsewhere, Israel remains one of the nations (South Africa is another) without any formal educational requirement for commissioning. In Israel, most officers (pilots, naval officers, and some specialists excepted) are chosen from the conscript intake by performance on intelligence and psychological tests and by leadership ability demonstrated during basic military and precommissioning training. In describing this system, Gunther Rothenberg, a professor of history both at Purdue University and at Monash University in Australia, also explains its relationship to Israel's security requirements, to the egalitarian ethos of Israeli society, and to the organizational nature and combat experience of the *Haganah* and the *Palmach*, the underground self-defense forces that were the Israel Defense Force's predecessors.

Note

1. Philip G. Altbach, "Patterns in Higher Education Development: Toward the Year 2000," in Robert F. Arnove, Philip G. Altbach, and Gail P. Kelly, eds., *Emergent Issues in Education: Comparative Perspectives* (Albany: State University of New York Press, 1992), 43–44.

Civil Education, Social Origins, and the German Officer Corps in the Nineteenth and Twentieth Centuries

Horst Boog

Whether or not it existed as a formal requirement for commissioning, the *Abitur* has generally been the yardstick for measuring the educational background of German officer candidates over the last two centuries. The *Abitur* was and still is the official leaving examination of a German *Gymnasium* (the designation for a normal German high school; the final year at a *Gymnasium* corresponds roughly to the beginning of the junior year at an American college or university). Normally taken by a student between ages eighteen and twenty after twelve or thirteen years of schooling, the *Abitur*'s successful completion would allow entrance to a German university. The state-controlled *Abitur* was introduced in Prussia in 1788 and, by 1834, its form had taken final shape. Various kinds of university entrance examinations had existed in previous years, but their purpose had also been to prevent young men of the lower social strata, who were subject to conscription, from escaping military service by becoming university students. The *Abitur*'s introduction was an expression of the absolute state's effort to organize its interior life.[1]

During the nineteenth century, the *Gymnasium* became the leading type of German high school. Its aim was to produce a broadly educated individual capable of independent judgment by developing, in harmony, all of his mental and physical abilities through the study of languages and literature, mathematics and the sciences, and natural and human history. These humanistic ideals had been propagated by the father of reform in German higher education, Wilhelm von Humboldt, and others, who believed that the classical languages and the humanities especially lent themselves to developing independent thinking in individuals. For this reason, the *Gymnasium* emphasized Greek, Latin, and the humanities. By 1837 the *Gymnasium's* curriculum was fixed and, from that point on, the subjects taught there were considered to be the foundation for higher education. They have remained so to this day, although the emphasis has shifted to more practical subjects such as the sciences and modern languages. In subsequent decades and as Germany industrialized, other types of high schools such as the *Realgymnasium*, the *Realreformgymnasium*, and the *Oberrealschule* appeared. They placed more emphasis on modern languages and the sciences.[2]

Prussian officers were never considered to be in need of the higher sanctifications offered by the *Gymnasium*, although the mere existence of the *Abitur* called for higher educational requirements for officers than had been customary until then. Not before the Franco-Prussian War of 1870–71, however, was this examination made a requirement for commissioning, and then it was only one among other requirements. Therefore, the *Abitur* cannot be regarded in isolation. Its role in the German officer corps must be examined in relation to other social, military, and personality requirements in an army confronted with the intellectual and technological challenges of an industrializing society and its accompanying social and political problems in an age of declining aristocracy and rising middle classes, as well as with the challenge to monarchical conservatism presented by liberalism and socialism. In this context, the German officer corps was caught between remaining a corps of class, status, and narrow military traditionalists and technicians, or becoming a democratically open corps of educated military professionals.

To explain the *Abitur's* relative unimportance for German officers throughout most of the nineteenth century, one must first examine Prussia's political and social situation in the years following its catastrophic defeat by Napoleon in 1806. Investigation into the reasons for the defeat revealed the need for reform. The officers of the old Prussian Army were found to be overaged, suggesting that changes in the system of promotion by seniority were required to introduce a standard of efficiency. Prussian military and political reformers such as Gerhard von Scharnhorst, Neithardt von Gneisenau, Hermann von Boyen, Karl vom Stein, and Prince Karl von Hardenberg, however, believed that the main reason for the country's collapse had been the separation of the people from the army. In the Prussian absolute state, when the army's ranks could not be filled with mercenaries from abroad, only the lower strata of the population became subject to military service. Exempted were the sons of the developing entrepreneurial middle classes which provided the king with the means to carry out mercantilist state policy, including war. Officer positions were a prerogative of the aristocracy. Reformers knew the aristocratic preserve had to be changed and the knowledge and education of the middle classes exploited for the military. Therefore an inner unity of the government, the army, and the nation had to be established. The liberal, democratic, and national feelings aroused in the masses in the wake of the French Revolution and against Napoleon worked toward this end, and the German Wars of Liberation from 1813 to 1815 served as a catalyst to establish this union temporarily.

General conscription, introduced on 3 September 1814, was viewed as the core of the political and military reforms. The reformers believed it would abolish the former differences, privileges, and prejudices within the nation and rally its people around the king to reestablish a strong Prussian state. General conscription was to be both an ethical principle and an honorable duty of the citizen. Even ordinary soldiers were to be treated like human beings and their dignity respected. Not only military *raison d'état,* but the love of king and nation was to be the basis of the new system, which was to be amalgamated with the old one. The slogan "With God for King and Fatherland" sealed the pact between king and nation. In practice, however, general conscription, except in wartime, was never fully applied because there usually was neither the need nor the money available to maintain so many soldiers.

Enthusiasm for the reforms was great as long as they helped liberate the country, but support for them had never been unanimous. Many a landed aristocrat, middle-class entrepreneur, civil servant, or military man regarded general conscription as a danger for German culture and the political order. To reconcile the hitherto exempted middle classes with conscription, the institution of the "One Year Volunteer" was introduced in 1814 and opened to the rich and better educated (at least six years at a *Gymnasium*) from the middle classes. Middle-class participants were thus able to feel somewhat coequal with the nobility, and with only one instead of two or three years of military service had a chance eventually to become officers (at least in the militia) if they volunteered and were able to keep a horse of their own and pay for their own equipment. Without going into the details of its further development, the "One Year Volunteer" institution actually represented the king's reluctant acceptance of the disinclination of his well-to-do subjects to be ordinary soldiers or even of their opposition to military service. In time, however, both the one-year volunteer program and especially its "reserve officer" replacement helped to create a better understanding of the military among the people which led, under new circumstances, to the militarization of German society.[3]

In practice, the reforms were not quite as liberal as the reformers and other segments of society wanted them to be. In fact, old conservative habits were maintained and new regulations preserving the exclusiveness of the officer corps were instituted. The reforms left the king's command and control over the army and his special relationship with his officers untouched and unhampered by other political institutions. Officers swore an oath of allegiance to the king not only as the head of the nation, but also as their supreme commander. Whoever wanted to become an officer had to apply first to the local regimental commander who screened the applicant. Then, after all other requirements had been fulfilled, including successful completion of training periods and examinations, the officer candidate had to be elected lieutenant unanimously by all officers of the regiment who, almost entirely from the aristocracy, preferred to choose candidates similar to themselves with respect to ways of thinking and conduct. Although the system of officer election had been introduced with a view toward getting efficient officers, the procedure soon became an instrument of conservative restoration. Additionally, the courts of honor that had been established within the officer corps in 1808 to investigate the conduct of officers during the war against Napoleon were instruments of self-purification, eventually developing into very conservative bodies that watched over not only the conduct of officers but their thinking as well. Both institutions, officer election and the courts of honor, maintained an exclusive *esprit de corps* among army officers. So did the king's marriage permit which was granted to an officer only if the bride belonged to the "right" family.[4] Finally the cadet schools, well established since the early eighteenth century, trained primarily sons of officers in the military spirit of order and obedience so that they might eventually become officers. The cadet schools also constituted a strong conservative element. Until they took on the character of a *Realgymnasium* late in the nineteenth century, their educational level was below that of a *Gymnasium*.[5]

The reformers themselves had thus laid some of the foundation for resistance to the educational renewal of the Prussian officer corps, and this resistance increased during the restoration period following the successful Wars of Liberation. Indeed, in 1808 King William III of Prussia had already instructed the Officer Examination Commission to be lenient toward officer candidates with respect to the requirements in French and mathematics because rigid adherence to them would bar from becoming officers the sons of the landed aristocracy and of officers of noble descent in small garrisons that lacked educational facilities.[6] In addition to formal education and knowledge of the sciences, the reformers had also believed that presence of mind, quick apprehension, orderly conduct in service, punctuality, and staunchness of character and personality should determine eligibility. This sprang from the conviction that the conduct of war was an art depending less on formal education than on practical knowledge of the sciences, the history of wars, and skills in fortification, tactics, strategy, and geography. The educational requirements to enter one of the army's schools for officer candidates (the *Kriegsschulen,* or war schools, created by Scharnhorst) still ranged far below the *Abitur.*

The decades of the restoration period before the Revolution of 1848 were marked by a growing aversion in the army to school education because this was more or less an attribute of the rising middle classes which were beginning to supplant the aristocracy's influence in society. As it was also declining materially, the aristocracy, which had hitherto furnished the king with officers and was still a buttress of the monarchy, had difficulty in meeting the costs of a high school education. Additionally, although many a senior officer was aware of a good education's importance for coping with the middle classes, the army was afraid of the explosive power of what was considered to be too much education and individual thinking in an officer corps that drew its strength from its conservative homogeneity as an elite, its tradition, and especially its "character." Character represented a rather diffuse set of values comprising mostly subjective qualities derived from the aristocratic officer corps of preindustrial times. These values included the "right" conservative convictions, honor, loyalty, a sense of duty, reliability, strong will, and fighting spirit—qualities that were attributed to the old nobility rather than to the middle classes. Although many of these qualities are more or less needed in any sound society, character soon became the prime quality expected of an officer and was always rated higher than formal education and abstract knowledge—qualities that were important in an industrializing, middle-class society. The two principles competing with each other, character and education, paralleled the political and social struggle between the aristocracy and the middle classes in the nineteenth century or that between traditional and irrational attitudes versus the rational way of life of modern times. Looking ahead, the overemphasis on character helped open the door for the Nazis' ideological manipulation and homogenization of an atomized (at least by its size) officer corps. During the Third Reich, the old Prussian virtue requiring an officer to disobey,[7] when justified by honor and circumstance, was only rarely, if at all, alive. Indeed, traditional character training enabled the regime to misuse individual officers already professionally disinclined to assert civil courage and moral responsibility.[8]

In the course of the nineteenth century, kings and senior officers had frequently admonished the army's directors of training and education to recruit vigorously the sons of former officers and to place more value on morality, conduct, obedience, and other soldierly qualities than on mere knowledge and abilities. Prince William, the later Emperor William I, wrote the Prussian minister of war in 1844: "Soldierly spirit is about to quit the army, if we, by our self-designed selection methods, drive out those classes which have hitherto been born and bred in this spirit and devolved it on succeeding generations."[9] Although not unaware of the necessity for officers in a modern army to possess scientific knowledge, he reacted to the minister of war's attempts to facilitate *Abiturientens* becoming officers by stating that "an enemy cannot be beaten with scientific knowledge," and pleading for lower educational requirements to enable to become officers the sons of those noble families that had helped to make Prussia great.[10]

In the 1850s and 1860s, Gen. Eduard von Peucker, the army's director-general of training and education and himself of middle-class descent (although later ennobled), raised the educational level for officers. The pressures of the middle-class Revolution of 1848 and of growing industrialization made giving greater place to middle-class educational values in officer training seem advisable in order to keep the officer profession at the top of the social hierarchy. In 1871, mainly due to Peucker's efforts, the *Abitur* became a desirable, if not obligatory requirement for an officer's career.[11]

The Prussian General Staff's successful performance in the Wars of Unification from 1864 to 1871 had proved the value of scientific and technical knowledge for officers, but at the same time had also revealed the officer corps' narrow military focus. For example, Gen. Helmuth von Moltke, the chief of the Prussian General Staff, attempted to exclude political interference as long as the fighting was going on. Advanced officer training had not followed Carl von Clausewitz's emphasis on the need to understand the relationship between war and politics, but had favored instead a high degree of military specialization and technical competence. As a result, the German officer corps lacked the educational preparation that would have enabled them to appreciate the broader strategic and political dimensions of warfare or to apply changes in science and technology most effectively. (This remained true until the end of World War II despite occasional attempts in the Weimar period to remedy these weaknesses by offering the cream of General Staff officers university liberal arts courses and, for exceptional officers destined for the technical branches, a program of study at the prestigious Berlin Technical Hochschule.) Victory in the Wars of Unification had also raised the officer corps' reputation and that of the military profession generally in society. General Staff officers achieved the status of demigods and, although only civil servants with a rank equivalent to colonel and above were considered suitable to be presented at court, the most junior army lieutenant could be presented if invited. In light of the officer corps' enhanced status, many Germans, especially academics, higher civil servants, and high school teachers were now anxious to become reserve officers.[12]

The Wars of Unification also fulfilled the political dreams of the nationally minded, liberal, upper middle class which now reconciled itself with the ruling monarchic-conservative system. In supporting the monarchy, upper middle-class Germans also accepted military values (although not always wholeheartedly) as guides for their own conduct as civilians. Their support for the monarch and the military system, both strong buttresses of the new national state, also stemmed from their view of the rising Social Democratic Party, the internationally minded political organization of the working class, as the enemy of the ruling political system as well as of the educated and propertied classes. The process of amalgamation of conservative monarchists and the upper middle class influenced also the conduct of the lower middle class and of the people generally, leading to what has been described as the German civilian population's militarization and feudalization in the last decades of the nineteenth century.[13] Carl Zuckmayer poignantly ridiculed this phenomenon in his play *Der Hauptmann [Captain] von Koepenick*.[14] While all Germans accepted military ways to some degree, the higher-ups tried to imitate the traditional aristocracy, the "feudal" class.

Despite the establishment of a constitutional monarchy in Germany, command of the army and officer selection remained the domain of the emperor and the military. Parliament did not have the right to interfere. When the *Abitur* was introduced following German unification as a qualification in officer selection, almost 50 percent of the officers came from noble families. In the infantry, 83 percent of the regimental commanders were aristocrats; in the cavalry 93 percent; but in artillery regiments only 50 percent. In the artillery, men from the middle classes with an appropriate higher education were pushing aside the aristocracy.[15] The handling of the *Abitur* qualification now mirrored the internal struggle to maintain class privileges. Social groups that stood for throne and altar—the aristocracy, officer families, land owners, the upper middle class, and higher civil servants—were considered eligible to provide officer candidates while the lower middle class, craftsmen, and workers were not. In 1900, thirty years after its introduction as a factor in officer selection, fewer than 44 percent of officer candidates possessed the *Abitur*. A significant percentage had obtained it at one of the cadet schools. (Although the cadet school curriculum had been adjusted to that of the more practically oriented *Realgymnasium*, it did not reach the level of a *Gymnasium's* curriculum.) Military authorities, however, pushed aside complaints about the low intellectual level of cadet school graduates, drawing attention instead to their practical abilities and proper social origins. To have the desired "character" was rated higher than theoretical school knowledge, so that, in practice, the *Abitur* remained just one of the qualifications for an officer candidate. Actually the regimental commander, to whom the candidate first applied, determined whether he had the "right" character or not, and that officer's opinion could not be disregarded. The officer selection commissions at district and Berlin headquarters used the same standard. In addition, the emperor himself could issue dispensations from the *Abitur*, and did so over a thousand times between 1902 and 1912. Finally, the cadet schools continued to furnish officer candidates without the *Abitur*, provided they met all the other practical, military, and character requirements.[16]

The expansion of the German Army required by the rearmament that stemmed from imperialism in the early 1890s challenged the restrictive policy of officer replacement. The nobility was increasingly unable to furnish the numbers of officers needed. Emperor William II, therefore, declared in March 1890 that the improved educational level of the German people opened the possibility for resorting to wider groups of the population to recruit officers:

> Nobility of birth alone cannot today, as in the past, claim the privilege of furnishing the officers for the army. However, the nobility of conviction, which has animated the officer corps at all times, must be maintained without change. And this is only possible when the officer candidates are taken from those groups of the people, where this nobility exists. Besides the scions of the noble families of the country and the sons of my brave officers, I see the future pillars of my army in the sons of those honourable middle-class families, who cherish the love of king and fatherland, a warm heart for the soldiers and a Christian way of life.[17]

The reference to Christian values not only meant what it said, but was also an expression of the traditional policy not to have Jewish officers.[18] Among the groups the emperor believed to possess the requisite nobility of conviction were academics, the educated higher professions, and the monied aristocracy. In the course of the feudalization and militarization of German society, these groups were now also considered to belong to the "desired circles" standing for the monarchy and representing the "right character." Altogether the "desired circles" supported the monarchy and were largely, although not completely, antirepublican, antidemocratic, antiparliamentarian, antiliberal, and antimodernistic. Despite the emperor's proclamation, however, the minister of war later thought it dangerous to open the officer ranks to these groups, and he proposed to abstain from further army expansion.[19]

The *Abitur* requirement was now used to distinguish between the "right" and the "wrong" middle classes for officer replacement because, meanwhile, a number of different types of *Gymnasien* had been established in response to the exigencies of an expanding industrial society. In addition to the old humanistic *Gymnasien*, the *Real-* and *Realreformgymnasien* and the *Oberrealschulen* had been created. They all offered the *Abitur,* and in 1902 all were officially recognized as being coequal despite their differences.[20] While the socially acceptable conservative and national-liberal, upper middle-class families put their sons into humanistic *Gymnasien,* the lower-middle and working classes preferred the newer *Realgymnasien* and *Oberrealschulen,* where practical subjects such as the sciences and modern languages were emphasized more than the humanities. Actually the army would have liked to obtain graduates from these latter schools because of the military profession's growing technical demands. The army's inspector-general of training and education, however, was afraid to accept *Abiturienten* from these schools as officer candidates without first examining their social and political attitudes. Accepting graduates from *Oberrealschulen* meant to him opening the "desired circles" "toward below" and to "importunate" people. Doubtlessly, he believed, the fathers selecting those schools did not intend to prepare their sons for a "higher" profession, but only for one to make a good living—a motive considered to be base. Here was the traditional prejudice of the old aristocrats and conservatives against the money-making bourgeoisie, which, as the *idée fixe* of the hero fighting the shopkeeper, dominated German war propaganda in World War I and World War II. (Conservative societies were considered to be much nobler than liberal-democratic ones.) The immediate reason for the caution displayed toward the graduates of the newer type *Gymnasien* was that their fathers were suspected of being Social Democrats or of not supporting the ruling system. Thus, the graduates of the humanistic schools continued to be preferred as officer candidates.[21]

Indeed, the *Realgymnasien* and *Oberrealschulen* with their emphasis on the "realities" also prepared students for an earlier specialization, which was counter to the general education offered by the Humboldt-inspired humanistic *Gymnasien* and always considered to be inferior by the educated upper classes in Germany. (This attitude later had a detrimental effect on how technology was handled by most of the generals in the highly technical *Luftwaffe*. Although more or less aware of the importance of technology, they had largely attended humanistic *Gymnasien* before World War I and did not sufficiently appreciate the inner regularities of technical and industrial processes to use it more effectively. Younger officers, who were more technical minded, were without influence.[22])

During World War I, when officer losses could no longer be replaced by men with the desired social and educational background, it was, nevertheless decided not to open the officer corps to replacements from the enlisted ranks and lower classes, although exceptions were made in cases of extraordinary performance and bravery. In general, and since most casualties occurred at the front among junior officers, the vacancies were not filled by promoting sergeants to officers, but by appointing them sergeant-lieutenants (*Feldwebelleutnant*) or officer-deputies (*Offizierstellvertreter*). Although these men fulfilled the duties of company-grade officers and wore outer insignia distinguishing them from the enlisted ranks, they otherwise remained sergeants and ranked behind the youngest, inexperienced lieutenant coming from school. It was feared that officers from lower-class families and without higher education would spoil the spirit and homogeneity of the officer corps as an instrument of the emperor's will.[23]

On the eve of World War I the German Army officer corps was composed as follows: about 75 percent had the desired social background (26 percent came from officer families, 39 percent were the sons of higher civil servants and academics, and 10 percent came from landowning families). Only 65 percent possessed the *Abitur*. This percentage compared unfavorably with the officer corps in Bavaria where possession of the *Abitur* had been obligatory since 1868, and almost every officer was an *Abiturient*. Because Bavaria was more of a middle-class and peasant country, the nobility was less numerous and played a lesser role than in Prussia. Only 13 percent of Bavaria's officers were aristocrats in 1913 as against 30 percent overall and 70 percent of the generals in the Imperial German Army.[24]

The Imperial German Navy was the beloved child of Emperor William II, but also a manifestation of industrialized German society. Only about 21 percent of its officers were of aristocratic origin in 1895 (and this percentage declined as World War I approached). Because the navy relied heavily on technology, about 83 percent of its officers held the *Abitur* just before World War I, and most were graduates of the humanistic *Gymnasien*. This figure was much higher than in the army, although the navy, as late as 1895, had not made the *Abitur* an entrance requirement for officer candidates because it wanted to conduct wars "with the sword and not with the pen." The navy believed it did not need scholars, but "people with practical capabilities, and a feeling of honour and duty."[25]

Although in need of graduates from *Oberrealschulen*, the navy assured the emperor that it would sift out socially unsuitable candidates. The pressure of circumstances requiring better education, however, led, in 1907, to granting officers holding the *Abitur* an earlier than actual date of commissioning (thus giving them, when due for the next promotion, the advantage of a longer time in rank) in order to attract adequately educated candidates. This caused a storm of protest especially from older flag officers who represented the highest concentration of aristocrats in the navy but mentally were still in the age of sailing vessels. The navy's inspector of training and education declared that it was undesirable to require a completed higher school education for naval officers at the expense of the values of chivalry and tradition of the German military profession. Admirals Prince Heinrich and Hans von Koester concurred. Disregarding contrary points of view, Adm. Alfred von Tirpitz, the head of the navy, pleaded for family origin as the basis of officer selection. In his opinion, those without the *Abitur* mainly came from "in every respect desirable" circles. Moreover, the navy was happy that the high cost of financial support for naval officer candidates would deter those from "undesirable" families.[26]

The emperor was caught in the dichotomy between educational and technological necessities and military-sociological traditions. On the one hand, he demonstrated his openness to modern technological developments by encouraging the navy to accept more candidates holding the *Abitur* and by establishing the first technical colleges on a university level in Germany; on the other hand, he asked the navy's leadership to increase the number of officer candidates from families that had been loyal to the monarchy and whose names were famous in the history of Prussia and the other German states.[27] As a result of this policy, the navy's engineer officers, although generally having a better formal education, continued to be discriminated against as noncombatant, second-rate officers with dirty fingers by the *Seeoffiziere* (the executive, or line officers) who exercised command and saw themselves as frontline officers risking their lives.[28] Although in the army the percentage of *Abitur* holders increased with rank—especially in the case of General Staff officers, it was just the opposite in the navy. In 1907 more than 76 percent of navy captains did not have the *Abitur*. In 1905–6 only 46 percent of the officers at the Naval War Academy were *Abiturienten* as compared to 77 percent in the entire naval officer corps.[29]

In the navy, the contrast between education and military professionalism on the one hand, and tradition on the other was greater than in the army, although naval officers were drawn from much broader social groups. In response to the perceived threats from new classes and modern ideas, the navy, without a long tradition of its own, overcompensated by being more Prussian than the Prussians. It adopted all the elements of the Prussian military system: officers' messes, honor courts, officer elections, the marriage license, seniority, the dueling code, the dispensation from standard entrance requirements, and the preference for education at home, as was customary in noble families, over formal, public education and the uniform measure of qualifications. Those with an academic or technical background were not considered coequals of officers with the right "character." The comparatively large number of senior navy officers of middle-class origins who were ennobled by the emperor is an indication of the eagerness with which the naval officer corps forgot its middle-class origins and created a traditional Prussian *esprit de corps* unmatched even by the army.[30]

The German Army's defeat in 1918 did not result in defeat for the traditional principles of officer replacement and advancement (despite a short period when requirements were loosened to remedy the wrong done during the war to the former sergeants who had been in officers' positions[31]) in the Weimar Republic's armed forces—limited by the Versailles Treaty to a 100,000-man army and a 15,000-man navy. In fact, the concomitant reduction of the officer corps to 4,000 in the army and 1,500 officers and upper boatswain grades[32] in the navy, and the change from a conscript to a volunteer force made it possible to select from the best officers with World War I service (General Staff officers above all) and only the best of the many postwar applicants. The reduction also facilitated the tacit reintroduction of the "desired families" concept from the former Imperial Army. Restoration of this criterion for officer selection was supported by the custom that the regimental commander should have first choice of applicants and by the practice of officer election by all the regiment's officers after the candidates had successfully passed the four-year training period. By the late 1920s, almost 90 percent of army officers came from the previously "desired" population groups. Fifty-four percent came from officer families, the educated middle class accounted for 28 percent, and the nobility for 24 percent (52 percent among generals). That the small officer corps, after a short lull, had no difficulty filling its ranks, explains why nearly 100 percent had the *Abitur*, although possessing it was not a *conditio sine qua non* for commissioned service. The officer corps in this "republican" army, so opposite in spirit from the Weimar Republic, sought to maintain the old values. The *Abitur* was a tool to preserve the old order. Nevertheless, there were more *Abitur* holders than ever before; the emperor would have rejoiced had he been able to achieve such high numbers of officers with good educations and the desired social origins in his army. General Hans von Seeckt and Adm. Erich Raeder, the heads of the army and navy, respectively, both stood for the priority of character and good breeding over education and training. In 1929, the navy even withdrew its officers from basic and advanced General Staff training. Character counted more than intellectual performance.[33]

The most conspicuous feature of the years from 1933 to 1945 was the rapid expansion of the *Wehrmacht*'s active (regular) officer corps from about 4,000 officers in 1933 to 24,000 in 1939 and 48,000 in 1944. By 1943 this expansion along with the activation of reserve and other kinds of officers, increased the army officer corps to about 250,000. By the same time the air force officer corps had grown to approximately 90,000, the naval officer corps to 13,000, and the SS to 15,700, for a total of approximately 370,000 officers in all services.[34] It stands to reason that this growth could not have been achieved by maintaining traditional standards of social origins, education, and training. Already, in the prewar years, about 1,500 sergeants, some 2,500 police officers, 1,800 retired officers and 1,600 officers of the former Austrian Army were transferred to the army officer corps. The *Luftwaffe*'s experience was similar. By 1939 about half of the German officer corps no longer came from desired families, although social origins as well as the *Abitur* were still the guidelines. The traditional pattern of officers drawn from the desired social groups, however, remained alive among the generals and admirals. Seventy-five percent met these two criteria. At the end of the war, 20 percent of generals were aristocrats as against only 7 percent in the active (regular) officer corps in 1943–44. Of the latter, 29 percent were the sons of officers.[35]

The homogeneity of the officer corps was loosened not only because of Adolf Hitler's war policy and the resulting need for more officers, but also for National Socialist ideological reasons. Conflicts arose between older, conservative officers and younger officers who had been indoctrinated by Nazi propaganda. In addition, the degree of homogeneity varied by service. The air force officer corps, since it was new and a mixture of many elements, had, in contrast to the army, no tradition of its own to defend. Political sensibility was, therefore, relatively higher in the army officer corps. In the *Luftwaffe*, technology counted most. The naval officer corps was brought up in a very conservative atmosphere—in the sense, not necessarily Nazi, of the national power state. The navy fought to the last for this reason, to preserve its honor, and to make up for the shame of the Revolution of 1918 that had originated partly from the rigidly conservative navy's own social defects.[36] The exploding size of the *Wehrmacht*'s officer corps broke social barriers and opened the way for the lower classes to rise socially, not necessarily by becoming educated but by becoming ideologically acceptable, brave "people's officers" whose character and honor demanded they fight for Hitler and his ideas. The intention was to create a socially "democratic" officer corps without a democratic mind. However, when defeat loomed and the last physical and mental reserves were to be mobilized by newly appointed political officers (NSFOs), the greater part of the officer corps fought on to defend their country, not for the sake of National Socialist ideology.

Until 1938, 50 percent of army officers were *Abiturienten* largely because admission to a university depended on completion of military service which had become obligatory again after the reintroduction of general conscription and because the *Abitur* was still a normal requirement for officer candidates.[37] Officer candidates with an *Abitur* had an advantage over those without the certificate because the period of service required to be eligible for regimental election was shorter for *Abiturienten*. Lack of formal education, however, could be made up by good military performance (*Bewährung*). Exceptions were made for those candidates, who, "by their personality and military abilities rise far above the mass of their comrades, so that they promise to become fully capable officers even without the normally required scientific education."[38] Speaking at the reopening of the General Staff college (*Kriegsakademie*) in 1935, the chief of the Army General Staff, Gen. Ludwig Beck, pointed out that "theoretical knowledge was only a first step to genuine ability and performance." Colonel Gen. Werner von Fritsch, the army chief, added that the *Kriegsakademie's* training objectives could only be achieved by "intimate contact with the daily practical work at the front" and not by "theories and regulations." Personalities are shaped by "exemplary action" rather than by "theoretical teaching." In addition to professional knowledge, it was believed to be critical that officers develop above all else the qualities of mind and character considered essential for a military leader, namely steadfastness of will, energy, judgment, readiness to take responsibility, audacity, and the self-assurance in conduct and action that came from confidence in one's own abilities. These maxims demonstrate again the emphasis on "irrational" values in German officer education, which, although stemming from pre-Nazi times and essentially having nothing to do with ideology, partly met with Hitler's preference for soldiers and officers with heart and "character" and with his dislike for "cool thinkers" of the General Staff type.[39] When the high losses and the devastation in Germany in the last years of the war destroyed any rational hope for victory, the German leadership appealed to these irrational notions to make the soldiers and officers "hold through." Here was an opportunity for Hitler to break up the last conservative strongholds of the old officer corps that had been built on social desirability and education and to create a National Socialist "People's Officer Corps" to his liking.

On 30 September 1942 the *Abitur* as a formal entrance requirement for officers was abolished. Since the age of conscripts had been lowered, officer candidates were too young in any case to finish school and obtain the *Abitur*. Moreover (a reflection once again of the attitude toward formal education and rational values) officers holding academic degrees and titles were forbidden to use them. Practice and character became paramount.[40] Also abolished in the fall of 1942 were two institutions that had served to maintain the officer corps' exclusiveness: the honor courts and the custom of officer election.[41] These had always been obstacles to Hitler's ideas. Abolishing them also dampened Hitler's fear that the officer corps might unite in spirit against him. The qualifications for officer candidates were changed so that every young German above the age of sixteen could, regardless of his social or educational background, become an officer if he had the "proper" personality, proved himself at the front against the enemy, was racially "pure," and believed in Hitler and his cause. New officer promotion regulations introduced in December 1942 and January 1943[42] struck a blow against the seniority system, a bulwark of officer *esprit de corps*. These regulations provided for accelerated promotion based on military efficiency and fighting qualities. Hitler wanted to have young, physically and mentally strong officers with good nerves—ideological fighters who would not start thinking about the general war situation, but had enough motivation to defend their country, which, to many at that time, seemed the highest priority task. Officers serving at the front rather than those on the General Staff benefited from the new promotion policy. Within half a year these officers were to be promoted to the rank corresponding to the position they held in battle, provided they proved themselves. Promotion based on battlefield performance further broke up the homogeneity of the officer corps because it favored egoism and could be manipulated politically.

The traditional underestimation of formal education and the widespread belief that the conduct of war and military leadership were arts that could be mastered by practice more than by knowledge had brought the German officer corps to a dead end. Of course, accumulating sufficient scientific knowledge to conduct a modern war required considerable time, something the Germans, in view of their unfavorable geographic situation, never seemed to have enough of. German battlefields were always nearby, calling for quick action and practical skill and, as Martin van Creveld has asserted, German officers excelled at the tactical level.[43] As a consequence of the preference for practice over abstract knowledge, however, German officers lacked strategic breadth and this proved fatal. It seems useless to ask whether the dividing line should have been drawn more in favor of formal education than practice. Germany's wars had to be short and successful to prevent the enemy from developing fully his war potential. Moreover, in most cases, they were prosecuted within continental limits. But as soon as either the emperor or Hitler disregarded and transgressed these limitations, the traditional recipe for officer training could not and did not work. Relatively speaking, the *Abitur* as a guideline, although not as a *conditio sine qua non*, secured a standard for the officer corps that was good enough to resist superior enemies in a continental environment. The rest of the story is bad politics.

The story of the *Abitur* and its significance for German officers is not complete without an epilogue for the postwar period. As a Cold War necessity and within the framework of the North Atlantic Treaty Organization (NATO), the Federal German Armed Forces, the *Bundeswehr*, was established rapidly ten years after the end of World War II. This could not have been done without employing former *Wehrmacht* officers. East Germany, which had been building an army even before the *Bundeswehr*'s creation, also exploited the experience of former *Wehrmacht* officers, although to a lesser degree than in West Germany because the East Germans wanted to fashion a Communist army. They were willing to risk having less educated and qualified officers in favor of Communist faithful drawn from peasant and workers families to create a radical break with the bourgeois past.[44] The West German Army preferred highly qualified officers—insofar as they were available at the beginning of West Germany's economic "miracle." The former *Wehrmacht* officers were screened both for their political views and for their conduct during the war. A special investigating committee, the *Personalgutachterausschuß*, was set up for former officers with the rank of colonel and above. This committee saved the *Bundeswehr* political trouble in the following years because it accepted only those officers who had not been Nazis and who had not violated any ethical responsibilities. For newcomers, the *Abitur* was again a prerequisite. In social composition the *Bundeswehr*'s new officer corps corresponded pretty much with the old pre-Nazi pattern. In 1967, 69 percent of

all officers had the *Abitur* as did 97 percent of the generals. The relatively low percentage of *Abiturienten* reflected the large number of former war officers
(*Kriegsoffiziere*) who had been commissioned during World War II for good military
performance, although they did not meet the educational requirements, and had been
accepted into the *Bundeswehr*.[45]

In the late 1960s and early 1970s a drastic change took place. Most former
Wehrmacht officers had retired. All of the young officers now usually had the *Abitur*,
and there were no longer any restrictions with regard to social origins. Only 6.9
percent of officers came from officer families in 1974; in 1977, the nobility accounted
for just 1.7 percent.[46] From the early 1970s the West German officer corps, socially
dominated by the middle classes in the broadest sense, could, for the first time, be
regarded as truly democratic in terms of social origins and mindset. The new concept
of *Innere Führung* had proved its value. By then, however, large portions of the
population were no longer convinced of the need to keep a large army. Many graduates of the *Gymnasien* preferred to make a living in civilian life rather than become
officers. This time German society had distanced itself from the army, essentially
because the German public mind and standards had been spoiled by economic success and Germans felt safe in the lap of NATO. The officer profession had to be made
attractive, and this coincided with the need for education and training to handle
modern military technology.

In the early 1970s, two universities of the German armed forces were established
for this purpose. Anyone who hoped to become a career officer had to obtain a
university diploma in certain desired subjects (e.g., engineering, teaching, electronics) in three-to-four-year courses beginning one or two years after entering the
Bundeswehr as an officer candidate. The *Abitur,* although still an entrance requirement, thus lost its former significance.[47] It had always been weighed against other
qualifications and had never been a *conditio sine qua non* for an officer's career, but
its importance had risen considerably in the twentieth century. It may appear surprising that, in the military, higher education counted relatively so little, since the Germans were always so proud of their achievements in philosophy and music and in
their higher education system. The reasons outlined in this essay to explain why the
German military undervalued formal education may be summed up as: the sociocul-
tural overlapping of preindustrial, aristocratic values with those of an industrialized,
middle-class society.

Notes

1. Michael Heafford, "The Early History of the Abitur as an Administrative Device," *German History,* no. 3, 1995; Friedrich Paulsen, *Geschichte des gelehrten Unterrichts auf den
deutschen Schulen und Universitäten vom Ausgang des Mittelalters bis zur Gegenwart,* 2 vols.
(Berlin and Leipzig, Germany: Walter de Gruyter, 1921), 2:97, 296, 348ff., 352.
    2. Paulsen, 2: Book 6, chaps. 5–6, and annex, chap. 2.
    3. Rainer Wohlfeil, "Vom Stehenden Heer des Absolutismus zur Allgemeinen Wehrpflicht
(1789–1814)," in Militärgeschichtliches Forschungsamt [hereafter MGFA], ed., *Deutsche
Militärgeschichte 1648–1939,* 6 vols. (Munich: Bernard & Graefe, 1983), 1:pt. 2:21ff., 100–53;

Manfred Messerschmidt, "Militärgeschichte im 19. Jahrhundert (1814–1890), Zweiter Teil, Strukturen und Organisation. Die Preußische Armee," ibid., 2:59ff., 87ff.

4. See also Herbert Schottelius and Gustav-Adolf Caspar, "Die Organisation des Heeres 1933–1939," ibid., 4:372.

5. Wohlfeil, "Vom Stehenden Heerdes Absolutismus," 139f., 143; Manfred Messerschmidt, *Die Wehrmacht im NS-Staat. Zeit der Indoktrination* (Hamburg: R.v. Decker, 1969), 386f., 428; Manfred Messerschmidt, "Militärgeschichte im 19. Jahrhundert (1814–1890), Erster Teil, Die politische Geschichte der preußisch-deutschen Armee," *Deutsche Militärgeschichte 1648–1939*, 2:10, 60, 67, 112, 123, 132, 147, 207, 211; Messerschmidt, "Strukturen und Organisation," 38f., 44ff., 56ff., 64, 87, 142, 146, 164, 191, 294, 272, 294, 310, 329.

6. Messerschmidt, "Strukturen und Organisation," 68.

7. Bodo Scheurig, "Insubordination als Gebot. Eine alte preußische Tugend und ihr späterer Verfall," *Frankfurter Allgemeine Zeitung*, 3 Aug. 1996.

8. Detlef Bald , *Der deutsche Offizier. Sozial- und Bildungsgeschichte des deutschen Offizierkorps im 20. Jahrhundert* (Munich: Bernard & Graefe, 1982), 110.

9. Messerschmidt, "Strukturen und Organisation," 84ff., 88ff.

10. Ibid., 102.

11. Bald, *Deutsche Offizier*, 111f.

12. Wiegand Schmidt-Richberg, "Die Regierungszeit Wilhelms II," *Deutsche Militärgeschichte 1648–1939*, 3:87.

13. Gerhard Ritter, *Staatskunst und Kriegshandwerk. Das Problem des "Militarismus" in Deutschland*, 4 vols. (Munich: R. Oldenbourg, 1964), 2:117–31; Bald, *Deutsche Offizier*, 28.

14. In Carl Zuckmayer's play, *Der Hauptmann von Koepenick* (1931), a journeyman shoemaker, just released from prison and without money, acquires a second-hand, captain's uniform, takes command of a squad of soldiers marching about in the street, heads for the town hall and asks for the community chest. The mayor, a reserve officer of lower rank, salutes and hands over the money. The false captain then orders the soldiers to continue marching and disappears with the money.

15. Messerschmidt, "Strukturen und Organisation," 63f.

16. Ibid., 109; Bald, *Deutsche Offizier*, 112f.

17. Manfred Messerschmidt, "Militär und Schule in der wilhelminischen Zeit," *Militärgeschichtliche Mitteilungen* 23 (1978): 55. Schmidt-Richberg, "Regierungszeit Wilhelms II," 85f.

18. After their emancipation in Prussia in 1812, German Jews sought constantly to become fully recognized as citizens by volunteering for military service. Although legally subject to conscription, the authorities did not want them for religious and other reasons—at least not Jewish officers. Despite the deep-seated antipathy against them in the military, Jews were accepted for military service in wartime but were discharged promptly when hostilities ended. In World War I, about 96,000 Jewish soldiers served in the German armed forces, suffering approximately 12 percent losses; 29,874 of them were decorated, 2,022 became officers (1,159 were medical officers and administrative civil servants equivalent to officers). If the number of Jews who became Christians is added to the total, then the number of Jews serving in the military is actually much higher. For more details, see Manfred Messerschmidt, "Juden im preußisch-deutschen Heer," in MGFA, ed., *Deutsche Jüdische Soldaten 1914–1945* (Freiburg: MGFA, 1981), 49–73.

19. Bald, *Deutsche Offizier*, 43; Schmidt-Richberg, "Regierungszeit Wilhelms II," 86.

20. Messerschmidt, "Politische Geschichte" 96, and "Strukturen und Organisation" 108; see also Franz Schnabel, *Deutsche Geschichte im neunzehnten Jahrhundert*, 8 vols., *Die moderne Technik und die deutsche Industrie* (Freiburg: Herder-Bucherei, 1964), 6:88ff., 109ff.

21. Messerschmidt, "Militär und Schule," 59.

22. Horst Boog, "The Luftwaffe and Technology," *Aerospace Historian* 30 (September 1983); idem, "Das Offizierkorps der Luftwaffe 1935–1945," in Hanns Hubert Hofmann, ed., *Das deutsche Offizierkorps 1860–1960* (Boppard am Rhein: Boldt, 1980), 315ff; idem, *Die deutsche Luftwaffenführung, 1935–1945. Führungsprobleme-Spitzengliederung-Generalstabsausbildung*

(Stuttgart: Deutsche Verlags-Anstalt, 1982), 36–76, 496ff., 549ff.

23. Schmidt-Richberg, "Regierungszeit Wilhelms II," 91.

24. Hermann Rumschöttel, "Das bayerische Offizierkorps 1866–1918," in *Das deutsche Offizierkorps 1860–1960*, 75–98; Bald, *"Deutsche Offizier,"* 45, 64, 69.

25. Holger H. Herwig, "Das Offizierkorps der Kaiserlichen Marine vor 1918," *Das deutsche Offizierkorps 1860–1960*, 148.

26. Ibid., 149ff.

27. Cited in Bald, *Deutsche Offizier,* 121ff.

28. The problem has been dealt with extensively by Werner Bräckow, *Die Geschichte des deutschen Marine-Ingenieuroffizierkorps* (Oldenburg and Hamburg: Stalling, 1974); see also Herwig, "Offizierkorps der Kaiserlichen Marine," 156ff.

29. Bald, *Deutsche Offizier,* 122f., 125, 128.

30. Ibid., 45, 64, 69, 74, 90, 97, 99, 113, 121ff.

31. Martin van Creveld, *Kampfkraft. Militärische Organisation und militärische-Leistung 1939–1945* (Freiburg: Verlag Rombach, 1989), 164.

32. Rainer Wohlfeil, "Reichswehr und Republik (1918–1933)," *Deutsche Militärgeschichte 1648–1939*, 3, pt. 6:93.

33. Bald, *Deutsche Offizier,* 48ff., 107ff. For more details, see David N. Spires, *Image and Reality: The Making of the German Officer, 1921–1933* (Westport, Conn.: Greenwood Press, 1984).

34. For the army, see Rudolf Absolon, "Das Offizierkorps des deutschen Heeres 1935–1945," *Das deutsche Offizierkorps 1860–1960*, 250; and Creveld, *Kampfkraft*, 190f. Including the medical, veterinary, administrative, legal and other officers, Creveld arrives at a figure of about 345,000 officers, 73,600 of them active (regular), and about 55,000 line (Truppen-) officers. For the air force, see Boog, "Offizierkorps der Luftwaffe," 289. The figures for the navy are based on Bernhard R. Kroener, "Die personellen Ressourcen des Dritten Reiches im Spannungsfeld zwischen Wehrmacht, Bürokratie und Kriegswirtschaft 1939–1941," in MGFA, ed., *Das Deutsche Reich und der Zweite Weltkrieg*, 10 vols. (Stuttgart: Deutsche Verlags-Anstalt, 1988), 5:pt. 1:975. The SS figures are taken from Bernd Wegner, "Das Führer-Korps der Waffen-SS im Kriege," *Das deutsche Offizierkorps 1860–1960*, 329.

35. Bald, *Deutsche Offizier,* 54ff., 60, 73, 90, 115; Kroener, "Personellen Ressourcen," 733.

36. Boog, "Offizierkorps der Luftwaffe," 324f.

37. Bald, *Deutsche Offizier,* 115f.

38. Ibid., 115.

39. Ibid., 108; cf. Messerschmidt, *Wehrmacht im NS-Staat,* 80, 169.

40. Reinhard Stumpf, *Die Wehrmacht-Elite. Rang- und Herkunftsstruktur der deutschen Generale und Admirale 1933–1945* (Boppard am Rhein: Boldt, 1982), chap. 3; Bald, *Deutsche Offizier,* 111f.

41. Messerschmidt, *Wehrmacht im NS-Staat,* 386f., 428.

42. MGFA, ed., *Untersuchungen zur Geschichte des Offizierkorps. Anciennität und Beförderung nach Leistung* (Stuttgart: Deutsche Verlags-Anstalt, 1962), 276, 286ff., 293ff., 303 ff., 307ff.; Creveld, *Kampfkraft*, 179f.

43. Creveld, *Kampfkraft,* 203ff.

44. Detlef Bald, "Sozialgeschichte der Rekrutierung des deutschen Offizierkorps von der Reichsgründung bis zur Gegenwart," in Bundesministerium der Verteidigung, Führungsstab der Streitkräfte I/15, ed., *Schriftenreihe Innere Führung, Reihe Ausbildung und Bildung*, No. 29 (Bonn: Ministry of Defence, 1977), 29f.

45. Bald, *Deutsche Offizier,* 109ff., 118ff., 130.

46. Ibid., 72, 83, 90.

47. Ekkehard Lippert and Rosemarie Zabel, "Bildungsreform und Offizierkorps. Zu den Auswirkungen der Neuordnung der Ausbildung und Bildung in der Bundeswehr auf das Rekrutierungsmuster des Offiziernachwuchses," *Schriftenreihe Innere Führung*, no. 29, 40–155.

The Evolution of Officer Education in the British Military Profession after World War II

Cathy Downes

Professions possess a unique body of knowledge, skills, and values vitally important to life's basic concerns that distinguishes them as occupational groups and that they are held responsible for maintaining, disseminating, and developing.[1] Most have met this responsibility by jointly sponsoring university-based programs of study and research followed by independent practitioner-centered self-development. When necessary, they have developed separate education establishments normally adopting university-level and type practices and philosophies. Through such institutions, professional aspirants receive distinctive formal and practical education, training, and socialization. Successful completion of these programs distinguishes *bone fide* members of a profession from members of the broader community.[2]

In the abstract, these statements represent a valid theoretical rationale for professional education and training. Nothing in the real world, however, is ever as blessedly uncomplicated as a theoretical rationale. Indeed, in the case of the British military profession, social culture, bureaucratic politics, and fiscal parsimony may explain more about how institutions of military professional development have evolved than any theoretical concept of the relevance and value of education in generating new members of the profession. At times, these factors have had greater influence upon systems of British officer education and training than changes in the nature of warfare or the demands of officership.

This paper explores this observation through three themes that have characterized the development and fortunes of the British entry-level officer education and training establishments in the four decades following World War II: the Royal Military Academy, Sandhurst, the Britannia Royal Naval College, Dartmouth, the Royal Air Force (RAF) College, Cranwell, and the Officers Training Wing of the Royal Marines Commando Training Centre, Lympstone (CTC).[3] These themes are: changing pedagogic roles for the service colleges with service-versus-civil provision of tertiary-level educational experiences; officer entry concentration and education and training courses/schemes rationalization; and shifting balances between preentry and postentry preparatory experiences.

Secondary School, University, or Something In-Between?

From their inception, United States service academies were designed as university-status institutions for officer aspirants. In contrast, the British service colleges struggled as late as the post–World War II period to define and to establish their pedagogic identity as tertiary educational institutions. This struggle focused around two role changes. The first concerned the postwar shift out of secondary- or postsecondary-level education into undergraduate-level, tertiary education intermixed with military professional studies and experiences. The second role change concerned the shift from undergraduate-level, tertiary education to a concentration on training in military professional skills, knowledge, and qualities.

From Secondary School to Undergraduate University

Prior to World War II, the Britannia Royal Naval College's key role was to provide secondary education for naval cadets. Established in 1905 as the Royal Navy's private secondary school, Dartmouth accepted thirteen-year-old boys for four years of secondary education in a naval environment before they went to sea as midshipmen. Older cadets could enter the college at age eighteen on the "Special Entry" scheme for a short naval studies course.

After World War II, however, continued provision of secondary school education was questioned and ultimately abandoned. The question was not so much whether Dartmouth should provide officer aspirants with a secondary or tertiary education. Rather, the issue was the social exclusivity of the college in a postwar Britain where wartime national experiences had challenged the rigidity of social class divisions and elitism.[4] In 1947, to overcome this exclusivity, the Admiralty announced the termination of the thirteen-year-old entry and its replacement with a sixteen-year-old entry scheme with a two-year program of instruction. Fees were waived. The Seaman Executive Branch was opened to Special Entry cadets and to candidates commissioned from the Lower Deck (enlisted ranks), and it was mandated that up to 25 percent of officers were to be drawn from the Lower Deck.

The sixteen-year-old scheme failed immediately. It took no account of the evolving civilian norms for educational development. It called for students to prepare for an entrance examination halfway through their secondary education to attend a course that did not qualify them to enter a university. This was at a time when students were increasingly encouraged to go on to university studies. Within five years, far from broadening the sources of recruitment, there were 50 percent fewer cadets entering the college.[5] Some twelve years of reviews followed with the Montague Committee, the Committee on Officer Structure and Training, and the Murray Committee, each producing modified entry and training schemes.

Although spurred by external concern over social exclusivity, the review committees could not help but focus as much upon *what* education as well as *when* education. If aspirants were to enter the college at age eighteen, having completed their secondary education, then what learning opportunities should the college provide? Where did the college fit into the hierarchy of pedagogic institutions—a college of further education, a polytechnic, a university? Each review gradually clarified the possibilities. By 1960, the Murray Committee had abandoned the Civil Service

Commissioners' examination for secondary school graduates and replaced it with civilian university entrance requirements as Dartmouth's educational entry standard. University teaching methods were introduced and at least one-and-one-half years of a three-year program of training were given over to university, undergraduate-level education.

Both Sandhurst and Cranwell also struggled with their own identity crises throughout this period.[6] By the early 1960s, to varying degrees, all the service colleges had reshaped their educational role and institutional status. They applied university entrance standards, adopted university-style teaching methods, gave teachers the more prestigious "lecturer" status, upgraded curricula, and grouped subject matter into academic or professional military categories. While adopting many of the attributes of university-level education, none of the colleges were able to expand, upgrade, and lengthen their courses to a full baccalaureate level (although RAF College Cranwell based its courses on civilian baccalaureate degree requirements).

As with the changes at Dartmouth, external support for similar changes in the other colleges reflected more a concern over recruitment than the need to provide officer aspirants with a tertiary education. The trend throughout the 1950s had been for more students to stay longer at school and complete their secondary education. This was followed in the early 1960s by a significant expansion in the national universities sector, opening up a great many higher educational opportunities for school-leavers. It became increasingly apparent that the armed forces were now directly competing not just with potential employers but also with universities and technical colleges for school-leavers. Tertiary-level education was not seen necessarily as a prerequisite to professional competence, but as a carrot for aspirant enlistment. As one naval officer put it, "The Navy does not want graduates with specific degrees . . . but the Navy needs the sort of people who aspire to degrees. The Navy must therefore enable future officers to satisfy their enhanced ambitions and to obtain a degree before or as part of naval training. This, or the Navy will not get the officers it must have."[7]

These sentiments reflected one of two perspectives on the most appropriate direction for the service colleges. This perspective was based in the conservative elements of the services, and often prevailed in the service staff directorates that sponsored officer education and training. The second perspective was based in the service colleges, especially in their civilian academic staffs. This latter perspective supported a change in pedagogic role for the colleges, not only in recognition of recruiting difficulties, but in response to the changing educational needs of the future military officer.

This latter perspective recognized the "neither fish nor fowl" status of each college and reflected a belief that the shift to a university-level status was the most logical development for these institutions. The perspective also recognized the embryonic state of academic fields of professional relevance—strategic studies, military history, civil-military relations, and military technology—in the wider university education sector and hence the need for the service colleges to make these fields available in their programs to potential military officers.

These differing perspectives on the value and purpose of tertiary education in recruiting and preparing officer aspirants led to tensions, disharmony, and conflict between the service colleges and their sponsors throughout the 1960s and 1970s. The dominant perspective would go on to define the role and function of officer entry institutions in preparing aspirant officers to enter the British armed forces.

From University to Military Training Establishment

By the early 1960s, most service colleges were providing mixed programs of secondary graduation-level remedial education and university-level undergraduate courses. However, the most important catalyst forcing the issue of tertiary education was the 1966 Howard-English[8] independent inquiry into the systems of officer education and training in the armed forces.

The Howard-English Report responded to the perspectives of both sponsors and colleges. On the one hand, the authors emphasized that the newly reestablished all-volunteer force would compete directly with universities for entrants, at a time when these institutions were rapidly expanding and offering broadened opportunities to school leavers. The report therefore recognized that providing tertiary-level educational opportunities would be essential to attract high quality officer candidates. On the other, the report recognized the upgrading of pedagogic programs that had taken place at the service colleges, and the need for future officers to be exposed to this type of educational experience.

To achieve a solution that met both sets of concerns, Howard and English proposed the creation of a joint services defense academy. After completing short courses of professional training and two to three years of commissioned experience, young officers, they suggested, should then undertake at least one year's academic study at such an academy. Sufficiently capable officers would continue on to complete their degrees.[9] While both bold in conception and introducing a joint perspective very much before the latter was fashionable or required, the Howard-English proposals failed to gain favor with either the colleges or their service sponsors.

From the colleges' perspective, a triservice academy, would, in one sweep, wipe out all the educational gains each had worked so assiduously to secure for their respective institutions. For the services, returning officers to educational courses, just when they were becoming useful in their units, squadrons, or ships, did not appear logical. Moreover, attendance for those officers requiring lengthy professional training, such as pilots, presented practical scheduling difficulties.

Although the Howard-English proposals were rejected, they had brought to a head the whole question of the feasibility of the service colleges providing university-level education. Each college sought to use the emphasis the Howard-English Report had put on university studies to entrench university-level studies in their own programs, including approaches to civilian universities to accredit those programs. These attempts were mostly unsuccessful. In Sandhurst's case, for example, civilian universities were willing to equate its program only with the university entrance, or "A" (Advanced) Level, qualification. In point of fact, the Howard-English vision, while recognizing how far the colleges had come, only demonstrated how far they still had to go to achieve full university status.

The service college sponsors were unconvinced that the proposed defense academy was the only way to compete for suitable officer candidates, particularly in view of the expanding population of university graduates. Nor were they willing to fund a new and expensive officer educational institution at a time when the national tertiary education sector was building universities, where potential officers could be educated. Their response reflected a "if you can't beat'em, join 'em" attitude. The sponsors opted not to support the emergent potential of the service colleges to provide officer aspirants with a university education. Instead, they choose to promote the development of university cadetship schemes that had been nurtured since the early 1960s by each service college, and the introduction of graduate officer entry schemes.

By the late 1960s, the tertiary education programs at all three service colleges were threatened. In 1968, the RAF made its choice and announced it would cease university education at Cranwell. From 1969, the RAF would rely on civilian universities to educate potential direct-entry permanent officers who would then be recruited into the service.[10] By 1971, the Royal Navy had modified entry schemes into Dartmouth to emphasize professional naval subjects and experience at-sea with the Dartmouth Training Ship in the first terms. After completing this training, civilian university graduates would be commissioned and sent to the Fleet. Naval university cadets would go to civilian universities, and return to the college for "top-up" professional courses after completing their degrees. Naval cadet (secondary school) entrants would undertake only two terms of "academic instruction" at the college before going to sea as midshipmen. By 1972, Sandhurst had been reduced to providing four-and-one-half months of "professionally relevant" academic studies for its nongraduate regular officer aspirants, once they had completed a twenty-eight-week course of military professional training.

From 1945 onward, entry-level education for Royal Marines officers had been dominated by professional military training. Although adopting university entrance standards in the 1960s, there were no attempts to develop the Officers Training Wing of the Commando Training Centre into a tertiary educational institution. The unique reputation of the corps proved sufficient to ensure a steady supply of officer aspirants without the carrot of a university education. In 1969, however, the corps followed the Royal Navy's lead and established its own university cadetship and graduate entry schemes.[11]

By the early 1970s, the die was cast for the service colleges as tertiary educational institutions, despite, in the case of Sandhurst, continuing and considerable resistance from its Academic Advisory Council. From the 1970s, the principal purpose and role of the service colleges was to provide professionally relevant military studies and training for officer aspirants. Such aspirants included secondary school graduates, civilian university graduates, and graduating university cadets.

The Trinity of Amalgamation, Rationalization, and Standardization

At the end of World War II, officer entry and training in each service was dispersed at different training establishments mainly determined by specialist branch, commission type and gender. From this start point, an almost inexorable press for amalgamation characterized the ensuing four decades. By the early 1980s, all entry-level officer education and training was concentrated at one single-entry point institution for each service. As the plethora of relocated and new schemes congregating in one location grew, a process of rationalization followed, with increasing standardization of training program content for all remaining entry types.

Amalgamation

In 1947 the RAF College Cranwell was reopened for flight cadets (pilots). Between that year and 1965, officer entry and education and training programs for equipment, secretarial, and general duties (navigators) officers were relocated from separate training facilities to the college. In 1966, the RAF Technical College at Henlow was closed and technical (engineering) officers' training was moved to Cranwell. Women's Royal Air Force and graduate officer training were collocated at the Officer Cadet Training Unit (OCTU), Henlow in the 1970s. And in the late 1970s, OCTU Henlow itself was closed and its training merged into Cranwell's Department of Initial Officer Training. Postcommissioning specialist training was also brought to the college, with a Department of Specialist Ground Training being established for Engineering and Secretarial Branch officers.

In 1939, the Royal Military Academy, Woolwich and the Royal Military College, Sandhurst had been merged to form the Royal Military Academy, Sandhurst. After the war this served as the entry point for all regular commission officers. Short service commission officers were trained at the Mons Officer Cadet School until 1972, when it was closed and its function collocated at the academy. A decade later, the Women's Royal Army Corps (WRAC) College, Camberley closed and WRAC officers' training was collocated at Sandhurst. In addition to these major incorporations, Sandhurst expanded the range of entry options available to officer aspirants by instituting preuniversity studies courses, pre- and post-university cadetship courses, and graduate entry courses. The academy also incorporated Rowallan Company (from Royal Army Educational Corps Centre, Beaconsfield) which ran courses for officer cadet aspirants who needed personal development to meet officer selection standards.

Although no other officer training institutions were merged with Dartmouth, the college expanded its entry options, particularly in the 1960s, to incorporate a broad spectrum of specialist entry schemes that had been previously maintained elsewhere or were new to the navy. In a fourteen-year period, some ten schemes were added to the college's standard Naval Cadet Entry "repertoire." These were a Supplementary List Aircrew Scheme (1960), an Upper Yardman's Scheme (1960), Supplementary List Seaman Cadet Entry Scheme (1961), Supplementary List Engineer Cadet Entry Scheme

(1961), Instructor Officers Scheme (1962), International Cadet Entry Scheme (1962), Graduate Entry Scheme (1964), University Cadetship Entry Scheme (1965), Queen Alexandra's Royal Naval Nursing Service Entry Scheme (1973), and the Women's Royal Naval Service Officer Entry Scheme (1974).

The Royal Marines officer entry system evolved in a pattern similar to Dartmouth's. In the immediate postwar period, Royal Marines officer entrants undertook up to thirteen separate training courses at a range of different training establishments. Over the 1950s, these courses were reduced in number and standardized into common core packages primarily taught at the CTC Lympstone. In the mid-1960s, the corps started to diversify its training schemes under the auspices of the CTC. In 1964 a short career commission scheme was introduced. In 1965 a Royal Marines Special Entry Flying Service List was introduced, followed in the late 1960s by a university cadetship scheme.

There are clear trends toward collocation and amalgamation on a single-entry point for all officer aspirants and tailoring entry and training schemes within the single-entry point to particular entry groups. However, there is less evidence that these actions were part of any larger, conscious plan to improve the systems of officer education and training. Rather, they appear to have been most often reactive responses to unanticipated events, or anticipated pressures (particularly to save money) either internal to the armed forces or in the broader society. Equally, it seems that some of the negative or even the positive consequences of these mergers were not entirely expected or necessarily intended.

For example, in the case of the Sandhurst-Mons merger, recruiting pressures prompted the collocation. While Sandhurst elevated its entry standards to compete with the universities, Mons offered young men an easier and quicker route to a commission. As a consequence, Mons increasingly attracted more officer candidates while Sandhurst attracted fewer. Because the number of those seeking regular commissions through Sandhurst declined, Mons graduates were able to convert to regular commissions with comparative ease and without disadvantage. It became untenable to maintain the two separate institutions. However, once the decision was taken to collocate short and regular commission training at Sandhurst, the possibility of achieving an undergraduate educational institution status was, in reality, lost because of the differential academic entry standards for short and regular commission entrants.

Rationalization and Standardization

Once the process of amalgamation was well established in each service college, it became increasingly apparent that some sort of order needed to be imposed upon the complexity of operation created by multiple-entry schemes. This process had two phases. Training subjects for each entry scheme were standardized. Then the number of schemes was reduced to one or a small number of common training programs. The process was most evident at Dartmouth which had admittedly the greatest diversity of entry schemes. By the early 1970s, common components of training for all entrants had been settled, with virtually all entrants undertaking common courses or naval general training and training on the Dartmouth Training Ship. With standardized training packages, the task of reducing the number of different types of entry was thus simplified. Where it was necessary for some groups of officer entrants to undertake more or less training than others, standardized training modules could be simply added or subtracted from a common core.

By the late 1970s, all the separate training programs for RAF branch officers, university cadets and graduates, and women officers, had been merged into one Single-Gate Initial Officer Training Entry Scheme at Cranwell. At Sandhurst, most standardization and rationalization took place in Victory College which had become the repository for all courses other than the Standard Military Course (for school leavers) and the Regular Careers Course. In 1972, the university cadetship and university graduate entry schemes were merged to create a Standard Graduate Course. The final merger came in 1992, when the separate commissioning courses for school leavers and graduates were combined into one common commissioning course.[12] A similar rationalization occurred in Royal Marines officer entry schemes.

A common feature of these rationalizations was an almost continuous reduction in the length of training programs, particularly the longer commissioning programs. By the mid-1970s, these courses had been reduced in length by between two-thirds and three-quarters in some cases. In 1985, the RAF College's initial officer training course was just eighteen weeks in length; Sandhurst's Standard Military Course was twenty-eight weeks, and Dartmouth's Naval College Entry (Supplementary List) course thirty-nine weeks. There were many shorter-length courses—a nineteen-week course for special duties officers (recruited from the Lower Deck), and a thirteen-week course for Women's Royal Naval Service officers.

This was partly a consequence of the decision to remove the tertiary educational content from many entry programs. However, it was also very much in response to financial pressures from and on sponsors to reduce costs. The 1970s at Sandhurst, for example, could only be regarded as the decade of "weekly adjustments" as the proponents of "streamlining" used the technique of continual institutional reviews and other tactics to ward off college defenders seeking to hold or even to restore course lengths. As with most decisions, it proved to be easier to shorten than to retain or expand training and education programs. Having accepted that more, if not ultimately the majority of, officer aspirants would enter the services having completed some form of university studies, entry-level training courses became restricted to the minimum sufficient to permit officers to proceed to their service units in what can be described as a "safe" but not necessarily a fully competent condition.

The service colleges, particularly Sandhurst, resisted the "salami-slicing" of course lengths, and continued to put up proposal after proposal to restore them. Decisions on officer education and training, however, tended to flow from an unequal debate between a service college and the sponsoring service staff directorate.

Also, there were only limited feedback mechanisms for measuring the effectiveness of newly appointed officers that might have indicated whether or not course lengths and content were satisfactory. Users of the service college product tended only to be informally and not always comprehensively consulted through various reviews. Therefore there were delays and gaps in information filtering back indicating that, in some cases, courses were too short to provide young officers with some skills, knowledge, and competencies regarded by their parent units as essential. Moreover, because the nature of junior officer appointments called more immediately upon technical and tactical skills, shortcomings in the skills and knowledge of tertiary-level educational experiences were not likely to become evident until some time into an officer's career.

In many instances, only when receiving units were able to provide feedback on the capabilities of newly commissioned officers were the issues, raised by the service colleges, validated and addressed. In the early to mid-1980s, consequently, there were a series of "claw-backs" in the length of entry officer training courses. This occurred, for example, at Sandhurst where the Standard Military Course was lengthened to a year, at the expense of the Regular Careers Course which was severely truncated.

Clearly, resource pressures were a key driver in the process of amalgamation and rationalization. It also reflected the resolution of competing pressures in favor of initiatives supporting a sustainable throughput of new officers to the services over those supporting precommissioning education. At the same time, it is evident that another philosophy threaded through the process. This philosophy represented the balance of judgements by the services and officer entry sponsors about how far initial officer education and training should go in preparing new entrants and what learning experiences should be provided by the profession proper.

Balancing Pre- and Post-Commissioning Preparatory Experiences

Both the British Army and the Royal Navy have a long historical precedence for limiting initial officer training to an absolute minimum. Not until the 1870s, for example, were army nontechnical officers required to attend a period of training at Sandhurst before joining their regiments. Moreover, because of the limited number of places at the college, a significant number of officers continued to "slip through" to their regiments by the circuitous route of securing militia commissions.

Both services strongly held that on-the-job learning was the best way to develop the military officer. Time spent in the artificial confines of a training institution was time not spent with the regiment or the ship at sea. It was felt that officers needed to be trained and socialized into their service culture and ways of functioning as young as possible, preferably in their mid-to-late adolescence. Indeed, the ranks of midshipman, second lieutenant, or ensign were recognized as in-service training ranks and young officers were treated accordingly.

This philosophy worked well when the demands of junior officership were comparatively uncomplicated and where the experiences of the young officer could be controlled and his more obvious mistakes preempted by experienced senior noncommissioned officers and tolerated by junior ranks. It also held true when officers were drawn overwhelmingly from social groups that stressed leadership development in early adolescent socialization and during secondary educational experiences.

By the turn of the century, however, it became increasingly difficult to sustain the philosophy of a solely service-based apprenticeship for young officers as warfare became more complicated and more technically sophisticated (always the case in the navy, but increasingly in the army). The costs of comparatively simple errors were seen to be rising and the services (particularly the army) were pressured to provide improved precommissioning preparatory experiences for officer aspirants. It became accepted that at least some components of the apprenticeship would need to be completed in a "safe" learning environment.

Despite this acceptance, and well into the post–World War II period, there continued to be an underlying, unresolved issue as to when an officer aspirant was ready to be released, as it were, to the service where the "real" learning could begin. The attempts of the service colleges to expand their programs to meet tertiary-level educational standards in the 1960s reawakened the power of this issue in a way in which the service colleges perhaps had not anticipated. The issue was not resolved in favor of service college expansion to provide an enlarged and higher level of educational experience for officer aspirants. Nor was it resolved in favor of getting officers to their receiving units before their adolescent socialization was complete.

From the 1950s through the 1980s, university-level educational opportunities significantly expanded in British society. Moreover, success in employment increasingly demanded tertiary-level qualifications. Students and parents alike placed a higher value on accessing such educational opportunities. It was increasingly evident to service recruiters that fewer and fewer officer aspirants were going to make themselves available prior to having achieved a tertiary-level qualification if the services were not prepared to offer such qualifications as part of their entry packages. By the 1990s, consequently, all permanent RAF officer aspirants have attained at least a baccalaureate degree, and nearly 80 percent of Sandhurst cadets and Dartmouth midshipmen have completed baccalaureate degrees.[13]

As a result, officer aspirants are undertaking commissioning training at later ages than their 1950–70s counterparts, and if their academic education and officer training are combined, are entering their services after some four to five years of postsecondary school education and training.

Conclusions

The aim of this paper has been to provide an understanding of the balance of factors, pressures, and tensions that have influenced the evolution of British officer education and training in the years 1945–85. In some senses, it is difficult to avoid the conclusion that the interplay of internal agendas has not always had the sole motive of producing the most appropriate learning experiences for the officer aspirant. Equally, external changes, such as the expansion of tertiary-level opportunities in the British education sector, have often preempted and narrowed the range of developmental options and choices for sponsors and the service colleges in developing officer entry programs.

By the 1990s, British officer entry training could be best described as vocational. It is assumed that those selected for commissioning have completed a suitable tertiary-level liberal educational experience elsewhere. Any aspirations to provide such experiences within the military establishment itself have ceased. Moreover, this situation is unlikely to change in the future, given the current Defence-wide philosophy of civilianization of activities not directly associated with the conduct of operations, particularly training.

There are some other features about the evolution of British officer education and training in 1945–85 that need to be highlighted. For example, throughout this period, to all intents and purposes, the systems for educating and training officers were almost under a state of perpetual review. There is no doubt that today and in the foreseeable future shorter and shorter change cycles will drive the need for more frequent evaluations of extant practices. In any era, however, a circumstance of continuous review must be regarded as not conducive to effective outcomes.

This sort of treatment leads to comparatively unhealthy attitudes on all sides. For those being reviewed, each subsequent review is judged with suspicion and instant resistance. Additionally, the incentives to generate improvements or changes internally are removed; sustaining the status quo becomes the only defensible action. For those conducting reviews, there is the pressure of all preceding reviews and the need to find something different to criticize or to institute. In such an environment, positions easily become polarized and unless the respective powers of reviewed and reviewer are disproportionate enough to force the issue, little is achieved as both sides resist each other. Even within shorter cycles of change, there is a need to permit a period of stability so that the effects of changes can be measured and evaluated.

Another feature of the evolution of British officer education and training is the apparent lack of a long-term strategy or development plan. Few of the changes that have been outlined here could reasonably be described as falling into a larger, overall vision of the direction of officer education—its desired outcomes. Many of the changes could be described as knee-jerk reactions to unanticipated circumstances that required immediate responses. Moreover, those responses, by focusing on a short-term problem, often could not cope with longer term, and unforeseen consequences. Furthermore, by choosing one response to a particular problem in isolation, some paths of future development, arguably, were prematurely closed off. Other paths of development became more valued in the resultant narrower field of possible options.

Notes

The content of this essay is the responsibility of the author and does not necessarily reflect the views of the New Zealand Defence Force.

1. Morris Cogan, "Towards a Definition of Profession," *Harvard Educational Review* 23 (1953): 36; and Richard H. Hall, *Occupations and the Social Structure* (Englewood Cliffs, N.J.: Prentice Hall, 1969), 76.

2. See Cathy Downes "To Be or Not to Be a Profession: The Military Case," *Defence Analysis* 1 (September 1985).

3. For convenience, the academy, the two colleges, and the training wing will be grouped as "the service colleges," unless treated and identified separately. These training establishments are not the sole education and training institutions for officer aspirants, although they are the commissioning sources for officers. Other service institutions include the Royal Military College of Science, Shrivenham, and until recently, the Royal Navy Engineering College, Manadon. Space limitations preclude addressing the role of these institutions.

4. As one observer noted: "Privilege of all kinds was under fire. Dartmouth which was in effect a dedicated, highly subsidised and expensive public school, was an obvious target. . . . The absolute monopoly of the private preparatory schools had been broken, but to many the College was still a citadel of privilege." Capt. J. J. S. Yorke, "The Entry and Training of Naval Officers," *Brassey's Annual*, 1959, 175.

5. E. L. Davies and E. J. Groves, *Dartmouth—Seventy-Five Years* (Portsmouth, England: Gieves & Hawkes, 1980), 18.

6. For more details, see Cathy Downes, *Special Trust and Confidence—The Making of an Officer* (London: Frank Cass, 1991), chap. 2.

7. Grey Rock, "The Royal Navy and Its Graduates," *Naval Review* 67 (July 1979): 220.

8. Professor (later Sir) Michael Howard and Doctor (later Sir) Cyril English.

9. Howard-English Report, quoted in Brian Bond, "The Labours of Sisyphus: Educational Reform at RMA Sandhurst 1966–1976," *Journal of the Royal United Services Institute for Defence Studies* 122 (September 1977): 38.

10. Sqn. Ldr. A. M. Newbould, "The Graduate Entry Scheme: A Milestone in Cranwell's Development," *Royal Air Force College Journal* 41 (July 1969): 29.

11. Downes, *Special Trust and Confidence*, 33–38.

12. House of Commons, Defence Committee, 9th Report, *Military Training* (London: HMSO, 1994), 12.

13. Ibid.

Combat Leaders First: Education and Junior Officers in the Israel Defense Force

Gunther E. Rothenberg

The *Zava Haganah le Israel,* the Israel Defense Force (IDF), differs in some respects from the other forces discussed in this volume. For one, while the Israeli soldier is an inseparable part of modern Israel, he is a historical oddity. During the two thousand years of their diaspora Jews rarely were warriors and Zionism, the late nineteenth-century Jewish nationalist revival, contained strong pacifist elements. The creation of various, for most of the time illegal, self-defense organizations by the Jewish population in what then was British administered Palestine in the post–World War I period, came as a reluctant response to political, social, and security needs brought about by Arab resistance to Jewish immigration.[1]

From the mid-1930s on, these organizations, especially the *Haganah,* grew in numbers to assume the character of an underground militia. They acquired the rudiments of a general staff, established larger territorial units, and introduced rudimentary training, including the famous section-leader course. During World War II these organizations, above all the *Haganah,* numbering about 40,000 men and women, formed part-time elite shock formations, the *Palmach,* about 3,000 strong, which—when there appeared a threat that the Africa Corps would penetrate through Egypt into Palestine—enjoyed a short-lived cooperation with the British Army.[2] In addition, initially reluctantly, the British Army accepted some 25,000 volunteers from the 600,000-strong Jewish population, and in late 1944 formed a Jewish Brigade Group, which, though its senior command, staff, and specialist positions were filled by British officers, provided more formal training in the handling of heavier weapons, larger size units, and the theory and practice of combined operations.[3]

When after 1945 pressure intensified for immigration and the establishment of a Jewish state, hostilities between Jews and Arabs increased. Britain declared its readiness to give up its mandate, and in November 1947 the United Nations voted in favor of creating separate Jewish and Arab states. Fighting escalated in December, forcing the Jewish command to commit larger *Haganah* and *Palmach* formations to secure Jewish settlements and their exposed communications against attacks by Arab irregulars. On 14 May 1948, the last British troops and officials left Palestine. The same day the State of Israel was proclaimed, regular forces from the neighboring Arab states entered the conflict. The War of Independence, December 1947 to January 1949, was well underway when on 26 May the new Israeli government decreed the establishment of a regular military establishment, the IDF, which absorbed, not without bitterness and even some bloodshed, the various former underground forces.[4]

147

Even so, the IDF retained many of their norms and standards, including leadership by example, the follow-me ethos, close personal relationships within units, and a strong streak of egalitarianism, manifested in little attention to military formalities and with emphasis on combat effectiveness, initiative, and personal bravery.[5] These characteristics have persisted to the present day.

Following the War of Independence, in the summer of 1949, there was a debate regarding the formal structure and character of the IDF. Rejecting calls for a small, highly trained and politically indoctrinated force on the *Palmach* pattern, backed by a territorial militia, Prime Minister Ben Gurion and his young Chief of Staff Yigal Allon, prevailed on parliament to enact the Defense Service Law of 8 September 1949, which repeatedly amended, still provides the framework for the IDF. Realizing that security would remain Israel's main concern, the law attempted to strike a balance between security requirements and limited manpower and financial resources. Based on a universal service obligation for all Jewish citizens and permanent residents, the force is divided into three components: the compulsory service (*sherut hova*) into which virtually all men and women aged eighteen are drafted, the career service (*sherut keva*), and the reserves (*miluim*). The first two, reinforced by reserve units doing their periodic training, constitute the active army, while the entire reserve is called up in an emergency to provide the bulk of operational forces.[6] Although the IDF is still described as a unified force, in practice both the Israeli Air Force (IAF), and the naval forces have assumed more independent roles than originally envisaged. Above all, since the 1960s the IAF has assumed a leading role in this triad.[7]

Immediately, however, the most pressing problem was dealing with the wave of postwar immigration that doubled the population during the next few years and enlarged the ranks of the IDF to almost 200,000 men and women. Many of the new immigrants were survivors of the concentration camps or came from backward North African and Middle Eastern societies, often very poorly educated, and all in need of considerable rehabilitation. Utilizing this manpower presented a serious problem. While the native born and the better educated recruits of European background were assigned to the technical services, most of the new immigrants were sent to serve in the infantry where their combat performance was less than acceptable. Finally, given the pressing social needs, there was little money left for the IDF where pay was low, living conditions harsh, and weapons, generally World War II surplus and wearing out. By 1950–51, the IDF was in danger of becoming a second-rate force.[8]

Among other problems that had to be addressed were the introduction of uniform standards and doctrine. The War of Independence had been fought with commanders of diverse backgrounds. The majority had come up through the *Haganah* and the elite *Palmach,* others had experience in the British, Canadian, American, and other foreign armies. Moshe Dayan recalled that as an underground force the *Haganah* had been compelled "to operate in small units. Only during the War of Independence had former platoon commanders found themselves commanding a company, and company commanders a battalion and even a brigade or more."[9] In the future they would have to lead even larger formations. In an attempt to standardize procedures and doctrine, Col., later Maj. Gen., Chaim Laskov had begun in 1949 to develop a new officer training course on the wartime British model—the aim was to train some 20,000 officers in eighteen months. There were considerable difficulties: opposition from the more independent minded *Palmach* veterans who objected to regimentation and, perhaps more important, a lack of suitable instructors and manuals. With most of the officers not having competed their secondary education, Laskov remembered that his first effort was "to teach many officers English so that they could read foreign military literature."[10] Still the crash program produced results. During the Suez War of 1956, the IDF fielded seventeen brigades. Though at the higher command levels there were some shortcomings, at the company and battalion level this army of hastily mobilized reservists exceeded expectations.[11]

This success was achieved even though during the years of austerity, perhaps until 1956, there was a substantial exodus of experienced officers needed to train the new formations from the army. One estimate put the shortage of junior officers for the standing army and the reserves at close to 10,000 lieutenants and captains, often men who had served in the pre-state underground as well as in the War of Independence.[12] Senior officers, by contrast, were willing to stay on. To halt the exodus, and within the limits of austerity and in the face of objections that he was destroying the egalitarian character of Israeli society, Yigal Yadin, IDF chief of staff, attempted to improve conditions for officers. He could not provide more pay, but tried to give career soldiers both psychic and material compensations—higher status, smarter uniforms, separate messing facilities, and family housing on military bases. Even then, compared with other military career establishments, the Israeli officer retained a simple lifestyle; patriotism, rather than material rewards, kept him in the service.[13]

At the same time, concerned about the difficult integration of immigrants from non-Western countries into Israeli society at large and the IDF in particular, Ben Gurion ordered Maj. Gen. Mordechai Makleff, chief of staff in 1952–53, to make a major effort to prove that an accelerated ten-month course, could turn the new immigrants into effective officers. About 150 candidates were chosen, and highly selected instructors assisted by sociologists and psychologists made considerable efforts to turn them into good leaders. The experiment was a failure and was not repeated. In all it produced fifteen sergeants, but not a single officer.[14] Clearly, there were no short cuts—producing substantial progress among this population group would be slow and take time. Only in the last few years have substantial numbers of officers, regular and reserve, emerged from this population group.

In 1953 IDF morale needed a boost. When Moshe Dayan took over as chief of staff, he set a new tone. "Officers of the Israeli army," he told a graduating cadet class, "do not send your men into battle. They lead them."[15] Commanders did not have to be insured against becoming casualties; the most important matter for any unit was to attain its objective. To improve combat performance Dayan ended the practice of assigning better educated soldiers to staff or office jobs. Instead he ordered that combat units—paratroops, armor, special forces, as well as the air force—were to have their pick of the best available manpower.

As the IDF stabilized, it rejected the system common to other armies under which officers are trained separately from the rank and file, and it discounted all civilian academic qualifications. It was, and remains, a firm principle that there would be no social or educational qualifications to enter the IDF officer corps and, except for a handful of academic specialists, as well as pilots and naval officers, all officers would enter the army as conscripts and initially train together with the men they might command in the future.

The IDF had begun to develop its new procedures for officer selection and procurement based on a series of tests, including a major medical and psychological work-up (later, of course, validated and further refined).[16] The IDF did not establish its own military academy; this would have been too great a departure from the egalitarian ethos of Israeli society. Still, there was one experiment. In 1952 Yadin proposed that Israeli high schools offer a special curriculum for future officers, but this was widely opposed by educators and political leaders. In the end only two high schools, the Beth Sefer Reali in Haifa and the Gymnasia Herzlia in Tel Aviv, allowed students to choose special military courses taught in army-sponsored boarding schools, but this instruction had to be taken as a supplement to the regular curriculum after day classes were finished. Moreover, graduates were not automatically commissioned, but entered the army as conscripts and underwent the common selection process leading to eventual commissioning.[17] The two boarding schools (*pnimiot zevaot*) still exist, but their influence has been marginal.

Initial screening of future officers takes place during the induction process common to all conscripts. At age seventeen, all young men and women register for service, and following their eighteenth birthday, high school graduates or not, are called up for compulsory service—three years for men and two for women. Deferments for university study in certain fields are given, but these do not lead automatically to a commission. Regardless of future branch or service, all inductees report to the basic training center, the *BAKUM*, at Tel ha Shomer near Tel Aviv for classification and assignment. Here they pass through a thorough medical and psychological screening which combined with measures of individual intelligence, knowledge, education, mastery of Hebrew, and projected motivation and adjustment to combat service results in an overall quality score, the *KABA*.[18] Volunteers for pilot training, special forces, and the like undergo additional screening. The *KABA* has fourteen quality groups—only those in the six highest groups can reach commissioned rank, while the next four groups qualify as potential noncommissioned officers (NCOs).

Inductees with a high *KABA* score are earmarked as possible junior officers and, except for future pilots, seagoing officers, and a few academic specialists, begin a three- to five-month basic training cycle if slated for combat service, and only one month if not. At the end of the cycle they are assigned to their units where they will again be evaluated for their potential on the basis of an efficiency report by their immediate superiors, peer ratings, and another look at their *KABA* score. If found qualified, the soldier now is asked to volunteer for officer training. Most, usually over two-thirds of those eligible, agree. Those who refuse may well do so because they obligate themselves to an additional year of service, thus falling behind their peers in starting their civilian careers. On the other hand, since most officers will become reservists after five years of active duty, and Israeli society, not unlike the Swiss, still regards military service as a vital rite of passage necessary for full citizenship and as a basic career requirement, there has not been a shortage of volunteers.[19]

For those who volunteer to accept the responsibilities of leadership, including the fact that the various IDF officer corps proportionately suffered four times as many casualties as the rank and file, there follows a demanding junior command course stressing leadership and technical skills. During this three- to four-month course, the participants are trained in small groups, subject to constant observation. The average pass rate is between 30 and 40 percent, with those who wash out being returned to their units and their active service obligation cut back to three years. Those who pass serve for several months as NCOs on active duty and may see combat service. After further screening, including recommendations by their unit commander, peer evaluation, and psychological tests, approximately at the end of their first year of service, they are classed as cadets and sent to officer basic training.

Since 1977, this process has been somewhat shortened. All officers still begin as privates, and take the junior command course, but their time as NCOs before they proceed to a branch-specific Officer Basic Course has been shortened. Courses by the various branches and corps all are conducted at Training Base 1 (*BAHAD ALEPH*) which thus serves as a unifying element for the IDF. Even so, there is some specialization. The Infantry Course (*KHEER*), designed for paratroops and infantry officers, lasts for six months; the Combat Arms Course (*AGAM*) for future armor, artillery, and combat engineers lasts for three months and is followed by another three-month course in the cadet's branch. Finally, for noncombat officers there is the basic course, *BASSISI*, lasting three months. Cadets are commissioned following successful completion of their courses.[20]

Courses are tough with a failure rate of up to 50 percent; they provide the young officer with confidence based on competence. Throughout, the lieutenants, not the sergeants, instruct and lead the men. In contrast with many other Western armies, when a new second lieutenant joins his unit he will not have to rely on the guidance of the senior NCOs. Unlike the British Army, an organization to which the IDF had considerable exposure, there does not exist a large, stable, and senior NCO corps in the IDF. For one thing, few NCOs will opt to join the career service (and then in technical or administrative positions), and for another, the IDF junior officer is better trained and more experienced than his NCOs. He has served as a private and as an NCO, he may already have seen combat, and he instructs the sergeants.[21] The IDF stresses unit cohesion and, as much as possible, follows the cohort principle with officers, NCOs, and enlisted men remaining with their unit, or at least within their branch, during their entire service.[22]

The time in grade for second lieutenants normally is one year, after which they are promoted and assigned positions as deputy commanders or executive officers. At the end of his second year of commissioned service, normally at age twenty-two, the young officer can either take his discharge and transfer to a reserve unit, or apply for a position in the career service where, with a total of five years active duty, he may expect promotion to captain. About 90 percent elect to be discharged and are transferred to a reserve unit where, with emphasis again on unit cohesion, they will serve until age forty to forty-five in combat specialties and an additional ten years in a noncombat assignment. Reservists attend professional schools and are promoted from platoon to company level (captain) and, if rarely, to battalion command (major).[23] The great majority of company commanders in the IDF are reservists; senior reserve officers normally are former members of the career service.

Those junior officers who do volunteer and are accepted in the career service, show a strong preference for service in elite units—paratroops, special forces, and the like, with pilots presenting a special case—and will often sign on for a few additional years in the career service, not because they intend to make the service their career (they fully intend to seek careers in civilian life), but out of a sense of obligation to the state. Such attitudes, accepting greater responsibilities and risks, appeal to members of the Kibbutz movement, a group comprising only 3 percent of the population, which has provided a disproportionate number of officers (and also has carried a disproportionate casualty rate—25 percent).[24] Also well represented are sons of prominent and highly educated families. Perhaps the most famous example of the latter was Lt. Col. Jonathan Netanyahu, the older brother of the current prime minister, who on his second tour in the *sayeret matkal,* the general staff special service force, was killed leading the assault forces during the raid on Entebbe Airport in 1976.[25]

It should be noted, however, that since 1982, there has been a marked drop, from 40 to 20 percent, in the willingness of Kibbutz sons to extend their service, reflecting the lack of consensus caused by the war in Lebanon.[26] Overall, the career service's prestige and esteem has eroded somewhat. Though still composed of highly proficient soldiers, there is a perception that it no longer is a dedicated brotherhood serving the nation, but just another group, and these days a well-paid one, motivated by career objectives.[27]

We must now briefly turn to the exceptions to the general rule that every IDF officer's career begins as a private taking a basic rifleman course. The most important exceptions are pilots. Pilot entrants are screened prior to and during the first days of the induction process. If they pass they are sent directly to flight school. Provided they survive the high washout rate, they are commissioned when they complete combat flight training.[28] Those rated highest are trained as fighter pilots, the next group as fighter-bomber pilots, followed by transport and helicopter pilots. Pilots normally stay with their unit throughout their five years of service and will continue in the reserve. If they intend to remain in the career service, most will take academic leave after the first five-year term to complete a degree. By this time they will be captains or majors and pass out of the junior grades. The training of future seagoing naval officers proceeds along similar lines.

Another small group following a different entry pattern is the academic reserve. While there have not been, and are not, any formal educational requirements for becoming an officer in the IDF, there are certain positions—legal, medical, and technical—for which the army requires officers with formal degrees. Conscripts studying in these fields, provided they have passed the initial officer selection process, can defer completion of their military obligation to continue their university studies. They will, however, have to do military service during the summer vacation, including the NCO course, and when they have obtained their degrees, finish a shortened Officer Basic Course.[29] Only then will they be commissioned and serve for three years as specialist officers, although rarely holding command. In fact academic reserve graduates were used in an experiment in the mid-1970s when a group were assigned to platoon command in line units. The outcome revealed that their intellectual level, higher than that of platoon leaders who had come up the normal route, did not translate into outstanding or even effective leadership, and confirmed the IDF's belief that leading a combat unit required qualities quite different from academic ability and civilian schooling.[30] The experiment was considered additional proof that IDF junior officers, up to and including company-grade, did not require an advanced civilian education. Leadership by example, efficiency, initiative, improvisation, and bravery are still considered the cardinal virtues.

Although in Israel, as in other Western nations, the values and expectations prevalent in society have become more at odds with the values and expectations that are indispensable to any effective military establishment, the IDF has managed to maintain high combat capability. At the same time, the IDF has retained, especially in its reserve components, an essential militia character reaching back to the days of the *Haganah.* To be sure, in the aftermath of the 1973 Yom Kippur War some questioned whether the old forms of discipline, suitable to the War of Independence and the small volunteer *Haganah,* remained appropriate for the new threat manifested by the Egyptian and Syrian armed forces. In the end, however, the basic egalitarian spirit prevailed. Because officers and men will often serve for decades within the same unit, resulting in strong group cohesion, differences of status and rank are blurred, officers and men will address each other by first or nicknames, salutes are seldom given, and except for general officers, titles are rarely used.[31] But group cohesion improves rather than degrades the unit's combat capabilities.

Finally, while the IDF and its officers no longer enjoy the status they had twenty-five years ago, at the same time most Israeli citizens continue to perceive genuine threats to their national security, a perception that continues, especially among junior officers, to promote retention of many of the old values—those of the combat leader and not those of the military manager.[32]

Notes

1. For a survey of the earliest defense efforts, see Yigal Allon, *Shield of David* (New York: Random House, 1970), 11–73. Cf. Edward Luttwak and Dan Horowitz, *The Israeli Army* (London: Allen Lane, 1975), 1–52, and Gunther E. Rothenberg, *The Anatomy of the Israeli Army* (London: Batsford, 1979), 13–38.

2. For the intricate story of the relationship between the *Haganah/Palmach* and the British, see Yehuda Bauer, *From Diplomacy to Resistance* (New York: Atheneum, 1973), passim.

3. Avraham Tamir and David Karmon, "The Legacy of the Jewish Brigade," in Yigal Allon, *The Making of Israel's Army* (New York: Bantam Books, 1971), 250–57; Luttwak and Horowitz, 24–26.

4. Luttwak and Horowitz, 38–39, 43–45; Rothenberg, 61–62.

5. Allon, *Making of Israel's Army*, 50–51.

6. Rothenberg, 71–73.

7. See the account in Ezer Weizman, *On Eagles' Wings* (New York: Macmillan, 1977), 173–98.

8. Rothenberg, 76–84; Luttwak and Horowitz, 99–108; Z. Schiff, *A History of the Israeli Army* (San Francisco: Straight Arrow Books, 1974), 52–56.

9. Moshe Dayan, *Story of My Life* (New York: Morrow, 1976), 162.

10. Schiff, 52–53.

11. Moshe Dayan, *Diary of the Sinai Campaign* (London: Weidenfeld, 1966), 103.

12. Rothenberg, 79.

13. Ibid.; Luttwak and Horowitz, 83–84.

14. Schiff, 54.

15. Dayan, *My Life*, 178.

16. On the development of psychological assessment in the IDF, see Reuven Gal, *A Portrait of the Israeli Soldier* (New York and Westport, Conn.: Greenwood Press, 1986), esp. 76–95, and on officer selection, 115–40.

17. Luttwak and Horowitz, 84; Edward B. Glick, *Between Israel and Death* (Harrisburg, Pa.: Stackpole Books, 1974), 74–75.

18. Gal, 78–90.

19. Ibid., 116–24.

20. Ibid., 121–22.

21. Ibid., 119–21.

22. Ibid., 153–56.

23. Ibid., 121–22.

24. Rothenberg, 118.

25. Max Hastings, *Yoni, Hero of Entebbe* (New York: Dial Press, 1979), esp. 99–108.

26. Gal, 68–69, 124; Meir Pail, "The Israeli Defense Forces: A Social Aspect," *New Outlook* 11 (1982): 40–44.

27. Gal, 38.

28. Ibid., 126.

29. Ibid., 35, 128.

30. Ibid., 34–35.

31. Yigal Allon, "Profile of a Commander," in *Making of Israel's Army*, 286–91.

32. Martin Levin and David Halevy, "Israel," in Richard A. Gabriel, ed., *Fighting Armies: Antagonists in the Middle East* (Westport, Conn.: Greenwood Press, 1983), 7–8; Jean Lartéguy, *The Walls of Israel* (Philadelphia, Pa. and New York: M. Evans, 1969), 201–5.

Section IV
Radical Political Change and Officer Education and Training

Radical political change has often resonated throughout systems for selecting, educating, and training officers. The French Revolution, for instance, so dramatically altered entry into the French Army's officer corps that it became the most egalitarian in Europe in the nineteenth century. After the Revolution, the middle and lower classes predominated in what had been largely an aristocratic preserve; indeed, two-thirds or more of the army's officers would now come from the ranks.[1] Ability replaced birth as the basis for officership, creating an officer corps much more in tune with French society and the nation's revolutionary ideals.

This section's essays examine officer education and training in the context of extreme political change or turmoil—in Japan following 1868 and 1945, in twentieth-century Russia, and in China after the Communist takeover in 1949. They reveal the overall sensitivity of officer formation to external political struggle, the tensions generated by competing demands for ideological conformity and professional skill, and the close connections between officer education and training and society as a whole.

Roger Dingman, an associate professor of history at the University of Southern California, argues that between the Meiji Restoration of 1868, which overthrew the shogunal system, and World War II, Japanese leaders succeeded in molding professional officers loyal to nation and to emperor. On the other hand, they did not tie these modern warriors to Japanese society, in part by failing to instill an understanding of, or tolerance for, politics, and did not equip them with the skills and broad outlook needed to govern an overseas empire. After the forced political revolution of 1945 and the reconstitution of Japanese armed forces in the early 1950s, the architects of a new military education and training system continued to rely on traditional ideas and rituals to produce self-disciplined officers with "noble spirit" and character, but designed a structure and content (notably social sciences and humanities courses) that would decrease the likelihood of Japanese officers becoming separated from their society. Dingman, however, questions whether instruction in the social sciences and languages has expanded sufficiently to prepare Japanese officers for their recently acquired role as international peacekeepers.

In Japan before 1945, the narrowness of officer education pushed the armed forces away from society. In Russia after the revolutions of 1905 and 1917, and again in 1991, the armed forces and society also drew apart. They separated, contends Russian military historian David Jones, because the two did not share a "myth" involving a common vision, cause, or view of history that legitimizes a state and its government and inspires a sense of national commitment. Such a myth, he writes, furnishes the motivation needed to attract officer candidates, to educate and train them effectively (especially in developing essential qualities of character and leadership), and to justify their role in society's and in their own eyes. Jones believes a unifying myth embodying elements of traditional nationalism and patriotism will probably reemerge in today's Russia, just as it did after 1905 and eventually after 1917, to raise the officer's status in society and to revitalize officer education and training.

In China after 1949, as in the Soviet Union after 1917, ideological imperatives clashed with professional military requirements. When the Communist party came to power, the People's Liberation Army had successfully combined ideological commitment and military proficiency since its origins in the mid-1920s. William Heaton, a former U. S. Air Force Academy professor and currently a Central Intelligence Agency expert on the Chinese armed forces, describes the regime's attempts after 1949 (mostly, although not exclusively, evident during periods of political crisis) to maintain the military's loyalty and revolutionary fervor. These efforts included conducting intense political indoctrination programs at the expense of training, closing military academies and schools in favor of selecting more politically reliable officers from the ranks, and assigning officers to civic action projects among the people. These and other measures have slowed, and at times halted, progress toward a professionally competent officer corps developed through a modern system of education and training and prepared for today's high-tech battlefield.

Note

1. Paddy Griffith, *Military Thought in the French Army, 1815–1851* (New York: Manchester University Press, 1989), 16; and Douglas Porch, *The March to the Marne: The French Army, 1871–1914* (Cambridge, England: Cambridge University Press, 1981), 17.

Paths to Officership: Military Education in Japan before and after 1945

Roger Dingman

Military education challenges those who give and those who receive it to answer one basic question: What does it mean to be a warrior? If considered narrowly, that puzzle can be simply resolved. Military education provides the warrior with the abilities and technical skills needed to survive and prevail in combat. But if military education is defined in broader terms, then answering the question of what it means to be a warrior is much more difficult. For the warrior is a complex creature. He is not simply a technician of death. Nor is he simply a manager of organized violence in the service of the state. He is the defender of the state and an embodiment of the values of the society that sends him into combat.

Military education must teach him *what* it is he is defending or fighting for as well as how to do so. The warrior-to-be must train and discipline not just mind and body but spirit as well. If he is to exercise the power of life and death over those he commands and those he opposes, he must know what he is about and why he is about it. And if he is to enjoy the support of the society that he serves and represents, he must resolve the warrior's paradox: How can one be simultaneously apart from society by virtue of one's license to kill, and yet a part of that same society?

The challenge of military education, then, is great. How has Japan tried to meet it over the last half-century? That question cannot be answered by focusing exclusively on developments in Japanese military education since 1945 because the structures used to provide it, the ethos of service to state and people it evokes, and the understanding of what its substance should be have their origins in the previous century. One cannot hope to see what is new in Japanese military education without comprehending what is, in fact, quite old about it.

The history of military education in Japan over the last century and a half can be seen as a series of responses to different challenges, each germane to its age but all concerned with the same basic question of what it means to be a warrior. In the first half-century, the challenge was to produce a warrior loyal to the nation who would defend, not overthrow, it. During the next fifty years, the problem was to create a warrior able to lead ordinary citizens in conquering and defending a vastly expanded imperial realm. In the years since 1945, Japan has struggled to produce a warrior capable of meaningful action in a world of alliances and nuclear weapons who is also acceptable to a population deeply divided over all things military.

How has Japan striven to meet these challenges of military education? What effect have the successes and failures of one era had upon the character of military education in the next? And what, in the end, does this century and a half of history suggest about the prospects for Japanese military education in the twenty-first century?

To answer those questions, it is essential to recall that modern Japan began as a revolutionary state. When Commodore Matthew Perry's "black ships" entered Edo Bay in 1853, two truths became painfully evident to the Japanese: they lacked the modern arms and know-how to keep out foreigners, and their government was woefully unable to devise a strategy for doing so. That realization was the taproot for revolution a decade and a half later when the leaders of four western feudal domains combined forces to overthrow the shogunal government and "restore" the emperor to power. That was easily accomplished in 1868, but carrying out the real revolution—turning a collection of feudal domains loosely united under a shogun into a modern nation able to defend and to expand its emperor's realm—was quite another matter.

No man better understood the nature of that challenge than Yamagata Aritomo. In 1869, this samurai-turned-subversive-turned-architect of a new order became the first Meiji leader to go to Europe. He returned a year later convinced that state, soldier, and society must change if Japan were to survive as an independent nation: His country needed strong armed forces, equipped with the latest weapons, and led by men who knew how to organize and to use both, if it were to fend off foreign foes. But Japan's most pressing need was for a new kind of warrior—a man unshakably loyal not to an individual domain leader but willing to die for emperor and nation.[1] Producing that warrior, and a society and polity that would support him, would take nearly two decades.

Yamagata began this new warrior's construction during the 1870s by making radical changes in Japanese military life. In 1871, he engineered the abolition of the feudal domains to which samurai owed loyalty and created a new imperial guard to counterbalance their power.[2] Between 1873 and 1875, Yamagata broke the samurai's monopoly on the knowledge and possession of deadly force, first by instituting conscription that potentially made every able-bodied male a soldier, then by forbidding the once-proud samurai to carry and use their swords.[3] Then, as if to say revolutionary change had gone far enough, he revised the system of military governance by separating administrative and command functions.[4] In that scheme of things only he, as minister of war (and his navy counterpart) would play the game of politics.

These radical changes in the definition of who might become a warrior were accompanied by equally significant changes in military education. Like military service itself, it had to be nationalized. The last shoguns had tried to accomplish that, but by the time Yamagata and his colleagues overthrew the shogunal regime, military education in Japan was neither centralized nor standardized. Although there were shogunal army and navy schools, individual domains continued to employ different foreign advisers—Americans, French, and Dutch—to instruct their samurai in the techniques of modern warfare. In an attempt to offset their power, Yamagata got the new Meiji regime to spend more on what had been the shogunal school for soldiers and sailors at Tsukiji in Tokyo.[5] He wanted students there to have the best instructors, not just foreign teachers of the same nationality. That meant replacing the French, who had just lost a war to Prussia, first in 1872 with British instructors for naval cadets, and then, somewhat later, with Prussian teachers for budding army officers.[6] To further improve their education and to reduce young warriors' tendency to attach themselves to an older master, military and naval institutions were divided. Aspiring young warriors, age eleven to sixteen, went to preparatory schools for two to five years of post-elementary training before entering the military or naval academy for three to four more years' education.[7]

These reforms did not produce the competent, national, and loyal force that Yamagata and the Meiji government wanted. Japanese expeditions to deal with unruly Koreans and the aboriginal Taiwanese in 1873 and 1874 ended in failure. Saigō Takamori, leader of the domain that provided most of the forces used to overthrow the shogunal government, revolted two years later. Yamagata's centrally controlled army required two years to cow the rebel force, and when Saigō committed suicide to atone for his defeat, he became a national hero. Worse still, the supposedly loyal imperial guard revolted in 1878.[8]

These disasters forced Yamagata and the Meiji government to take a variety of steps to insure the warrior's loyalty. Some were designed to inculcate emperor-centered nationalism in the soldier- or sailor-to-be. In 1878, Yamagata issued his Admonition to Soldiers, which extolled loyalty, bravery, and obedience while warning against involvement in politics. Four years later, he persuaded the Meiji emperor to issue a rescript to soldiers and sailors. It commanded those who bore arms to cultivate the virtues of loyalty, propriety, valor, righteousness, and simplicity; and it ordered them to eschew politics. In 1890, to further strengthen the loyalty of male subjects who might someday bear arms, the emperor issued a rescript on education that taught those same values to all Japanese.[9]

Other measures imposed structural constraints on the possible involvement of young officers or officers-to-be in politics. In 1886, the Council of State decided to move the naval academy from Tokyo to Etajima, a tiny isolated island near Hiroshima. The ostensible reason for the shift was to put cadets close to the naval station at Kure. Its practical effect, however, was to move sons of samurai, whose unruliness forced the navy to commission civilian instructors to transform them into strict disciplinarians, to a spot hundreds of miles from the capital.[10] In 1887 and 1888, the Army and Navy War Colleges, open only to a handful of top graduates from each service academy, were established.[11] Concentrating the study of strategy, history, and the art of leadership at that higher level further depoliticized junior officers in Meiji Japan.

Despite Yamagata's efforts to standardize military education while nationalizing and depoliticizing it, significant differences between army and navy paths to officership emerged. The military academy at Ichigaya never enjoyed a monopoly on educating and training junior officers. Later Prime Minister Tōjō Hideki's father, for example, rose from the ranks and retired as a lieutenant general. Domain of origin and sponsorship by a ranking, serving officer strongly influenced where the aspiring cadet began his military education. He might go straight to the central preparatory school attached to the military academy at Ichigaya; or he might progress from regional preparatory schools (relics of the domain schools) to that central facility before capping his military education with eighteen months' study at the academy. The latter was the course followed by the younger Tōjō and most other aspiring army officers of his generation.[12] But the navy dropped domain sponsorship as a prerequisite for admission to its academy when the move to Etajima was made. From 1888 onward, successfully passing rigorous, competitive and comprehensive, academic and physical examinations was the starting point on the path to becoming a junior naval officer. One simply had to get into Etajima and succeed there to become a sublieutenant, second class in the Meiji emperor's navy.[13]

The aspiring junior officer's educational experience also differed considerably according to service. The military cadet's studies were shorter and much more likely to be punctuated, or even interrupted, by training with troops in the field. By the turn of the century, those in the regional preparatory schools gathered to compete in tough exercises that culminated in a "great march" meant to separate the men from the boys. The military cadet was also not assured of a commission until he had completed six months as a probationary officer in the field.[14]

The Etajima cadet, by contrast, spent three years in one isolated place with only brief summer and New Year's holidays. His experience was much more academic than that of the army cadet. Not until his final year did professional subjects—gunnery, torpedoes, navigation, seamanship, and engineering—take more of his study time than academic disciplines. A single foreign language, English, remained preeminent among the latter. The naval cadet had to understand it because, thanks to the presence of British instructors, he was taught professional subjects from texts written in English. His hands-on training came in his last year of study, eight months of which were spent on a training cruise that took him to both nearby East Asian and more distant Western countries. The capstone test for the potential junior naval officer was successfully completing the last four months as a midshipman aboard a ship in regular naval service.[15]

These variations in education interposed subtle but important differences between army and navy junior officers by the turn of the century. While both cultivated the same virtues, served the same emperor, and faced lives of genteel poverty, the relationship of each to society had become quite different. Graduates of Ichigaya or other army schools were less likely to have shed domainal loyalties; they remained more attached to their own and their subordinates' rural roots than did junior naval officers. They were definitely a part of the society from which they had come. Etajima products, by contrast, were self-conscious elitists. Four years of intensely competitive education and isolation from society at the naval academy as well as at sea, together with steady exposure to things foreign, changed them in ways that set them above and apart from Japanese society.

Regardless of these differences, the junior officer became what Yamagata Aritomo most wanted him to be—a respected and trusted member of society rather than a threat to it or to the Meiji polity.[16] Moreover, his performance in battle—first in the 1894–95 war against China, then a decade later against Imperial Russia—left no doubt of his professional competence. In 1905, three and a half decades after Yamagata began its reformation, no one doubted that Japan had met and mastered the challenge of military education.

No one, that is, except Yamagata's protégés and their navy counterparts. They confronted a new, threefold challenge. In one sense that challenge was societal. The military and the people had to become as one if Japan were to meet the demands of modern total war. In another sense, it was technical. The early twentieth-century warrior would have to master a much wider array of more complex weapons—the machine gun, the tank, the submarine, and the airplane—than his predecessors if he were to prevail in battle. But in its most fundamental sense, the challenge that Japanese military educators faced during the four decades between the end of the Russo-Japanese and Pacific Wars was international. Japan, the aggregate of feudal domains, had become a nation state. After 1905, however, that nation would become a continental and maritime empire whose warriors needed tools for understanding the peoples whom they conquered and presumed to defend no less than the latest weapons of war.

How did Imperial Japan's leaders try to meet this challenge through changes in military education? Their response to its societal aspect appears to have been schizophrenic. On the one hand they renewed emphasis on cultivating character. The Japanese officer had to be spiritually superior if he were to prevail over foes that, like China and Russia in the past, could be expected to be materially superior. Military educators at Ichigaya and Etajima developed such a character in their charges in three ways. Cadets had to listen reverently and attentively to *kun'iku*—moral instruction given by academy superintendents and distinguished visitors at the beginning of and regularly throughout the school year. Memorizing and reciting the Imperial Rescript to Soldiers and Sailors supplemented that form of moral instruction. Giving praise and honor to predecessor role-models was another form of character education. The heroes of the wars against China and Russia, for example, were enshrined at Etajima in the Great Hall whose walls listed the name of every naval officer who had died in the line of duty.[17]

Self-scrutiny provided yet another means of character development. By the early 1930s, naval cadets were required to answer five questions at the end of every evening's study period: Had they always acted sincerely, spoken and acted carefully, and worked vigorously? Had they spared themselves any pain or, worse still, been lazy? Presumably the answer to the first three questions was yes, and that to the last two vigorously no.[18]

The officer-to-be, especially if he wore an army uniform, was expected to reach out and to lift up Japanese society. Beginning in 1908, under the leadership of Tanaka Giichi, arguably Yamagata's most important protégé and like him, eventually both war and prime minister, the Japanese Army set out to make itself "the school of the people."[19] In 1910, the army introduced the Imperial Military Reservists Association, half of whose members were civilians who had never served in the military, as its first effort to educate civilians. Over the next decade and a half, it established first the Greater Japan Youth Association and then a nationwide system of youth training centers. These organizations, aimed at the 80 percent of the Japanese male population with no more than an elementary education, were designed to produce "physically and spiritually fit citizens" who could readily become soldiers. In 1927, the conscription system was modified to shorten the term of service and to allow the best conscripts to enter an army officer candidate school and become reserve officers. The youth training centers, the presence of serving military officers in high schools, and the fact that no less than six generals served as ministers of education before 1945, all attest to the success of the "militarization" of civilian education.[20]

Its side effects on junior officer education and training, however, were far less healthy. Because so much emphasis was placed on civic education and civil-military relations at the secondary school level, those subjects had no place in the curricula at Ichigaya and Etajima. Aside from whatever insights they might draw from the bit of Japanese history they were offered, cadets got no higher education in politics. That may have been appropriate in the early, shaky years of the Meiji regime. By the 1910s and 1920s, however, when political parties contended for power at home and new, challenging ideologies—democracy and Leninism—flooded in from the West, it left young officers who might someday govern with little understanding of or tolerance for politics. What did not happen in the classrooms at Ichigaya and Etajima in these earlier years goes a long way to explaining "radical" junior officers' participation in abortive coups during the 1930s.[21]

Japanese military educators were much more successful in meeting the technological challenge that confronted them. In aviation matters, for example, they followed the pattern set in the early Meiji years. In 1911 and 1912, both services sent men to Europe and America for pilot training, but their own subsequent attempts to train aviators and Japanese engineers' efforts to build military aircraft yielded little progress. Thus, in 1919, a team of French aviators came to train army pilots, and two years later a British group arrived to do the same for navy men. By 1925, the army had enough trained men and aircraft to establish an independent aviators' corps that was coequal to other specializations within the service. Two years later the navy followed suit, creating an aviation department in 1927 and a tender squadron the following year.[22]

Increasing weapons complexity and variety led both services to develop specialized schools for post-academy training of junior officers. In the decade after World War I, the navy added a submarine school to the five other specialized institutions it had developed for engineers, doctors, paymasters, and gunnery and torpedo experts.[23] The number of army schools, apart from the three primary officer preparation institutions and the prestigious staff college, jumped from ten in 1918 to seventeen a decade later.[24]

These specialized institutions and the unique training experience provided by the powerful Kwantung Army in Manchuria[25] undoubtedly enhanced the professional competence of Japanese junior officers, but they also accelerated factionalization within the armed services.[26] The intense learning experience they provided helped bond individuals to seniors within the same subspecialty and created loyalties that, on occasion, transcended the importance of service ties. In some instances—the Kwantung Army for one, naval aviators for another—these ties rivaled those of domain or type of basic officer training in the army or Etajima class in the navy as determinants of subsequent thought and behavior. Technically trained specialists who lacked civic education commensurate with the military's role in the Japanese polity all too easily became "radicals" willing to act independently in either the military or the political sphere.

Japanese military education leaders failed utterly to confront the challenges of internationalization that possession of empire presented. In 1918, Communists began their takeover of Siberia; the following year students in China and Korea rose up to protest Japan's colonial rule. These events, together with the acquisition of Germany's Central Pacific islands in 1919, should have been a wake-up call for broadening junior officer education in the social sciences. But neither Ichigaya nor Etajima heard the call to begin the warrior's acculturation to the challenges he would face on the frontiers of empire.[27] Foreign language training remained a small and specialized element in both academies' curricula. The navy continued to employ Britons to teach words and manners, but they had too little time with students to open their minds to British and American ways of thinking.[28] The army literally rationed language education. While those preparing for Ichigaya were allowed to study French, German, and Russian, only Ichigaya cadets could try Chinese or English—the language of the people whose land they would occupy for fifteen years and the tongue of their most formidable foes in the Pacific War.[29]

Prior to 1945, then, the junior officer in Imperial Japan was a warrior at best imperfectly educated for the tests of battle that lay ahead. Competitive long marches or a ten-mile swim together with intramural sports put him in superb physical condition.[30] Spiritual education gave him a strong character and an even stronger sense of the superiority of the empire he served. He was also trained to use the most modern weapons in ways that would make him a formidable foe in the Pacific War. But his education increasingly set the young Japanese officer apart from and above the society he was pledged to defend. Although they would never have been so crude as to sing it, Etajima graduates shared the ethos that a popular army cadet song proclaimed: "In the muddy stream that is the world, . . . [shit] rises steeply to the heights of Ichigaya. The world is a muddy stream, and only we are pure."[31] Officers who saw themselves that way proclaimed their ignorance of their own nation and of the world. Little wonder, then, that from 1937 onward they proved singularly ill fitted to conquer, govern, and defend an ethnically diverse empire that was nearly six times as populous as the prewar French and Dutch empires and almost three-quarters the size of the British Empire.[32]

The Pacific, or Greater East Asia, War of 1937–45 destroyed the empire Japan's young officers fought to conquer and defend, shattered the polity and military system in which they served, and, at least for a time, ended military education in Japan.

The war strained the imperial military system to the breaking point, first by flooding it with too many men who could not be trained thoroughly enough to prevail in battle, and then by forcing it to become so harsh as to sever the bonds of trust and respect that existed between the military and society at large. That the *kempeitai*, the military police established by Yamagata Aritomo in 1881 to keep young officers from meddling in politics became, sixty years later, dreaded enforcers of ideological conformity and obedience to military authority in Japanese society at large is one measure of that change.[33] By the summer of 1945, Japan's military leaders were fighting to save their emperor from vanishing with the empire they had created.

Defeat was followed by a political revolution comparable to the one the architects of the Meiji Restoration had brought about ninety years earlier. The victors occupied Japan and set out to demilitarize and to democratize the nation. Although the emperor kept his throne, the army and navy that had served him were abolished and their officers purged from public life.[34] The occupiers forced Japan to accept a new constitution. Its ninth article proclaimed that "the Japanese people forever renounce war as a sovereign right of the nation and the threat or use of force as means of settling international disputes." Moreover, the new constitution pledged that Japan would never maintain "land, sea and air forces, as well as other war potential." Finally, in this reconstituted polity, "the right of belligerency of the state" was not to be recognized.[35]

This forced political revolution occurred at a time when the nascent Cold War was deepening in ways that soon altered dramatically the desires of Japan's most powerful conqueror, the United States. Committed to disarming Japan in 1945, five years later Washington saw retaining bases in Japan as essential to its grand strategy of containing the Soviet Union. By January 1951, when the United States was fighting a very real war to do so in nearby Korea, John Foster Dulles came to Tokyo convinced that some measure of Japanese rearmament within the framework of a Japanese-American security treaty was the essential prerequisite for a peace treaty. Tokyo would have to find some way around Article Nine of its new constitution if Japan were to recover its independence and sovereignty. Thus when Prime Minister Yoshida Shigeru signed peace and security treaties at San Francisco in September 1951,[36] he took on a formidable challenge: creating a new warrior whose existence and values would be acceptable to society and whose role and capabilities would be appropriate to a new, post-imperial international order.

Meeting that challenge was particularly difficult because war, defeat, and forced revolution left Japanese deeply divided on questions of national defense. Some concluded the warrior was too dangerous; Japan should not re-create military forces lest they threaten society and polity as had their imperial predecessors. Others insisted the warrior was obsolete. Embracing Article Nine and exaggerating the implications of the nuclear revolution, they insisted that the destructiveness of modern weapons made possession or use of military force senseless.[37] Still others, among them depurged politicians, held that the warrior was essential. Possessing armed forces able to defend the nation was a right implicit in sovereignty. The new constitution, they argued, had to be modified to remove any ambiguity on that point.[38] Pragmatists took yet another view: Japan needed a new warrior who would be both deterrent to the Soviet Union and communism, and placebo to guarantee America's continued defense of the nation.

Prime Minister Yoshida Shigeru espoused the last point of view. Between 1950 and 1954, with characteristic initial reluctance that became shrewd single-mindedness, he presided over the creation of a new national defense force. It was a hybrid creature, born both of Washington's desire to build up domestic police power when the outbreak of war in Korea took away most American occupation troops, and former Imperial Army and Navy officers' hopes to restore the Japanese armed services. The new defense force first appeared as the National Police Reserve in July 1950, but from the outset its American sponsors saw it as the nucleus of a four-division Japanese army and coastal defense navy. Thus Gen. Douglas MacArthur's chief of staff designed its command structure so that its senior administrator would be a civilian directly responsible to the prime minister.[39]

Initially the force was supposed to be made up entirely of new recruits. But in March 1951, despite his lingering and deep distrust of the prewar military, Yoshida agreed to the entry of three hundred, depurged, younger, former Imperial Army and Navy officers into the new military. Not until 1952, when the force became the National Safety Agency, were ten more senior officers allowed to assume commands.[40]

By that time Washington's vision of what a reconstituted Japanese military force should be had matured and grown considerably. What began as an emergency defense force against Soviet subversion or invasion turned into a junior partner of and eventual replacement for American forces. Thus the United States pressed Japan—first by providing a security advisory group, then by providing aging ships, and finally with a full-fledged mutual defense assistance agreement—to commit itself to creating a force of nearly 300,000 well-equipped and trained men. In 1954, this force, given legal sanction by the Diet after a bitter debate resulting in a proviso that its troops not be dispatched overseas, became the Japan Self-Defense Force (SDF). It was to be comprised of ground, maritime, and air components.[41]

Where were new junior officers, future leaders for this new force, to come from? How were they to be educated and trained? And what should they be taught? Yoshida gave "serious consideration" to those questions and tried to resolve them in much the same way as he had those about the wisdom of reconstituting Japanese armed forces. In dealing with the latter, he had relied heavily on the advice of military men who had been his subordinates during his prewar term as ambassador to England.[42] Once a National Safety Academy was authorized in 1952, he sought to civilianize military education by naming as its president an academic whose pedigree, prewar study in Britain, and politics pleased him.[43] By the time Yoshida left office late in 1954, the school, renamed the National Defense Academy, was preparing to move from old army barracks to new but still incomplete buildings on a bluff overlooking Tokyo Bay near Yokosuka, once Japan's but now America's, biggest naval base in the western Pacific.[44]

The unfinished buildings were a metaphor for what Obaradai, as the school was then called, was to become. It was not the only gate to junior officership in the new Self-Defense Force; officer candidates who came up from the ranks and those who were university graduates could gain a commission by passing through officer candidate school. But it was clearly meant to be the nation's *only* defense academy to educate the military elite of the future. Those who came to it were already an elite; like Etajima entrants before the war, they had to be high school graduates (or equivalents as certified by the Ministry of Education). They had to pass rigorous academic and physical qualifying examinations.[45] They came from all over Japan, although home for a disproportionate number (nearly one in four) of successful candidates for admission was Kyushu—the more rural, less economically developed southern island that had produced so many distinguished army and navy officers in Imperial Japan.[46] The ratio of applicants to entering cadets between 1955 and 1965 ranged from eleven, to fifteen to one—a figure roughly equivalent to that for the Imperial Naval Academy during 1896–1905 when officership was a popular dream for many Japanese youth.[47] The new cadets were thus, in some measure, a self-selecting elite.

The school they entered differed from the prewar military and naval academies in having a mixed civil-military administration dominated by civilians. Its first president was a distinguished Keio University law professor and former dean of students, Maki Tomō. His deputy held a doctorate in science and had previously directed the Defense Agency's Technical Research Institute.[48] In marked contrast to the pattern of prewar academy administration, when senior serving officers rarely stayed long as superintendents at the academies, President Maki remained in office for a dozen years.[49] He and his deputy outlasted a succession of short-term major generals (and a single rear admiral) who ran the military training program at Obaradai during its first decade.[50] Together, an administration, faculty, and staff (only a quarter of which was military) that was more than twice the size of the incoming cadet student body sought to shape a new warrior for the new, peaceful and democratic Japan.[51]

In some respects, the notion of how that was to be done changed little from prewar days. Military educators still sought to develop a warrior with "a self-disciplined noble spirit and . . . moral character." He was also to be made into a "responsible member of . . . state and society."[52] The new warrior, like those brought up by Yamagata Aritomo, was to avoid participating in politics. The oath taken by entering cadets originally required simply an avowal of that abstention; later, when omitting any mention of the constitution became a kind of political statement, cadets were required to pledge to support and to defend that still controversial document.[53] Neither were the sterling characters produced at the Defense Academy to be allowed to identify too strongly with one branch of the service as had their imperial predecessors. The new team of civilian and military instructors was committed to using "every possible means" to promote camaraderie and a spirit of "understanding and cooperation" that would preclude the reappearance of factionalism within or between the new Self-Defense Force's ground, maritime, and air branches.[54]

The curriculum these young men would follow presumed that the warrior in the age of nuclear deterrence best be trained as an engineer. Its academic component resembled prewar Etajima's, and the structure and character of its military training segment echoed Ichigaya's. Even so, an Obaradai education, unlike that of the prewar service academies, was first and foremost an engineer's education.[55] All students had to take courses in the humanities, social sciences, and basic science. Moreover, as administrators and faculty settled into their jobs, the basic social science courses were redesigned and repositioned in the curriculum to emphasize the principles of democracy, the constitution and laws of Japan, international law, and physical and human geography. Additionally, every cadet had to study two foreign languages—one of which had to be English. More than half of the courses, however, were in science or engineering. Every graduate, moreover, majored in one of six applied scientific or engineering fields.[56]

The new warrior's education did not end at Obaradai. After completing his studies, he became a sergeant or chief petty officer at the officer candidate school of the individual service to which he had been assigned in his second year at the Defense Academy. There he competed with students from the enlisted ranks or others, more like himself, who had just graduated from a civilian university.[57] He also learned more of particular service traditions; at the postwar Etajima maritime officer candidate school, students practiced the same, daily self-scrutiny of character as prewar naval cadets.[58] This cap to study and training at the Defense Academy was meant to make the officer-to-be more competitively minded and less elitist than his prewar counterpart. Only after succeeding at officer candidate school did the new cadet return to the Defense Academy where, in an echo of prewar ceremonies over which members of the imperial family had presided, the prime minister handed him his commission.[59] With that, the junior officer was ready to begin his "real" professional education on active duty.

The basic pattern of junior officer education that crystallized at the National Defense Academy a decade after its move to permanent quarters in 1954 changed but twice over the next thirty years. Those changes reflected the influence of a man second in importance only to Yoshida Shigeru in shaping Japan's post–Pacific War defense establishment—Nakasone Yasuhiro. Born in 1918 the son of a wealthy lumberman, he graduated from Tokyo University in the year of Pearl Harbor, entered the powerful Home Ministry as a civil servant, passed through officer candidate school, and then went off to Indonesia as an Imperial Japanese Navy paymaster. He lost a younger brother, a kamikaze pilot, in the war and returned home bitter over Japan's defeat and occupation by American forces. Reentering the bureaucracy, he became supervisor of the Tokyo Metropolitan Police Board before running successfully for the Diet in 1947.

In the national legislature, Nakasone gained notoriety by always wearing a black necktie in mourning for the defeat in World War II, displaying (as was then forbidden) Japan's national flag, and opposing both Article Nine of the new constitution and what he insisted was the one-sidedness of the U.S.-Japan Security Treaty.[60] Never one to shy from controversy, in the years to follow he voiced the desires of conservatives within the ruling Liberal Democratic Party to revise the constitution, spoke of the need for self-reliant national defense, and suggested that possession and, if need be, use of tactical nuclear weapons was constitutional.[61]

In January 1970 Nakasone was named director-general of the Defense Agency. In a gesture meant to suggest continuity with the past, he changed the Self-Defense Force insignia from the dove of peace to the cherry blossom, the symbol worn by those in the Imperial Japanese Army and Navy. His pressure to dramatically increase Japan's defense spending (something that Washington, in accordance with the Nixon Doctrine, welcomed) worried socialist legislators and cautious Finance Ministry bureaucrats alike.[62] The greatest danger to Nakasone's hopes for enlarging and improving the Self-Defense Force, however, came on 25 November 1970 when the novelist Mishima Yukio, who had just harangued SDF troops for their weak patriotism and failure to accomplish anything over the preceding twenty years, had an aide ritually behead him on the parade ground at Ichigaya. That event sent shock waves through the Japanese body politic and stirred fears about the revival of prewar militarism.[63]

Nakasone responded to the challenge by trying to alter the Self-Defense Force's image. He published, for the first time, a defense white paper—one that stressed the need for an "autonomous" national defense in which the treaty with the United States contributed to, but did not control, the nation's security.[64] The SDF flag was also modified; the dove of peace returned to perch above the cherry blossom, as if to say that continuity with the past was not dangerous.[65] Nakasone also allowed the new, third president of the Defense Academy, Inoki Masamichi, a politically conservative academic who had been working on a multivolume biography of Yoshido Shigeru, to enter the political arena to defend self-defense.[66] In his *Kuni o mamoru* (To defend the nation), published in 1972, President Inoki left no doubt of the folly of unarmed neutrality, of the need for larger and better equipped forces, and of the difference between bad (old) superpatriotism and legitimate love of country in the contemporary world.[67] What he and others like him wrote probably contributed to a decided shift in public feeling about the Self-Defense Force. By 1972, one year after Nakasone moved on from the directorship of its parent Defense Agency, polls showed that nearly three out of four Japanese thought maintaining the Self-Defense Force was necessary.[68]

Under Nakasone, President Inoki introduced major curricular reforms that made the Defense Academy more than a school for military and naval engineers. The humanities and social sciences, previously housed in a single, liberal arts department, metamorphosed into three departments: general education, administrative science, and international relations. When cadets were asked to state their preference for majors, they rushed to the latter two subjects. Within two years, the ratio of applicants to acceptances in administrative science and international relations jumped from thirty-seven- to nearly sixty-to-one.[69] In the late 1970s, Japan's warriors-to-be clearly came to the Defense Academy with a broader conception of the range of skills that officership demanded and left strengthened by a curriculum designed to develop those skills.

Eight years later, in 1982, Nakasone Yasuhiro became Japan's sixteenth postwar prime minister and the only one to have previously served as director-general of the Defense Agency.[70] Once again he came to office with an ambitious agenda: strengthened defense abroad and administrative reform, especially educational reform, at home. Now he reached out across the Pacific to help achieve the first goal, cultivating a personal friendship with President Ronald Reagan and pledging to assume a greater share of the cost of defending Japan. He produced deeds to match his words. In his first year of office, he raised defense spending 6.5 percent,[71] temporarily getting it above the unwritten cap of one percent of the gross national product and setting what he hoped would be a precedent for keeping it at or above that level.[72] Educational reform proved slower and more difficult to achieve, even if what was proposed was only incremental change.[73]

Nakasone's drive for educational change prompted the establishment of a study group within the Defense Agency to develop better Self-Defense Force human resources.[74] Working first with Defense Academy President Tsuchida Kuniyasu who, like the prime minister was a former Imperial Navy paymaster, and then with his successor at Obaradai, Natsume Haruo, a long-time educational administrator within the Defense Agency, the study group recommended four basic changes in what cadets would experience at the academy.[75] Collectively the proposed changes sought to modernize majors, tailor education to individual needs, and reduce the disruptive impact of professional military training upon the academic program. When implemented in 1989, the reform more than doubled the number of scientific and technical majors open to students while reducing required courses to the same number as those in civilian universities. Rather than spreading language learning between two tongues, the new curriculum especially emphasized the development of a working knowledge of spoken English. The ratio of compulsory to optional courses was changed to allow cadets more freedom in shaping their education. Basic hands-on training in military specialties was shifted away from time-consuming training at individual service facilities to the Defense Academy itself.[76]

These reforms suggested that Japanese leaders had gone a long way toward meeting the challenge of post-imperial military education. Japan had a small but powerful, well-trained, better educated, and less threatening military than at any other time in the preceding century and a half. That force, and the young officers its educational institutions produced, were just what was needed for the Cold War world.

By 1993, when the first cadets educated under this revised curriculum received their degrees, the Cold War was over. For the first time since 1948, a man who did not owe loyalty to the conservative Liberal Democratic Party was Japan's prime minister.[77] The Japanese economy, which had enjoyed seemingly limitless growth for decades, slumped into protracted recession. Before these cadets' eyes, two other changes they would have thought impossible when they entered the Academy, occurred. Some of their predecessors were sent abroad as peacekeepers in combat zones to man aircraft and minesweepers in the Persian Gulf and to help enforce a truce in Kampuchea. Suddenly, from 1 April 1992 onward, women cadets came to the Defense Academy hoping to join the officer ranks.[78] The very definition of who the warrior of the future was and how he or she would relate to Japanese society and to the world was changing before these young cadets' eyes.

What conclusions can be drawn from the foregoing review of military education in Japan over the last one hundred and fifty years? What do those chapters from the past suggest about the probable shape of the next, twenty-first-century chapter in the history of Japanese military education?

Military education in Japan was not, and will not be, something constructed on a *tabula rasa*. The plate, to mix metaphors, is always full. Neither Yamagata Aritomo, nor his successor generation, nor those who struggled to give birth to a post-imperial Self-Defense Force and an officer education program appropriate to it worked free of the constraints and lessons of the past. On the contrary, each generation learned how very difficult it is to blend the traditional warrior ethos with the demands of present and future so as to shape an outlook appropriate for the warrior in a new age.

Military educators, if they are to succeed, must look beyond military traditions, administrative structures, and the particulars of curricula to confront, with and for their students, the warrior's paradox. If the history of military education in Japan teaches anything, it is that the separation of the warrior from the society he is to serve is fraught with peril for both. Yamagata tried to insulate the junior officer from politics, only to have him emerge later upon the national stage unable to understand, and bent upon destroying, the more complex political system that had evolved. The succeeding generation of army and navy officers, who came to see themselves as an elite better and purer than the rest of society, set it on course to disaster in the Greater East Asia War. Nakasone Yasuhiro, like the founders of the postwar Self-Defense Force, was forced to recognize that even the most rational plans to reform that force and the education of its future leaders must be implemented with sensitivity to society's attitudes toward the military. The warrior must be of the people if he is to be supported by the people and to emerge from his professional education and training able to work for the people.

Military education must change, constantly and carefully, as society itself evolves. What does the history of that change in Japan over the past century and a half suggest about its possible pace and direction in the century that lies ahead? Historians are trained to analyze the past, not predict the future. But if a prediction based on analysis of that past presumes the absence of some transforming event of the magnitude of the Greater East Asia War may be hazarded, then three basic points emerge.

First, military education in Japan will change, but at differential rates in its various aspects. Military educators will continue trying to develop "noble spirit" in the youth who come to the National Defense Academy. They will, like their predecessors, look to ideas, rituals, and traditions of the past in those efforts. The vestiges of "spiritual education" as practiced at Etajima and Ichigaya will not vanish. Nevertheless, Japanese youth are exposed to an ever-widening diversity of values in a popular culture deeply penetrated by foreign influences; those who seek to produce uniformity of character in the future officer will find that task increasingly difficult.

Second, technical and technological change will be more readily accommodated in the curriculum and practical training that cadets receive. The Japanese officer, sprung from a society fascinated by applied technologies in electronics, robotics, and by the literature of futurology, is likely to demand and to get the most sophisticated tools for his or her trade. The long-standing relationship between industrial/economic development and the military in Japan seems likely to continue, and to provide military educators and their students with what they need to develop and to operate the implements of national defense.

Third, change will come more slowly in the areas where it is most needed: English language and social sciences instruction. With respect to the former, the Nakasone reforms of the late 1980s pushed Japanese military education in the right direction. The Japanese junior officer must be able to speak English well if, as seems likely, his nation is to play a greater role in international peacekeeping under United Nations or some other collective auspices. Today's Obaradai cadets have the benefit of videotapes and computer-assisted English-language instruction,[79] but changing college-level language training through new technologies is too little, too late. More fundamental changes at the secondary level, difficult to achieve with any speed, given the Japanese educational bureaucracy's ability to resist, must, however, accompany them if future officers are to be able to understand and to communicate with the wider world in which they will serve.

The same might be said of social science education. Despite their nation's status as a global economic colossus, most Japanese remain rather insular in their norms of behavior and outlook. Defeat in the Greater East Asia War a half-century ago did not rob the language, the educational system, or the society at large of notions of superiority. The Japanese officer of the future needs to know and to understand much more about the richness and diversity of other societies and cultures—beyond what is seen on videos, television, and films—if he or she is to serve effectively as national defender and contributor to international peace in the next century. One can only hope that the next generation of reform-minded military educators will offer cadets the opportunity to major in an even wider array of social science subjects—cultural anthropology, history, economics and psychology—as well as public administration and international relations. One hopes also that they will find ways to breach the bureaucratic walls that have heretofore separated military and civilian education in ways that will further "internationalize" both.

The challenge confronting military education in Japan at the end of this century is, then, not so very different from that at its beginning: it is to produce a warrior who knows who he is and where he stands in relation to his society, and to equip him for action in the diverse and constantly changing world beyond Japan.

172 Forging the Sword

Notes

1. Roger F. Hackett, *Yamagata Aritomo in the Rise of Modern Japan 1838–1922* (Cambridge, Mass.: Harvard University Press, 1971), 51–54.

2. Robert J. C. Butow, *Tōjō and the Coming of the War* (Stanford, Calif.: Stanford University Press, 1961), 3.

3. Peter George Cornwall, "The Meiji Navy: Training in an Age of Change" (Ph.D. diss., University of Michigan, 1970), 44.

4. Hackett, 76, 81; Butow, 5. These structural changes were later amended to make an inspectorate general of military education the third essential component of military administration.

5. Cornwall, 45–46, 62–64, 77; Umetani Noboru, *The Role of Foreign Employees in the Meiji Era in Japan* (Tokyo: Institute of Developing Economies, 1971), 92.

6. Cornwall, 77; Butow, 5. Umetani, 40–45, details the activities of two of the most prominent foreign military/naval educators in Japan during the 1870s, Albert Charles du Bosquet and Archibald Lucius Douglas.

7. Cornwall, 80–81; Hata Ikuhiko, *Rikkaigun sōgō jiten* (Dictionary of the [Imperial Japanese] Army and Navy) (Tokyo: Tokyodaigaku shuppan kai, 1991), 312, 429.

8. Hackett, 56, 77–81; Meirion and Susan Harries, *Soldiers of the Sun: The Rise and Fall of the Japanese Army* (New York: Random House, 1991), 28; Ivan Morris, *The Nobility of Failure: Tragic Heroes in the History of Japan* (New York: Farrar Straus Giroux, 1975), 217–74. Morris analyzes Saigō's rise from defeat to national hero.

9. Hackett, 82–86, 128, 133.

10. Cornwall, 81, 107.

11. Hata, 313, 429; Cornwall, 86; Butow, 5.

12. Butow, 7–8; Mark Peattie, *Ishiwara Kanji and Japan's Confrontation with the West* (Princeton, N.J.: Princeton University Press, 1975), 21.

13. Cornwall, 81. There was one exception to this admissions procedure: graduates of the army's military preparatory schools could enter Etajima. See ibid., 118.

14. Butow, 7–8.

15. Cornwall, 119, 127–30, 133; Arthur J. Marder, *Old Friends, New Enemies: The Royal Navy and the Imperial Japanese Navy; Strategic Illusions, 1936–1941* (Oxford, England: Clarendon Press, 1981), 268.

16. Two measures of the junior officer's standing in society are noteworthy. One was the rapid increase in the ratio of applicants to appointments to the academies. In 1884, that ratio was 2.8 to 1 for Etajima; by 1905 it had risen to 13 to 1. See Cornwall, 115. A second measure of status was the attractiveness of young officers to prospective brides. Despite their poor prospects for high income, academy graduates were prized bridegrooms.

17. Marder, 271–72, 276; Hillis Lory, *Japan's Military Masters: The Army in Japanese Life* (New York: Viking, 1943), 99.

18. Marder, 275.

19. Leonard A. Humphreys, *The Way of the Heavenly Sword: The Japanese Army in the 1920s* (Stanford, Calif.: Stanford University Press, 1995), 14.

20. Richard J. Smethurst, *A Social Basis for Prewar Japanese Militarism: The Army and the Rural Community* (Berkeley: University of California Press, 1974), xiv, xvi, 18–19, 25–28, 35–43. The army's "education" of society also extended to women, following the establishment in 1932 of the Greater Japan National Defense Women's Association and publication of a special magazine directed toward them. Cf. Smethurst, 44, 166; Humphreys, 94, 176, 178.

21. The Harries, 170–75, present a scathing indictment of Ichigaya officer education that is tinged with overtones of prewar British racism. Ben-Ami Shillony, *Revolt in Japan: The Young Officers and the February 26, 1936 Incident* (Princeton, N.J.: Princeton University Press,

1973) is the best treatment of junior officers' roles in the most spectacular coup attempt of the 1930s.

22. Takenobu Yasaburō, ed., *Japan-Manchukuo Year Book, 1938* (Tokyo: Japan Year Book Publishing Co., 1938), 121–23.

23. Takenobu Yasaburō, ed., *The Japan Year Book, 1921–1922* (Tokyo: Japan Year Book Publishing Co., 1922), 303; Takenobu Yasaburō and Karl Kawakami, eds., *The Japan Year Book, 1928* (Tokyo: Japan Year Book Publishing Co., 1928), 186; Lory, 100–101.

24. *Japan Year Book, 1921–1922,* 291; *Japan Year Book, 1928,* 173.

25. On the development of the Kantōgun, see Alvin D. Coox, *Nomonhan: Japan against Russia, 1939,* 2 vols. (Stanford, Calif.: Stanford University Press, 1985), 1:1–11; Yamamoto Masao, "Pearl Harbor: An Imperial Japanese Army Officer's View," *Journal of American–East Asian Relations* 3 (Fall 1994): 253–54, captures the essence of the cadet's Manchurian field training experience.

26. Humphreys, 33–37, 108–13, 181, 205–7, analyzes the nature and growth of factions within the Japanese Army during the 1920s. For a corresponding treatment of factionalization in the navy, see Asada Sadao, "The Japanese Navy and the United States," in Dorothy Borg and Okamoto Shumpei, eds., *Pearl Harbor as History* (New York: Columbia University Press, 1973); and idem, *Ryōtaisenkan no NichiBei kankei Kaigun to seisaku kettei katei* (Japanese-American relations between the wars: The navy and the course of policy decision-making) (Tokyo: Tokyo daigakushuppan kai, 1993), esp. 149–210.

27. Lory, 99.

28. Marder, 276, notes that naval cadets had but four hours of English instruction weekly, only one of which was with a native speaker. The absence of any mention of such in the memoirs of an officer who later became an expert in dealing with Britons and Americans suggests its relatively slight impact upon naval cadets. See Rear Adm. Yokoyama Ichirō, *Umi e kaeru Yokoyama Ichirō kaisō roku* (Return to the sea: The memoirs of Rear Admiral Yokoyama Ichirō) (Tokyo: Hara shobo, 1978), 8–13.

29. Hayashi Saburō and Alvin D. Coox, *Kōgun: The Japanese Army in the Pacific War* (Quantico, Va.: Marine Corps Association, 1959), 23. Lory, 98, notes that Ichigaya cadets had, after Pearl Harbor, the opportunity to study Dutch or Malay.

30. Marder, 268–69; Butow, 8.

31. Humphreys, 51.

32. Peter Duus, "Introduction: Japan's Wartime Empire Problems and Issues," in Peter Duus, Ramon H. Myers, and Mark R. Peattie, eds., *The Japanese Wartime Empire, 1931–1945* (Princeton, N.J.: Princeton University Press, 1996), xii–xiii.

33. Hackett, 85; Ienaga Saburō, *The Pacific War: World War II and the Japanese* (New York: Pantheon, 1978), 113–14.

34. Japan Defense Agency, *Defense of Japan 1993* (Tokyo: Japan Times, 1993), 336; Richard B. Finn, *Winners in Peace: MacArthur, Yoshida, and Postwar Japan* (Berkeley: University of California Press, 1992), 30, 34, 78–79, 82–85.

35. Finn, 89–104, is the best recent account of the tangled origins of the 1946 constitution. Article Nine is quoted in Masuhara Keikichi, *A Review of Japan's Strength* (Tokyo: Defense Agency, Japan, 1956), 1.

36. Howard B. Schonberger, *Aftermath of War: Americans and the Remaking of Japan, 1945–1952* (Kent, Ohio: Kent State University Press, 1989), 240–65; Finn, 245–306.

37. This view was espoused publicly by the left wing of the Japan Socialist Party, and privately by a former army chief of staff, Gen. Kawabe Torashiro. See Finn, 273, 307, and William Joseph Sebald, *With MacArthur in Japan: A Personal History of the Occupation* (New York: Norton, 1965), 262.

38. Finn, 116–19, 248–50. Funada Naka, then director of the Defense Agency, told John Foster Dulles in 1956 that it was "regrettable" the new constitution with Article Nine remained

in force. See David W. Mabon, ed., *Foreign Relations of the United States, 1954–1957*, vol. 23, pt. 1, *Japan* (Washington, D.C.: Government Printing Office, 1991), 159–60.

39. Finn, 263–65; Jieitai jūnenshi henshan iinkai, *Jieitai jūnen shi* (A ten-year history of the Self-Defense Forces) (Tokyo: Bōei chō, 1961), 19–66; *Defense of Japan 1993*, 337–38; James E. Auer, *The Postwar Rearmament of Japanese Maritime Forces, 1945–1971* (New York: Praeger, 1973), 39–41, 49–61, 72–89, details Imperial Japanese Navy officers' survival and planning for the creation of a new force and the emergence of the Maritime Safety Agency. Frank Kowalski, an U.S. Army colonel who served as advisor in the creation of the National Police Reserve and Self-Defense Force, details his experience in *Nihon sai gunbi* (Japan's rearmament) (Tokyo: Saimul shuppan kai, 1969).

40. Finn, 265.

41. *Defense of Japan, 1993*, 338; *Jieitai jūnen shi*, 68–70; Auer, 94–95.

42. Auer, 72; Ōtake Hideo, ed., *Sengo Nihon bōei mondai shiryō shū* (Documents on Japan's postwar defense problem), 3 vols. (Tokyo: Sanichi shobo, 1992), 2:568.

43. Bōei daigakkō (Japan), *This is Obaradai: The Japanese Defense Academy* [hereafter *Obaradai*] (Yokosuka: Japan Defense Academy, 1962), 2–3; Kamesaka Tsunesaburō, ed., *Who's Who in Japan, with Manchoukuo and China, 1939–1940* (Tokyo: Who's Who in Japan Publishing Co., 1939), 525.

44. *Obaradai*, 2; Matsuda Akira, *Bōei daigakkō sono kyōiku no honshitsu* (The National Defense Academy: The essence of education there) (Tokyo: Orijinu shuppan senta, 1989), 96–114, surveys the problems the academy faced at this time of transition.

45. *Obaradai*, 7.

46. Thomas M. Brendle, "Recruitment and Training in the SDF," in James H. Buck, ed., *The Modern Japanese Military System* (Beverly Hills, Calif.: Sage Publications, 1975), 72–73, points out that as late as 1973 Kyushu, with roughly 10 percent of Japan's population, produced 25 percent of Defense Academy cadets. The over-representation of Kyushu in the cadet student body persists to this day. Author's interview with Rear Adm. (Ret.) Hirama Yoichi, Library Director, Bōei daigakkō, 10 Mar. 1997, Yokosuka, Japan.

47. Brendle, 85; Cornwall, 115.

48. *Obaradai*, 3; *Who's Who in Japan, with Manchoukuo and China, 1939–1940*, 525. Yoshida reportedly chose Maki so as to exclude prestigious Imperial University professors who had collaborated with the military before the war. See Maeda Tetsuo, *The Hidden Army: The Untold Story of Japan's Military Forces*, trans. Steven Karpa, ed. David J. Kinney (Chicago: Editions Q, 1995), 54.

49. Hata, 312–13, 429–30; *Bōei nenkan 1996* (Defense annual, 1996) (Tokyo: Bōeinenkan kankō kai, 1996), 626–29.

50. *Obaradai*, 4.

51. *Jieitai nenkan 1960nen* (Self-Defense Force yearbook 1960) (Tokyo: Bōei nippō, 1960), 80.

52. *Obaradai*, 8; Maeda, 59–60, points out that Maki consulted former Etajima president Adm. Inoue Shigeyoshi about the proper character of military education before assuming the presidency at Obaradai. He also retained as vice president and director individuals who were convinced that cadets must be inculcated with the "military spirit" prewar educators at Ichigaya and Etajima had considered essential.

53. *Obaradai*, 1; Japan National Defense Academy, *The National Defense Academy* (Yokosuka: National Defense Academy, n.d., ca. 1987–93), 1.

54. *Obaradai*, 9.

55. Ibid.; Maeda, 58, notes that the Obaradai cadet devoted four times as many hours to the study of the sciences as his prewar counterpart at Ichigaya.

56. *Obaradai*, 9–12; *Jieitai nenkan 1960nen*, 80–81; *Jieitai nenkan 1965nen* (Self-Defense Force yearbook 1965) (Tokyo: Bōei sangyō kyōkai, 1965), 618–47.

57. *Obaradai*, 19; *Jieitai nenkan 1965nen*, 648–49. If the 1961 prime ministerial address printed here is typical, postwar commencement addresses bore a striking resemblance to prewar *kun'iku*.

58. Marder, 275, points out that in the 1970s the five points were still recited by maritime OCS candidates at Etajima.

59. *Obaradai*, 19.

60. *Japan Biographical Encyclopedia and Who's Who*, 2d ed. (Tokyo: Rengo Press, 1960), 1014; Charles Moritz, ed., *Current Biography Yearbook 1983* (New York: H. W. Wilson Co., 1983), 262–63; "Nakasone Yasuhiro," in Nakasone Yasuhiro et al., *Warewarera Taishōko* (We children of the Taishō era) (Tokyo: Dokumon shoten, 1961). In 1956, Nakasone penned lyrics for the "Constitutional Revision Song" which proclaimed "As long as this constitution survives, the unconditional surrender continues." See Maeda, 138.

61. *Current Biography Yearbook 1983*, 264; *New York Times*, 6 Mar. 1970.

62. *Current Biography Yearbook 1983*, 264.

63. *New York Times*, 25–27 Nov., 12 Dec. 1970.

64. *Kodansha Encyclopedia of Japan*, 9 vols. (Tokyo: Kodansha International, 1983), 2:82.

65. I have inferred this point from Japan Defense Agency, *Defense of Japan 1989* (Tokyo: Japan Defense Agency, 1989), 131.

66. *Japan Biographical Encyclopedia and Who's Who*, 377. Inoki was a political conservative on a strongly Marxist-influenced faculty at Kyoto University. His *Hyōden: Yoshida Shigeru* (Yoshida Shigeru: A critical biography) was published by Yomiuri shimbun sha in 1981. Matsuda, 125–52, offers an extremely critical assessment of what he perceived as Inoki's attempt to turn the Defense Academy into a bastion of anti-communism.

67. Inoki Masamichi, *Kuni o mamoru* (To defend the nation) (Tokyo: Jitsugyō no Nihon sha, 1972). Others who wrote in the same vein included the former director of the National Defense Council, Kaihara Osamu. See his *Watakushi no kokubō hakusho* (My defense white paper) (Tokyo: Jiji tsushin sha, 1975).

68. Japan Defense Agency, *The Defense of Japan 1976* (Tokyo: Japan Times, 1976), 53.

69. *Obaradai*, 6; *Defense of Japan 1976*, 115–16.

70. *Current Biography Yearbook 1983*, 264; *Boeinenkan 1996*, 631, 634.

71. *Current Biography Yearbook 1983*, 265; *New York Times*, 28 Nov. 1982. One fascinating relic of the "Ron-Yasu" relationship is a signed Nakasone painting of Mount Fuji which the prime minister gave to the president in honor of his birthday in 1986. The painting is housed at the Ronald Reagan Presidential Library in Simi Valley, Calif.

72. S. Javed Maswood, *Japanese Defence: The Search for Political Power* (Singapore: Institute of Southeast Asian Studies, 1990), 54–59.

73. Hayao Kenji, *The Japanese Prime Minister and Public Policy* (Pittsburgh, Pa.: University of Pittsburgh Press, 1993), 46–67, renders a mixed verdict on the outcome of Nakasone's educational reform effort.

74. *Defense of Japan 1989*, 131.

75. Nichigai assoshiatsu, *Gendai Nihon jinmei jiten* (A biographical dictionary of contemporary Japan), 3 vols. (Tokyo: Nichigai assoshiatsu, 1987), 2:1372; *National Defense Academy*, 3.

76. *Defense of Japan 1989*, 131–32.

77. *Bōeinenkan 1996*, 636.

78. *Defense of Japan 1993*, 353–56.

79. Author's observation of Bōei daigakko foreign language training facilities, 10 Mar. 1997.

The Persistence of Memory: Radical Political Change and Officer Training in Twentieth-Century Russia

David R. Jones

> Time present and time past
> Are both perhaps in time future
> And time future contained in time past. . . .
>
> T. S. Eliot, "Burnt Norton" from *The Four Quartets*

Eliot's insight is especially applicable to Russia, a country that treasures, if not sanctifies, its history. It is also especially relevant to discussions of "radical change" in any sphere of social life. The continuity of underlying historical processes often has unpleasant ways of reasserting itself, frequently mocking the ambitions of radicals and revolutionaries, and more recently the pretensions of "postmodernists" and "deconstructionists." Human beings cherish tradition, either consciously or unconsciously, and tend to do so in a "mythic" sense that is often the despair of historians as well as their academic detractors who attempt a more "professional" and "objective" study of the past.

I shall consider the role of "myths" later in this paper. Here I will say only that "myths" have played a highly significant role in Russian and Soviet history over the last century, and especially in the formation of the command cadres of the Red, later the Soviet and today the Russian Army. First, however, I shall address my two primary concerns: junior officer training and radical political change. In discussing the first—training—I will concentrate on issues of motivation, not the details of the contents of changing curricula. More narrowly, I will focus on the role of motivation in developing such qualities as "character," loyalty to the state, a sense of duty and professional dedication, and that most elusive of military qualities, "leadership." For it is difficult to inculcate these highly esteemed (at least in military circles) qualities with even the most sophisticated educational techniques unless the future officer is "motivated," be it "self-motivated" or otherwise.

The Military Academy, Politics, and Professionalism

The relationship between political change and military education, of course, is neither a uniquely Russian nor a new problem. The historian Richard Preston made this point on the occasion of the U.S. Air Force Academy's 25th anniversary in 1989. Since World War II, he noted, "the extent of social and political change has been just

176

as great and has produced problems that are just as difficult" as in Europe after 1850. Furthermore, such changes "tend to set the military academy even further apart just at a time when many of these same things call for closer relations between the officer and civilian society." This reflected the fact that while military academies sought to provide a sound general curriculum and meet changed professional demands, they simultaneously adhered to "military traditions and social structures that were threatened by social conditions and political needs." Today as well they usually uphold "standards of conduct quite different from those that prevail outside of the academies," and this "obviously imposes greater strain on the academies than they have known so far."[1]

This last problem entails more than the attempt to preserve past traditions and structures for their own sake. Rather, it reflects the dual nature of "professionalism" when this term is applied to officers. First, it commonly denotes the trend that emerged during the nineteenth century toward formally training officers to become what Russians sometimes call the "military intelligentsia."[2] Such officers were to replace the traditional, rough-and-ready commanders who had learned (or not learned) the art of leadership by practice since, as Preston and others have chronicled, advances in technology then demanded that officers receive a more formal, technological training. This was notably true in arms like the artillery and engineers, but it also affected such traditionally aristocratic and "nontechnological" branches as the infantry and cavalry. These technical, troop management, and other combat skills distinguish the distinct "professional group" we know as officers.

Second, Russian (and other) officers often stress that their "profession" differs from most civilian professions because when practiced in its proper environment, it entails the constant threat of death and the responsibility to lead others into the jaws of death.[3] Contrary to some opinions, an officer's career is not merely one way to obtain a trade, education, or managerial skills. These may well be acquired in passing, but pursuit of a military career still remains a decision to involve oneself with the "technology of death." Put differently, "body bags" are inseparable from combat and as long as wars are fought, the traditional, although possibly socially outmoded and "primitive" virtues of courage, self-sacrifice, honor, and duty will assume special significance within any effective military establishment. Although such virtues are commonly scorned by those civilian critics who see soldiers as barbaric vestiges of a "feudal," "aristocratic," and more recently, a "sexist-chauvinist" past, these virtues remain vital elements for successful leadership by even the most educated, sophisticated, and "professional" junior commanders.[4]

The psychological attributes of what is certainly one of humanity's oldest professions, that of the soldier, have changed remarkably little since organized armies first marched to war in the fourth millennium B.C. Furthermore, the development of these attributes has remained a primary goal of what Preston calls the "process of character building" in all officer training systems. That their students and teachers feel separated, if not alienated from their civilian counterparts, is hardly surprising and to a degree, probably inevitable. Yet the extent of that gap can vary considerably and according to the common wisdom, usually is widest during periods of radical political change.

Russia and Radical Political Change

Since 1900 Russian soldiers have experienced repeated bouts of such "radical" change. These include five cases: first, Russia's defeat by Japan in Manchuria in 1904, the outbreak of revolution in 1905, and the birth of the "Duma Monarchy" in 1906–7; second, the Revolutions and the Civil War of 1917–20, and the birth of Soviet Russia; third, the so called "Stalin Revolution" of 1928–32 that brought the collectivization of agriculture and rapid industrialization; fourth, the purges that convulsed Joseph Stalin's Russia from 1934 (after the Kirov murder) to 1941, and disrupted the armed forces from 1937 to 1941; and fifth, the confusion initiated by Mikhail Gorbachev's reforms in 1986 which ended in the traumatic transformation of the Soviet Union into the Commonwealth of Independent States (CIS) in 1991, and which was complicated militarily by the Afghan War and, more recently, the Chechen conflict.

Some might add to this list the wartime period of 1941–45, and that of the "de-Stalinization" which followed Nikita Khrushchev's famous "secret" speech of 1956. After all, as ex-partisan and Belorussian writer Ales Adamovich concluded, by waging the Great Patriotic War the Soviet "people facilitated the complete victory of Stalin's absolutist tyranny" and this therefore might represent another period of radical change.[5] Perhaps so, but despite the turmoil, at the time few saw it in this light. To a lesser extent, this is true as well of the eras of Stalinization (cases three and four), and that of de-Stalinization (late 1950s–early 1960s). Whatever the traumas, in all these instances the regime remained intact, as did the officer corps and training system that supported it.

The first, second, and fifth cases listed above are very different. Not only did all three bring (or seem likely to bring) radical political change, but all involved a seemingly fundamental restructuring of the military system, its educational component included. These cases share other similarities as well. In each the existing regime initially appeared to be losing control or "stagnating" politically and, in cases two and five, failing economically. Again, in each the regime faced an articulate opposition led by a section of the intelligentsia and part of the political elite, but one from which the majority of officers held aloof. In each, the army and regime were humiliated by defeats which highlighted the armed forces' professional deficiencies and indicated the need for another round of military, if not necessarily political, reform. For these reasons, I shall focus largely on cases one, two, and five.

Of these, case five, the restructuring of junior officer training in today's Russian Federation, is of most immediate interest. But unfortunately, this is still in flux and despite the deluge of information that pours from the former Soviet Union, we probably understand Russia's social-political dynamics (and so her likely future) as little as our predecessors did in 1905, 1917, 1936, or 1986. Today's predictions thus may well appear absurd tomorrow. Yet if "time future" remains imprisoned in "time past," as Eliot suggests, then history may provide clues to the future, or at least an indication as to where we may find such clues.

At the same time, superficial similarities between the cases selected should not obscure their substantial differences. Unlike 1917, in 1905 the regime survived. Again, in 1991 Boris Yeltsin edged the USSR from the stage in what is perhaps the classic modern example of B. H. Liddell Hart's "strategy of the indirect approach" as applied to politics. Similarly, both 1917 and 1991 witnessed the senior commanders (M. V. Alekseev unwillingly, D. T. Yazov with premeditation) acting politically with disastrous effects for themselves and their armies. Despite this, the Soviet Union died with a whimper rather than the bang that brought down Tsar Nicholas II. Finally, if the Tsarist Army collapsed in 1917–18, the USSR's has survived more or less intact, at least externally, the transition from USSR to CIS. True, some units, bases, factories, schools, and academies were lost to successor states in the Baltic, Ukraine, the Caucasus, and Central Asia. But as in 1905–6 the military's core remains, battered but still functioning, in a world vastly different from that of a decade earlier.

The 1917 Revolution and Officer Training

The drastic differences between the outcome of 1917–18 and that of either 1905 or 1991 are in part due to the intentions of the actors involved in the Bolshevik takeover. Whatever the cynicism of later Communist leaders, those of 1917 were ideologically committed to destroying the existing military system. Equally important, before V. I. Lenin and his Bolsheviks seized power in October, this ideological precept was a practical tactic of revolution as well. Marxists long had viewed armies as the protectors of the status quo and Lenin himself dismissed Russia's *ofitsery* as "the spoiled and darling sons of the capitalists."[6] Here as in other fields, he and his colleagues consciously sought to create a *tabula rasa* by destroying and demobilizing the "old" army before they built a "new type" of armed force for their "socialist" society. As for the Tsarist officer-training system, this network naturally had expanded dramatically during World War I. With the brief exception of the General Staff Academy, the system disintegrated rapidly after 1917 as the majority of its cadet-junker students, along with many of their instructors, took up arms against the Bolsheviks/Communists.

Most Western scholars long accepted Soviet assertions that the Red Army's officer training network therefore was built from scratch. At best the new regime supposedly inherited only some physical plant and equipment, along with a handful of old instructors.[7] In creating their new army the Bolsheviks relied on the traditional radical model of a "militia army" and naturally sought to staff it with reliable, revolutionary commanders. They first had relied on leaders elected from the ranks to disrupt the old military system and to transform it into a force for revolution. Despite its oft recounted drawbacks in terms of professional efficiency, this process was positive in that it created a pool of politically reliable commanders in the form of sympathetic Tsarist junior officers, noncommissioned officers (NCOs) and leaders of the Red Guard militia. It thus provided a considerable proportion of the junior commanders for the Communists' Red Army as proclaimed in January 1918. But these proved inadequate for the needs of the expanding Civil War and the Soviet government soon was forced to create its own system of *Kraskom* or "Red Commander" courses.

Over the next two decades this beginning developed into a far-flung and ramified system of "military-educational institutions" or "*VUZy*." For Soviet scholars the result was a new corps of politically conscious and committed "commanders" (the term "officer" only resurfaced in 1943) who were "blood of the blood and bone of the bone" of the working and toiling classes. As proof they proudly cited figures on the kraskoms' membership in the Komsomol and Communist Party of the Soviet Union (CPSU), pointed to the large numbers with "worker and peasant" origins, and compared this to the Tsar's "noble" officers.[8] Given these characteristics, along with the emphasis placed on the political education or "indoctrination" of the new commanders, the Red Army's training network indeed appears very different from that of its predecessor.

In fact, this difference is more apparent than real. While Tsarist officers indeed were "nobles," this was because Peter I's Table of Ranks made every officer a "nobleman" on being commissioned. By 1914, however, 40 percent of all officers below the rank of colonel, and 35 percent of that year's newly commissioned and "ennobled" infantry officers, had been born peasants! Again, if the massive effort at indoctrinating commanders was a novel aspect of the new Soviet system, before long the young Red Army had its own decorations and ceremonial banners, and increasingly resembled not only contemporary "capitalist" and "imperialist" forces, but those of the derided Imperial regime as well.

No one was more aware of this than the Red Army's founder, Leon Trotsky. As early as 1923 he attacked the reemergence within his "socialist" army of regimental histories, a form of literature traditionally used by "reactionary" regimes to foster the soldiers' sense of heroic distinctness from the rest of society.[9] Needless to say, by then the militia model had been abandoned in favor of the "mixed" cadre-territorial system, and by 1938 even this doctrinaire (and once economically necessary) concession had gone. Thereafter, the USSR reverted to a conscript army of the old type. More striking still, after 1935 this latter was headed by "marshals," and after 1940 by "generals" as well. Similarly, by that time its junior and middle commanders again enjoyed such privileges as the services of batmen, their own officers' messes and *voentorg* commissariat stores, ever-increasing levels of pay, their own "courts of honor," and so on.[10]

As William Faulkner once observed, the "past isn't dead; it's not even past." More practically, this evolution showed that building a "new type" of army was easier said than done. Apart from the *kraskoms,* during 1917–20 the Red Army included some 50,000 former Tsarist officers or *voenspetsy* (military specialists). Use of the once discredited and hated "class enemies" began in late 1917, accelerated during 1918, and was confirmed by the Eighth Party Congress in March 1919. Thereafter, former officers rapidly filled positions that ranged from company through front commander, and worked as teachers in the military school system. True, they served under the watchful eyes of political commissars, but the latter assured that the rank-and-file obeyed the former officers as much as they prevented the officers' treason. Equally important in the long term, tens of thousands of former NCOs, including such prominent figures as the later Marshals S. M. Budenny, S. K. Timoshenko, R. Ya. Malinovskii, and G. K. Zhukov, also were recruited. But during the formative stage it was the ex-officers who exerted a dominating influence on the development of the combat doctrine and training system of the young "proletarian" army.

For example, during the Civil War a commission of ex-generals analyzed the combat experience of that conflict, and of the preceding world war. Again, of the one hundred authors of the field regulations of 1929, seventy-nine were veterans of the Imperial Army. The same was true of 198 of that year's prominent 243 military writers. As for the military schools per se, there even former enemy "Whites" trained the *kraskoms*. Not surprisingly, the curriculum was as much directed to studying the practical "lessons learned" in tactics and operations since 1914 as it was to creating leaders to liberate the world's oppressed toilers, a fact that in time had major implications for the ideological stance of the command cadres of the Red and Soviet Armies, and for their view of their place in Soviet and post-Soviet society.

To explain the emergence of old patterns merely in terms of the use of Tsarist officers begs the question of why so many were recruited by a regime so obviously hostile to their conceptions of duty and service. The real answer, as even committed revolutionaries recognized at the time, was the Red Army's need for cadres of "professionals" who understood the deployment and employment of the modern technologies of death, and who possessed the requisite if "noble" skills of leadership. However much the officers' professional reputation suffered during 1914–17, and however much the victorious revolutionaries wished to rely on the elected "proletarian" commanders of 1917–early 1918, the first campaigns made it abundantly clear that political enthusiasm and reliability alone could not bring victory. But unfortunately, with military skills came the military forms that promoted their exercise.

Officer Training and the 1905 Revolution

Before examining further the political and social attitudes of Soviet officers we must look briefly at case one, that of 1905.[11] If most historians, Soviet and Western alike, naturally focus on 1917 and its aftermath, this nonetheless obscures the equally significant changes introduced after 1905. These too followed defeat in war and domestic revolution, but in this case the revolutionaries were as unsuccessful as the generals. Political, social, and military reforms thus were implemented within a surviving if altered, but not a new regime. They therefore were politically less extensive, and militarily more explicitly "professional" in nature than those introduced by Lenin and Trotsky after 1917.

After 1905 the military reforms in general, and the reform of both the military schools and overall training system in particular, emerged after acrimonious debate and organizational confusion. By 1908–9 the issues involved assessing and acquiring the developing technologies of death, and adapting tactical and operational practices accordingly. At the lowest level, this brought an increased emphasis on marksmanship and small-unit tactics, at the higher tactical and operational levels a stress on mobility and "meeting" or "encounter" battles, and at the top the demand for a "unified military doctrine." The extent to which these and other innovations could be realized obviously depended on the quality of the command staff at all levels. Consequently, the authorities took steps to clear out "deadwood" that had accumulated at all levels of the military hierarchy by instituting "attestation commissions" to review the qualifications and performance of its officers.

To provide junior commanders of a higher and consistent quality, the War Ministry restructured and streamlined its military schools. This effort included introducing revised curricula and updated teaching methods, transforming the old "junker" institutions into full military schools, creating more and better equipped specialist technical facilities and academies, and expanding the base from which junior officers were recruited. Despite the sometimes successful opposition of conservatives and other self-interested parties, these measures had their effect. In 1914 Russian officers, especially junior officers, entered action considerably better prepared for modern combat than had those of 1904.

Unfortunately, the warfare they anticipated was not the warfare they had to wage. Nonetheless, and despite the fact that much remained to be done, the reforms implemented after 1905–6, and especially under War Minister V. A. Sukhomlinov after 1908–9, had had a considerable impact. But their implementation had demanded that the service chiefs lobby intensively to secure the necessary budget credits in times of fiscal restraint. This meant that whatever the alleged "apolitical" nature of the officer corps, its leading representatives became very involved in the political maneuvers of the new parliament or "State Duma." Similarly, its junior members were fully aware of the Duma's influence on their conditions of service and personal prosperity. This, and the fact that the supposed professionalism of the military reforms masked their real social content, make the period from 1905 to 1914 instructive in today's context.

Myths and Motivation

The relevance of these earlier cases for today becomes clear if we turn to another aspect of recruiting and training junior officers. This is "character formation" which, Preston argues, is prominent among the "complex and persistent" problems that face military educators, even in times of normal political evolution. "Motivation," he maintains, is "part of the process of character building, an element in all officer training" that is rightly stressed in all military academies, past and present.[12] Without motivation, officer candidates will decline in number and quality, thereby diminishing the prospect of developing leaders with character. So it is small wonder that there is as vast, complex, and usually technical a literature on officer motivation as there is on any other aspect of military education.

In preparing this paper I discussed this issue with a number of officers of the Royal Canadian Navy. This group included an admiral once responsible for personnel as well as recent officer-graduates. Morale in today's Canadian Forces probably is at least as low as in the post-Tailhook United States Navy, and both services—albeit for different reasons—share some of the problems faced by Russia's armed forces. This aside, my conversations suggested that the junior officers' most prominent concerns include: first, the availability of high-quality professional training, the opportunity for a rewarding career, and a reasonable level of financial security; second, the degree of respect accorded their profession by other professional groups in particular, and by society as a whole; and third, a cause justifying their service in their own minds and in the eyes of society.

These three groups of motives demand comment before applying them to junior officer training in Russia, past or present. Whereas the first set is self-explanatory, groups two and three are less so. I have noted that since the professional officer is a specialist in the technologies of death, and since he may lead men into situations where death, his own included, is likely, his training must be different, and indeed may appear somewhat "primitive" when compared to that given in other professions. To be sure, politicians and public relations experts disguise this fact by lauding modern military leaders as "managers" rather than "commanders," and today's soldiers as "peacekeepers" rather than warriors. Yet Vietnam and Afghanistan, the Middle East and Bosnia, demonstrate that when a junior commander finds himself at the sharp end of war, he must "lead," not "manage."

Because the virtues required for such leadership often are scorned in today's world, the officer's professional counterparts on "civvy street" may well regard him as a throwback, if not a lout. To some extent this may be inevitable, but there is one condition by which the soldier's unique skills can gain some degree of respect from his civilian colleagues, from the politicians on whom his career and conditions of life depend, from his fellow military men and, indeed, in his own eyes. This is through the existence of what we may call the appropriate "myth."

Social scientists long have recognized the importance of military "myths." One type, described by Richard Holmes, meets the "legitimate need to de-fuse deep-seated cultural and psychological taboos against killing," and therefore "is an inseparable part of military training." Such a "myth" of necessity is crude and involves the "almost obligatory dehumanization of the enemy." This process becomes "particularly pronounced when there are radical ideological differences between the opponents, differences which . . . may portray the enemy as the foe of civilization and progress, as a 'Godless Communist' or a 'Capitalist-Imperialist bloodsucker.'"[13] There also are the more positive "traditions" (or myths) such as those preserved by the regimental histories, colors, and battle honors so disliked by Trotsky. These too are important to provide models that sustain an officer under stress; to strengthen the cohesion between him and his fellows, as well as with his men; and to ensure that they all come to grips with their now "dehumanized" foe.

However important both these forms of military myth are, neither has real resonance outside of military circles. But there is a type of "myth" that is far more compelling and complex—one offering a vision or cause that in certain circumstances justifies the officer's use of his skills on behalf of his society, his nation, or the community of nations as a whole. Such myths usually involve a view of history, the essentials of which are shared by the officers and their community, that underlies and justifies the existence of a state or nation, and which legitimizes its government. The history embedded in these myths is not necessarily false, but it usually is abbreviated and simplified so as to be acceptable to as broad a stratum of society as possible. When such a myth is strongly entrenched, a national "will" and confidence usually results that provides the officer with his cause and justifies his career in both his own and others' eyes. This same myth buttresses the more purely military myths just mentioned, and helps define the nature of the military establishment and of its service to the state. It thus is an essential, motivational prerequisite for all effective programs of military training, that of officers included.

When this complex of myths is tested or undermined by radical social change, the officer's position suffers. For example, troops employed against people who are not successfully "dehumanized" may be seen as "murderers." This usually means that their morale and sense of purpose will fall, perhaps terminally in an organizational sense. Morale will decline similarly if "sacred" regimental and service traditions are suddenly condemned and banned by new considerations of morality, financial or political expediency. As the condition of today's Canadian Forces demonstrates, when myths at both levels are assaulted, a state simultaneously loses its self-confidence in general, and fiscal restraints meanwhile place prospects for remuneration and advancement in doubt, the results are serious problems in recruiting, training, and retaining new generations of officers.

Russian Myths and the Officers: 1905

Considering the number and extent of modern Russia's political upheavals, one might expect that her armed forces had frequently found themselves left "mythless," and that the military educational system suffered accordingly. Perhaps paradoxically, the opposite was true of the revolutions in 1905 and 1917. In retrospect it is clear that each, albeit in its own way, eventually provided the myth needed to give emotional content to the professional changes that followed. More surprising still, the myths that emerged from these revolutions were curiously complementary, and this despite their apparent ideological contradictions.

This seeming paradox is explained by an important change of emphasis within the myth underlying service in the Imperial Army brought by the Revolution of 1905. Before that event this myth was expressed in the formula "Faith, Tsar, and Fatherland," with the Tsar being the most potent of the three symbols. That symbol, however, already had lost much of its attraction for society as a whole thanks to the rise of the anti-autocratic opposition after the mid-1890s. Officers then found themselves with increasing frequency leading their men against the "internal enemy," as strikers and other protesters were known. Accordingly, their popularity fell among the educated public, the military became an unpopular choice of career, criticism of officers mounted in the press and, in spite of numerous personal and family ties, a widening gap developed between "society" and its officers.

Military defeats in Manchuria and the use of troops to repress the Revolution of 1905 momentarily transformed this gap into a chasm. Stung by reproaches from their civilian counterparts and even their relatives, many officers contemplated and some did retire. The unrest gripping civilian students meanwhile affected the cadet corps and military schools; the number of applicants for training as officer candidates fell and again, some students left. Reflecting on these events, thoughtful officers yearned for a nationalistic union between society and its military such as they perceived existed in Germany and Japan.[14] Then Nicholas II's October Manifesto and the creation of the State Duma changed matters abruptly. Many officers saw this institution as a means to reconcile liberal society with the military so that together, acting in a spirit of patriotic dedication, they could build an army capable of upholding Russia's national interests.[15]

Thanks to a similar upsurge of nationalist sentiment in much of civilian "society," such dreams were not unjustified. If officers were alienated by the political radicalism dominating the first two Dumas, the situation changed with the election of the Third Duma. Its nationalist-liberal majority was happy to fund reforms aimed at improving officers' training and conditions of life. This upsurge also brought an influx of nationalist youths into military schools. In these schools, courses appeared to familiarize future officers with the basics of the day's political theories while some school directors made serious efforts to encourage contacts between their own and civilian students. Other reforms raised the standards of the junior officers' training and service within the ranks, improved their material conditions and career prospects, and renewed their sense of professional pride. In addition, many junior officers and officer candidates became involved in the various youth movements springing up across Russia, of which the most successful was the Boy Scouts.

By 1908–9, then, a new nationalist spirit infused both the military-educational system and the ranks of junior officers. Equally important, the latter now enjoyed considerable respect from a public that shared their nationalist sentiments and, not infrequently, sense of national commitment. But this new spirit also brought a revision of the old tripartite creed of "Faith, Tsar, and Fatherland." If the Tsar retained his legal position as commander-in-chief and remained one potent symbol of "Fatherland," he now shared this honor with the Duma. This presented no problem as long as government and parliament agreed, as when significant military reforms were initiated during 1908–9. Unfortunately, this unified nationalist front collapsed when competition to control future military policy developed between the Tsar and his ministers on the one hand, and the Duma's "experts" on the other.

At that point the majority of nationalist officers sought to avoid choosing between Tsar and Duma, withdrew from party politics and in this sense, became "apolitical." By 1912–13 the gulf again was widening between the military professionals and society. True, it was bridged momentarily by the patriotic emotions roused by the outbreak of war in 1914. Nevertheless, while Russia's nationalist youth flocked to wartime officer schools, the political feuding between Tsar and Duma grew ever more bitter. When the February Revolution of 1917 forced the army to choose between these two symbols of Fatherland, Gen. M. V. Alekseev and the high command opted for the Duma, and a new era opened for Russia.

Russian Myths and the Officers: 1917–1918

In 1917 the nationalist officers' dual focus on Tsar and Duma seemed to doom their cause. As events unfolded the army, along with the rest of Imperial society, fragmented and disappeared beneath the tide of social revolution. From the empire's ruins arose the new Soviet Republic with its openly internationalist "founding myth." Proclaiming itself the world's only truly "democratic" state, Lenin's Russia stood as the vanguard of a socialist revolution destined to destroy capitalism everywhere. Thus soldiers and commanders of the new Red Army in 1918 vowed "before the laboring classes of Russia and the whole world" to serve "the great ideal of liberating the toilers of the world" and, if necessary, to die "for socialism and the brotherhood of man."[16] This and similar rhetorical pronouncements then became the ideological basis for a resurrected military-educational system.

During the mid-1920s, however, the prospects for "world revolution" dimmed, and Stalin began building "Socialism in One Country." Consequently, a Red soldier's status as "citizen of the Soviet Republic" gradually took precedence over that of "son of the toiling people." Meanwhile, during 1917–20 the explicit exponents of "Russia—One and Undivided" were the anti-Bolshevik Whites. If the Communists opposed this with the slogan of "national self-determination," in practice they too sought to preserve Russia's empire, though for other motives. Nonetheless, overt expressions of the nationalist spirit of 1905 were suspect in the army of international labor and so went underground. Yet from the first a handful of former officers, generals included, had discerned the germ of a new and powerful Russian state within Lenin's Soviet Republic. True, the overwhelming majority of their fellows preferred to follow Gen. A. A. Brusilov's counsel and remain neutral until the new Russia emerged from the chaos. Some even managed to do so, or found "apolitical" work with groups such as Brusilov's commission for examining "lessons learned." Even so, the majority were eventually conscripted by one side or the other.[17]

For many ex-officers, Brusilov and his fellow "neutrals" included, the Soviet-Polish campaign of 1920 was the moment of truth. Even before this event a perceptive English observer had reported that the Red Army's morale was largely "drawn from its patriotism, and whatever Government were in power, it would make no difference if the enemy were thundering at the gate."[18] Then, thanks to Marshal Josef Pilsudski's

Poles, in early 1920 the Communists found themselves leading Russians in a crusade against an ancient enemy from the west. Brusilov and his colleagues immediately renounced neutrality and appealed to their fellows, regardless of recent allegiances, to join the Red Army in beating back the common foe. Although this strident nationalism, not to say chauvinism, worried Lenin and Trotsky, they welcomed the reemergence of those ex-officers who had escaped conscription, and the defection of White officers to the cause of Red Russia.

The result was a true *smychka* (merging), if not in the form of the worker-peasant union celebrated by Soviet ideologues. Rather, it was the union of the ex-officers' nationalist sentiments with the cause of defending the "Socialist Fatherland," which under Stalin became the "Soviet Motherland." In this form the founding myth of the Soviet Union—that of the "Great Proletarian Socialist October Revolution"—was transformed into one suitable for training command cadres for a Red (after 1945, Soviet) Army that increasingly was Russian in its "spiritual" essence. If anything, Stalin's military purges of the late 1930s increased the effectiveness of this socialist-nationalist myth in recruiting officer candidates and training junior officers. After the mid-1930s the military's understated nationalism was validated by the ever-more deafening chorus celebrating "Soviet patriotism." Meanwhile the threat of war, long a constant theme in the Soviet media, became a reality. These factors combined with the social and financial advantages offered the command cadres, as well as the frequent promotions brought by the purging of some 80 percent of commanders above colonel, to offer career opportunities and social prestige to junior commanders that had seemed impossible only a few years earlier. Consequently, by 1939 the Red Army had acquired all the requisites for effectively recruiting and training its command cadres.

If some officers undoubtedly found the USSR's founding myth tarnished by the horrors of collectivization, the disappearance of colleagues during the purges, and Stalin's sudden friendship with Hitler, all hesitation vanished when German tanks drove eastward on 22 July 1941. The impact of these events on most Russians as well as on many of their non-Russian Soviet fellows, is obvious in the works of a multitude of novelists, poets, artists, composers, and memoirists. For almost all, the "Great Patriotic War" of 1941–45 was a Rubicon, a defining event of their lives. "June of the year forty-one drew a line across life and time," recalls one Soviet veteran. "On the one hand there was what had been before 22 June. On the other there was what began that day, that which was defined by the terrible word 'war.'"[19] The horrors of German occupation quickly convinced millions of Russians and non-Russians, Communists and non-Communists alike, that Hitler was no alternative to Stalin. Despite the latter's brutal treatment of the Volga Germans, Crimean Tatars, Chechens, and other Caucasian peoples, even most non-Russians joined their Russian counterparts in defending the embattled USSR in what became a true *vsenarodnaia voina* or, as translated by an American Stalinist and later Maoist, a "war of the whole people."[20]

Even Gen. A. A. Vlasov, who organized an anti-Communist liberation army from Russian prisoners-of-war, recognized this fact. In vain he pointed out to his German sponsors the unpalatable truth that the "majority of the Russian people—especially the educated classes—now regard this war as a war of German conquest. . . ."[21] Given a lack of alternative leadership, Russian and Soviet patriots had no rallying point other than Stalin and his government. Even the most skeptical Russian nationalist saw the Stalinist Soviet state again legitimized by an assault from a traditional enemy. The victory of 1945 therefore provided both state and party with a "sustaining myth" to buttress and validate the flagging "founding myth" of 1917. As Nina Tumarkin demonstrates, the Soviet rulers' success in shamelessly manipulating the Great Patriotic War's history is still evident in the aged and marching veterans, the remnants of the same youths who "won the war" in 1945, who annually celebrate "Victory Day."[22]

The changed ethos of the Soviet Army is evident by comparing its oath of 1947 with the Red Army's of 1918. Soldiers now pledged as citizens of the USSR to serve "my people, my Soviet Homeland, and the Soviet Government." Similarly, the officers' old credo of "Faith, Tsar, and Fatherland" had reemerged as "Communist Party, Soviet Government, and the People."[23] More importantly, the war solidified the place of Russian nationalism within the military's ethos. When the Communist party set out to restore the supremacy of Marxist dogma and social discipline after 1945, it did so under "the cover of Soviet patriotism" in an attempt "to ignite a xenophobic hatred of the outside world." Arguing this point, Merle Fainsod adds that Russian became Soviet patriotism as a result of "a many-sided effort to mobilize support for the regime and to combine the specifically Russian nationalism of the war period with an ideological commitment to Marxism-Leninism." In this amalgam one "basic ingredient was pure love of country, with particular accent on the leading position of the Great Russian people in the Soviet family of nations."[24]

This development paralleled the beginning of the Cold War, a conflict most expected to become a "hot" war. In preparation, the Soviet government retained its massive military machine, manned by conscripts and supported by a powerful military-industrial complex. This was a far cry from the militia system envisaged by the revolutionaries of 1917–20, and its commanders proudly termed themselves "officers," wore the once-hated epaulets, bore the once-banned personal ranks, and once again were drawn from cadres prepared and refined by an extensive network of military-educational institutions. As in 1914, graduates departed with expectations that an officer's career guaranteed both professional respect and social privilege.

The extent of the Soviet military-educational system was vast by Western standards. Indeed, it was almost a separate armed service in itself. Apart from the military faculties attached to some civilian schools, by 1990 it included 163 junior officer schools supported by extensive facilities for the teaching staff and students, support personnel, libraries, laboratories, and so on. Since this network existed to produce qualified "professional" commanders, the curricula were at least as narrowly specialized as those of similar institutions under the old regime. Moreover, despite the emphasis placed on the "*kraskom* tradition" and a "Marxist-Leninist world view," both instructors and graduates of all nationalities were inspired by a Russian nationalism that was as dogmatic and exclusive as that of their Tsarist forebears.[25]

As in other places and times, Soviet military schools were largely alienated from academe. Contemporary Russian critics charge that "certain historical reasons" caused the Soviet military system as a whole, and its educational network in particular, to develop "as an enormous isolated organization protected from any effective outside influence and, therefore, [as being] incapable of self-development and self-refinement."[26] Ironically, such charges echo those of radical critics of the Tsarist Army in the early 1900s and suggest the presence of the same gulf between society and its armed forces that the Bolshevik military program had sought to bridge in 1917. But if thinking officers of the late Imperial era were troubled by their alienation from civilian "society," their Soviet successors seem to have been less so, at least in public. Until challenged by *perestroika* in the late 1980s, they appeared confident in their sense of nationalist mission, a concept which had survived intact through two revolutions.

Russian Myths and Officers Today: Case Five

Demands for radical reform and the USSR's collapse left Russia's army facing its third major political crisis of the century. Although its present and future status again are in question, this case both differs from and resembles those already considered. Thus there is not even the apparent *tabula rasa* of 1917. Rather, today Russia has a battered, recently humiliated, but surviving military establishment that more resembles the Imperial Army of 1905–6 than the disintegrating corpse inherited by Lenin and Trotsky in the gray dawn of January 1918. Again as in 1905, the present army prefers to refrain from direct participation in the political process, but instead finds itself pushed into action by "circumstances and the ill-considered steps of politicians."[27]

Consider briefly the observations of the well-known commentator Aleksei Arbatov. In early 1995 he noted sadly that "the army itself is not so much an actor in, but rather a victim of, the Chechen adventure. After August 1991 and October 1993, the army . . . suffered its third and greatest blow" when the politicians "irresponsibly" chose to send in the troops rather than seek a negotiated solution. But if the politicos bore primary responsibility, Arbatov also charges that the "unnecessary casualties and hardship" suffered in Chechenia were a direct result of "the incompetence of top military leaders, the foolish boasting of the defense minister, ill-prepared and uncoordinated action and the traditionally indifferent and merciless attitude toward soldiers."[28]

Arbatov's comments differ little in essence from the charges made by critics of the Imperial regime and its army's performance in Manchuria during 1904–5, and its actions in restoring order during 1905–7. Fueled by scandals of conscript abuse and military incompetence in peace and war, this tone became dominant among civilian critics after the mid-1980s. Their mounting attacks since then have left the army and its officers increasingly humiliated in the eyes of nationalist Russians and their opponents alike. More mundane factors likewise have lowered public perceptions of the value of a military career, as well as the officers' sense of self-respect and personal well-being. Even adequate housing often is unavailable and pay is late, assuming it arrives. In the military schools directors face major obstacles in simply feeding and clothing their staffs and students, let alone in obtaining fuel to heat quarters or classrooms. Frequently they get by only through leasing school facilities to civilians and selling the staff's services on the open market. Efforts to acquaint junior officers with the art of command, the costly "lessons learned" in Afghanistan and Chechenia, or to prepare them for their new role as international "peacekeepers," suffer accordingly.

Resolution of the problems of officer recruitment and training obviously waits on the anticipated "restructuring" of Russia's military system. Furthermore, few deny the need for substantial military reforms in general, or of measures to "democratize" the military structure. Such calls were sounded long before the USSR's demise but, a decade of debate later, little has materialized and reform is more urgently needed than ever. Thanks to the scandals revealed in the late 1980s and the events in Moscow of 1991 and 1993, as well as the recent Chechen intervention, this issue has become a constant in the daily politics of Yeltsin's Russia. But if it remains near the top of the political agenda, the chances of immediate action are far from bright.

The problem, as Arbatov rightly notes, is that the issue of military reform "concerns the entire Russian state policy."[29] In fact, it goes much deeper still and involves the very nature of the state whose policies the armed forces must implement or support. This becomes clear if we consider the practical problems of recruiting the rank-and-file soldiers, be they conscripts or "volunteers," and the officers who are to lead them. As Gen. (Ret.) Yu. Ya. Kirshin observes, success in both depends on overcoming the military's alienation from society in general, and in this way creating "motivation so that an absolute majority of young people choose army service." Idealistic young officers then might decide to remain in service despite depressed living conditions. Yet here as in other areas, the Soviet Union's collapse has left an immense void. Kirshin points out that with the USSR disappeared ideals such as "the defense of 'the socialist fatherland,' 'the socialist way of life,' 'the achievements of October,' etc., which served the CPSU and which were directed toward maintaining the military's high moral spirit. . . ."[30]

Whatever the Soviet system's defects, Kirshin argues that these ideals "undoubtedly played an important role in rallying the Soviet Army." Meanwhile, although the "indoctrination of army command cadres loyal to Russia and its ideals" has become more vital than ever, even the "designation of the Russian Army as the Army of Russia and its functions" remain undefined.[31] In the Soviet period, such definitions would have been enshrined first and foremost within the official "military doctrine." While such now exists in the form of the "Basic Statement of the Military Doctrine of the Russian Federation," both military and civilian specialists agree in rejecting this formulation as inadequate. In late 1996 the Army therefore awaited nervously a new edition that hopefully would give its future efforts more precise direction.

This situation is made more pressing, and the task of indoctrination especially confusing, by the vague responsibilities assumed by Moscow within the new CIS, an organization that lacks its own armed forces, and by possible new threats on frontiers once deemed secure. Consequently, Kirshin maintains that the "most complex situation now exists in this sphere, inasmuch as many problems, on whose resolution will depend the Russian Army's moral spirit and its place in the political life of the country, not only have not been resolved, but also have not even been posed as practical tasks."[32]

While politicians squabbled and dallied, the high command recognized fully the need to find alternative formulas for expressing their traditional nationalist sentiments. Their search naturally has been for purely Great Russian forms that ignore the "internationalist" rhetoric of past Soviet slogans. Faced with the Communist party's declining prestige during the late 1980s, some prominent military men already had begun seeking "Russian" symbols to inspire their subordinates. Some saw a return to Tsarist traditions as one possible alternative. Thus military journals increasingly celebrated the successes of Imperial Russia's arms and by 1990, the once prestigious *Voenno-istoricheskii zhurnal* had become a magazine more suited to military historical "buffs" than to serious scholars.

Numerous books similarly celebrate the traditions and glories of Imperial Russia's army and fleet in general, and of their military-educational system in particular.[33] This trend was paralleled by official measures such as the navy's adoption of the old Andreevskii flag, the banner of the Imperial Fleet. Yet these were mere stopgaps and frequently led to amusing contradictions. For example, one missile cruiser proudly bears the name of the Communist aparatchik and Soviet defense minister, "Marshal" Dmitrii Ustinov, but sports an Imperial flag and has the Romanov's double-headed eagle on its wardroom wall. This last symbol now is widely displayed throughout the Russian Federation's armed forces, but some officers refer to it with ironic scorn as the "Chernobyl mutant."[34]

In the wake of the USSR's collapse, the military leadership also flirted briefly with building a new mission statement around protecting the millions of Great Russians left stranded in the newly independent states of the "Near Abroad."[35] If this course suited the military's own nationalist inclinations, it proved politically untenable and in view of the armed forces' declining capabilities, impossible in practice. Lacking other options, the military authorities have had little choice but to rely on past formulas. Thus Adm. F. Gromov, writing on the occasion of his service's 300th anniversary, could only exhort his seamen to emulate their predecessors by preserving "the combat core of the Navy, its spirit, and three centuries of tradition."[36]

Conclusions

"Time the destroyer," Eliot told us, "is time the preserver." Today the truth of his dictum is generally more evident than at any moment since 1945. Under the watchful eye of the Soviet gendarme, the nationalist fires of Eastern Europe and Eurasia long had seemed completely extinguished or, at the very least, safely banked. Some even argued that the Great Russians' nationalism had been coopted and even emasculated through the Soviet regime's assertion of "Soviet Patriotism" and "Fraternal Internationalism." As a consequence, as late as the mid-1980s many political commentators agreed in assigning these sentiments to Trotsky's historical dustbin.

Later events, and especially those since 1989–91, have unmasked this illusion. Rather than Russian nationalism having been coopted by Marxism-Leninism, the opposite was true. In spite of the ideological chasm supposedly separating the Imperial from the Soviet regime, Russian nationalism had resurfaced during the later 1920s to become the dominant element in the myth inspiring the Soviet military elite. This credo's essence had emerged from the Revolution of 1905, gone underground during that of 1917–18 and, on reemerging, adapted successfully to the rhetoric of the new regime. Supposedly an alien element in Marxism-Leninism, this nationalism survived in part due to the revolution's need for professional military skills to ensure its survival. In time the Soviet armed forces became a bulwark of such sentiments and their training programs, especially those for officers, a major conduit by which the nationalist ideal passed from generation to generation of Communist military commanders.

At present, the Russian state and its armed forces are searching for a new "founding myth" to justify their place on the future world stage. When this myth does take shape this same nationalism, in one form or another, will retain its place of honor. Western analysts must never forget that revolutionary enthusiasms aside, Russia has been a conservative nation and has always seen its "future contained in time past." This past can provide Western analysts with a vantage point from which to view today's problems. In so doing, they also will come closer to an understanding of their Russian opposites. For whatever the defects of their historical memory, Russians instinctively obey the imperative of the poet Olga Berggolts: "Let no one forget; Let nothing be forgotten."[37]

Notes

The basis for the arguments made in this essay are outlined more extensively, and with further citations, in David R. Jones, "Continuity and Change in the Russian Military Tradition," *RUSI Journal for Defence Studies* 120 (June 1975): 31–34; and idem, "Russian Military Traditions and the Soviet Military Establishment" in K. M. Currie and G. Varhall, eds., *The Soviet Union: What Lies Ahead? Military-Political Affairs in the 1980s* (Washington, D.C.: Government Printing Office, 1985), 21–47. On the development of military education in particular, see David R. Jones, "Military Academies" in G. N. Rhyne, ed., *The Supplement to the Modern Encyclopedia of Russian, Soviet and Eurasian History,* 2 vols. (Gulf Breeze, Fla.: Academic International Press, 1994), 1:30–40.

1. Richard A. Preston, *Perspectives in the History of Military Education and Professionalism,* Harmon Memorial Lectures in Military History No. 22 (USAF Academy, Colo.: U.S. Air Force Academy, 1980), 2, 26–27.

2. See, for example, L. K. Erman, *Intelligentsiia v pervoi russkoi revoliutsii* (Moscow: Mysl, 1966), 14–15.

3. E. Messner et al., *Rossiiskie ofitsery* (Buenos Aires: Institut N. N. Golivina, 1959), 37–39.

4. Ironically, these same civilian critics fully appreciate these same outmoded values when displayed by a police officer or fireman! Otherwise, in an era when managerial skills are valued most highly in peacetime, the need for leadership abilities seems to be rediscovered at the beginning of each new conflict. This was the case with the Russians in Afghanistan, and apparently in Chechenia as well. On the first conflict, see Anthony H. Cordesman and Abraham R. Wagner, *The Lessons of Modern War,* 4 vols. (Boulder, Colo.: Westview, 1990–96), 3:135–36. More generally, see the discussions in Robert L. Taylor and William E. Rosenbach, eds., *Military Leadership in the Pursuit of Excellence* (Boulder, Colo.: Westview, 1984).

5. Ales Adamovich, "Kuropaty, Khatyn, Chernobyl," *Literaturnaia gazeta,* 15 Aug. 1990, 6.

6. V. I. Lenin, *Collected Works,* 4th ed. (Moscow: Progress, 1965), 28:195.

7. For a fuller exposition of these arguments, see David R. Jones, "Armies and Revolution: Trotsky's Pre-1917 'Military Thought,'" *Naval War College Review* 27 (July–August 1974): 90–98, and "From Imperial to Red Army: The Rise and Fall of the Bolshevik Military Tradition," in Carl Reddel, ed., *Transformation in Russian and Soviet Military History* (Washington, D.C.: Government Printing Office, 1990), 51–74.

8. Typical are L. Spirin, "V. I. Lenin i sozdanie sovetskikh kommandnykh kadrov," *Voenno-istoricheskii zhurnal* 1 (April 1965), and V. S. Staritsyn, ed., *Sovetskii ofitser* (Moscow: Voenizdat., 1970).

9. Leon Trotsky, *The New Course* (Ann Arbor: University of Michigan Press, 1965), 99–105.

10. For more on this, see David R. Jones, "Motives and Consequences of the Red Army Purges, 1936–1937," in D. R. Jones, ed., *Soviet Armed Forces Review Annual 3: 1979* (Gulf Breeze, Fla.: Academic International Press, 1980), 256–64.

11. These reforms are detailed, and their effectiveness analyzed, in David R. Jones, "Russia's Armed Forces at War, 1914–1918," in Allan R. Millett and Williamson Murray, eds., *Studies in Combat Effectiveness*, 3 vols. (London: Allen & Unwin, 1987–88), 1:249–328.

12. Preston, 27.

13. Richard Holmes, *Acts of War: The Behaviour of Men in Battle* (New York: Free Press, 1985), 366.

14. For example, E. I. Martynov's *Iz pechal'nogo opyta russko-yaponskoi voiny* (St. Petersburg: V. Berezovskii, 1906), and "Citizen" in the series "Obshchestvo: armiia," nos. 3 and 4, *Voina i mir. Zhurnal voenno-obshchestvennoi* (Moscow: V. G. Zhdanovich, 1906).

15. Typical is Maksim Lipkin's ode "Po povodu otkritiia Gosudarstvennoi dumy" in the Warsaw Military District's newspaper *Ofitserskaia zhizn,* 8/21 May 1906, 1.

16. The first oath's text is printed in *Dekrety Sovetskoi vlasti,* 13 vols. (Moscow: Politizdat., 1959), 2:156–57.

17. On the development of officers' attitudes during 1917–20, see David R. Jones, "The Officers and the October Revolution," *Soviet Studies* 28 (April 1976): 207–33, and "The Officers and the Soviets, 1917–1920: A Study in Motives," *Sbornik of the Study Group on the Russian Revolution No. 2* (Leeds, England: Study Group on the Russian Revolution, 1976), 21–33.

18. Mrs. Philip Snowden, *Through Bolshevik Russia* (London: Cassell, 1920), 87.

19. Aleksei M. Sobolev, *Razvedka boem. Zapiski voiskogo razvedchika* (Moscow: Moskovskii rabochii, 1975), 5.

20. Anna Louise Strong, *Peoples of the USSR* (New York: Macmillan, 1944), 231. On the deportations, see Robert Conquest, *The Soviet Deportation of Nationalities* (London: Macmillan, 1960).

21. Nikolai Tolstoi, *Victims of Yalta,* rev. ed. (London: Corgi, 1979), 52.

22. On this postwar "myth," see Nina Tumarkin, *The Living and the Dead: The Rise and Fall of the Cult of World War II in Russia* (New York: Basic Books, 1994).

23. A. I. Odintsov, *Uchebnoe posobie po nachal'noi voennoi podgotovke* (Moscow: Voenizdat., 1971), 4–5.

24. Ibid., 114–15.

25. Jones, "Military Academies," 30, 39–40.

26. Yu. Ya. Kirshin, "The Army in a Democratic Society and Democracy in the Army," *Journal of Slavic Military Studies* 6 (March 1993): 3.

27. Ibid., 6. For a general review of the military in this period, see Pavel K. Baev, *The Russian Army in a Time of Triumph* (London: Sage, 1996).

28. Aleksei Arbatov, "Army Reform in the Midst of Disaster," *Moscow News* 65 (20–26 Jan. 1995): 4. The references to August 1991 and October 1993 refer to the attempted coup against Gorbachev and Yeltsin's later dispersal of the parliamentary opposition, respectively.

29. Ibid.

30. Kirshin, 16.

31. Ibid., 3–4.

32. Ibid.

33. For example, Yu. Galushko and A. Kolesnikov, *Shkola Rossiiskogo ofitsera: Istoricheskii spravochnik* (Moscow: Russkii mir, 1992), and S. V. Volkov, *Russkii ofitserskii korpus* (Moscow: Voenizdat., 1993).

34. A paratroop officer used this term in an interview aired on National Public Radio's *All Things Considered* in early October 1996. This writer noted the mixture of symbols during a reception on the cruiser *Marshal Ustinov* in Halifax, Nova Scotia in the late spring of 1992.

35. See, for example, "Principles of the Military Doctrine of Russia" (Draft), *Voennaia mysl,* Special Issue (May 1992): 4, as translated in Joint Publications Research Service JPRS-UMT-92-008-L (16 June 1992), JPRS Report, *Central Eurasia, Military Thought*, May 1992 Special Edition, 2. The "Near Abroad" is comprised of the non-Russian states of the former USSR.

36. Fleet Adm. Felix Gromov (Commander-in-Chief Russian Federation Navy), "After Three Centuries," U.S. Naval Institute *Proceedings* 122 (October 1996): 45.

37. Quoted in Harrison E. Salisbury, *The 900 Days: The Siege of Leningrad* (New York: Harper, 1969), xv.

Junior Officers in the Chinese People's Liberation Army*

William R. Heaton

In traditional China, military service was not a respectable occupation. According to a Chinese proverb, "Good metal is not used for nails, good men do not become soldiers." Young men were discouraged from seeking careers in the military. Persons seeking to become government officials prepared for their careers by obtaining a classical education. Such education emphasized stability, models from the past, and moderation. Martial qualities or skills were rarely sought or glorified.[1]

With the advent of modern military forces, attitudes began to change soon after the turn of the twentieth century. The Chinese Nationalist leaders of the Republic created in 1911 sought to introduce a military academy system in the early 1920s. As a key part of this effort, Nationalist authorities, with Soviet assistance, established the Whampoa Military Academy in May 1924. President Sun Yat-sen told the first cadet class that they would form the "backbone of the revolutionary army."[2]

Chinese Communist troops inherited this tradition. Samuel B. Griffith, a noted observer of the People's Liberation Army (PLA), wrote that Whampoa's curriculum "emphasized ideological indoctrination with the avowed aim of instilling in the young cadets a sense of dedication to the revolution." Revolutionary leaders believed that "highly motivated, idealistic, and politically conscious junior officers could successfully convey this spirit to the men they commanded." Since the emergence of Communist-led armed forces in the 1920s, leaders have emphasized the need to ensure both political loyalty and military success. The performance of Communist forces in subsequent battles against the Nationalists and the Japanese fulfilled the leadership's expectations.[3] Thus the development of junior officers, particularly since the founding of the People's Republic of China (PRC) in 1949, has underscored the importance of maintaining both military skills and political loyalty. Tensions between these two objectives have characterized the history of modern China's armed forces.

Junior Officers and Development of the PLA

In addition to waging a civil war with the Nationalists through most of the 1930s, the PLA participated in the "War of Resistance Against Japan" from 1937 to 1945. These struggles, including the epic "Long March," have been recounted elsewhere.[4]

*The views in this essay are the author's and not necessarily those of the United States Central Intelligence Agency or any other agency of the U.S. Government.

Most junior officers did not survive these early struggles; those who did eventually became leaders of the People's Republic.

During the civil war, the Chinese Communists' main objective, notes Jane L. Price, was to "build a unified political movement that could bring them to power."[5] Training institutions were established, including the Red Army School in 1931; the Red Army Academy in 1933; the Anti-Japanese Military and Political University (Kang-da) in 1937 (this institution trained 100,000 students by the end of World War II); and a variety of other academies and specialty schools. About 70 percent of the instruction was military training, with an emphasis on guerrilla warfare. The remaining 30 percent consisted of political instruction. Mao Zedong, who personally headed the party's education apparatus, insisted that politics were "in command."[6]

When the Red Army completed the Long March in 1935, it had about 20,000 soldiers. By the time of the Japanese attack in 1937 at the Marco Polo Bridge near Beijing, the number had increased to 45,000. By then the army was called the Eighth Route Army, and the typical soldier was in his early twenties. Foreign observers reported that these soldiers were disciplined and well led. Despite an intense Japanese counterinsurgency and continuing harassment by the Nationalists, the army grew to 910,000 by the end of World War II.[7]

The rapid expansion of the armed forces required an increasing number of junior officers. Most of those selected were comparatively uneducated men drawn from the ranks primarily for their political enthusiasm or combat skills. The Communist leadership, increasingly concerned about the political reliability of such soldiers, conducted a massive "rectification" campaign during 1942–44. This involved a systematic educational program to remold the ideological outlook of party members, including junior officers. All party members were required to study the writings of Mao and other party leaders. Officers were to become models of obedience to the party, while also showing individual initiative; they conducted self-criticism sessions designed to improve their political awareness. Many observers credit the rectification campaign with giving the party a renewed sense of unity and revolutionary fervor. Price points out that this campaign developed the key techniques for party political education in the early decades of Chinese Communist rule.[8]

The Consolidation of Communist Power

After the PRC was established in 1949, China was essentially governed by military control commissions until people's congresses could be established. This period culminated with a new constitution and the election of the National People's Congress in 1956. In 1950, meanwhile, Chinese "volunteers" had intervened on the Korean peninsula and fought United Nations (principally United States) forces until the armistice of 1953. Thus, for the PLA, the challenges of the early 1950s included both consolidation at home and fighting the United States abroad.

Many PLA junior officers served effectively in the Korean War. Citing military histories and personal accounts, Griffith observed that Chinese troops displayed excellent discipline and were "courageously led at the small-unit level. . . ." Additionally, Chinese combat techniques consistently improved.[9]

The Chinese were seriously deficient, however, in modern weapons and logistics support, resulting in huge losses of men. In the spring of 1951, for example, a major Chinese offensive failed because of these deficiencies.[10] PLA morale dropped sharply, and many enlisted personnel defected to UN forces. The party intensified political indoctrination among the officer corps, which remained loyal for the most part; few officers defected. Many were captured, however, and reported a serious decline in morale among both officers and men.[11]

Setbacks in Korea helped military leaders recognize that modernization was essential. Assisted by the USSR, the Chinese sought to improve the PLA rapidly. They reduced the military's overall size, imported large quantities of military equipment, and strengthened the armed forces' air and naval components.

Widespread illiteracy posed a significant obstacle to modernization. In 1951, about 80 percent of the soldiers in the PLA were illiterate, and many junior officers could barely read. A campaign to improve literacy was undertaken between 1952 and 1954, and formal military training was introduced in 1953. Because the PLA lacked a noncommissioned officer (NCO) corps, junior officers were expected to play the central role in all training—literacy, military, and political.

Professionalizing the Officer Corps

In 1952, the PLA took its first steps toward creating a professional officer corps. General Xiao Hua, director of the General Political Department (which runs the PLA's political commissar system), asserted that the military needed to master modern military techniques and the use of modern weapons and equipment.[12] With help from Soviet advisers, a new system of military academies was established.

Prior to 1955, officers were designated by position, not rank, and wore no rank insignia. A system of ranks and awards for officers was introduced in that year. Junior officers included second lieutenants, lieutenants, captains, and senior captains. The maximum age for lieutenants was fixed at thirty and that for captains at thirty-five. Officers not promoted were transferred to the reserves.

In 1955, following the enactment of military conscription, about 500,000 soldiers were conscripted, a number that remained fairly constant until the reforms of the early 1980s.[13] With such large numbers of new recruits entering the PLA, much of the work of junior officers was related to training the conscripts.

Junior officers also were expected to become specialists in the new weaponry being provided by the USSR. They were encouraged to study aviation, shipbuilding, mechanics, civil engineering, meteorology, and a variety of other technical subjects. Others were assigned to learn foreign languages in order to participate in intelligence work, particularly preparing to serve as military attachés in Chinese embassies abroad.

Resistance to modernization quickly developed. Some veteran officers asserted that the changes were unnecessary and could hurt the armed forces. Political officers also may have opposed the changes, possibly because their technical proficiency was falling behind that of new officers.[14] During 1956–58, a major campaign within the PLA sought to restore the primacy of ideology. All junior officers were required to study Mao's writings in order to raise their political awareness.[15] These campaigns were apparently unsuccessful; by 1958 and the initiation of the Great Leap Forward, a gulf was evident between the party and elements of the officer corps. John Gittings, an authority on the PLA, noted that either because of or despite the ideological campaign, the relationship between the PLA and the party was deteriorating.

The Great Leap Forward and Its Aftermath

The Great Leap Forward (1958–61) was marked by an effort to return officers to the ranks to re-create the spirit of the revolutionary war. Belt-tightening measures were a key part of this effort. Military training and military construction were reduced on the grounds that they were too expensive. Officers and men were required to arrange for their own training equipment. Young officers, fresh from training school, were criticized for "bad attitudes" when they asked for greater quantities of materials and equipment.[16] Officer salaries were reduced. Dependents were sent home.

The "people's war" concept espoused during the Great Leap emphasized that all workers and peasants could be soldiers and all soldiers could be workers and peasants. Militia forces were considered the backbone of the "people's war" strategy. In the PLA, senior officers, as well as the few junior officers who had not risen from the ranks, were returned for a year to basic units where they performed manual labor alongside ordinary troops. Because of China's economic troubles, the PLA had to supply much of its own food and to assist in civil construction projects and harvests.

Just as the Great Leap proved disastrous for China's economy, so did it disrupt the armed forces. Officers who had embarked on professional training were suddenly whipsawed by new political indoctrination movements. Initial drives for professionalism, including technical modernization and the introduction of ranks, gave way to renewed emphasis on political reliability. By the end of the Great Leap the PLA was neither militarily effective nor politically unified.

Because of the damage done by the Great Leap Forward, Beijing reversed many economic and social policies. In the early 1960s, the regular armed forces were ordered to concentrate once again on their military mission, and the militia was redirected to concentrate on production. But the growing tension with the Soviet Union and the 1962 war with India spurred Beijing to resume treating the militia as a key component of national defense.[17]

Two factors underscored the priority given to professionalizing the military: China's detonation of a nuclear weapon in 1964 and, following the suspension of Soviet assistance, its need to rely on its own defense industries to produce major military items. China, however, was not about to return to the type of professionalism introduced in 1955. A leading advocate of military professionalization, Peng Dehuai, commander of Chinese forces during the Korean War and subsequently defense minister, was replaced because of a dispute with Mao over the Great Leap. His successor, Lin Biao, advocated socialist education for the PLA and launched a movement to study Mao's works. Officers were expected to be both "red" and "expert." Lin strengthened the political commissar system in order to reassert the party's leadership. He also introduced a new movement within the armed forces to ensure party control through ideological indoctrination.

At the company level, this new movement meant dismissing unreliable party members and restoring party branches in all platoons. Party organization was then revamped in the companies. Junior officers were expected to be members of the Communist Youth League or the party itself, and the Youth League became responsible for recruiting new soldiers. As party organization solidified in basic units, junior officers assumed responsibility for conducting military and political training.[18]

The program experienced some success. The "Socialist Education Campaign" of the mid-1960s encouraged the entire nation to learn from the PLA. Ordinary soldiers such as Lei Feng were made national heroes whom the populace was urged to emulate. The PLA's prestige appeared to rise substantially.

In 1964, China's detonation of a nuclear weapon conveyed an image of national power and intense national feeling. The political loyalty of the armed forces seemed assured. Yet modernization and technical specialization remained elusive. Many junior officers had no more than vague notions of modern equipment. With Soviet aid, China had developed a large navy and a large air force, but skill and logistics lagged.[19] At the basic unit level, ground forces were able to maintain combat skills but could do little beyond that. So long as China did not need to project power abroad, this level of combat capability was generally seen as sufficient.

The Effects of the Cultural Revolution

Chinese generally characterize the Cultural Revolution (beginning in 1966 and continuing through Mao's death in 1976) as ten years of disaster for China. It was no less so for the PLA. The military suffered from the same factionalism that disrupted the party and the nation. In the Wuhan Incident of July 1967, for example, an independent division with the support of the military region headquarters arrested two central authorities.[20] This "mutiny" was suppressed, but local disorder persisted. In 1968 Beijing ordered main-force units to help provide administration throughout the country. Revolutionary Committees were formed, with the military playing a prominent role.

Most military academies, along with other colleges and universities, were closed during the Cultural Revolution.[21] As had been the case during the revolutionary war of the 1930s and 1940s, officers were chosen from the ranks. Most had a high school education or less. Political connections were emphasized, and strong patron-client relationships developed within the military.[22] Promotions depended on political qualifications. Ranks and insignia were abolished. Headquarters became bloated. Troops devoted about 70 percent of their time to political training; military training was relegated to the remainder.

Only after border clashes with the Soviet Union in 1969 did China's leadership take pause and begin to relieve the military of many of its civil administration duties. This retrenchment gained momentum after an apparent coup attempt in 1971 by Defense Minister Lin Biao, Mao's designated successor. Lin, who died in an airplane crash together with some other top military leaders while trying to flee the country, allegedly resisted a plan to restore party committees within the armed forces that signified the reassertion of civil authority.[23] By 1973, the PLA was moving out of local administration, and the party was beginning to reassert its role. Not until 1977, however, did military professionalism begin to reemerge.

At the time of Mao's death in 1976 and the winding down of the Cultural Revolution, the state of junior officers in the PLA could be assessed as follows. First, most had been selected and promoted on the basis of ideological credentials. (Despite this, the turmoil of the Cultural Revolution had destroyed party organization within the military.) Few junior officers were well educated and few had technical skills. Second, junior officers spent most of their time in ideological indoctrination, civic action, and production. Less time was spent in military training. Third, these officers were imbued with outmoded strategic and inadequate tactical doctrine. They were taught the strategy of "people's war," in which the PLA was to lure an enemy deep into China, trade space for time, allow the enemy to become overextended, and then defeat it. Air and naval doctrines were similarly inward oriented.[24] Fourth, many junior officers became entangled in local politics when their superiors became involved. Fifth, many had little meaningful work because "staffs were bloated, headquarters were overmanned, and authority and responsibility were diffused."[25] Sixth, many officers were despised by the civilian population because the PLA had taken over property during the Cultural Revolution. The military's slowness in returning much of this property—including land, buildings, and homes—heightened resentment among ordinary citizens. The PLA, once an honored and glorified institution, was now anathema to many, and junior officers took some of the slings and arrows.

Deng and Reform

Most observers credit the late Deng Xiaoping with launching the effort to revitalize the PLA, although he was not able to implement his ideas until he consolidated his position in 1978. By then, most military leaders agreed that the PLA was ill suited to fight a modern war. The reforms associated with Deng picked up steam after the PLA encountered difficulties attempting to invade Vietnam in 1979.

Deng's reform agenda included these key elements: creating the foundation for a modern national defense by strengthening the Chinese economy; revitalizing the officer corps by improving education and professionalism; adopting a new strategy, "People's War under Modern Conditions," that more realistically matched China's vital interests with its military capabilities, and adjusting tactics to fit reality; reorganizing all of the services; cutting the size of the military by one million, bringing it down to 3.2–3.5 million, and using the savings to invest in modernization; obtaining modern equipment by purchasing technology and weapons from abroad, and pushing defense industries to enter the forefront of advanced technology.[26]

The impact of these measures on the junior officer corps was first reflected in recruitment and training. The number of military schools and academies expanded rapidly. Deputy Chief of Staff Zhang Zhen told interviewers in 1984 that more than 100 such institutions had been restored.[27] Two categories were established: command schools and technical schools. The command schools had three course levels: junior, intermediate, and senior. Junior officers at the platoon level attended the junior-level course. A few outstanding officers at the company level were permitted to attend the intermediate course. The technical schools were intended mainly for more senior officers, although junior officers with medical and other technical qualifications could be admitted.

In 1982 the Military Commission directed that by 1985, over 70 percent of junior officers should have completed training in military schools and colleges. All air force pilots and naval ship commanders were to complete the requirement. By the time the new military service law was enacted in 1984, nearly 1 million officers had already received training in the new academies.[28] Lonnie Henley's ten-year review of professional officer education identifies further refinements. The PLA sought high school graduates for the most part, believing that college graduates were more difficult to train (except for those in a few selected technical fields). The basic officer course was offered for one to four years: one year for college graduates, two years for enlisted persons seeking to advance to the officer corps, and three to four years for high school graduates.

A majority of officers remained in the ground forces, but some were selected for specialized training. A number of company-grade officers were selected to attend basic-level political academies for training as political officers. Others were selected for air and naval training at service-operated schools. Still others were chosen to receive special training in particular fields such as strategic nuclear forces, logistics, and medical support.[29] Some of the schools for ground forces were operated by China's military regions (in 1979 there were eleven military regions; the number was reduced to seven in 1984). Henley estimated that the various academies could train about 15,000–20,000 officers per year, approximately two-thirds of the PLA's perceived need for 30,000 per year.

The Military Commission also ordered significant curriculum changes. In the fall of 1980 at a conference on military academies, China's top military leaders had concluded that military educators must "emancipate our thinking and boldly change the content of the training and education programs" in the schools and academies.[30] In time, Deng Xiaoping's military writings (his *Selected Works* was published in 1983) supplemented Mao's. In one essay, Deng stated that all officers should attend military academies, that they should study modern warfare—including strategy—and that promotions should be related to their mastery of the material. He also wrote that officers should study the works of Marx, Lenin, and Mao, as well as history, science, geography, foreign languages, mathematics, physics, chemistry, industry, agriculture, and other disciplines. Deng argued that military education should prepare officers not only for modern combat but also for civilian jobs when they retired.[31]

To make room for the greater emphasis on technical fields, the time spent on ideological instruction was reduced (except in the political academies where ideology was the specialty). Lessons denounced the idea that "millet plus rifles" (a slogan common during the Great Leap and the Cultural Revolution) were sufficient. Students were to learn from the world's "modern scientific and cultural achievements and the fresh experiences gained from modern military campaigns."[32]

How successful were these reforms? Henley predicted in 1987 that improvements in military education would be of "great benefit." Because of the PLA's improved technology, better educated and more specialized officers would be needed to employ modern weapons. Furthermore, such officers would be needed to develop the strategy, doctrine, tactics, and training necessary to conduct modern warfare.[33] In 1988, ranks and rank insignia—banned since 1965—were restored. Problems persisted, however. One source noted that the academies felt no obligation to develop training materials for field units, and that links between operational units and the academies were not formed.[34] Moreover, key challenges arose to confront the PLA in the late 1980s and early 1990s.

Shocks to the System

Two events substantially altered the PLA's course. The first, in 1989, was the student movement that culminated in the Tiananmen suppression. Some junior officers in the Beijing Military Region participated in the demonstrations at Tiananmen Square that preceded the 4 June crackdown. Furthermore, some officers hesitated to order troops to enforce the government's martial law decree issued in May. Only when outside units were brought to the capital could the popular movement be crushed. Senior officers in the Beijing garrison were blamed for the initial failure and were replaced. The government was embarrassed by the international publicity given to its use of armed force to suppress the populace, and the PLA was discredited in the minds of many Chinese because of its role in the crackdown. Leaders concluded that the PLA urgently needed political revitalization to ensure loyalty to the party. Consequently, the armed forces once again focused on political indoctrination. A major propaganda campaign was launched and political training was upgraded.[35]

Just as political revitalization was gathering steam within the PLA, another shock occurred: the Gulf War. The Chinese clearly believed Iraq would defeat the United States because Baghdad was pursuing a Chinese-style strategy of "people's war under modern conditions." The Chinese expected that the U.S.-led coalition would be overextended, fail to dislodge the Iraqis from Kuwait, and eventually give up and go home. Chinese press accounts clearly show that Beijing was stunned by the coalition's rapid victory. Practically overnight, Chinese leaders discovered that their military tactics, as applied by the Iraqis, collapsed against superior Western tactics and technology.[36] For Beijing, this meant that its own forces, despite improvements in training and equipment over the previous two decades, were ill prepared to defend the nation.

Reports of deteriorating junior officer morale accompanied the twin shocks of Tiananmen and the Gulf War. In 1991, investigative journalists of the *Jiefangjun bao* (Liberation Army Daily) reported that:

> Sharp disagreements existed between junior and senior officers over the value of ideological indoctrination. Senior officers argued that junior officers did not understand the army's traditions and past travails, and accused them of depending on assistance from higher headquarters and resorting to makeshift solutions instead of engaging in grassroots political work.

> Political commissars claimed that academy-trained officers lacked ability and initiative and did not set good examples for their units. According to the commissars, the officers did not care about their soldiers, and relations between officers and men were often tense.

> For their part, young officers told the investigators that unit leaders were jealous of their academic backgrounds, and did not display confidence in their ability to lead. One platoon leader stated that his company commander had not allowed him to perform his duties for two years. Another claimed that he was bypassed and ignored by his commander, harming unit cohesion.[37]

Leadership Response: Politics and Professionalization

The events of June 1989 apparently took a heavy toll on company-grade officers. One U.S. attaché noted that such officers were mostly college-educated lieutenants and captains who had been in universities recently enough to empathize with student demands. Thus their political reliability was uncertain.[38] The *China Daily* stated that the PLA would again select some new officers from the ranks, at least in part because of a shortage of academy-trained officers.[39] But another purpose probably was to ensure greater political reliability.

In the wake of the Gulf War shock the leadership once again was forced to review junior officer qualifications and preparation. In May 1991, the New China News Agency reported that PLA armies and divisions were creating teams for the professional training of junior officers. The item stated that superior junior officers should be selected for training as instructors.[40] In 1992, the General Staff Department modified its 1990 general training program to place more emphasis on military operations under high-tech conditions. The program was to be completed by 1995.[41] One of the goals was to develop an NCO corps that could augment junior officers commanding small units.

The PLA also reviewed the selection process for junior officers. While reaffirming selection criteria based on political integrity, ability, and merit, the PLA said that collective discussions should be used to select officers rather than having one person make the decision. The *Jiefangjun bao* warned that until the quality of officers was improved, unit training could not be upgraded. The General Staff Department instructed that junior officers in various departments should be assigned temporary duty to lower combat units in order to be "tempered" and improved.[42]

In a speech to leaders of the Beijing Military Region on the party's seventy-third anniversary, General Secretary Jiang Zemin emphasized the training of young officers, stating that they should "have high political quality, very strong work capability, and very good work style." A related report stated that PLA academies and schools had trained more than 700,000 officers over the past fifteen years, and that the quality had improved significantly.[43] In 1996, the Central Military Commission apparently restored a requirement that all officers attend a military academy before commissioning.[44]

Several recent accounts assert that junior officer training and professionalism are making significant progress. In June 1996, the *Jiefangjun bao* reported that 145,000 officers had graduated from college through inservice education. The army paper observed that the General Staff Headquarters, the General Political Department, and the General Logistics Department had jointly developed a program to develop science and technology over the next five years.[45] A July 1996 report from the Nanjing Artillery College, which trains junior artillery officers, established the slogan, "A commander without technical knowledge is not a qualified commander." The college claimed that 83 percent of its new graduates were competent when they were sent to the field.[46] Around the same time, China's Strategic Missile Corps claimed that a historic change concerning the "knowledge structure" of its officers had occurred, with the result that "overall quality had improved by a large margin." Most lower level officers had been trained in military academies and schools, and more than 50 percent of the officers were "scientific and technological officers."[47]

Despite such efforts, criticism of junior officer training persisted. In an article in *China Military Science*, Lt. Gen. Tan Naida, political commissar of the Jinan Military Region, argued that junior officer training should be radically revised. Academies, he wrote, provided inefficient and substandard education: "The contents of military education are outdated and one-dimensional, the starting point is too low, and there is no system: the necessary coordination among military academies at all levels is absent." In Tan's view, the primary-level military academies should "make attaining college undergraduate standards [their] basic educational goal; they should insist on providing a regular, basic college-level science education." He urged that China's goal should be to prepare to fight and win high-tech wars because China's future opponents would be armies with high-tech weaponry.[48]

At the same time, the PLA leadership was not satisfied with political attitudes among many junior officers. In his 1 August 1996 speech on the PLA's anniversary, Chief of Staff Zhang Wannian stated that the PLA must strengthen party leadership within the military. He asserted that because "hostile Western forces" did not want to see a strong China, "they are doing all possible to advance their strategy of 'Westernizing' and 'dividing' socialist China." To combat Western efforts to overthrow the Communist party and destroy China, the army must "intensify" ideological indoctrination.[49] Zhang denounced factions and cliques, and demanded that the PLA "liberate thinking, break with convention, and boldly employ outstanding young cadres to develop and create a large number of cross-century military officers."[50]

In the spring of 1996, the PLA conducted maneuvers adjacent to Taiwan in an attempt to exert pressure on voters in the island's presidential election, especially to intimidate advocates of Taiwan independence. China also intended the exercises to send a message to the United States about the dangers of encouraging "separatist" activities on the island. An article in *Jiefangjun bao* on the exercises claimed that the PLA had improved its combat ability under high-tech conditions but noted serious gaps in carrying out operations. The article implied that the PLA had far to go before it would genuinely be able to wage combat when faced with high-tech weaponry.[51] In an inspection tour of units in northeastern China, Military Commission Vice Chairman Zhang Zhen also reiterated the importance of realistic training under modern conditions. Zhang emphasized leadership at the squad and platoon level, observing that only realistic training would ensure the "healthy growth of the younger generation."[52]

The Air Force

Conditions among junior officers in China's air force are not very different from the ground forces, according to a 1995 RAND Study.[53] The air force has twenty-six schools and academies—ten for flying and sixteen for related subjects (AAA, SAM, communications, engineering, logistics, etc.). The numbers have changed considerably over the years; there were seventeen flight schools at the time of the Cultural Revolution. In the mid-1980s, as a consequence of Deng's military reforms, some schools were consolidated. For the first time, new pilots were expected to graduate from air force academies with college degrees.[54]

Because the air force does not have an NCO corps, officers are trained for hands-on operations. For example, maintenance officer cadets at the Engineer College and maintenance technical training schools complete from two to four years of hands-on training before they join operational units.[55]

The air force uses the PLA "5-3" tier system for training and education. This system consists of five tiers of specialized or technical training and education for officers: secondary specialized, specialized college or equivalent, university or equivalent, master's degree program, and doctoral program. Also, a three-tier system for professional military education has been established: primary command and leadership training, intermediate command and leadership training, and higher command education.[56]

According to the RAND study, the air force selects about 3,000 qualified high school students for pilot training annually, of whom about 1,300 graduate. Some college graduates also are recruited, but the success rates are not as high. In 1986 China moved from a three-phased process to a four-phased system of pilot training. For example, fighter and ground attack pilots attend basic flight school for twenty months, a flying academy for twenty-eight months (where they are expected to graduate with a degree), a transition program for one year, and then four or five years of training in an operational unit.[57]

Fighter pilots fly about 100–110 hours per year (A-5 ground attack pilots fly up to 150 hours) and bomber pilots average about 80 hours per year. The air force began awarding pilot ratings in 1986—special, first, second, and third grade—after the pilots completed various stages of training. The rating received depends on time-on-station, flying hours, flying in bad weather, and special missions. The air force awards aircrew ratings to navigators, communications and gunnery personnel, and instructor pilots.[58]

Pilot training suffers from important deficiencies. The air force appears unable to conduct: combat training at night and in poor weather; two- and four-ship attack training against low-altitude targets; mobility training, including frequent scramble exercises, "intertheater" airstrike combat training, and massing aircraft for defense or for an offensive strike; combined combat training among various air defense branches and services; and dissimilar aircraft and simulator training.[59]

Key Issues Today

Quality of junior officer recruits

For the past twenty years the PLA has been largely unable to attract the most qualified people to serve as officers. Talented and educated Chinese, for the most part, want to avoid military service because the PLA has a bad public image (especially since the Tiananmen upheaval) and because they can make more money elsewhere. Chinese youths with skills and ambition go into business or abroad as researchers. The PLA obtains some officers via the compulsory service laws, by appeals to patriotism, and by offering scholarships to academies. For the most part, however, it attracts those who want to get ahead but are less able to function effectively in the civilian economy.[60]

A December 1992 article in *Jiefangjun bao* observed that improvements in China's economy have caused PLA soldiers to feel disadvantaged. Many wanted to quit to get better paying jobs. According to the article, "Servicemen must be made to understand that the disparities in benefits between them and their civilian contemporaries was only a temporary phenomenon," and that they must "make sacrifices."[61]

Reports of violent incidents involving junior officers reflect the lack of quality. Such accounts, although infrequent, suggest serious problems. In November 1993, for example, the Hong Kong journal *Cheng Ming* reported that a shooting incident broke out in a PLA antichemical warfare unit in Anhui Province. The soldiers were from Hebei and Sichuan provinces; the company officers were from Hebei. One weekend the soldiers requested leave to visit a local city. The battalion commander disapproved all leave requests because the company was scheduled to participate in a live-fire training exercise the next week. Despite this, the company commander allowed three platoon leaders and forty soldiers from Hebei to take leave. The Sichuan soldiers found out and protested in the mess hall. The company commander fired a pistol in the air, infuriating the soldiers who then beat him and the unit political officer to death. In the ensuing chaos, three soldiers were killed and several officers were injured. An inquiry identified the causes as "anarchism, lack of military discipline, and opposition to unit leadership."[62]

Corruption and factionalism

Widespread corruption affects the senior officer corps more than junior officers, but some junior officers also have been tainted. With reform and the opening of the Chinese economy in the 1980s and shortfalls in military budgets, the PLA began to operate businesses of its own on a fairly large scale. Today it runs more than 20,000 businesses, including prominent hotels and restaurants in Beijing. Many of these enterprises have been extremely profitable, but business activities probably have aggravated corruption.[63]

In a recent study of China's officer corps, June Dreyer notes that "factions with various different bases such as place of birth, family connections, educational institution, and field or group army affiliation not only continue to exist, but thrive." Such factionalism particularly pervades the senior ranks, but, as junior officers rise in ranks, they too become implicated.[64]

Shortfalls in expertise

Both the Blasko and RAND studies maintain that junior officers lack the conceptual and technical skills necessary for command on the modern battlefield. Blasko points out that ground-force officers are good at basic research and memorization but have difficulty with applications. Most junior officers do not know how to drive automobiles and know little about computers. Chinese researchers, including those in military academies and "think tanks," write about modern battlefield concepts, including information warfare, but little of this filters through to the field.[65] The RAND study shows that Chinese Air Force officers fly barely enough hours to maintain proficiency.[66] The air force emphasis is on not losing or damaging equipment, and pilots are very cautious. Observers of Chinese military exercises indicate, for example, that the Su-27s purchased from Russia by China are not employed to their capabilities. These observers also note that the military has problems with equipment maintenance, probably because the responsible junior officers lack technical ability.

Conclusions

Most Western specialists on the Chinese military agree that it will be ill equipped to fight a modern war against an adversary with advanced weapons—such as the United States or Japan—over the next decade and beyond.[67] At the same time, many observers also believe that the Chinese have little incentive to change because most of their potential opponents in the region similarly lack modern capabilities. China probably can achieve its basic stated objective to protect its economic interests 200 miles beyond its coasts through gradual conventional-forces improvements.[68]

Junior officers are one component—albeit a key one—of China's overall capabilities. The corps of junior officers has been shaped in a cauldron of revolution, upheaval, and debate over the roles of military professionalism and political reliability. Observers generally credit junior officers in the PLA's ground forces as being average but not superior leaders. Much of their time is spent training because of the large number of new recruits in units each year. Most of the training is rote; officers are not motivated to change the system, to innovate, or to take risks. Leadership probably would not be bad in small-infantry attacks, but is not adequate to the modern battlefield. Officers in other Chinese services have greater technical skills and different missions but are much like their ground force compatriots in leadership qualities.

When the PLA was founded nearly seventy years ago, its creators envisioned a core of young leaders, strongly motivated by ideology, who could lead China and the Communist party to victory. Today, although some younger officers seem nationalistic, most appear disillusioned by the atrophy of the system. Whether the party can re-create the dedication and altruism of the revolutionary generation—let alone modern fighting skills—is doubtful. Perhaps the key question today is whether the regime can maintain the military support it needs to survive.

Notes

1. For a detailed study on the education and training of Communist elites in the pre-Communist period, see Jane L. Price, *Cadres, Commanders, and Commissars: The Training of the Chinese Communist Leadership, 1920–45* (Boulder, Colo.: Westview, 1976), 7.

2. Samuel B. Griffith II, *The Chinese People's Liberation Army* (New York: McGraw-Hill, 1966), 14–15.

3. Ibid., 15.

4. The best known source is Edgar Snow, *Red Star Over China* (New York: Random House, 1944, 1948).

5. Price, 122.

6. Ibid., 137–64; Griffith, 34.

7. Griffith, 47–77.

8. Price, 184–85.

9. Griffith, 169.

10. John Gittings, *The Role of the Chinese Army* (New York: Oxford University Press, 1967), 133–34.

11. Griffith, 171.

12. Gittings, 132.

13. Ibid., 149. Gittings indicates the PLA probably numbered about 2.5 million in 1955. Other sources estimate the number to be about 3.5 million. Terms of service were: ground forces, three years; air forces, four years; naval forces, five years.

14. Ibid., 156.

15. Ibid., 168–69.

16. Ibid., 188.

17. China defeated India decisively in the border war. In part this reflected China's superior preparation and conditioning to fight at high altitudes. It also reflected the poor preparation of the Indian troops who were moved to the front. The ability of China's ground forces to fight effectively reflected good leadership at the unit level, just as in the Korean War. See Neville Maxwell, *India's China War* (London: Cape, 1970).

18. Gittings, 242–62, provides an excellent account of developments during this period.

19. For the size of China's armed forces, including air and naval components, throughout the period between the Anti-Japanese War and the Great Leap Forward, see ibid., 303–5.

20. Harvey W. Nelsen, *The Chinese Military System: An Organizational Study of the Chinese People's Liberation Army,* 2d rev. ed. (Boulder, Colo.: Westview, 1981), 35–37.

21. Lonnie D. Henley notes the number of military schools dropped from 140 to 40; see "Officer Education in the Chinese PLA," *Problems of Communism* 36 (May–June 1987): 56.

22. A classic study by William Whitson, *The Chinese High Command* (New York: Praeger, 1973), examined carefully the concept of field army loyalties in China. Although some have challenged Whitson's hypothesis that field army identification explained the balance of military power in China's high command, most acknowledge the value of his work in examining personal and organizational ties. See Robert E. Johnson, Jr., "China's Military Modernization: A Systemic Analysis," in Charles D. Lovejoy, Jr., and Bruce W. Watson, eds., *China's Military Reforms* (Boulder, Colo.: Westview, 1986), 118–19.

23. Nelsen, 43.

24. Lonnie D. Henley, "China's Military Modernization: A Ten Year Assessment," in Larry M. Wortzel, ed., *China's Military Modernization: International Implications* (New York: Greenwood, 1988), 99–100.

25. Henley, "Ten Year Assessment," 99.

26. Henley notes that this approach rejected doctrinaire Maoist sentiment that favored a highly politicized, low-technology military. See ibid., 111–12.

27. "Interview with PLA Deputy Chief of Staff Zhang Zhen" (in Chinese), *Liaowang*, no. 31, as carried by *Xinhua* (New China News Agency), 29 July 1984, in *Foreign Broadcast Information Service (FBIS) Daily Report (China),* 30 July 1984, K5–6.

28. William R. Heaton and Charles D. Lovejoy, Jr., "The Reform of Military Education in China: An Overview," *China's Military Reforms,* 93. Note that the system for recruiting and training officers came under the direction of the Central Military Commission of the Chinese Communist Party, not the Ministry of National Defense.

29. Henley, "Officer Education," 58–59. Henley's article has an excellent chart providing details about the various military academies.

30. *Xinhua,* 7 Nov. 1980, in *FBIS Daily Report (China),* 10 Nov. 1980, L4. For further details on the issues that military education faced, see William R. Heaton, "Professional Military Education in the PRC," in Paul H. B. Godwin, ed., *The Chinese Defense Establishment* (Boulder, Colo.: Westview, 1983), 89–120. The first group of academy cadets joined PLA units as junior officers in 1982. See *Jiefangjun bao* (Liberation Army Daily), 16 July 1991.

31. *Xinhua,* citing *Jiefangjun bao,* 4 June 1983, in *FBIS Daily Report (China),* 5 June 1983, K33.

32. Radio Beijing, 4 Oct. 1984, in *FBIS Daily Report (China),* 9 Oct. 1984, K8–9.

33. Henley, "Officer Education," 71.

34. Heaton and Lovejoy, 95.

35. Michael T. Byrnes, "The Death of a People's Army," in George Hicks, ed., *The Broken Mirror: China After Tiananmen* (Chicago: St. James Press, 1990), 133–51. General Byrnes presently serves as U.S. defense attaché to China.

36. This point was elaborated in an interview with Lt. Col. Dennis J. Blasko, U.S. Army, 9 Oct. 1996, National Defense University, Washington, D.C. Also see his chapter, "Better Late Than Never: Non-Equipment Aspects of PLA Ground Force Modernization," in C. Dennison Lane, Mark Weisenbloom, and Dimon Liu, eds., *Chinese Military Modernization* (Washington,

D.C.: AEI Press, 1996), 141–42. Lieutenant Colonel Blasko served as a military attaché in Beijing and Hong Kong in the early 1990s.

37. Report from the United States Defense Liaison Office (USDLO), Hong Kong, 7 Apr. 1992. This report summarizes the reports on junior officer morale in *Jiefangjun bao*. Consecutive articles in *Jiefangjun bao* on 23 and 24 July 1991 detailed some of the problems. The new junior officers initially brought enthusiasm to their units but often did not mix well with the troops. The reports generally were negative about the leadership skills of these officers. Morale among junior officers also may have been low because they were not authorized accompanied tours. See ibid., 4 and 10 Aug. 1990.

38. Report from USDLO, Hong Kong, 21 Apr. 1990.

39. *Renmin ribao* (China Daily), 21 Apr. 1990. This article is cited in the above report.

40. *Xinhua*, 31 May 1991, in *FBIS* (OW3105080091), 31 May 1991.

41. For an extensive discussion of PLA training practices and issues, see Dennis J. Blasko, Philip T. Klapakis, and John F. Corbett, Jr., "Training Tomorrow's PLA: A Mixed Bag of Tricks," *China Quarterly* (June 1996).

42. Report from USDLO, Hong Kong, 8 Feb. 1994, citing several issues of *Jiefangjun bao*.

43. Report from USDLO, Hong Kong, citing *Jiefangjun bao*, 2 July 1994.

44. Reuters, 22 July 1996.

45. Zhang Handong and Li Ping, "Three Headquarters Set Out All-Army Goals for Inservice Education in Science and Culture During Ninth Five-Year Plan" (in Chinese), *Jiefangjun bao*, 2 June 1996, in *FBIS Daily Report (China)*, 11 July 1996, 50–51.

46. *Xinhua*, 30 July 1996, in *FBIS* (OW01811496), 30 July 1996.

47. Zhang Jiajun and Sun Jilian, "People's Liberation Army Strategic Missile Corps Modernized" (in Chinese), *Jiefangjun bao*, 2 July 1996, in *FBIS* (OW2708093196), 2 July 1996.

48. Lt. Gen. Tan Naida, "On a Cross-Century View on Military Personnel and Developing Qualified Servicemen" (in Chinese), *Zhongguo junshi kexue* (China Military Science), No. 4 (20 November 1995), 57–66, in *FBIS Daily Report (China)*, 30 May 1996, 41–47.

49. Zhang Wannian, "On Sharply Reinforcing Our Ideological and Political Construction to Ensure Absolute Party Leadership of the Military—Commemorating the 69th Anniversary of the Founding of the PLA" (in Chinese), *Qiushi* (Seeking Truth), 1 Aug. 1996, 2–7, in *FBIS* (96CM0571A), 1 Aug. 1996.

50. Ibid.

51. Commentator, "Improve Training Quality with Live Operation as Criterion" (in Chinese), *Jiefangjun bao*, 29 May 1996, in *FBIS Daily Report (China)*, 11 July 1996, 49.

52. Zhang Haiping, "During His Inspection Tour of Units Garrisoning the Three Northeastern Provinces, Zhang Zhen Stresses the Need to Carry through the Requirement of Stressing Politics with Tangible Results" (in Chinese), ibid., 27 June 1996, in *FBIS* (HK1508042096), 27 June 1996.

53. Kenneth W. Allen, Glenn Krumel, and Jonathan D. Pollack, *China's Air Force Enters the 21st Century* (Santa Monica, Calif.: RAND, 1995).

54. Ibid., 126–27.

55. Ibid., 129.

56. Ibid.

57. Ibid., 130.

58. Zhang Wannian, "On Sharply Reinforcing Our Ideological and Political Construction to Ensure Absolute Party Leadership of the Military—Commemorating the 69th Anniversary of the Founding of the PLA" (in Chinese), *Qiushi*, 1 Aug. 1996, 2–7, in *FBIS* (96CM0571A), 1 Aug. 1996.

59. Ibid., 133.

60. Interview with Lt. Col. Blasko, 9 Oct. 1996.

61. Report from USDLO, Hong Kong, citing *Jiefangjun bao*, 3 Feb. 1993.

62. Report from USDLO, Hong Kong, citing *Cheng Ming*, 1 Jan. 1994.

63. Tai Ming Cheung, "China's Entrepreneurial Army: The Structure, Activities and Economic Returns of the Military Business Complex," in Lane et al. This chapter details the business activity and notes its corrupting effect. Anyone who lives in China very long sees the fruits of corruption. For example, on any given day, PLA officers and their dependents can be seen going on shopping expeditions using expensive Mercedes with official tags.

64. June Dreyer, "The New Officer Corps," *China Quarterly* (June 1996): 331. Dreyer's study covers the issues of training and education, technological proficiency, loyalty, factionalism, fiscal accountability, and regionalism.

65. Interview with Lt. Col. Blasko, 9 Oct. 1996.

66. Allen et al., 183.

67. Blasko, "Better Late Than Never," explains why the PLA ground forces are unlikely to become modernized soon. Similar assessments apply to the other services, as can be found in the reports of U.S. air and naval attachés in China.

68. For divergent views on China's future capability to project power see Paul H. B. Godwin, "Force Projection and China's National Military Strategy," and Chong-Pin Lin, "The Power Projection Capabilities of the People's Liberation Army," in Lane et al.

Section V
United States Officer Commissioning Programs since 1945

Postwar transformations in the United States' international role and in American society produced a new context for selecting and preparing the nation's future officers. Its most important features included: the establishment of a four-year college or university degree as a virtual prerequisite for commissioning; the liberalization of service academy and Reserve Officers Training Corps (ROTC) curricula;[1] the emergence of ROTC and officer-candidate schools as sources of the great majority of active-duty officers;[2] the integration of women into the regular military establishments and their increasing percentage of the total officer corps;[3] the need for a greater variety and number of specialists for large standing forces with diverse tasks; and the turn to an all-volunteer force after the Vietnam War. This context forms the background for the very specific aspects of United States cadet and junior officer formation discussed by the authors of this section's six essays.

The first three focus on different dimensions of service academy education and training. Although a modest trend toward a more liberal general education characterized the curriculum at the U.S. Military Academy from the turn of the century through World War II, cadets at West Point and at the U.S. Naval Academy received a completely prescribed course of instruction, heavily favoring science and engineering. Malham Wakin, a permanent professor and U.S. Air Force Academy faculty member for over forty years, describes the different curricular path taken by the Air Force Academy within a decade of its founding in 1955. The first of two key initiatives reduced the prescribed curriculum, but kept the remaining core evenly balanced between science and engineering on the one hand, and the social sciences and humanities on the other (a broad general education Wakin considers vital for future officers). The second initiative, made possible by the smaller core curriculum and other measures, enabled every cadet to choose an academic major (today there are twenty-five majors). West Point and Annapolis followed the Air Force Academy's lead, but critics of this liberalization have lamented the reduction in the required scientific and technical curriculum and a tendency among cadets to choose majors less obviously and traditionally related to the profession of arms (principally majors in the social sciences and the humanities).[4]

This criticism reflects long-held concerns in the services since World War II, but especially after the Vietnam War, that many officers have become occupational specialists and "military managers" rather than leaders embodying the warrior spirit. Moreover, junior officers (and too often their seniors) are believed to be largely ignorant of such professional fundamentals as individual service traditions, history, doctrine, and operations of their service, or military history, tactics, and strategy in general. Since 1980, partly in response to these perceived trends (some think most apparent in the Air Force), the Air Force Academy has sought to improve its professional military studies instruction. Philip Caine, a retired Air Force Academy professor of history and the first permanent professor charged to develop a coherent military studies program, outlines its origins and discusses the numerous pedagogical and bureaucratic difficulties encountered as the program was defined and implemented.

In 1976, women cadets first entered the three major service academies. Mandated by Congress, their admission was another milestone in the American woman's struggle to gain full equality in her society and a reflection of the effort to achieve all-volunteer military forces. From the beginning, women cadets experienced blatant and subtle sexual harassment and discrimination. In 1994 at West Point, for example, several varsity football players inappropriately touched female cadets during a pep rally. D'Ann Campbell, a military historian and vice president for academic affairs at the Sage Colleges, describes the incident and the swift and effective response to it by academy officials and both male and female cadets. Her essay clearly illustrates the complex challenges facing women in a service academy environment, but that, as at West Point, the potential for gender issues to divide and disrupt may be overcome by placing them within the larger framework of leadership development.

The academies have drawn much more attention from scholars than have other commissioning sources such as the ROTC and officer-candidate schools. The articles by Vance Mitchell, a military historian and retired Air Force officer, and Michael Neiberg, an assistant professor of history at the Air Force Academy, chart some of this virtually unexplored territory. Mitchell, viewing postwar Air Force officer accessions from the service's institutional perspective, shows that the Air Force's goal of a college-educated officer corps was held back by quantitative pressures, including volatile force levels and the controlling need for pilots and navigators. Thus, the Aviation Cadet and Officer Candidate School programs—neither requiring a college degree—lasted through the early 1960s. The Air Force ROTC and the Officer Training School (admitting only college graduates) moved the Air Force toward the educational standard it desired, but were also affected by fluctuating force levels and other problems exacerbated by the Vietnam War.

On the nation's campuses, the war in Vietnam stirred intense criticism of the ROTC. Neiberg contends that the most significant challenge came not from the radical student movement protesting the military and the war, but from civilian faculty questioning the quality and content of ROTC curricula. Although no doubt stimulated by the war, he writes, this scrutiny was part of a process of ROTC curricular reform that had been underway since the early 1950s and that had involved negotiation and compromise between the services and the universities. By the 1970s, the result was a liberalization of ROTC curricula—a reorientation from military training to the professional education appropriate for both a university community and for the complex national security environment in which ROTC graduates would have to function.

The history of U.S. Marine Corps officer formation has also been relatively neglected. Drawing on his personal experience in officer selection and junior officer training assignments from the 1950s to the 1970s, retired Marine Colonel Herbert Hart illustrates how several of the larger forces shaping those programs in all the services—higher educational standards, the increase in female officers, and the need for more specialists—affected the Marine Corps.

Notes

1. David R. Segal suggests there has been "a long-term secular trend from purely technical to more broadly liberal education" for officers in several major Western nations since the eighteenth century. See "Military Education: Technical Versus Liberal Perspectives," in Ernest Gilman and Detlef E. Herold, eds., *The Role of Military Education in the Restructuring of Armed Forces* (Rome: NATO Defense College, 1993).

2. In 1995, the ROTC commissioned 6,400 active-duty officers, the academies 3,070, and officer-candidate schools 2,700. See Report of the CSIS Study Group on Professional Military Education, *Professional Military Education: An Asset for Peace and Progress* (Washington, D.C.: Center for Strategic & International Studies, 1997), 22.

3. In 1997, the percentage of women officers ranged from seven and one-half in the Marine Corps to nearly sixteen in the Air Force.

4. See, for example, David A. Smith's article in this volume, and M. D. Van Orden and Stanton S. Coerr, "Reverse Engineer the Academy: Toward Restoring Service Integrity," *Strategic Review* 25 (Fall 1997).

The Evolution of the Core Curriculum at the U.S. Air Force Academy

Malham M. Wakin

Historical Development of the Core Curriculum

The roots of the core curriculum at the United States Air Force Academy may extend as far back as 138 years prior to the academy's founding. Sylvanus Thayer, superintendent of the U.S. Military Academy at West Point from 1817 to 1833 developed a four-year, totally prescribed curriculum taught in small, homogeneously sectioned classes, with daily recitations and frequent examinations. In his remarkably well-researched book, *Neither Athens Nor Sparta,* John Lovell describes the "seminary-academy vision" which drove Thayer's conception of "the modes of discipline and character building that were employed, the pedagogical techniques that were utilized, and the academic curriculum that was prescribed."[1] That curriculum was essentially an engineering and science curriculum although it included general history, moral philosophy, law, and geography. Lovell also points out that as civilian colleges began to move away from prescribed curricula at the beginning of the twentieth century, West Point continued the prescribed "Thayer system" and became a "bastion of the status quo."[2] Through the twentieth century until the Air Force Academy opened in 1955, there were changes made in subject matter and particularly significant developments in the social science offerings but the essential characteristics of the totally prescribed curriculum and pedagogy continued. These characteristics were eventually carried over to the U.S. Naval Academy after its founding in 1845.

As late as 1953, the curriculum at West Point was totally prescribed except for a choice of which foreign language would be studied. According to Lovell, a faculty curriculum review committee, commissioned by the superintendent to examine the feasibility of introducing choices of fields of study, defended the traditional prescribed curriculum with two arguments:

> First, because the Academy was preparing its students for a common professional career, it was possible to identify a single "best" program of studies and practical experience that would provide the requisite preparation. Secondly, there was the traditional concern that any significant departure from a uniformly prescribed program would carry the risk of destroying the sense of organic unity among the corps of cadets.[3]

The core curriculum at West Point through the nineteenth century could be described fairly as predominantly a technical curriculum buttressed by a few courses in the humanities and social sciences. After World War I, more studies in English and literature were added along with classes in economics and government, and by World War II expanded offerings had been developed in history, comparative government, and international relations. After World War II, studies in psychology and leadership were incorporated. Thus, prior to the Air Force Academy's establishment, the trend at West Point was clearly in the direction of a more general education.

This trend influenced the members of the Air Force Academy Planning Board which was organized at the Air University in the fall of 1948 to recommend structure and curriculum for the yet-to-be-established Air Force Academy. The Planning Board devoted the second volume of its three-volume report to a proposed curriculum described as one which should "be designed to offer a broad general education as well as a sound background in aeronautical science and tactics." Planning Board members emphasized education as opposed to specialized training in the technical duties of junior officers. They recommended three main divisions of study: the Division of Humanities, the Division of Science, and the Division of Military Studies. Their version of humanities included courses in English, psychology, philosophy, foreign language, history, geography, economics, government, international relations, and what was then called "great issues." The Division of Science was to include courses in mathematics, physics, chemistry, electrical engineering, engineering drawing, thermodynamics, aerodynamics, and mechanical engineering. The Division of Military Studies was to offer courses in military studies, military law, administration, tactics, physical education, and military hygiene. The breakout, as the Planning Board saw the core curriculum, included 66 semester hours of humanities courses (33.9 percent), 76 semester hours of science courses (39 percent), and 53 semester hours of military studies (27.2 percent). The board members hoped this curriculum would produce a graduate who would "be an individual with the breadth, regardless of his specialty, to represent the Air Force advantageously in any educated group, at home or abroad, socially or officially."[4]

It is worth pointing out that the nine officers charged with developing this curriculum plan consulted thirty distinguished educators. As they were completing their report in the spring of 1949, then Secretary of Defense James Forrestal appointed the Service Academy Board to study the entire issue of service academy education. Frequently referred to as the Stearns-Eisenhower Board (its chair was Robert L. Stearns, president of the University of Colorado, and its vice-chair was Gen. Dwight D. Eisenhower, then president of Columbia University), it also solicited the services of a number of distinguished educators. The Service Academy Board reaffirmed the need for an Air Force Academy, generally approved of the existing national military academies, and also confirmed the Planning Board's findings. In its interim report (4 April 1949), the Stearns-Eisenhower Board stated:

> The mission of the Service Academies is to provide undergraduate instruction, experience, and motivation to each student so that he will graduate with the knowledge, character and the qualities of leadership required of a junior officer. This program should provide a basis for continued development throughout a lifetime of service to the Nation and a readiness for military responsibilities of the highest order.[5]

The language of this mission statement carried over to the 1954 legislation authorizing the establishment of the Air Force Academy, and was repeated in a 1955 Air Force regulation and in subsequent mission statements used in academy catalogues.

As the Stearns-Eisenhower Board was completing its final report, Lt. Gen. Hubert R. Harmon, who was to become the first superintendent of the new Air Force Academy, was made the special assistant for academy matters to the Air Force chief of staff. His staff included Col. William S. Stone who was also destined to be a future academy superintendent. This group refined the curriculum proposals made by the Air Force Academy Planning Board and the Stearns-Eisenhower Board and also sought reviews from civilian educators. Faculty members from Purdue University and the Massachusetts Institute of Technology evaluated the science course proposals while the faculties of Stanford and Columbia examined the humanities and social science courses.[6]

In July 1955, when the first cadets arrived at Lowry Air Force Base in Denver, Colorado, the interim site of the new academy, the faculty and staff that had been assembled to launch the institution shared at least one important characteristic relevant to curriculum development and pedagogy. Most of the officers were either graduates of the U.S. Military Academy or had served as faculty members at West Point or Annapolis. Not surprisingly, then, the initial curriculum they distilled from the work of the three planning groups was completely prescribed (like that at West Point), the class-size goal was twelve students, the students were to be homogeneously sectioned by ability and resectioned during each semester, and recitations from each student were to be graded almost daily. The academic courses in the core (as distinguished from the airmanship and physical education courses) totaled 137-2/3 semester hours (46 courses). Of these, 63-2/3 semester hours were in science and engineering and 74 in the social sciences and humanities. Aside from a focus on airmanship studies, the attempt to balance the technical and nontechnical courses was the most obvious difference from the curricula at the other service academies. Maintaining a relatively equal number of courses in the social sciences and humanities, on the one hand, with the number of science and engineering courses, on the other, has characterized all curriculum adjustments from 1955 through 1996. In some years the balance has tipped in one direction or the other by as many as two courses, but the specific courses and course content have undergone continuous evaluation and frequent revision.

By 1957, Brig. Gen. Robert F. McDermott, dean of the faculty, had instituted an enrichment program to challenge gifted cadets and those with previous college work to go beyond the standard core curriculum. He convinced the Academy Curriculum Committee and the Academy Board that, contrary to service academy tradition, cadets should be granted transfer credit for college-level courses successfully completed prior to matriculation at the academy, they should be permitted to take validation exams for courses in the core in which they had previous backgrounds, and they should be allowed to take elective courses outside the core curriculum by substituting for the validated courses or by overloading if their grades were sufficiently high. By academic year 1959–60, the academy catalogue described a total of 142 "prescribed" and "special" academic courses, plus navigation, physical education, and military studies. Of the 142 academic courses, 45 were in the core curriculum. The graduation requirement included 135 semester hours of academic courses and 48 semester hours for all airmanship, military studies, and physical education courses for a total of 183 semester hours. As part of the enrichment program, a cadet could earn a major by transfer, validation, substitution, or overloading to pick up 17 semester hours in one of four major subject areas (corresponding to the four academic divisions). All of these curriculum changes took place before the first class graduated, and, miraculously, the North Central Association of Colleges and Secondary Schools accredited the Air Force Academy bachelor of science degree in time for the graduation of the first class.

The graduates of the Classes of 1959, 1960, and 1961 all completed sufficient navigation courses and flight missions as part of the core curriculum to earn navigators' wings. By 1960–61, however, the navigation program as a core requirement for all cadets was discontinued and the total academic course core was pegged at 144½ semester hours, airmanship at 25½ semester hours, and the athletic program at 12½ semester hours. These, plus a 2½-hour flying option, made the total graduation requirement 185 semester hours. The enrichment program was expanded to include cooperative master's degree programs in engineering (with Purdue) and international affairs (with Georgetown). The first students to participate in these cooperative graduate programs were in the Class of 1963. The cooperative arrangements were pursued after attempts to obtain authorization to grant master's degrees at the Air Force Academy itself did not succeed.

The success of the enrichment program led naturally to a majors-for-all curriculum, and a major became a graduation requirement beginning with the Class of 1966. Note that transfer credit, validation, electives, substitution in the core, majors, and cooperative graduate programs, were all significant departures from the completely prescribed curricula that had been the standard in service academy education for many decades. Nevertheless, the commitment to a broad liberal education as so clearly mandated by the early planning boards continued to be reflected in a substantial and balanced core curriculum. In 1966, there were 36 core academic courses (103 semester hours) and 17 different majors available which could be earned by completing 16 majors' courses. By 1971–72, the core academic courses totaled 99 semester hours, leadership and military training 27 semester hours, physical education and athletics 14½ semester hours, and majors' courses (including a 5-hour airmanship elective and other electives) equaled 46½ semester hours—all for a total graduation requirement of 187 semester hours.

With the Class of 1975 the cooperative graduate programs were discontinued. In 1976, based on the twenty-year curriculum review, a major change in scheduling the academic year resulted in equalizing the length of the two semesters, reducing the cadet preparation load by one course each semester, eliminating 2½-semester-hour courses, adding two courses to the core, and reducing the semester hours for majors and elective courses.

After the most recent curriculum changes, the core curriculum for the Class of 2000 still maintains the balance between technical and nontechnical courses. The academic core courses total either 31 courses or 33 depending on choice of major; there are 25 majors available, and a major can be earned by taking from 11 to 14 courses beyond the core depending upon the major selected.

Over the past forty-one years, although the balance between technical and nontechnical courses in the core curriculum has been maintained, the specific courses have changed and the pedagogy has also changed. The newest version of the core curriculum includes nine courses in the basic sciences (chemistry [2], mathematics [3], computer science, physics [2], and biology); six courses in engineering (mechanics, electrical engineering, aeronautics, energy systems, astronautics, and engineering systems design); eight courses in humanities (English [3], foreign language [2], history [2], and philosophy); and six courses in social sciences (economics, management, political science [2], law, and behavioral science). In addition, all cadets must elect two core-option courses and must take ten semester hours of core military studies and six semester hours of physical education. By the mid-1960s most departments had abandoned homogeneous sectioning. The tradition of small class size has been continued with the average today stabilized at approximately fifteen students. Over the years, the amount of testing has decreased considerably from the early practice of daily grades, but the institution as a whole continues to do more testing than many civilian schools. Of course, over the years, pedagogy has also evolved with available technology both in the classrooms and in the laboratories.

Although I have skipped over some changes that have taken place in the core curriculum, I have provided the essentials of its history. It has continuously reflected the basic parameters laid down by the planning groups of the late 1940s. It moved very quickly from the original, totally prescribed core to a substantial core with electives and majors. Since the mid-1960s, the core has occupied roughly the same proportion of the graduation requirement—from 90 to 112 semester hours (if graded military studies and physical education courses are included) out of a total 136-to-154-semester-hour graduation requirement that has fluctuated over time with the status and choice of majors.

The question of appropriate balance in the core between technical courses and courses in the social sciences and humanities has surfaced historically in several contexts. The case of the core curriculum during different periods of time at the U.S. Naval Academy is instructive. During the deliberations of the Service Academy Board (1949), the subcommittee panel on the social sciences observed in its report to the board:

> In our opinion, the limited time devoted to the Social Sciences at the Naval Academy permits the attainment neither of the objectives set forth in the directive of the Service Academy Board . . . nor that minimum standard of general education in the area that should be required of officers in the military establishment?[7]

At that time the Naval Academy core courses in the social sciences were thirteen semester hours fewer than those in the West Point curriculum and twenty semester hours fewer than in the proposed Air Force Academy curriculum.

In the mid-1980s, John Lehman, the secretary of the navy, called for a rebalancing of the core curriculum at the Naval Academy. In a memorandum for the chief of naval operations, Lehman noted the pressures involved in "meeting the exacting technical standards set by the highest engineering accrediting boards" and others, but expressed concern that in meeting those pressures "some of the broader educational objectives have suffered." In his view:

> The current faculty ratio—for example, 70 professors of Engineering and only one of Philosophy—reflects both the pursuit of technical specialization . . . and the 1975 policy directive affirming that 80 percent of the midshipmen must major in Science or Engineering. The circumstances supporting a quota of 80 percent Engineering and Science majors have now passed and the quotas are no longer justifiable.[8]

The navy secretary also pointed out that not more than 20 percent of midshipmen were receiving a true core curriculum in the humanities and social sciences and directed a strong core be established in these areas. Lehman's memorandum accompanied *SECNAV Instruction 1531.2* in which he quoted John Paul Jones: "It is by no means enough that an officer of the Navy should be a capable mariner. He must be that, of course, but also a great deal more. He should be as well a gentleman of liberal education, refined manners, punctilious courtesy, and the nicest sense of personal honor." The "guidance" in this instruction sounded very much like the positions adopted by the Service Academy Board of 1949–50, and included among many other points that "the mathematics and science core and the humanities and social science core shall be equal in required semester hours."

The Relevance of the Core Curriculum

Samuel Huntington, in his classic 1957 study of the military profession, emphasized the need for officers to have a broad liberal education. His reflections are worth noting:

> The military skill requires a broad background of general culture for its mastery . . . just as law at its borders merges into history, politics, economics, sociology, and psychology, so also does the military skill. Even more, military knowledge also has frontiers on the natural sciences of chemistry, physics, and biology. To understand his trade properly, the officer must have some idea of its relation to these other fields and the ways in which these other areas of knowledge may contribute to his own purposes . . . he cannot really develop his analytical skill, insight, imagination, and judgment if he is trained simply in vocational duties. The abilities and habits of mind which he requires within his professional field can in large part be acquired only through the broader avenues of learning outside his profession. . . . Just as a general education has become the prerequisite for entry into the professions of law and medicine, it is now almost universally recognized as a desirable qualification for the professional officer.[9]

I find Huntington's views to be remarkably consistent with those offered in the reports of the academy planning boards of the late 1940s and early 1950s, succinctly summarized in the following excerpt from a 1955 Air Force regulation specifying the Air Force Academy's mission:

> Mission. The Academy provides instruction, experience, and motivation to each cadet so that he will graduate with the knowledge and the qualities of leadership required of a junior officer in the United States Air Force, and with a basis for continued development throughout a lifetime of service to his country, leading to readiness for responsibilities as a future air commander. . . . Objectives. As a minimum goal the Academy will accomplish its mission by providing each cadet with:
>
> A basic baccalaureate level education in airmanship, related sciences, the humanities, and other broadening disciplines.
>
> A knowledge of and an application for airpower, its capabilities and limitations, and the role it plays in the defense of the nation.
>
> High ideals of individual integrity, patriotism, loyalty, and honor.
>
> A sense of responsibility and dedication to selfless and honorable service.[10]

Armed with all of this advice and direction, one should be able to review the courses in the academy's core curriculum to ascertain the relevance of each course to the ideal preparation for future Air Force officers. We might test each core course against two elemental criteria. Some courses will need to be in the core because they prepare cadets to be educated, participative citizen-leaders in a democratic republic; other courses will be needed to prepare them for their specific role in the military profession and an arm of that profession which is highly technical. Courses in the liberal arts and sciences are needed to satisfy the first criterion; courses in aeronautics, astronautics, electrical engineering, military history, and the more specific military studies courses are needed to fulfill the second criterion. Some courses, like law, defense policy, ethics, literature, and political science may well serve both criteria. We have not yet mentioned the overall academy experience which is also a part of the core curriculum but not delineated in terms of courses. The teacher-student relationships at the academy, the immersion in the principles of the honor code, the plethora of programs promulgated by the Character Development Center, the daily professional application in every classroom, every military training program, and every athletic endeavor comprise the environmental glue of the core curriculum. And why is this core so crucial at this or any other military academy?

One needs only to reflect for a moment on the relationship between the military profession and its parent society to see the critical relevance of the core curriculum. Like the other public professions, beneficiaries of academy educational and training opportunities are clearly obligated to reciprocate by serving a parent society that grants the military profession considerable autonomy and authority to act on its behalf. If we review the authority to act bestowed upon the officers in the U.S. military hierarchy (not to mention the large portions of the national treasury), we get some sense of the remarkable trust American society places in them. Officers have almost absolute power over their subordinates; they govern directly those lives through the greater portion of every working day. They may lead the youths of this country, who are bound by their leadership, to their deaths. Within the scope of military law, they may promote and demote, assign to worldwide billets, and place their subordinates in harm's way. They have licensing authority and court authority, and they have at their fingertips the most destructive weapons ever devised; indeed, they may unleash the weapons which could destroy all of humanity in a very short period of time. May we not conclude that where such trust is given, when the authority to act on behalf of the entire society is bestowed, when national survival and individual human life may so frequently be at stake, when large portions of the national treasury support their existence—the obligations weighing on the leaders of the military profession are many and they are severe and they require the kind of preparation that the core curriculum described here can provide.

Notes

1. John P. Lovell, *Neither Athens Nor Sparta?: The American Service Academies in Transition* (Bloomington: Indiana University Press, 1979), 18.

2. Ibid., 24.

3. Ibid., 93–94.

4. U.S. Department of the Air Force, Air Force Academy Planning Board, *Air Force Academy Planning Board Study*, Vol. 2, *The Curriculum* (Maxwell Air Force Base, Ala.: Air University, 1949).

5. Cited in U.S. Department of Defense, Service Academy Board, *A Report and Recommendation to the Secretary of Defense* (Washington, D.C.: Government Printing Office, 1950), 4.

6. James P. Tate, "Origins of the United States Air Force Academy Curriculum," in *The United States Air Force Academy's First Twenty-Five Years* (United States Air Force Academy, Colo., 1979), xix.

7. Service Academy Board, *Report and Recommendation to the Secretary of Defense*, app. E, 58.

8. John Lehman memorandum for the chief of naval operations, "Curriculum and Admissions Policy at the U. S. Naval Academy," Washington, D.C., n.d.

9. Samuel P. Huntington, *The Soldier and the State* (Cambridge, Mass.: Harvard University Press, 1957), 14.

10. Cited in William T. Woodyard, "A Historical Study of the Development of the Academic Curriculum of the United States Air Force Academy" (Ph.D. diss., University of Denver, 1965), 26–27.

A U.S. Air Force Academy Dilemma: Professional Military Studies

Philip D. Caine

On the surface it might seem strange that a subject as basic as professional military studies could be anything other than a key area of the education received by a cadet at an institution such as the United States Air Force Academy. And yet, although it is an essential portion of the curriculum on the one hand; on the other, military studies is widely misunderstood and has, over time, been an area with which no one seems very comfortable. Both the Air Force secretary and chief of staff have recently underlined the importance of military studies in the publication, *Global Engagement,* which stresses the need for each new officer to have a thorough knowledge of combined air and space operations.[1] Moreover, in 1998 the Air Force will test a six-week Air and Space Basic Course designed to "instill in each new officer the concept of being an airman first; an understanding of what air and space power brings to the fight; and the ability to advocate air and space power's contribution to national security." If successful, this program will be required of all officers entering active duty.[2] The same stress has been laid on professional military studies by the other services as well, all focusing on their precommissioning programs—the service academies, reserve officer training corps, and officer training schools—as the primary places for emphasizing the subject. Yet the record of professional military studies is not very good, too often characterized by frequently changing courses, poor instruction, and second-class status in the education of future officers. I have chosen to consider the situation at the Air Force Academy since it is one of the best examples of the problems and issues associated with the evolution of a military studies curriculum and also the program with which I am most familiar.[3]

Unlike most academic subjects taught at the Air Force Academy by the faculty under the direction of the dean, professional military studies, or military art and science, or professional development (among the various names it has been called over the years) has never had a very comfortable home. Military studies has not been recognized as a separate academic discipline, but rather included as part of the commandant of cadets' military preparation program and generally lumped together with military training. As such, it has far too often been of little interest or concern to many on the faculty and staff and, unfortunately, sometimes to the cadets as well. In 1975, the Air Force Academy's 20th Anniversary Study Group recognized only the military training curriculum; not until five years later did the 25th Anniversary Review Committee refer to a broad field of military studies.[4] Even then, there was a definite overlap, in the minds of the committee, with military training. Still today, many members of the staff and faculty at the Air Force Academy can neither define military studies nor explain the differences between it and military training, often using the two terms interchangeably.

But there is a distinct difference that has been critical throughout the history of service academies. It centers on the definitions of training, on the one hand, and education on the other. Military training encompasses courses and activities with defined, practical objectives, designed to teach a skill or concept that is needed to perform a task, display a knowledge of facts, or employ a concept in everyday life. The level of training can generally be measured by demonstrated ability to perform the tasks for which the training was designed. Much that makes the military unique falls into the area of military training. At one end of the huge training spectrum are relatively narrow and regimented areas such as drill, procedures for filling out reports, and the rote memorization required of new cadets. At the other end are highly skilled activities such as parachuting and soaring.

Education is harder to define. It is the process by which we acquire that knowledge enabling us to pursue a profession, to make moral decisions, to ask why and how, to think critically and autonomously. We speak of the "educated man" and we see a person who is conversant with the world around him, able to bring diverse forms of knowledge to bear on a situation and equipped to function effectively in a complex, changing world in which he must make decisions involving unfamiliar circumstances. Part and parcel of education is that body of knowledge equipping the civil engineer to think through a complex construction problem, the chemist to invent a new plastic, or the economist to devise financial alternatives for a developing nation.

So, too, must military officers be educated in their profession just as the physicist is in his. At the service academies, the portion of the education spectrum that is primarily responsible for building this foundation is professional military studies. At the Air Force Academy, for example, such courses as Foundations of the Military Profession, Air Power Theory and Doctrine, Joint and Multi-National Operations, and Combined Arms Operations not only help to build character and the ability to make rational and moral decisions, but they are also essential to interpreting combat situations, working out strategy and tactics, understanding national priorities, and responding to new and rapidly changing circumstances.

Given the mission of the service academies, it follows that professional military studies should be the cornerstone of that portion of the core curriculum (the courses required of all cadets regardless of major) which sets these institutions apart from the normal college or university. The professional part of the core curriculum includes the relatively few courses generally unique to service academies and essential in the military preparation of officers. At the Air Force Academy these include military history, aeronautical engineering, leadership concepts, language and expression, and ethics. Together with military studies, they form what I call the professional core.

There is another far larger group of courses placed in the core curriculum so that academy graduates will be familiar with the wide variety of areas with which they might come in contact during their careers as well as have the academic background needed to perform in the diverse situations expected of college graduates. Among these are literature, physics, many engineering courses, chemistry, psychology, foreign language, economics, and a number of others. While their presence in the core curriculum is no less valid than the first group, they are not unique either to the Air Force Academy or the other service academies. This total core curriculum, augmented by essential honor and character education, ensures the well-rounded graduate who is ready to be both an officer and an educated citizen.

All might be well if what I have just described was the way the curriculum worked in practice, but that, unfortunately, has not been the case over the years. While the core curriculum, taught under the auspices of the dean of the faculty, has generally done a good job of fulfilling its part of the equation, the portion of the curriculum allotted to professional military studies has not been so successful. These courses, taught under the control of the commandant of cadets, have had a stormy existence at best due, at least in part, to the confusion between professional military studies and military training. It is, therefore, necessary to examine military training, its place in the service academy experience, and the impact it has had on professional military studies.

I shall use four examples. First, because of its nature, military training is relatively easy to modify, so its specific content changes frequently, both as a result of changes in personnel and issues in the military at large. For this reason, military training has become the academy sponge soaking up the ideas and whims of commanders, the various staffs in the Pentagon, or other significant sources of input, such as the media. Altering military training enables an academy to comply without having to make substantive changes in the academic arena. In this guise, military training performs an essential, but often overlooked, service for the institution in addition to its generally understood training function. Unfortunately, the close identification of military studies with military training over time has led to the idea that military studies can do the same thing; its courses have changed frequently, tending to dilute its academic credibility.

Secondly, a significant amount of military training during the academic year takes place at times not thought suitable for academic classes, such as early morning or early evening, as well as on Saturday. The heretofore blurred line between training and military studies has often doomed the latter to these same times with the resultant message to cadets that it is not "academic" and not as important as courses taught during prime time.

This situation has been compounded by a third factor—using nearly any officer to conduct military training, and often to teach military studies on the assumption that every officer is an expert on military training, and by extension military studies, by virtue of being a military professional. The result has frequently been officers teaching military studies courses with little actual qualification for the duty. While this can be tolerated in military training, it can often lead to a poorly taught or sometimes wasted period in military studies, further tarnishing its reputation.

The fourth element in this equation is the commandant of cadets. Normally and appropriately a very successful, rapidly promoted officer from the operational line of the Air Force, the commandant has seldom had anything to do with education programs. Training is generally second nature to him by virtue of his assignments, but military studies is almost always a relatively unfamiliar and therefore uncomfortable area. Consequently, commandants have tended to stress military training and to give little attention to military studies. That tendency has been exacerbated by every new commander's perceived need to make changes in organization and programs as well as the operational service's general reluctance to value stability and continuity. More often than not, the result has been a well-intentioned but dramatic change in the commandant's policies and programs with little understanding of or appreciation for long-term stability. While some of this can be absorbed in military training with little difficulty, it can be a disaster for military studies.

The overall result has been that since World War II, and probably since the inception of the service academies, the entire area of military studies has been proscribed by four critical questions: what should be taught, who should teach it, how to achieve stability and academic credibility, and how to award credit for these courses? For those areas of study universally recognized as part of the academic curriculum, these have not been very significant questions. The subject matter is generally clearly defined and widely accepted, as is its presence under the jurisdiction of the dean. Those teaching academic subjects generally possess a specialized and adequate graduate education. Moreover, the academies' permanent professors provide the stability that is the norm in higher education, as well as ensuring academic credibility for their various departments. Finally, course credit is given in the accepted manner for the particular academy, either in semester hours or some other type of course unit.

Not so with military studies. The courses in this discipline have varied greatly from year to year. There are no standard qualifications for teaching them, other than the requirement to be an active-duty military officer. Academic credibility has been elusive and stability has more often been a goal than a reality. Moreover, there is no agreed upon formula for determining course credit. These four questions, then, have taken an inordinate amount of time and energy year after year and have helped keep military studies in a generally chaotic and ever-changing condition. Michael Howard, the eminent military historian and active participant in developing British officer education, laid out the problem well when he said: "The moment I myself came up against the problem of actually organizing education in the Services, I realized, for the first time, the enormous complexities which must be dealt with and the number of compromises which one must make, however reluctantly, with one's ideal."[5]

The best way to illustrate some of the complexities and compromises that characterize a professional military studies curriculum is to consider the experience at the Air Force Academy since 1980. It is an excellent example of the evolution of a military studies program from one that was poorly taught, randomly scheduled, and frequently changing, into a program that has become an accepted part of the academic curriculum and that is presently available as a minor and scheduled to be an accredited major for the Class of 2000.

In 1980, to its credit, the Air Force Academy's 25th Anniversary Review Committee identified and addressed the most critical problem in the military studies area: "Quality is difficult to maintain in the absence of some continuity in personnel. The area of military studies has suffered from lack of continuity in personnel and [therefore] of credibility with the cadets." Among the committee's key recommendations were a permanent professor to serve as the deputy commandant for military instruction and several tenured faculty positions in military studies.[6] Thus, in late spring 1980, the superintendent offered the position of deputy commandant for military instruction to the deputy head of the History department and agreed that, in the following two years, three tenured positions would be made available. In supporting these changes, the commandant endorsed the stability in both program and personnel that the superintendent envisioned.

The first step in building a military studies program along the lines envisioned by both the 25th Anniversary Review Committee and the superintendent was to establish a military studies division completely separate from military training. This division was charged with developing a quality professional military studies curriculum that would equal the courses taught by the faculty in every way. Each offering had to be a solid, college-level course dealing with meaningful topics essential to cadet professional military education and one that could become part of the cornerstone of the professional core curriculum. Establishing some continuity in personnel and creating a separate division provided a firm base for dealing with the four questions historically plaguing military studies. It is important to note that the entire military training program was also in line to benefit from these developments because of the close tie between military studies and military training. Since the deputy commandant for military instruction was also in charge of military training, the foundation was laid for stability in both personnel and programs in that area as well, significantly decreasing some of the problems that have historically resulted from the close association between military studies and military training.

By the summer of 1981, the mechanism was in place to address the first two questions: what should be taught and who should teach it? The 25th Anniversary Review Committee had commented about both gaps and overlap in the professional curriculum. One charge given to the new permanent professor that was to prove of inestimable value over the next few years was to establish a curriculum responsive both to cadet and Air Force Academy needs and stable enough to weather the changes in superintendents, commandants, and other key personnel.

To design a comprehensive military studies curriculum that the entire institution would support, the superintendent formed the Professional Development Review Committee, made up of several faculty department heads, a representative from the athletic department, and chaired by the now permanent, deputy commandant for military instruction. This group's task was to integrate the entire professional military curriculum into the academy's curriculum as a whole, with a focus on professional military studies. In reality, this committee accomplished little in terms of curriculum development. The major stumbling block, ultimately dooming the group to failure, was the faculty argument that military studies was not a defined academic discipline, such as economics, and therefore had no established or universally accepted body of knowledge. This was compounded by the possibility that a group of so-called academic courses would not be under the dean's jurisdiction which, in turn, raised the question of whether it was proper to give them either academic credit or class time during the academic day. Despite the efforts of some members, the group was disbanded after about a year. In the course of the committee's meetings a number of senior faculty officers learned a great deal about the commandant's programs, but most of the issues that compromised the group's effectiveness have continued to be problems to the present.

With respect to what should be taught, attention first turned to the new cadet's transition from basic cadet summer to the academic year and to the military in general. A new freshman course was created called The Foundations of the Military Profession. The objective was to help the new cadets understand what the military profession entailed and to make it clear that their experience at the academy was focused on becoming professional military officers. This was joined by a new sophomore course, Command and Control of Air Power; its evolution illustrates the problems that confronted the construction of a viable professional curriculum. In the mid-1970s, the superintendent had agreed with the Air Force's chief of public affairs that there should be a course dealing with that area in the curriculum. The institutional solution was to create a military studies course called Leadership Communication Skills, which hardly fit the definition of military studies. Still, the deal had been struck, and not until 1983 was this course finally replaced by a new one for sophomores focusing on air power; the course on communication skills was properly made the responsibility of the English department.

Another early objective was to construct a computer-assisted war game to support the second class (junior) course concentrating on resource allocation in the battle zone. Although it was to be a few years in the making, this game eventually became a model for similar programs throughout the education community. The course themes for the first class (seniors) were a much needed grounding in military theory and a force analysis of the Soviet Union. Thus, by 1983, a four-year course flow had been established which took the cadet from looking broadly at the military profession, through the organization and employment of air power and the allocation of military forces in a real world scenario, to military theory and a case study of the Soviet Union. This general course structure has survived six superintendents and ten commandants without major change.

The second question was who would teach the military studies courses. Since the inception of the Air Force Academy, the policy has been that every classroom teacher have at least an appropriate master's degree. Unfortunately, only about half of the officers teaching military studies in 1980 had a master's and many of those degrees were from marginal institutions. About that same percentage of officers was otherwise qualified to teach the courses. During the next two years, conforming to the master's degree policy became a primary goal and, by sending officers to graduate school, hiring others who already possessed the appropriate academic credentials, and shuffling personnel within the commandant's area, the goal was met. However, both the problems of subject area and university quality remained. Since there was no master's degree available in military strategy or doctrine, for example, officers were hired with a wide range of subject area expertise. While history and political science emerged as the two fields that best prepared officers for instructor duty in military studies and secured entrance for them to first-class schools, the problem of subject area for graduate education has yet to be solved. The primary venue for training new instructors had been the Academic Instructors Course taught by the Air Force's Air University. Unfortunately, this course was designed to teach enlisted personnel how to instruct at the Air Force's many technical schools and had little that was applicable to college teaching. Thus the academy designed a comprehensive instructor training program. This three-week course covered every aspect of lesson preparation including sessions on the cadet honor code and the academy grading system as well as several hours of practice teaching. The latter activity was monitored by a designated master teacher and videotaped so that the instructors could easily see both their strong and weak points. By the fall semester 1982, military studies was finally being taught by a group of college teachers who at least met the minimum education requirement established for the faculty.

The third major question, how to provide program stability and academic credibility, cannot be discussed without understanding the tempering influence the permanent professors have on the policies and actions of the dean. Although the commandant can pretty much dictate what he wants, the faculty at the Air Force Academy is organized in such a way that the dean's every action related to the curriculum is subject to the influence, and in some cases, the veto of the permanent professors. This results in a stability that has almost no parallel anywhere else in the armed forces. Continuity is the reason permanent professors exist, so they are very reluctant to make or approve radical change. Change that does come is very slow, generally taking years to complete rather than days or months as is the norm in operational organizations within the services. Thus, while the academic curriculum has remained very stable over the years, the lack of anyone under the commandant comparable to the permanent professors under the dean has made it possible for the commandant to institute radical and short-term change in his programs with little difficulty.

In 1982, however, an important change increased stability and continuity in the military studies program. Prior to the establishment of the military studies division, the officer in charge of military studies and military training was a line officer, on a three-year tour, with no particular qualification in academy education. With the separation of military studies and military training into two divisions, the commandant agreed to allow the deputy commandant for military instruction to hire an officer from the faculty with superb academic qualifications to head the military studies division on tenure status. This was a critical decision in that the commandant acknowledged that one could not become qualified to teach and to administer military studies and military training simply by service in the Air Force; to do so required expertise developed from both specialized education and experience. The commandant was now committed to quality teaching and to developing a curriculum that would compare favorably with that taught by the faculty under the dean. Within two years, the cadets rated the instructors in the military studies division as some of the best at the academy. Putting an officer on tenure status reporting directly to the commandant's permanent professor (the deputy commandant for military instruction) also ensured that long-range policies could be established. Thus, a firm foundation for more than a decade of program and personnel continuity was laid.

The fourth question, how to determine credit for military studies, continued as an issue for nearly fifteen years. While it may not seem on the surface to be as critical as the preceding three, it significantly impacts each of the others because the cadet perception of the place of military studies courses in the total curriculum is directly related to this issue. At the Air Force Academy, academic courses have always been counted both by semester hours for academic credit and as course units for figuring cadet academic load. Although semester hours varied from course to course, each full-semester course taught by the faculty was awarded one course unit. A full load for a cadet was considered to be six course units, or six courses. Although professional military studies had carried semester hours of credit since the academy's inception, these courses were never awarded course units. They were actually treated in the same way as a number of military training activities, such as basic cadet training, soaring, and survival training, which were awarded from one to five semester hours of credit but no course units. The reality of this distinction was that cadets, and often members of the staff and faculty (reasoning that if a military studies course was as important as an academic course then the academy would give it course-unit credit that would count in the total course load), considered military studies courses to be less important than academic courses taught by the faculty. This situation had both a psychological and a real-world impact. During the semester a cadet took a three semester-hour military studies course, he or she was actually taking seven full courses rather than the six counted by course units. The unfortunate part of this was that the cadet was taking a one-course overload while the system said he or she was taking only six courses, or a normal load. For those cadets who did not find academics difficult, this situation had little or no impact (other than not making sense), but to marginal cadets it often spelled academic disaster since they were not credited for being overloaded. The same issue that doomed the Professional Development Review Committee has been at the root of this problem—a basic unwillingness on the part of many faculty leaders to give course-unit credit to an academic endeavor not controlled by the dean of the faculty. Doing so would be tantamount to legitimizing the existence of an academic program outside the dean's area.

Despite the problems I have outlined, military studies at the Air Force Academy has made enormous strides in the past sixteen years. In the spring of 1996, a military art and science minor that would be available to cadets in the Classes of 1998 and 1999 was approved. This is to be followed by a major in military doctrine, operations, and strategy for the Class of 2000. Still, the procedures for implementing this major have not yet been worked out and the sticky issue of whether it will be under the commandant or the dean of the faculty remains unresolved.

Three issues germane to this paper remain to be discussed. The first is who should be responsible for the military studies curriculum, the second is where this discipline should fit in the total academy program, and third is the relationship of military studies at the Air Force Academy to an officer's total professional education.

At the Air Force Academy, professional military instruction, or its equivalent, is under the commandant of cadets because that office is assumed to be primarily responsible for cadet professional and military development. Since, over the years, military studies has been seen as part of the total military development program, it has appeared natural for that curriculum to remain with the commandant. There is also a perceived need for the commandant to have a formal teaching element in his area of responsibility. But are these valid arguments, or should military studies be with the rest of the academic courses under the dean of the faculty?

There are two reasons to keep military studies with the commandant. First, a teaching mission ties the commandant and his people more closely to both the faculty and the cadets. An academic division provides a valuable link, that otherwise would be lost, between personnel assigned to the commandant and those serving on the faculty. There is a tendency at each of the service academies for members of these two primary mission elements to drift apart, compete with each other for cadet time, and often blame one another for perceived problems at the institution. Having a vital stake in the core curriculum and the teaching mission helps the commandant identify with the faculty and understand their concerns. Unfortunately, most commandants do not understand this situation when they first arrive; indeed some are in the position for such a short time that they never recognize the problem's validity. This issue is also directly related to the need to have a permanent professor on the commandant's staff. The permanent professor is actually the prime agent for interaction between these two primary mission elements, emerging as the commandant's main spokesman with the faculty. Most commandants find this odd because it is completely out of their experience in the line of the service. Until they understand the place of the permanent professors in the total operation of the academies, they will not understand the essential role to be played by their own permanent professor.

There is a general belief at the Air Force Academy in areas outside of the faculty, that the air officer commanding (AOC—the active-duty officer assigned to each of the forty cadet squadrons) is the primary agent for interaction between officers and cadets. This is often not true. The average instructor sees each cadet much more often than the AOC because the primary interaction between officers and cadets usually takes place in the classroom, not to mention the time spent in academic advising and counseling sessions outside of class. The commandant needs this valuable channel to cadet concerns, attitudes, questions, and issues. Without a teaching mission, only the AOCs have direct access to cadets and the very valuable academic instructor channel is lost.

The second reason military studies should remain under the commandant is that the separate academic departments tend to be parochial. Department heads tend to view the institution and the academic program through their own discipline, and there is a tendency for each department to put its stamp on any course it teaches. Although this is desirable for courses in their own discipline, it can be fatal to military studies which, by its very nature, crosses disciplinary lines. For example, a military studies course that emphasizes history will soon become a history course and cease to be one in military studies. Putting military studies courses under one academic department, or the alternative of a consortium of departments, would practically guarantee that its nature and soon its very existence would be subsumed by those departments. Also, without a permanent professor to lead the military studies department, military studies courses would have no dedicated spokesman, making them especially vulnerable whenever course or personnel cuts become necessary.

There is a counter argument that any academic effort should be accountable to the dean. This problem was solved at the Air Force Academy by having a permanent professor with outstanding academic experience and ability, and with whom the dean was comfortable, assigned to the commandant and responsible for the quality of the professional military curriculum. Further, academic procedures for military studies courses followed faculty guidelines and had to be endorsed by the curriculum committee, made up of representatives of the faculty, commandant, and director of athletics. This system worked well; it ensured that the academic content and the teaching of the military studies courses met the dean's standards and enabled the commandant to maintain a teaching mission. Critical to this entire process, however, is the presence of a permanent professor on the commandant's staff. Without this position, the mechanism will not work.

The second remaining issue is where professional military studies should fit in the total academy experience. The cadets are immersed in the academic program, and this is where professional military studies plays its key role. Very few academic areas are directly and completely focused on the profession of arms. I believe there are only two—military history and military studies. These two disciplines must build in each cadet the foundation for a lifetime of service. While military history provides a base from which to understand the profession and its roots, military studies is principally responsible for ensuring that each graduate has come to grips with the mission of his or her service, the unique foundations of the profession, and how military forces operate. In short, it is the military studies curriculum that makes the service academies unique. Therefore, it must be the cornerstone of the professional and total core curriculum if the academies are to fulfill their responsibility and mission.

The question that naturally follows is whether establishing a major in military studies is a good idea. Although such a development is viewed by many as a milestone in military education and a victory for reason and academic credibility over those who opposed recognizing any academic course not under the purview of the dean, it also has the potential for seriously changing the focus of military studies and its purpose at a service academy. The natural tendency in any academic department is to put the best resources into the advanced course to be taken by the special few who have identified with that department's major. Unfortunately, this development often occurs at the expense of the basic or core courses. Military studies is no exception. The problem here is that, unlike any other discipline, military studies is the essential core area for every cadet. It is the glue that holds diverse career fields and academic majors together as part of the operational mission of the service. That is the reason for the existence of the subject area in the first place. Any dilution of the quality of the four basic or core military studies courses to enhance those taken by cadets in the military studies major could have grave consequences for the entire institution and the purpose for which it exists. Such developments could also undermine the reasons for keeping military studies under the jurisdiction of the commandant. Any loss of focus on the profession of arms and the foundations of the service in the name of more challenging or sophisticated courses for cadets in the military studies major could divorce the military studies division from the commandant and find it losing its applicability across the academic spectrum.

The final issue, the relationship of military studies to an officer's total professional education, is critical both for the academies and the services at large. Each precommissioning source is responsible for building a foundation for the officer's continuing professional education. Here I am not addressing professional education in terms of a career field or academic discipline, I am talking about the profession of arms. Service academies have only one reason for their existence—to prepare young men and women for a career of service to the United States in the armed forces. Each of the services has one overriding mission, the defense of the nation. It necessarily follows that every cadet and officer, regardless of his or her academic major or career field, must understand how they are preparing to contribute or are contributing to that mission. Unfortunately, many do not. This shortcoming is one of the major reasons for the proposed Air and Space Basic Course referred to previously. During their four years at an academy, cadets can too easily focus on their academic major and subsequent career field, and lose sight of why they are at an academy and what they are preparing to be.

The professional curriculum, anchored by military studies, has thus become more important than ever in cadet education. It builds the base for understanding the military profession and the responsibilities of service so that graduates, throughout their careers, will realize the importance of keeping current and continually broadening their knowledge and understanding of their chosen profession. As the academies address the military profession's unique nature, such areas as service doctrine, how force is employed, the responsibilities of the military in society, service heritage, and the other myriad issues in the professional core must be brought together in one understandable whole with which the cadet can identify and commit to as a professional. This is the responsibility of the military studies curriculum and those who teach it. A great deal of progress has been made in this cornerstone of military education. Nevertheless, there are still very basic issues to deal with and to solve before the total professional curriculum, and military studies in particular, provides the optimum professional education for every academy graduate.

Notes

1. U.S. Department of the Air Force, *Global Engagement: A Vision for the 21st Century Air Force* (Washington, D.C., 1997), 19.

2. U.S. Air Force News Release, April 1997.

3. For further specific information on the development of professional military studies, professional military training, cadet summer programs, and cadet navigation training at the Air Force Academy, see Brig. Gen. Philip D. Caine, "CWI, 1980-1992, End of Tour Report," December 1992, Papers of Philip D. Caine, USAF Academy Library Special Collections, USAF Academy, Colo.

4. U.S. Air Force Academy, "20th Anniversary Study: Curriculum and Cadet Way of Life, Final Report, 11 July 1975" and "Report of the 25th Anniversary Review Committee, 1 March 1980," USAF Academy Library Special Collections.

5. *Education in the Armed Forces: Report of a Seminar Held at the Royal United Services Institute for Defence Studies, November 15, 1972* (London: Royal United Services Institute for Defence Studies, 1973), 14.

6. U.S. Air Force Academy, "Report of the 25th Anniversary Review Committee," 24-25.

The Spirit Run and Football Cordon: A Case Study of Female Cadets at the U.S. Military Academy

D'Ann Campbell

Newsweek magazine interviewed me about a 1994 incident involving a spirit run and football cordon during a pep rally at the United States Military Academy (USMA), West Point, New York, at which several women cadets reported that they had been inappropriately touched. The interviewer explained that she had visited West Point after the incident and had found some women cadets telling her that they were treated as equals and others who argued that they continued to be discriminated against. She asked me, "Which is true?" I answered "both." Luckily, she was willing to explore shades of gray and the history of the "long gray line." I will argue in this paper, as I did to this reporter, that the spirit run and football cordon on 20 October 1994 can provide an excellent case study of the status of the academy's women cadets.

A review of the basic features of the USMA will help set this incident in context. The oldest of the academies, West Point has provided a key training environment for future senior officers. It has a four-year curriculum similar to other accredited colleges but its pedagogy is based on the principle of demanding that each of the 4,000 cadets accomplish far more than is humanly possible. Practically every minute of the day the cadets are monitored by a complex, hierarchical system of cadet officers, closely supervised by regular Army officers. The cadets spend several hours a day on athletic training and military training as well as taking five to six courses each semester in an academic program that heavily emphasizes engineering (five courses), sciences (every semester), and mathematics, regardless of the cadet's overall major. All cadets must engage in varsity or (more commonly) intramural sports. The Army has observed that, like Dwight Eisenhower, many of its great generals were student athletes, so it competes vigorously for high school athletes.

When I arrived as the Visiting Professor in Military History in 1989, the academy began requiring a mandatory five hours' sleep each night (from midnight to 5 A.M.). It even relaxed the dinner hour to better accommodate a variety of athletic and military training schedules. It also made it possible for the plebes (freshmen) to eat at least one meal a day. Although hazing was officially not allowed, the plebes were tormented daily, with food deprivation a favorite technique.

The hierarchical system is rigorous, with primary emphasis on cadet class—the lowly plebes yearning to become "yearlings" (sophomores); the "cows" (juniors) eyeing impatiently the privileges enjoyed only by the "firstees" (seniors). Within each class, the cadets are elaborately organized by rank and unit. Parades and marches practically every day instill pride in the system and constantly reinforce a hierarchy that specializes in handing out punishments. Very trivial offenses are identified and punished immediately; serious offenses can lead to expulsion. The usual punishment, however, is a tour—forced marching back and forth while wearing a backpack and carrying a rifle in a seventy-five-yard area for an hour. This marching has been characterized as "valueless," a chore that guarantees the offender will fall even further behind in assigned duties and classwork, making the next few days even more frantic.

The administration and faculty have their own hierarchical structure. A lieutenant general serves as superintendent on a five-year stint, while brigadier generals serve as dean of the college (no fixed term mandated) and as commandant of the Corps of Cadets (dean of students, typically a two-year rotation). The department heads, all colonels, stay until they retire. They basically control West Point—I was astonished at the way the colonels could resist policies issued by the generals in command (or by Washington). Each academic department has a few lieutenant colonels (titled, "permanent associate professors," PAPS) who serve five to eight years and then return to the field. All have Ph.D.s. Most recently, a handful of civilian, tenure-track professors has been hired by mandate of Congress. (In five years, 25 percent of the faculty will be civilians.) However, the great bulk of the teaching is done by captains and majors, who rotate through West Point in a five-year cycle. The academy selects likely captains, sends them at Army expense to civilian graduate schools for two years—usually for a master's degree—then brings them to West Point for three years of teaching. They are mature soldiers—role models for the cadets—with the academic credentials of a third-year graduate student. Very few have ever taught before—or will ever teach again. In addition to the academic officers, West Point assigns another group of "tactical officers"—also bright captains and majors—to supervise the cadets directly. (West Point is now requiring master's degrees for these tactical officers.) Just as the academic instructors, the "tac" officers typically serve a three-year tour of duty and then rotate to the field. In the strange Army way of doing things, the "firstee" cadets have been at West Point longer than most of the professors.

West Point, like the military generally, demands everyone to be in top physical shape at all times—civilians excepted! Cadets and officers are running almost any hour of the day or evening. Most captains and majors skip lunch to workout as a group in the gymnasium. All officers including generals and colonels have to pass a rigorous physical fitness examination twice a year that includes a two-mile run, pushups and situps. It is not possible to develop a crash course of reducing and working out to pass the test. It is common for the officers to comment directly to another officer or a cadet if he or she thinks the person is adding weight or has not been seen working out lately at the gym. This physical activity often bonds groups of officers or cadets as they run together and share runners' pain and runners' high.

Since men in general have more upper-body strength than women, they are required to do more pushups to pass the test. Likewise in running, men are generally faster than women. What would be an "A" time for a woman would be an "F" for a man. Only when it comes to situps are the numbers similar. In fact the year I left, women were required to do a few more situps than men in the two-minute timed test. My suggestion that some aerobic forms of conditioning be timed and measured was not welcomed. Since most women can outperform most men in aerobic dancing, I wanted this physical fitness and endurance test worked into the curriculum. Otherwise, women cadets, many of whom represent the top athletes in the country, consistently find themselves coming in second best. The handful of women who can meet the men's standards is given special respect.

The varsity football team is first among equals. When Army plays Navy and to a lesser extent, Air Force, cadets spend all week on pep rally antics and activities. Each Friday during football season, all cadets wear BDUs (battle dress uniforms, a comfortable outfit that everyone prefers). The varsity athletes eat at special tables, have tutors to help them stay abreast of their academic assignments when traveling, and are seen as role models. However, even they must pass the rigorous academic curriculum; they cannot substitute for required courses in electrical engineering, physics, or calculus.

The physicality of the Army stems from World War II when a system of intelligence testing shifted the brightest—and strongest—soldiers into the Army Air Forces and jobs such as finance. The dumbest were left over for the infantry, even though the top generals recognized that only the infantry could win the war. Because World War II soldiers had grown up in the 1930s, the severe effects of the Depression's deprivations were very evident. Class effects of urban unemployment and rural poverty meant that millions from the working class had mediocre health and physique. (Draftees in poor health were rejected.) At the same time, the class stratification of the educational system meant that these working-class youths had lower than average test scores. The result was that soldiers who scored low on the tests—and went into combat—were also physically the weakest. To combat this perverse result, the Army began to emphasize physical qualifications for the infantry—and eventually for every branch. The qualifications selected were, however, male-oriented. The Army deemphasized or even ignored physical skills in which women do well, such as eyesight, hearing, swimming, speed in learning bodily maneuvers, gymnastic ability, aerobic skills, and tolerance of pain or of food and water shortages. Rather it stressed upper-body strength and emphasized classic male sports such as football. It is not clear how much the Army's stress on physicality—indeed, its definition of physical fitness—drew upon concepts of masculinity. In World War II, when the standards were being developed, there was widespread fear—even loathing—of the entry of women into the masculine domain of soldiering.

When I stayed a second year at West Point to work with the leadership on issues of subtle forms of discrimination, the academy already had over fifty hours of training workshops on human resources including sexual harassment, racial equity, discrimination, and sex education. The administration took great pains to explain that these were not gender issues, but leadership issues, and that cadets who did not take the workshops seriously would have problems becoming effective leaders. The focus was not on blatant forms of discrimination and harassment because the message for years had been "zero tolerance" of any such activity. Officers and cadets were watched closely, and were quickly dismissed if they were found guilty of either.

I worked closely with the administration to create a modified version of "Chill in the Classroom" workshops. We had cadets design and produce videotapes on instances of inadvertent, unintentional ways officers could and did "chill" or create a negative learning environment. When West Point representatives gave papers at conferences at Harvard University and the Massachusetts Institute of Technology on these workshops, the other academics marveled how much further ahead the academy was than civilian colleges which struggled with women breaking into the ranks of typically male professions. The next year, the academy's Human Resources Committee added a graphic, live drama on date rape featuring cadet scripts and cadet actors and actresses, and modeled after workshops sponsored by Cornell University.

By 1993, the USMA had raised discussion of human resources issues to the next higher plain by adopting two bedrock values as the foundation of the Corps of Cadets: honor and integrity, and consideration of others. By adding the second dimension, consideration of others (in terms of religion, race, or gender), the focus was not just on the honor code but also on the responsibility for all cadets to treat each other with dignity and to be sensitive to unintentional forms of discrimination. For example, a cadet might pass a group of minority cadets, perhaps African American, Asian, or Hispanic, who are watching television or eating together, and say: "What are you guys plotting now?" Such a comment can be hurtful and would violate the "consideration of others" value. Today, members of the Human Resources Committee are busy creating with cadet input and actors and actresses a series of video vignettes on "Chill in the Foxhole" or "Chill in the Military Training Program" to be used in leadership classes.

The military academies integrated because of pressure from Congress. The first class of female cadets entered West Point in 1976 and graduated in 1980. By 1985, the admissions office was not treating women's applications differently than those from men. Women were admitted as approximately 10 percent of each class and about that proportion of qualified applications came from women.

In the beginning, resistance to women was fierce. A much higher percentage of women cadets left the academy before graduation than men. The autobiographies and other data, collected under Project Athena by the USMA's institutional research branch, showed blatant forms of discrimination were the norm for the early years and that women cadets were reluctant to report possible date rape, other sexual assault, or sexual harassment to their cadet chain of command. As Lt. Gen. Andrew Goodpaster, then the school's superintendent, related, some of the hazing of female cadets was "crude and obscene toward women." For example, female cadets "who seemed to be the most squeamish" were picked out to kill live chickens by biting their necks.[1] If unfairly treated, women would either "tough it out" or quit. Such treatment was a key reason that retention rates were lower for women in the early years. However, in the first half of the 1990s, women have graduated in the same proportion as men. In fact, in 1993, the women's retention rate was a few percentage points higher (79 percent for women, 75 percent for men). How the women cadets reacted after the groping some experienced in the football cordon in October 1994 and whether they felt comfortable in reporting possible inappropriate behavior may serve as a weathervane for future high retention rates for women at West Point. A 1985 woman graduate talking to reporters in November 1994, explained, "If this [groping incident] had happened to me while I was a cadet, I would probably not have reported it."[2]

What has been slow to form are networks of women. In general, women do not seem to be willing to mentor or to nurture other women. Indeed, many try to out-macho or out-masculinize each other, and the men. Some women feel that to be accepted they have to be smarter and more physically fit than men and to deny their feminine characteristics. Although some attempt has been made in the past few years to form networks of women cadets, women officers are even less likely to form networks or be seen off post with other women officers. I find it ironical that the USMA's leadership training classes are trying to inculcate traits—such as allowing each member of a squad, company, or regiment the opportunity to be all that he or she can be by nurturing their strengths—that have traditionally been seen as female.

The Spirit Run and Football Cordon, 20 October 1994

Let us now turn to the spirit run and football cordon and describe what happened, how the cadets and officers reacted, what the short-term and long-term implications might be, and what this incident reflected about the status of women cadets at the USMA. Each Thursday before a game, one of the four cadet regiments always runs as a unit to the football practice field as part of a spirit run. Once at the field, they run through a gauntlet (called cordon) of football players. This is the reverse of what takes place at most college campuses where the *athletes* run the gauntlet. Since some cadets are excused from this exercise because of other varsity sports or leadership training commitments, probably 550–600 cadets from the First Regiment participated in the spirit run that rainy Thursday to spur on the football team. They arrived earlier than expected and the players had to quit practice, forming their cordon quickly, although chaotically. The distance between the football players on either side of the cordon was apparently closer than usual which made it possible simply to extend one's hand and graze a cadet in the chest as he or she passed. The team usually added to the rally with "high fives," "low fives," and back-slapping.

After the run, some of the football players allegedly bragged in the showers about having felt some women's breasts. Meanwhile, back in the dorm, a handful of women went to their immediate cadet supervisor, usually a male, and reported that something inappropriate had just happened on the playing field. The report went quickly up the Cadet Corps' chain of command. Many involved in this chain were football players. These leaders did not block the report or try to put pressure on the women cadets to drop the matter. As Stephanie Arnold, deputy brigade commander from Ohio, concluded, "The chain of command system worked, and the academy did what was right. . . . It makes me proud to be a woman here but more than that, proud to be a cadet. I'm part of an organization that doesn't tolerate abuse of other people or each other."[3] The women cadets expressed concern that their complaints not become the basis for a witch hunt that could damage the team's morale. As Arnold explained, "They got the maximum punishment you could give. . . . We're not on a witch hunt."[4] The women emphasized that they simply wanted the behavior stopped; they did not want fellow cadets suspended from the academy. Three of the women interviewed concluded that the touching of their breasts had been accidental; others thought it was intentional. No one could identify any of the men, but they could say approximately where in the line they had been inappropriately touched. One woman jumped the chain of command and "e-mailed" the commandant of cadets with her concerns. E-mail often breaks down hierarchies; in no other way can a lowly cadet make direct contact with a general officer. E-mail, however, is not anonymous because a person's name or identity number is at the top of the message.

The next morning, the commandant of cadets, Brig. Gen. Freddy E. McFarren, began an investigation. West Point's leaders wanted to take the women's concerns seriously and also their plea not to overreact. All women in the regiment were interviewed that day (about fifty to sixty, 10 percent of all cadets who participated in the pep rally). Eventually fifteen of the women interviewed reported that they had been intentionally touched; three other women who had also been touched believed that the touching had been unintentional. In the meantime the football players were also interviewed. Some players, under the honor code, came forward to report the bragging they had heard in the locker room after the incident. Eventually three came forward to admit that their hands had touched women's breasts, but each maintained that the encounter was unplanned and unintentional. In all, seven black and white football players (sophomores and one junior) went before a regimental-level disciplinary board. Three, including the star tight-end, received stiff penalties. Their punishment was thirty-five demerits, eighty hours of walking in dress uniform toting a pack and a rifle, ninety-day suspension from the football team (effective immediately), and grounding (inability to leave the post). If a second offense were ever added to their record, they would probably be dismissed from the academy. Since cadets can work off only five hours of walking tours each weekend (on the weekend), a sentence of eighty hours of walking tours would take these three suspended football players sixteen weeks or well into February to complete. All three decided to stay at West Point and to accept the punishment. However, in the past, others who have received fewer punishment tours chose to leave the academy. From the perspective of those punished, the punishment would seem quite stiff.[5] But, as a junior cadet commented, "The actions taken were pretty normal, and they [academy officials] acted like they said they would."[6]

Some critics agreed that West Point had acted quickly but felt that the perpetrators should have been dismissed. Yet, according to Col. Donald J. McGrath, the USMA's director of public affairs, "The punishment is sufficient to insure that they become the kind of leaders with proper character that we're developing in West Point."[7] To have tried to expel them would have been much more difficult; the investigating team could find no one who could positively identify the offenders. Given the level of evidence they had, the investigators advised the superintendent that the academy would have a hard time proving the intent required for dismissal. In addition, there was a time factor to consider. A USMA disciplinary board, necessary for harsh disciplinary action, would have taken three months to gather all the evidence and reach a verdict. In the meantime, the players would have been allowed to stay on the football team. The investigating team and Army lawyers believed that the defendants would probably be found not guilty. By going with the lower-level regimental board, the officers sitting in judgment could consider whether bragging in the locker room was inappropriate behavior and a violation of the bedrock value, consideration of others. This regimental board could reach a swift judgment and could send a clear, quick message to the Corps of Cadets.

None of the newspapers printed the names of the eighteen women who came forward to report that they had been inappropriately touched; these eighteen constituted roughly one-third of the women in the First Regiment. A handful of newspapers, however, did print the names of the three football players punished and one published photographs of two of them.[8] One possible explanation why the *New York Times* and other newspapers did not print the names or publish the punished football players' pictures is that the editors wanted to be careful not to violate the cadets' rights. Yet, there is even a more basic explanation. West Point wanted the media and the general public to focus on the women victims and the swift and stiff penalties the academy had imposed. Publishing the photographs could have distracted the general public from these issues, drawing attention to what the USMA considered a secondary matter. The women cadets touched were white, the football team members who reported the inappropriate locker room comments were mostly if not all white, and the three football players punished were black. The officer in charge of the regimental board was also African American.[9]

Unlike at many civilian colleges and universities, race is not a major tension point at the USMA. Although subtle discrimination still exists, the blatant comments and actions that create tension on the civilian campus do not exist at West Point. In this respect, the USMA is years ahead of its civilian counterparts. As two cadets explained, "West Point cadets are trained to be leaders. Gender, race and everything you put aside"; "You can't have those hang-ups."[10] The Army is also in advance of many civilian employers. This accounts in part for the fact that 50 percent of enlisted women are African American. They find that the Army provides them with opportunities not available in the civilian sector. The *New York Times* editors knew the race of the victims and those accused, and made the wise decision that race did not play a primary role in the groping incident or in the response by the USMA's leadership. Yet the general public could easily have concluded that because the players were black, they were being given special treatment. Some might have said the punishment was too light; others that it was too harsh.

The three football players were not eligible for the next game (Army vs. Air Force). In that game, the USMA lost because the team was not able to develop a passing offense without their star tight-end. Yet, there was no hue and cry from the male cadets or the football team that the "women" were to blame because they overreacted to the incident. Knowing the possible consequences of the loss of three players, the team still wrote a "Memorandum for the Corps of Cadets" thanking the Corps for the support they had given the team "throughout the year and specifically through this ordeal." The memo continued, "We apologize to you for this incident and assure you it is not representative of the Army Football Team. . . . We do not want this incident to damage the bond between the Corps and this team."[11]

West Point's leaders were pleased with the outcome. They were delighted that several women felt that they could come forward and report behavior that was inappropriate and that should be stopped. The male cadets were pleased too. "I'm glad the female cadets felt they could come forward," said one junior cadet.[12] Indeed, after this incident some women reported other instances of alleged hazing or harassment that were then investigated. The academy's leaders were proud that male football players had come forward to report other football players. (Team members are customarily very close and form a wall around themselves.) Superintendent Lt. Gen. Howard Graves labeled the incident an "alarming lapse," but concluded that, "The identification of football players had been done by other members of the team so we don't see the team circling the wagons to protect a small group."[13] The USMA football coach, Bob Sutton, called the apparent incident "a serious problem" and concluded, "If someone did it, it's wrong. It's a wrong judgment, a wrong decision."[14] General Graves labeled the football cordon a leadership builder, but explained that it has now been suspended.[15]

Academy leaders were relieved the reaction of the rank-and-file cadets was that the players involved were jerks or "morons" and should have been punished, and that the women were right in filing complaints.[16] One male cadet put it bluntly: "I find myself wondering what are they? Brain dead?. . . There's more than fifty hours of instruction here on how to treat each other with respect. After the first hour, you get the message: This stuff won't be tolerated."[17] The women cadets urged the academy to get back to business and stated publicly their respect for the majority of their male colleagues at the USMA.

Although this football pep rally incident received the media's immediate attention (the superintendent even met with officials from the *New York Times*), media interest died quickly. The same pattern occurred in 1991 at the U.S. Air Force Academy in Colorado Springs, when several cadets were disciplined and others dismissed for watching or participating in sexual intercourse. The Naval Academy, however, remains chained to the urinal of bad publicity. Indeed, well over twelve dozen articles have appeared about that notorious episode. But even stories reporting positive actions by the Naval Academy such as appointing the first woman to lead the midshipmen had a paragraph reminding readers of the handcuffing incident. Why did the media treat these similar events so differently?

I maintain that the Army was able to enforce for a longer period of time and in a more effective format than the Navy, a zero tolerance policy. Cadets were warned to take their leadership training classes seriously or be disciplined or even removed from the academy. The Navy is a few years behind in getting the zero tolerance message across and in creating training programs that incorporate sexual harassment and discrimination under general leadership issues. Leaders of all the service academies have also learned from the "Tailhook" incident and its aftermath. "The parallel with Tailhook is so striking and the aftermath of Tailhook has been so strong that I can't imagine a military officer being that close to the parallel and not feeling obligated to take strong action," explained Larry R. Donnithorne, author of a key book on West Point's warrior leadership and a USMA graduate and former member of the faculty.[18] Because of the superficial similarities between Tailhook and the USMA spirit run (both had gauntlets through which women ran and were touched), the media asked West Point officials to comment. A *New York Times* headline captured the USMA response: "Cadets See Aberration, Not Tailhook Scandal." Specifically, as a male junior cadet explained, "The academy is dealing with it and the guys who have done it. . . . Tailhook was blatant and they tried to cover it up."[19] A female cadet agreed with her male classmate, "The two situations are completely different." "The Spirit Run," according to her, "was an unpremeditated incident involving individuals who made a grave mistake. Nobody said 'those women deserved it.' Stuff like that just isn't tolerated around here."[20] Lieutenant General Graves agreed: "There's a difference between a cordon of cadets trying to build spirit and a gauntlet of people who were drunk and who engaged in this kind of behavior before."[21]

In the spirit run and groping incident, the USMA leadership reacted quickly from bottom to top in both the Corps of Cadets' and the academy's chain of command. In a united front, West Point's leadership decided to send a message to all cadets that no cadet—even a football player—can be a cadet in good standing if he or she violates the "consideration of others" value. "I have to give the Academy and General Graves an absolute 'A' plus for handling that," said public relations expert Richard Yarbrough, vice chairman of West Point's voluntary Civilian Public Affairs Committee.[22]

The academies in general are ahead of civilian institutions in making more than fifty hours of training on social issues and social dynamics an integral part of the curriculum, and they do not limit such programs to freshman orientation or dorm workshops. However, as valuable as such programs are, they must be reinforced yearly and developed in breadth and depth. The academies are also proactive—unlike most institutions that whip together some sort of ad hoc program on date rape for a fraternity after a serious incident has occurred. Moreover, they have done a better job in reducing racial tension and racial discrimination than most colleges or universities. At the academies, race is not an issue providing "special treatment" for or treating minorities as second-class citizens who are not required to meet the standards set for white students. Although subtle forms of discrimination may never be eliminated, I was stunned by the lack of tension and general acceptance of minority cadets by their peers and by the officers. Finally, athletics is important to most colleges and universities, and does not always clearly take a back seat to academics. Unlike many civilian institutions, no one at West Point is allowed to graduate without passing the rigorous mathematics/sciences/engineering core curriculum, the military leadership program, and the physical fitness testing. If football players put on weight while at the academy, they must shed it and pass the same rigorous physical fitness tests as other cadets before being commissioned. According to Colonel McGrath, "We are not in the business of developing football players, we're in the business of developing leaders, both men and women."[23] Yet anyone who has been at a service academy (or military base or post or even the Pentagon) the week before the Army versus Navy or Army versus Air Force football games knows that football is taken very seriously.[24] Under these circumstances, it is even more remarkable that the USMA chose a process that led to swift punishment and suspension from the football team for the offenders.

In sum, I would argue that the spirit run and cordon incident highlights both the complexity of issues facing female cadets at West Point and the quick and effective response of the administration. As a *New York Times* editorial, entitled "Wisdom at West Point," concluded: "The Superintendent of the Academy, Lt. Gen. Howard D. Graves, moved rapidly, intelligently and with an openness that should constitute a watershed in the armed forces; treatment of what in recent years has seemed an endemic sexism."[25] Blatant discrimination and harassment are not tolerated and will no longer typically be experienced by female cadets. A 1993 General Accounting Office report stated that 76 percent of the women attending the service academies said that they had experienced some form of harassment. Colonel Patrick Toeffler, director of policy, planning, and analysis at the USMA, has analyzed the data to distinguish between overt or blatant and subtle forms of discrimination. He reported that, "The majority of women [at the USMA] said they heard disparaging comments such as women should not be there."[26] However, subtle forms of discrimination and harassment are omnipresent and are much harder to root out of a closed culture such as the Corps of Cadets and its alumni. Subtle forms of discrimination can actually be more damaging to a woman's overall self-esteem and self-confidence and thus will continue to be a serious problem—even if reduced in numbers of incidents and depth of treatment at the academies and on other college and university campuses. In addition, men and women will continue to fall in love. Issues concerning fraternization will continue to be delicate and potentially damaging to overall leadership and morale. Alcohol abuse will continue to be a key factor in date rape. The realistic goal of the academies is to reduce, not to eliminate, inappropriate behavior to make clear that punishment will be swift and harsh for violators, to develop a reporting system that will protect victims and whistleblowers from blatant and subtle forms of recrimination, and to handle without over or underreacting any violation of the bedrock values of the cadet code of conduct.

Notes

An earlier version of this paper appeared in *MINERVA* 13 (Spring 1995).

1. Robert Hardt, Jr., "Just the Latest in the Long Gray Line of Sex Scandals," *New York Post,* 2 Nov. 1994, 5.

2. Interview of Capt. Lorelei Coplen, USMA Class of 1985, *New York Daily News,* 2 Nov. 1994, 7.

3. Dale Russakoff, "West Point Hopes to Send Message in 'Groping' Investigation," *Washington Post,* 4 Nov. 1994, 1.

4. Carol Trapanl, "Cadets: Ruling is Valid," *Poughkeepsie Journal,* 5 Nov. 1994, 1.

5. Brig. Gen. Freddy E. McFarren, Memorandum for Corps of Cadets, "Update 2 to Incidents During Spirit Run/Football Cordon, 20 October 1994," 3 Nov. 1994, MACC-O-SP.

6. Trapanl, 1.

7. Raymond Hernandez, "Three Suspended For Groping At West Point. Congresswoman Calls Penalties Too Lenient," *New York Times,* 5 Nov. 1994, A26.

8. "Guilty Cadets Punished," *Times Herald Record,* 5 Nov. 1994, 4.

9. The regimental tactical officer for the First Regiment just happened to be African American; another regimental tactical officer was not pulled in to conduct the hearings. However, it is also indicative of the Army's general policy of color-blindness that there are black cadets and black officers at the USMA throughout both the cadet and officer chains of command. Also, the actual board hearings are confidential, so it is difficult to determine which football players testified at the hearings.

10. Trapanl, 1.

11. Undated memorandum for the Corps of Cadets from the Army football team signed by the team members with the highest cadet ranks.

12. Trapanl, 1.

13. Wayne A. Hall, "Point Tackles Scandal," *Times Herald Record,* 1 Nov. 1994, 3.

14. Kevin Gleason, "Sutton Calls Groping Case 'Serious,'" ibid., 3 Nov. 1994, 77.

15. Hall, 3.

16. Russakoff, *Washington Post,* 4 Nov. 1994, 6.

17. Ibid.

18. Wayne A. Hall and David Kibbe, "Quick Disclosure Applauded," *Times Herald Record,* 2 Nov. 1994, 32.

19. Joseph Berger, "Cadets See Aberration, Not Tailhook Scandal," *New York Times,* 1 Nov. 1994, B2.

20. Virginia Breen, "Sex Harass Probed at West Point," *New York Daily News,* 2 Nov. 1994, 7.

21. Frances X. Clines, "West Point Cadets Face Investigation in Sex Harassment," *New York Times,* 1 Nov. 1994, A1.

22. Hall and Kibbe, 3.

23. "5 Army Football Players Are Accused of 'Groping,'" *Washington Post,* 2 Nov. 1994, 1.

24. Army lost to Air Force the week after the three football players were suspended, but by the end of the season, the Army was back in full force and beat Navy.

25. *New York Times,* 2 Nov. 1994, A22.

26. "5 Army Football Players," *Washington Post,* 2 Nov. 1994, 1.

U.S. Air Force Non-Academy Officer Commissioning Programs, 1946–1974

Vance O. Mitchell

Since its establishment as an independent military service in 1947, the United States Air Force has used a number of programs to satisfy its officer procurement needs. Over the years, these programs have included the military academies as well as three non-academy sources. This paper examines the non-academy programs during their formative years and evaluates their importance to the Air Force in providing for the service's future leadership.

To satisfy its officer procurement needs in the immediate post–World War II period the newest military service called upon five programs. Although an air academy was still a decade away, agreements negotiated in 1948 between Secretary of the Air Force Stuart Symington and the other service secretaries pledged the Army and Navy to send 25 percent of each West Point and Annapolis class to the Air Force. These agreements, in effect until the Air Force Academy's production equaled the other two service academies, netted the Air Force about 4,200 officers between 1949 and 1968.[1]

Beyond the Army and Navy contributions, the Air Force looked to its three non-academy sources. These non-academy programs were not merely procurement supplements. Over the next twenty-five years they produced at least 90 percent of the incoming officers. In 1970, over 64 percent of Air Force general officers came from one of these sources. Moreover, they produced new officers cheaper than did the academies and they broadened the procurement base. This last point addressed legitimate concerns that an officer corps too heavily laden with academy graduates would be dangerously inbred.[2]

The smallest of the non-academy programs was Officer Candidate School (OCS). OCS shrank almost to nothing following World War II and did not begin to revive until 1947. In July of that year, the program was assigned a quota of five hundred graduates annually, obtained from both civilian life and from the enlisted ranks. Those commissioned through OCS, which included both men and women, would be assigned to nonrated (nonflying) duties.[3]

The Aviation Cadet program, the main source of rated (flying) officers and destined to be the largest of the non-academy programs during most of the late 1940s, was actually closed for almost two years following the war so great was the glut of fliers produced during that conflict. Only 778 individuals, almost exclusively recent West Point graduates and nonrated officers, earned pilot's wings in all of 1946 and 1947. These individuals trained as student officers, not as cadets. Navigator training ceased in the spring of 1946, not to begin again for four years.[4]

The pace quickened with the end of demobilization in June 1947. With its force levels more or less stabilized, the Air Force laid the ground work for a long-term flight training program that would train three thousand pilots and one thousand navigators annually. Given an expected elimination rate of about 30 percent, over five thousand trainees had to enter flight training per year to meet production goals. The Aviation Cadet program was immediately thrown open to both enlisted personnel and civilian applicants and given widespread publicity. Recruiting teams visited colleges to interview prospective applicants and give a cursory physical exam (eyesight and hearing) to whomever seemed genuinely interested. Those who passed the physical exam could then apply for flight training. The publicity program and the recruiting teams achieved good results and the Aviation Cadet program met its production goals until the Korean War forced further adjustments.[5]

The Air Force Reserve Officers Training Corps (AFROTC), the remaining non-academy commissioning program, was a joint venture by the service and the nation's universities and colleges. AFROTC, called Air ROTC for the first year of its existence, got underway at seventy-eight campuses in the fall of 1946. Confidence in the program was high, and plans called for units on 150 campuses and as many as 8,000 graduates annually within a few years. Although AFROTC did not fulfill those early expectations, it nevertheless remained a substantial program that in 1949 commissioned 3,300 second lieutenants, its largest number before the Korean War.[6]

The problem with the AFROTC was that its graduates were beyond the reach of the active-duty force. As the name implied, the ROTCs of all services supplied officers to the reserve, not the active-duty components. Active-duty service was voluntary except during a national emergency, and no national emergency existed in the late 1940s. With the nation's economy entering a twenty-five-year period of almost unbroken growth and an academic degree to help them along, AFROTC graduates usually elected to remain in the reserves and pursue more lucrative civilian careers. Fewer than 800 out of 5,500 AFROTC graduates in 1948 and 1949 chose active duty. The same problem existed in all ROTC programs, and in 1948 a committee chaired by Gordon Gray, assistant secretary of the army, began drafting legislation to remedy the situation. The legislation, the ROTC Act of 1949, had still not been enacted when the Korean War brought a declaration of national emergency and voided its need.[7]

The non-academy commissioning programs met the quantitative requirements of the period, but that was only half the equation. All personnel procurement programs had to deal with qualitative issues as well. The qualitative aspect was more difficult to satisfy because it was more complex, and quality has traditionally been sacrificed to make up for numerical shortfalls, particularly in times of national emergency. As a result, the ranks have always been manned, albeit at times with much less-than-ideal individuals.

If the service had an unlimited pool from which to select commissioned personnel it would have chosen those with high physical profiles, emotional stability, self confidence, high scores on qualifying exams, and, not the least, a college education. In fact, a college-educated officer corps was early on identified as an Air Force objective. That objective had little to do with piloting aircraft. Flying an airplane rested on mechanical aptitudes—gross motor skill development, hand-eye coordination, eyesight, and the ability to rapidly translate a variety of external cues into control surface inputs; in short, good physical coordination and quick reflexes. These were attributes largely unrelated to intelligence and largely unaffected by academic achievements.[8]

The need for college-educated officers arose from other factors. In 1944, an Army Air Forces (AAF) study group headed by Maj. Gen. Laurence Kuter, assistant chief of air staff for plans, had examined some of the changes that the service needed to make in the postwar era. Among the problem areas identified was the narrow skill spectrum of the regular officer corps. Over 98 percent were pilots whose careers had focused almost exclusively on flight operations. The postwar military was going to make much broader demands on its leadership.[9]

General Henry H. "Hap" Arnold, the commanding general of the AAF, effectively summarized those demands during a 12 January 1945 staff meeting. After offering his vision of aerial warfare in the future he turned his thoughts toward the postwar officer corps, warning that:

> The phase in which exclusive pilot management was essential is drawing to a close. ... Regulations limiting the responsibilities of non-rated personnel must be changed. Every opportunity must be given to skills and abilities needed for a well rounded organization if the United States is to maintain its air leadership.[10]

Arnold's words preceded by fifteen years the classic pronouncement by Morris Janowitz, the dean of American military sociologists, that modern military leadership must "strike a balance between the three roles of heroic leader, military manager, and military technologist."[11] There would always be a place in aerial warfare for the fighting leader, embodied in the pilot, but a modern military also required officers able to master a wide variety of skills, both technical and nontechnical. In the Air Force that would soon translate into over 250 individual specialties grouped into approximately 20 career fields.[12]

That many of the 250 specialties were academically based was sufficient justification to have a college-educated officer corps, but there were other reasons as well. Well-educated officers were likely to have better perspective on and insights into complex issues such as civil-military relations, national policy-making, and America's role in the postwar world. At rock bottom, a college education suggested that an individual had ambition, the will to overcome obstacles, a capacity to solve problems, a capability to deal with adversity, and the ability to get along with people. None of these presumed qualities guaranteed success, but collectively they did indicate an enhanced potential that could translate into growth and achievement given time and opportunity.[13]

Unfortunately, plans for a college-educated Air Force officer corps in the late 1940s were more utopian than realistic. The average age of Air Force officers at that time was only about twenty-six years. The vast majority of these young people had scarcely been out of their teens when plucked from civilian life by World War II. Few had finished college, and a significant number had barely completed high school when called to the colors. Since entering military service they had not had an opportunity to improve their academic education. As a result, hardly more than a quarter of postwar Air Force officers had the coveted college diploma. The other military services had likewise seen their officer academic levels decrease from the prewar years, but the Air Force, for reasons of its relatively larger growth, suffered the most.[14]

The educational deficiency highlighted a qualitative problem in officer procurement, one destined to haunt the service for many years. Of the sources of officers, three presented no problems as far as academic levels were concerned: All academy and AFROTC graduates had degrees and a majority of those from OCS had similar achievements. Yet academy, and OCS graduates were few, and only a small number of those from AFROTC chose active duty. As a result, the combined annual contribution of these programs to the active-duty force probably did not much exceed one thousand officers.

Most new officers of the period, possibly as high as 70 percent, came from the Aviation Cadet program, now on its way toward annually producing almost four thousand pilots and navigators. The program's quantitative success had, however, come at the expense of educational standards. In 1948 and 1949, only 2 percent of cadets had a baccalaureate. Even those recruited from the nation's campuses tended to quit college prior to graduation. Two years of college was the stated educational requirement to become an aviation cadet, but high school graduates were accepted if they passed an examination demonstrating knowledge equal to two years of college. The lack of college-educated aviation cadets and the overwhelming numerical contribution of that program meant that a college-educated officer corps was far out of reach in the late 1940s.[15]

Concerns about the low educational achievements of newly commissioned officers were forced into the background with the outbreak of the Korean War in June 1950 and the return of quantitative pressures that accompanied the resulting military buildup. By the end of 1950, the pilot training goal had more than doubled to 7,200 annually, and navigator training had also increased, although the exact figure is unknown. Just to satisfy pilot production, over ten thousand qualified applicants had to enter training each year. Overall, the Air Force began building toward 143 wings and an officer corps of over 150,000, almost a three-fold increase over the 54,000 officers in active service at the beginning of the war.[16]

The build-up was facilitated by the recall of reservists to active duty and a declaration of national emergency. The national emergency required that all ROTC graduates (veterans excepted) serve at least two years of active duty. By 1952, AFROTC was providing six thousand officers annually, but, consistent with the attitudes of AFROTC graduates before the war, no more than 17 percent wanted to be rated. The vast majority wanted to avoid the longer active-duty commitment that came with flight training and to return to civilian life in the shortest time possible.[17]

Since flight training was voluntary, AFROTC graduates could not be ordered into training, and, as before, the Aviation Cadet program had to meet the demand for rated officers. Throughout the war there were always enough cadets to satisfy requirements, but the pool of qualified applicants sometimes got dangerously low. When that happened, quality was sacrificed to avert shortages. In 1952, the minimum passing score on the qualifying exam was lowered from six (out of nine) to three. The following year, the educational requirements were reduced from two years of college to a high school diploma for civilians and from a high school diploma to a graduate equivalency degree for enlisted men.[18]

The end of the Korean War in August 1953 ushered in a new era in American history when, for the first time, the United States did not demobilize following a conflict. The nation found itself technically at peace, but with a sizable peacetime draft and a large standing military due to the Cold War. At this critical juncture the Air Force began exploring ways to convert AFROTC, now the largest commissioning program, from a nonrated source into a rated one. How to make that conversion was still being worked out when the Air Force personnel system got a sudden, unpleasant preview of the future.

Unlike previous periods of national peace, the "peacetime" Cold War was destined to be a protracted period of volatility in which the military was alternately subjected to force increases and force reductions. In April 1953, and even before the guns fell silent in Korea, the newly installed administration of President Dwight D. Eisenhower ordered the Air Force to halt its growth at 120, rather than 143, wings. A 120-wing force meant an officer corps of about 130,000, only 1,500 more than the number projected to be on active duty at the end of the year. Officer procurement had to be reduced, and quickly. The recall of reservists to active duty was sharply curtailed and OCS production was slashed from two thousand to five hundred annually, but the continuing need for rated officers meant that the Aviation Cadet program had to be left alone. That left only the AFROTC from which to make cuts and the reduction campaign focused on that program. The focus was appropriate because AFROTC was the main problem.[19]

The AFROTC, unlike the other non-academy sources, was a long-term program that took four years to produce a second lieutenant. This already inflexible arrangement was further complicated by the contract signed by cadets upon entering the advanced (junior and senior) portion of the program. The contracts, legally binding documents, guaranteed commissions upon graduation. The problem facing the Air Force was that in June 1954 the AFROTC, swollen by draft pressure to 125,000 cadets on 209 campuses, would turn out 13,000 second lieutenants, the overwhelming majority destined for nonrated duties where they were not needed.[20]

The Air Force's first impulse was to renege on its deal with AFROTC cadets. H. Lee White, assistant secretary of the air force for management, requested permission to deny commissions to all 1954 AFROTC graduates. John A. Hannah, assistant secretary of defense for manpower and personnel, refused the request; all AFROTC contracts must be honored. Lieutenant General Emmett O'Donnell, air force deputy chief of staff for personnel, then recommended that many of the graduates be sent directly into the reserves. Again, Hannah refused; everyone had an active-duty obligation. He was, however, receptive to a third idea, disenrolling all cadets before their senior year if they did not volunteer for flight training.[21]

Acting quickly, the Air Force in July 1953 gave the cadets, then in summer camp between their junior and senior years, the option to accept flight training or to resign from AFROTC. Many resigned, but there was an immediate outcry from parents, colleges, and members of Congress over the Air Force's arm-twisting methods. Just as quickly as the choice had been imposed it was removed and those who had resigned were welcomed back into the program. The paperwork was tidied up by attributing the resignations to "administrative error."[22]

Instead, only those 1954 graduates who met specific requirements, mainly those agreeing to flight training, got commissions. The remainder, about 4,600, received "Certificates of Completion" entitling them to commissions in the reserves *after* completing two years on active duty as enlisted men. Again, there was an uproar, but the decision stood. The certificates of completion plus the separation of thousands of reserve officers from active duty allowed the Air Force to stay under the 130,000 officer ceiling, but at a huge public relations cost. The host colleges sharply criticized the whole episode, and certificate recipients were understandably bitter. So were those reservists involuntarily separated from active duty just to make room for thousands of others.[23]

The experience of the AFROTC class of 1954 was repeated in subsequent years, albeit handled in a more enlightened manner. Beginning in 1955, physically qualified cadets had to volunteer for flight training before they could be enrolled in the advanced portion of the program. Since the cadets in question had not as yet signed contracts, there was no legal impediment to forcing the choice upon them. AFROTC had become the main source of rated officers among the commissioning programs, academy or non-academy.

In 1957, and in a time of sharply reduced officer procurement, declining flight training, and another mandatory force reduction, the Air Force made a further change in the AFROTC contracts. Those completing the advanced program would still receive commissions, but the change released the service from any firm commitments with regard to active-duty assignments. Students destined for rated duties signed a statement of understanding that their entry into flight training would be contingent on service needs when they graduated. Those headed for nonrated duties acknowledged that their assignments were based on the same consideration and might not be related to their college majors. This was apparently the Air Force's last major effort during the period covered by this paper to "cut itself some slack" and make AFROTC as responsive as possible to sudden changes while at the same time avoiding criticism.[24]

In addition to the four years AFROTC took to produce officers and the limitations imposed by the student contracts, the Air Force's ability to manipulate the program was circumscribed by the nation's colleges and universities. All other commissioning programs, academy and non-academy, were firmly under military control. Not so with AFROTC, whose base of operation was the nation's campuses. The host institutions provided classrooms, granted academic credit, bestowed professorial status on instructors, and encouraged students to enroll. In return, the hosts received a government stipend and shared in the management of the program. Even before the on-campus turmoil of the Vietnam War, however, there were clashes of interests and questions about AFROTC management and just how much authority the host institutions really had.

Speaking for the Air Force, Maj. Gen. Turner C. Rogers, the AFROTC commandant in the mid-1950s, saw host-service management in terms of how it should work. In his words, this was "a joint venture—a working partnership—between the educational institution and the Air Force. . . . This implies close collaboration between the partners in the solution of ROTC problems and the formulation of ROTC policy."[25]

Not so, responded critics within academia. Calling it a working partnership was merely a verbal fig leaf to hide near-total military control. The military developed the curriculum, controlled who enrolled in the program, and appointed the AFROTC staff personnel virtually free of university influence. The officers who headed ROTC units functioned more like foreign ambassadors than faculty members. They ran their departments with little university input and entered into campus affairs only when their interests were involved. Their instructions came from the parent service, and it was to the parent service, not the host institution, that they owed their loyalties.[26]

The tension generated by these opposing views had risen more than a little over the certificates of completion incident and the Air Force had to give ground on the next AFROTC reform effort, an attempt to cut operating costs by eliminating the weaker units. Plans to terminate these units were first drawn up in 1953, but the furor over the certificates of completion had relegated the idea to the back burner for over a year. Then, true to the concept of joint management, the Air Force in January 1955 convened a panel of civilian educators to advise and consent on a plan to eliminate those units that did not meet minimal production. The panel, chaired by Dr. Everett N. Case, president of Colgate University, agreed with the service's position that minimum production should be twenty-five officers annually who were qualified for flight training. The panel made only a few suggestions, including that all units be given until the end of academic year 1955–56 to meet the minimum requirement. This was agreed to.[27]

The trouble came when, in early 1956, David C. Smith, assistant secretary of the air force for manpower, personnel, and reserve affairs, identified twenty-nine units to be closed. Unwilling to see their students' plans dashed, the affected institutions replied with a barrage of protests. Unwilling to further damage AFROTC, Secretary of the Air Force Donald Quarles directed that an individual unit be closed only if the affected host institution concurred. Only eight of the twenty-nine agreed. Another, Yale University, asked that its unit be closed even though it was not on the list. This meant that the Air Force had very little to show for two years of work.[28]

In 1961, the Air Force returned to the problem of AFROTC's operating costs by planning for a two-year program open only to juniors and seniors. Officials hoped to trim the AFROTC enrollment from approximately one hundred thousand to as few as

13,000, all of whom would be under contract. Again, there was no serious opposition, but the host institutions reserved the right to choose between two-year and four-year programs. After a four-year delay caused in part by problems encountered in gaining congressional approval for a two-year program, the colleges and universities made their choices. In a defeat for the Air Force, 155 chose to offer *both* two-year and four-year programs, 8 opted to retain the four-year program, and only 16 elected the two-year option. Those selecting both programs did so because giving students the widest possible choice was in the best interest of the institution, but it at least temporarily negated yet another cost-cutting effort.[29]

In the meantime, the other non-academy commissioning programs, both firmly under military control, responded much as the Air Force wished. The OCS, restricted to enlisted personnel since 1951, remained a steady, if small, source, producing about 450 officers annually between 1953 and 1957. In 1958, its quota was raised to 600 for the remainder of the decade. Most OCS graduates were still bound for nonrated duties, but a few now entered flight training after being commissioned. With ten qualified applicants for each opening, meeting production goals was not difficult.[30]

The Aviation Cadet program also responded well, but unlike OCS it was anything but steady. After 1954, the AFROTC produced mainly rated officers and the Air Force suddenly had two major sources for rated officers at a time when both the size of the officer corps and the number entering flight training were declining. Cuts had to be made and, not surprisingly, the Aviation Cadet program, the vast majority of whose graduates did not have college degrees, bore the brunt. In 1954, over 11,000 cadets entered flight training. In 1957, the figure was less than half that, and falling.[31]

Yet, the Air Force still had to maintain the Aviation Cadet program. In 1957, AFROTC graduates commissioned during 1954, the year the program was converted to a rated source, came to the end of their active-duty commitments. Confirming earlier fears, about 72 percent separated from active duty. Driven at least in part by this poor retention the Air Force increased the active-duty commitment for rated officers by one year. This better recouped the money spent on flight training, but the effect on AFROTC was quick and negative. By 1959, production had slipped below four thousand and the program no longer met production goals.[32]

The problems with the AFROTC meant that the Aviation Cadet program even in its reduced state still produced over 30 percent of the new officers, but, as before, at a penalty in terms of a college-educated officer corps. Whereas most AFROTC graduates left the service, retention of both aviation cadets and OCS graduates was good, even excellent. This gave retention an obvious pattern. Young officers with college degrees opted for civilian life where the compensation was greater than that the military could offer. Those without degrees stayed in the service because they were less competitive in the civilian arena, despite an economy that was in its second decade of unparalleled growth. This retention pattern once again frustrated attempts to achieve a college-educated officer corps.

By 1958 it was obvious that the manifold problems with AFROTC and the inability of the Aviation Cadet program to attract college graduates required a new approach. An entirely new, short-term commissioning program was needed, one that accepted only college graduates, provided officers for both rated and nonrated duties, and was firmly under military control. Such a program, Officer Training School (OTS), began taking shape in late 1958. The first class entered training in December 1959. The goal for the remaining six months of the fiscal year was a modest three hundred graduates.[33]

OTS was short, requiring just under three months to produce a commissioned officer. It was also intensive with training conducted six days a week. The academic portion, roughly 210 hours, covered a span of subjects from effective writing to the conflict of nations to the responsibilities of a commissioned officer. There were also rigorous programs of inspections, physical training, and drill and ceremony. The uncharitable might conclude that the academic portion was little more than a protracted series of briefings, but given the material covered and the time available brevity had to rule.

OTS was also different. Both men and women were accepted, but the number of female trainees remained small for a number of years. Although it relied heavily on the coercive effects of the draft to furnish qualified applicants, the program itself was noncoercive to the extreme. Standards were set and enforced, but trainees had considerable latitude in how they went about their daily tasks. No harassment of trainees discharging their responsibilities was allowed, and only those who failed to meet standards were subjected to the harsher forms of discipline. Hazing was forbidden on pain of dismissal. Trainees were even permitted automobiles and could wear civilian clothes in the limited free time available during the last half of the program.[34]

In 1961, enlisted personnel who had earned their college degrees at government expense began receiving their commissions through OTS. This obviated the need for OCS which closed in 1963. By 1963, OTS was producing several thousand new officers a year, many of them destined for rated duties. The obsolescent Aviation Cadet program, already reduced to a mere shadow of its former self, closed in 1965.[35]

The closures left two non-academy commissioning programs to meet the challenges of the Vietnam War. OTS had no quantitative problems for most of that period. Indeed the twin prods of the war and the draft resulted in as many as three qualified applicants for every training slot. The program met production goals that ran as high as seven thousand annually, although year-to-year volatility in those goals made effective planning virtually impossible and brought protests from the school's harried staff.[36]

The main problems with OTS lay elsewhere. OTS graduates were, in general, of lower quality that those from the academies and AFROTC. Some of the qualitative deficiency could be attributed to the brevity of OTS itself which did not adequately prepare its graduates for what awaited them. Time and experience usually took care of this problem, but time and experience could not compensate for the questionable motivations of some who entered the program.

The majority of OTS applicants had little interest in making the military their career. About 80 percent sought nonrated assignments to avoid the longer active duty commitment. Some agreed to accept flight training only to get into the program and secure a commission, but had no serious intention to earn their wings. The ramifications of such shallow commitment were vividly displayed in pilot training where the elimination rate for OTS graduates in some years was a horrendous 47 percent. Some resigned from pilot training upon arriving at their training base or shortly thereafter. A few were candid in admitting that they had never intended to follow through on their end of the deal. When, in 1969, a lottery system based on an individual's birthday established the priority for draft calls, as many as 15 percent of those in OTS at the time resigned, mainly those with lottery numbers exempting them from involuntary service. OTS produced many fine officers during this period, but in having a program so attractive to those looking for an easy path to a commission the Air Force had a built-in problem that could only be ameliorated by further adjusting production goals to make good the inevitable shortfalls.[37]

OTS's problems were, however, small when compared to the travails of AFROTC in the last half of the 1960s. The Vietnam-related troubles of AFROTC began in academic year 1967–68 when forty-five units reported some sort of anti-AFROTC activity. The activity was not violent. Mostly, the protesters addressed issues relative to the previously mentioned joint management of the program by the service and the host institution, but now given a sharper edge by the growing antiwar sentiment. Both students and faculty questioned the quasi-independence of AFROTC units, the quality of both the curriculum and the instructors, and whether academic freedom was possible in a military classroom. More extreme elements charged that teaching people to kill had no place in an educational institution.[38]

AFROTC came under heavier pressure in academic year 1968–69 when even conservative students and faculty members began to question program content and the quality of the instruction. In December 1968, in a harbinger of things to come, the Boston College faculty voted to deny academic credit to AFROTC after those students currently enrolled had completed the program. In response, Theodore C. Marrs, assistant secretary of the air force for reserve affairs, asked the host institutions themselves to recommend changes that would improve the program. Like other critics, the hosts suggested that the curriculum and the instruction be improved by allowing the civilian faculty a greater role. None thought that AFROTC and academia were incompatible.[39]

These recommendations and the deteriorating position of AFROTC on the nation's campuses prompted Lt. Gen. Albert P. Clark, commander of the Air University, to order changes. Henceforth, only officers with advanced academic degrees would be assigned to AFROTC. The senior officer in each unit would no longer be an elderly colonel looking for a soft assignment before retirement. The hosts were urged to reject senior officers deemed unacceptable, a right always present but seldom exercised. Clark also authorized host institutions to substitute academic courses for related AFROTC subjects.[40]

Whether too-little-too-late or simply irrelevant given the times, General Clark's initiatives never had a chance. In academic year 1969–70, AFROTC was the target of over 1,000 "hostile acts" that included 37 violent demonstrations and 193 instances of property damage or personal injury. Interestingly, the number of hostile acts dropped sharply following the start of the draft lottery in December 1969. Although American involvement in Vietnam lasted until early 1973, hostile acts declined to 479 in academic year 1970–71 and 169 the following year.[41]

More significant, however, were the campaigns on some campuses, mainly centered in the more elite colleges and universities, to deny AFROTC academic credit and to strip unit staffs of faculty rank. The military gave ground by reducing drill, offering to accept civilian instructors in some courses, recommending that AFROTC be called a "program" rather than a "department," and suggesting titles other than "professor" for staff members. However, AFROTC must have academic credit, a condition deemed essential to the program's survival, and the staff must have faculty status as required by the legal statutes governing all ROTC programs. Unfortunately, negotiations to resolve these issues were seldom successful. Between 1969 and 1971, thirteen AFROTC units were closed either by mutual agreement between the host and the Air Force or by failure to reach suitable contracts.[42]

This turmoil did not seriously hurt AFROTC numerically for most of the period. Enrollment declined from 70,000 in 1966 to 44,000 in 1969, but that was due as much to progressively fewer schools requiring participation in basic ROTC and a slow, but steady, increase in the number of institutions offering only the two-year program than to antiwar sentiments. Indeed, war and conscription resulted in stiff competition to get into the advanced portion of the program with as many as four applicants for each billet. Between 1966 and 1969, AFROTC met its production goals for the first time since the late 1950s.[43]

The real downturn came after the draft lottery was announced in December 1969. AFROTC lost over 25 percent of its enrollment within a few months, production goals were no longer met, and there were concerns about the program's future. Broadening access to the program, increasing financial assistance to cadets, and lowering the minimum acceptable scores on the qualifying exam slowed but failed to halt the decline. By 1973, enrollment stood at only 20,000, including 1,800 women who gained entrance into the program on an equitable basis in 1969. The fact that those still in AFROTC were now better motivated and the program was still producing almost 4,000 new officers annually led some officials to predict a resurgence within a few years. Others, however, were unsure how the end of the draft in the early 1970s would affect on-campus procurement and how well AFROTC would fare in the world of the all-volunteer military. Thus did AFROTC end the period 1946–74 with its future in some doubt.[44]

This paper closes with two observations. The first is an assessment of how well the non-academy commissioning programs met the demands placed upon them during the period. Given this paper's relation of what must seem to be an endless series of problems and crises one might be tempted to answer "no," but that would be too hasty. Problems, not successes, generate staff work, staff work generates paperwork, and paperwork is the life-blood of most histories. Hence the problem-oriented focus of this paper. Moreover, the fact that many of those commissioned during the period were draft motivated is not relevant to a final judgment. Regardless of their attitudes toward the military or their reasons for seeking a commission, those who honorably discharged the trust placed in them fulfilled their obligations. Only the minority who willfully reneged on their part of the bargain need be judged harshly.

So, did the non-academy commissioning programs meet the needs of the service? Yes, they did, but their success in quantitative terms was uneven and short lived. As soon as the programs achieved a desired level of production, requirements changed and adjustments had to be made to meet new demands. Such volatility, of course, still exists today. Witness the massive force reductions that struck the military in the late 1980s and early 1990s.

Figure 1 is an attempt to break through the mountain of numbers and show the numerical contribution of each program from the mid-1950s to 1972. The U.S. Air Force Academy's contribution, beginning with its first graduated class in 1959, has been added to show the entire officer commissioning picture. If nothing else, the illustration shows the volatility of the period and the overwhelming numerical contribution of the non-academy commissioning programs.

Qualitatively, at least in terms of academic achievements, the service was plagued for much of the period by either the inability to attract college graduates into its ranks or to retain them once they had completed their minimum service obligations. Only with the beginning of OTS in 1959 and the termination of OCS and the Aviation Cadet

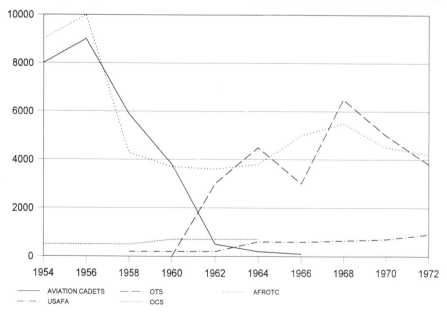

Figure 1. USAF Officers Commissioned by Program, 1954–1972
(in Two-Year Intervals).

program a few years later was it possible to say that a college-educated officer corps was on its way to becoming a reality. That reality was, however, not achieved until the retirement of all those career officers who did not hold degrees. The problems of poor retention among college-educated officers and of poorly motivated individuals obtaining commissions would be ameliorated by the end of the draft and the better financial compensation that accompanied the transition to an all-volunteer military. Overall, the best evidence that the triad formed by the Air Force Academy, AFROTC, and OTS met both quantitative and qualitative requirements is that they are still in place today.

The second observation offers an explanation of why personnel policies have always been, and certainly always will be, contentious and difficult to manage. The Air Force was made up of "things" and people. Things, be they pencil erasers or airplanes, responded to human guidance in predictable ways and without value judgments. Not so with people, who were emotional, exercised value judgments, made demands, petitioned Congress, weighed options, and made decisions in ways advantageous to themselves. Dealing with, say, three thousand OTS graduates was also to deal with three thousand personal agendas pulling in as many different directions. Call it greed or enlightened self-interest, it was this factor, the human factor, that troubled the waters and over which the Air Force had only limited control. The late Maj. Gen. Haywood S. Hansell clearly had that in mind when, in a conversation with the author, he voiced a terse, but accurate, summary of the period: "You know, we always did better with airplanes than we did with people."

Notes

1. Schlatter to SECAF, subj: Allocation of West Point and Annapolis Graduates to the Air Force, 15 Apr. 1949; SECNAV to SECAF, subj: Allocation of Annapolis Graduates to the Air Force, undated; and SECNAV to SECAF, subj: Allocation of Annapolis Graduates to the Air Force, 5 Aug. 1949; all located in Deputy Chief of Staff, Personnel (DCS/P) 210.511, Assignments USMA and ROTC to USAF, book 1, 31 Dec. 1953, Record Group (RG) 341, Modern Military Branch (MMB), U.S. National Archives and Records Administration (NARA), Washington, D.C.

2. The 1970 statistic about Air Force general officers is based on a 50-percent sampling of flag officers on active duty on 30 June of that year. The source was the biographical sketches published by the Air Force public affairs staff.

3. Exec Asst of the Mil Pers Division (AC/AS-1) to Johnstone, 3 Oct. 1947, Mil Pers file 1, RG 18, MMB, NARA; Dir of Training Historical Summary, 1 July 1949–30 June 1950, 7, Air Force History Support Office (AFHSO) 140.01, Washington, D.C.; "The Development and Functions of OCS and OTS, 1942–1951," USAF Hist. Study No. 99, Air University Hist Div, 1953, 114–15 (AFHSO 101-99), 114–15.

4. *Air Force Statistical Digest, 1947*, table 56; Memo for the SECAF, undated and unsigned, Directorate of Personnel Planning (D/PDP) 353.2, Pilot Trng, 31 Dec. 1952, book II-S, RG 341, MMB, NARA; Air Training Command (ATC) Hist, 1 Jan.–30 June 1947, 42–44, and 1 July–31 Dec. 1946, 77, AFHSO 220.01; *Army Air Forces Statistical Digest*, 1946, table 12.

5. Presentation before the Bureau of the Budget by Maj. Gen. Robert Nugent, 11 Oct. 1949, Dir of Military Personnel (D/DPD) 250.01 (1948–50), Speeches, 1 Jan.–30 Sept. 1947, RG 18,

MMB, NARA; USAF/CC to multiple addressees, 11 Sep. 1947, Mil Pers file 1D, 1 July–30 Sept. 1947, ibid.; Dir of Trng Hist Summ, 1 July 1949–30 June 1950, 6; "Development of AAF and USAF Training Concepts and Programs, 1941–1952," USAF Hist. Study No. 93, Air University Hist Div, 1953, 393–94 (AFHSO 101-93); Seventh Meeting of the Air Board, 33. The Air Board minutes are maintained in the AFHSO.

6. Air Defense Command (ADC) Adjutant General to USAF/CC, subj: Requirements for Air ROTC Instructors, 2 Feb. 1948, ADC Hist, July 1947–November 1948, documents 1:215, AFHSO 410.01; Statement by Gen. Carl Spaatz to the Finletter Cmte, November 1947, ibid., 6:8; *Air Force Times*, 18 Aug. 1948, 3; Fourth Interim Rpt of the Air Board, 3 Dec. 1948, Air Board 4 Dec. Mtg file, box 7, papers of Maj. Gen. Hugh Knerr, Library of Congress (LOC).

7. Air ROTC Hist, app. 38, AFROTC historian files; Transcript of the AF Advisory Cmte on ROTC Affairs, 14–15 Feb. 1949, 55–59, AFHSO 132.31-9; Draft speech, subj: Future Ofcrs in the Armed Services, DCS/P 210.002, Gen Ofcrs, 31 Dec. 1949, book 1C, RG 341, MMB, NARA; Exec D/PMP to multiple addressees, subj: Review of AFROTC Ofcr Production Plans, 19 Jan. 1949, Directorate of Personnel Training (D/PTR) 326.6, ROTC, November 1948–August 1949, ibid.

8. Neal E. Miller, *Psychological Research on Pilot Training*, AAF Aviation Psychology Program Research Report No. 8 (Washington, D.C.: Government Printing Office [GPO], 1947), 48–50; Philip H. Debois, *The Classification Program*, AAF Aviation Psychology Program Research Report No. 2 (Washington, D.C.: GPO, 1947), 14.

9. AC/AS-1 to AAF/CV, Special Qualifications for Technical Specialists, 30 Aug. 1946, decimal 210, Commissioned and Warrant Ofcrs, Misc 1946–1947, tab A, RG 18, MMB, NARA. The 1944 studies are in the microfilm collection of the AFHSO under decimal 140.xxx.

10. Deputy CG/AAF to multiple addressees, subj: Principles of Future AAF Action, 29 Jan. 1945, decimal 210, Misc 1945, vol. 1, ibid.

11. Morris Janowitz, *The Professional Soldier: A Social and Political Portrait* (New York: Free Press, 1971), 21.

12. D/PTR Hist Summ, 1 July 1949–30 June 1950, 3, accession (ACC) 67A-0575, box 1, RG 341, Washington National Records Center (WNRC), Suitland, Md.; Survey, subj: Officer Career Programs: Principles, Policies, and Procedures, 26 Sept. 1949, DCS/P 319.1 (1949), RG 341, MMB, NARA.

13. D/PDP to D/PTR, subj: Educational Requirements for Regular Commissions under AFR 36-5, 10 Nov. 1949, D/PDP 210, Ofcr Personnel 31 Dec. 1952, box 1-C, ibid; Noel Parrish, "New Responsibilities for Air Force Officers," *Air University Quarterly* 1 (Spring 1947): 41–42.

14. Chairman of Air University Board of Visitors to USAF/CC, subj: Rpt of the Visitors Fourth Mtg, 5 June 1948, DCS/P 210.2, Promotions 1 July–31 Dec. 1950, book IV-C, RG 341, MMB, NARA; Edwards to Armed Forces Staff College, 26 Oct. 1948.

15. D/PTR Hist Summ, 1 July 1949–30 June 1950, 7–8; Draft speech, "Future Ofcrs in the Armed Services"; "Development of AAF and USAF Training Concepts and Programs, 1941–1952," 393.

16. ATC Hist, 1 July 1950–30 June 1951, 1:130–32 and inclusion 1, and 1 July–31 Dec. 1951, 1:41–42; Hist, Training Div, D/PTR, 1 July–31 Dec. 1950, 1, AFHSO K141.31; Hist, HQ USAF, 1 July 1950–30 June 1951, 117, AFHSO K168.01-1A; Gerald Cantwell, "From Flying Club to Total Force: The Air Force Reserve, 1946–1981" (unpublished manuscript, Washington, D.C.: AFHSO, 1988), chap. 5, 75; Chief Programming Div, Deputy Chief of Staff, Operations (DCS/O) to multiple Air Staff agencies, subj: Preparation of Mil Pers Programs, FY 1953, 18 Oct. 1951, D/PDP 320.22, Ofcr Strength, 1 Jan.–31 Dec. 1951, book IV-S, RG 341, MMB, NARA; Alfred Goldberg, *A History of the United States Air Force, 1907–1957* (Princeton, N.J.: Van Nostrand, 1957), 69.

17. D/PTR to SECDEF Dir of Manpower Requirements, untitled, 27 Oct. 1953, D/PDP 326.6, ROTC, 1 Aug.–31 Dec. 1953, book VI-C, RG 341, MMB, NARA; Hist, Pers Procurement

Div, 1 Jan.–30 June 1951, 6; D/PTR Hist Summ, 1 Jan.–30 June 1951, 1; *Air Force Times*, 17 Mar. 1951, 1.

18. ATC Hist, 1 Jan.–30 June 1952, 1:47; *Semi-Annual Report of the Secretary of Defense and the Semi-Annual Reports of the Secretary of the Army, Secretary of the Navy and the Secretary of the Air Force*, 1 Jan.–30 June 1952 (Washington, D.C.: GPO, 1953), 212; D/PTR to DCS/P, subj: Aviation Cadet Procurement, 23 Jan. 1953, and D/PMP to DCS/P, subj: Aviation Cadet Procurement, undated, DCS/P 220.001, Aviation Cadets, 31 Dec. 1953, book I-C, RG 341, MMB, NARA; *Air Force Times*, 27 June 1953, 1 and 24 Oct. 1953, 1.

19. Robert Frank Futrell, *Ideas, Concepts, Doctrine: A History of Basic Thinking in the United States Air Force, 1907–1964* (Maxwell Air Force Base, Ala.: Air University, 1971), 208–10; Director of Personnel Planning (D/PDP) to DCS/P, subj: Ofcr Recalls for the Remainder of FY 53 and FY 54, 6 May 1953, DCS/P 210.455, 1 Jan.–31 Dec. 1953, RG 341, MMB, NARA; D/PDP to Procurement Div, D/PTR, subj: Ofcr Candidate School, 8 July 1953, D/PDP 326.6, ROTC, 1 Jan.–31 July 1953, book V-S, ibid.

20. D/PTR to Asst SECDEF for Manpower and Pers, subj: AFROTC Program Information for the Asst SECDEF (M&P), 12 Mar. 1953, ACC 60A-1055, box 20, chronological file 1 Mar.–31 Mar. 1953, RG 340, WNRC; Air University (AU) Hist, 1 July–31 Dec. 1953, 1:51–52; *Semiannual Rpt of the SECDEF, 1 Jan–30 Jun 1952*, 213.

21. Dir of Manpower Requirements to Asst SECAF, untitled, 14 May 1954, DCS/P 210.8, Separations, 1 Jan.–30 June 1953, book IV-C, RG 341, MMB, NARA; Press briefing by the Hon. John A. Hannah, Asst SECDEF, 22 July 1953, ACC 60A-1055, box 20, chronological file 15–30 Sept. 1953, RG 340, WNRC; Memo for record, subj: Results of Mtg with the OSD, 29 July 1953, D/PDP 326.6, ROTC, 1 Jan.–31 July 1953, book V-S, RG 341, MMB, NARA.

22. AU Hist, 1 July–31 Dec. 1953, 1:49–50, and 2:exh. 42; Commander, 14th AF to DCS/P, untitled, 12 Aug. 1953, D/PDP 326.6, ROTC, 1 Aug.–31 Dec. 1953, book VI-C, RG 341, MMB, NARA.

23. AU Hist, 1 July–31 Dec. 1953, 2:exh. 42; Paper, Hist and Current Review of the ROTC Program, undated, D/PDP 326.6, ROTC, 1 Aug.–31 Dec. 1953, book VI-C, RG 341, MMB, NARA; D/PTR Staff Hist, 1 Jan.–30 June 1954, 17; D/PDP to Asst SECAF, subj: Establishment of AFROTC Graduates, 19 Apr. 1954, chronological file 19–30 Apr. 1954; Ltr Senator Clyde Hoey to SECAF, untitled, 26 Apr. 1954, chronological file 3–14 May 1954; Memo, Asst SECAF to D/PTR, subj: Active Duty Tour for AFROTC Graduates, 1 July 1954, chronological file 15–30 June 1954, all in box 21, ACC 60A-1055, RG 340, WNRC.

24. AU Hist, 1 Jan.–30 June 1957, 1:41–42, and 1 Jan.–30 June 1958, 3:sup. doc. 23.

25. AU Hist, 1 Jan.–30 June 1958, 1:75.

26. Gene M. Lyons and John W. Masland, *Education and Military Leadership: A Study of the R. O. T. C.* (Princeton, N.J.: Princeton University Press, 1959), 174–75; President of Williams College to D/PTR, untitled, 22 Feb. 1956, Williams College folder, box 1, ACC 63A-1536, RG 341, WNRC; AU Hist, 1 Jan.–30 June 1958, 1:85.

27. Dep D/PTR to DCS/P, subj: Conference Regarding Criteria for Withdrawal of AFROTC Units, 18 Aug. 1953, D/PDP 326.6, ROTC, 1 Aug.–31 Dec. 1953, book VI-C, RG 341, MMB, NARA; Memo, Asst SECAF to DCS/P, subj: Disestablishment of AFROTC Units, 30 June 1954, 15–30 June 1954 chronological file, box 21, ACC 60A-1055, RG 340, WNRC; AU Hist, 1 July–31 Dec. 1955, 1:57, and 3:sup. docs. 29 and 41; *Air Force Times*, 1 Jan. 1955, 1.

28. AU Hist, 1 July–31 Dec. 1955, 1:56–59, and 1 Jan.–30 June 1956, 1:40–41.

29. D/PTR Hist Summ, 1 Jan.–30 June 1962, 12; AU Hist, 1 Jan.–30 June 1962, 1:302, and 1 Jan.–30 June 1964, 1:130; Dir of Pers Training and Education, 1 Jan.–30 June 1965, 34, and 1 Jan.–30 June 1966, 42; *Air Force Times*, 30 Dec. 1964, 11.

30. ATC Hist, 1 Jan.–30 June 1954, 1:158; 1 July–31 Dec. 1956, 1:80; 1 July–31 Dec. 1957, 1:110; 1 Jan.–30 June 1958, 1:39.

31. ATC Hist, 1 Jan.–30 June 1954, 1:68, 70–71, 141–42; 1 July–31 Dec. 1954, 1:33–36,

and 1 Jan.–30 June 1957, 1:46–47; Paper, subj: Pilot Procurement Problems, SECAF file, box 7, Papers of Gen. Thomas D. White, LOC.

32. D/PTR Hist Summ, 1 July–31 Dec. 1956, 10; Asst SECAF to President of the University of New Hampshire, untitled, 30 Sept. 1957, 1957 chronological file, box 8, ACC 65A-3152, RG 340, WNRC; AU Hist, 1 July–31 Dec. 1957, 1:36.

33. AU Hist, 1 July–31 Dec. 1958, 1:40; D/PTR to Chief of Staff, USAF, subj: Ofcr Training School Program, AFCCS read file, box 12, White Papers, LOC; ATC Hist, 1 July–31 Dec. 1959, 1:80–81.

34. This and the preceding paragraph are based on the author's experience as a member of OTS class 61-B, June–September 1960.

35. D/PMP Hist Summ, 1 Jan.–30 June 1959, 34, and 1 July–31 Dec. 1963, 2:3; D/PTR Hist Summ, 1 Jan.–30 June 1959, 2, and 1 July–31 Dec. 1963, 56; *Air Force Times,* 21 Nov. 1959, 6; 11 Nov. 1961, 55; and 3 July 1963, 5; Maurice G. Stack, "The Aviation Cadet Program in Retrospective," *Air University Review* 16 (July–August 1965): 89.

36. *Daily Staff Digest,* 5 Aug. 1966, item 11; 15 June 1967, item 1; and 23 Feb. 1968, item 2; ATC Hist, 1 Jan.–30 June 1966, 119–21, and 1 Jan.–30 June 1968, 1:156.

37. ATC Hist, 1 Jan.–30 June 1963, 1:95, and 1 July–31 Dec. 1966, 1:187–88, 241–42.

38. Ibid., 1 July–31 Dec. 1968, 1:50.

39. Ibid., 1 July 1968–30 June 1969, 81–82, and sup. docs. 14 and 42.

40. Ibid., 1:116–17, and sup. doc. 40; *Daily Staff Digest,* 14 July 1969, 6; *Air Force Times,* 19 Feb. 1969, 1, and 8 Apr. 1969, 10.

41. AU Hist, FY 1972, 1:22.

42. "Report of the Special Committee on ROTC to the SECDEF," 22 Sept. 1969, 48, AU Education Ofc file ED 93, Maxwell Air Force Base, Ala.; *Daily Staff Digest,* 18 Dec. 1970, item 5; AU Hist, 1 July 1969–30 June 1970, 1:53–60, and 1 July 1970–30 June 1971, 1:23; *Air Force Times,* 14 May 1969, 5; 17 Sept. 1969, 19; and 23 Sept. 1970, 23.

43. Dir of Pers Training and Education Hist Summ, 1 Jan.–30 June 1966, 41; AU Hist, FY 1969, 1:74, and 1 July 1969–30 June 1970, 1:61; *Daily Staff Digest,* 21 June 1967, item 2.

44. AU Hist, FY 1969, 1:76–77; FY 1970, 1:62–65; 1 July 1970–30 June 1971, 1:65, 67–68; FY 1972, 1:36, 39–42; *Daily Staff Digest,* 7 Feb. 1966, item 1, and 20 June 1969, item 7; Lawrence J. Korb, ed., *The System for Educating Military Officers in the U.S.,* International Studies Paper No. 9 (Pittsburgh, Pa.: University of Pittsburgh, 1976), 71; Women in the Air Force (WAF) Newsletter, January 1973, 1.

The Education and Training of ROTC Officers, 1950–1980: The Military Services and the Universities

Michael S. Neiberg

As the Cold War developed in the 1950s, the United States military committed itself to a larger standing armed force than had been envisioned in the immediate postwar years. To staff the officer corps of that military, the services turned to a previously marginalized program initially designed to give men rudimentary military training during their time on campus. After 1950, the Reserve Officers Training Corps (ROTC) program became indispensable to the active-duty officer corps of the Cold War military. The program thus entered a new period in its history with a mission much more central to the overall mission of the American armed forces.

As it did so, ROTC underwent rapid transformation on many levels, including changes to its academic curriculum. The military services and the universities, the two institutions responsible for ROTC management had separate goals for the on-campus portion of the education of the American junior officer candidate, but those goals overlapped in important ways. This paper examines the evolution of the ROTC curriculum on American college campuses from 1950 to 1980.[1] I argue that significant reform of the program resulted from a consensus among educators and military officials that ROTC was critical to a military that would share the values of American society. This reform process, not the student movement, was responsible for the reorientation of the ROTC curriculum from technical training to professional education.

The debates of the 1950s, 1960s, and 1970s produced an ROTC curriculum that more closely resembled civilian course offerings. For educators, these curricular changes allowed ROTC to fit in with the regular course offerings of their civilian faculty and therefore helped to assure ROTC's acceptance in the university. For the military, the evolution and steady improvement to the ROTC curriculum made possible the accession of large numbers of better educated officers. This coincidence of interests produced a consensus favoring steady reform of ROTC that far outweighed radical calls for the program's abolition.

Consensus and Substitution: The Era of the Cold War

The highest ranking officials of both the military and higher education agreed that ROTC could staff the American officer corps with a talented group of young people whose exposure to civilian life during their academic careers would provide

264

for more diversity than that which would emerge from the service academies. Most university presidents and chancellors firmly supported ROTC, not because they were militarists (indeed many were deeply suspicious of the military), but because they believed that ROTC represented an opportunity for civilian higher education to influence the military.

From the university's perspective, civilian participation in military training programs reinforced two traditional positions that had also influenced ROTC in the prewar years. First, civilian participation would provide a critical contribution to the growing professional military. Many academics believed that a firm grounding in the demands of citizenship, not Universal Military Training, would best prepare the citizen-soldier for military service. The universities, of course, believed that they could play a great role in such preparation.

Second, and perhaps more important, the universities took great pride in preparing young men for all walks of life. Why, they asked, should not the military profession be included? As American interests became more international and as the military became a more important profession in America, in many respects just like other professions, why should not the universities take an active role in preparing junior officers? The military fit in nicely with how Harvard University President Nathan Pusey understood the mission of American universities in the 1950s:

> It was our national policy in the postwar years to endeavor to ensure peace through the creation of a world order built on cooperation in many fields of activity and on enlarged international understanding. The nation's colleges and universities had a role of fundamental importance to play in this, for only they could prepare the highly trained people, specialists and laymen . . . needed in the public and the private sector, in government and in business, for the implementation of the policy.[2]

Few administrators could find logical reasons to exclude military officers from this role. Instead they welcomed the chance to participate in educating military professionals in much the same way that they welcomed the chance to influence the development of doctors, engineers, and bankers.

For its part, the military needed to recruit large numbers of experts in such diverse fields as public relations, engineering, and financial administration in both the officer and enlisted ranks. Enlisted men were no longer the typical laborer or foot soldier "assumed to be ignorant"; they had expert knowledge that had to be put to best use. Having college-educated men in the officer corps, planners assumed, would create a corps of leaders familiar with a diverse body of knowledge and technical skills as well as men inclined to be solid leaders.[3]

"Leadership" to military planners of the 1950s meant something more akin to management based upon psychology and group dynamics than what Morris Janowitz called the "hero" model of the past. With this shift in leadership definition came a concurrent shift away from authoritarianism as a means of structuring the military; instead, military leaders were expected to invoke a sense of "solidarity" among the men they led. Successful manager-officers therefore needed to be more sophisticated than the old "hero" types. The universities were the obvious training ground for the sort of well rounded, educated men the military believed it needed to manage the increasingly complex and technical men and machines of the Cold War military.

Despite the general level of harmony, differences between military and university culture, values, and mission manifested themselves often in the debates over the goals and character of the ROTC curriculum. In the 1950s, the military preferred an officer preparation program that emphasized *training*, the teaching of the practical knowledge a man needed to become a junior officer the day after his graduation from college. The obvious person to conduct such training was, of course, a uniformed officer. University administrators, while conceding that some training was necessary, argued for *education* as the model of officer preparation. By education they meant the teaching, partly by civilian faculty, of courses in psychology, history, engineering, and other fields.

To the military, the changing nature of warfare and America's new global role meant that large numbers of men had to be ready to serve in future mobilizations. As a result, they successfully contended in the early 1950s that the ROTC curriculum had to consist of technical and professional classes in order to reduce as much as possible the postcommissioning training junior officers needed to assume their roles. Therefore, ROTC courses were more specialized, more vocational, and less generally applicable to civilian professions than were most university courses.

The courses ROTC students took, whether at summer camps, on cruises or on campus, were strictly military in subject matter and quite squarely vocational in application. In the early 1960s, the Army described the goals of the national ROTC curriculum model as "not designed to educate the student, but to train him in the basic skills required of a second lieutenant."[4] This viewpoint, of course, stood opposed to traditional modes of instruction in higher education, especially in liberal arts colleges. As long as the Cold War seemed to threaten national security, however, higher education officials by and large accepted the military's argument that cadets and midshipmen needed to learn the nuts and bolts of military service on campus in order to be useful immediately upon commissioning. Over time, however, the dissonance between pedagogical goals resulted in conflict.

After World War II, according to Janowitz, progressively less discipline characterized the junior officer's daily life. Such does not appear to have been the case for the junior officer's training, however. ROTC classes were well known for strict, sometimes arbitrary discipline intended to prepare cadets for the rigid environment of the military. To assure a firm grounding in the basics of such regimentation, the ROTC curriculum devoted considerable time to "personal appearance, wearing the uniform, military courtesy, military discipline, leadership, drill, and customs and courtesies of the service."[5] Success in mastering these subjects largely determined individual success and advancement in the ROTC program. University of Texas Army ROTC cadets, for example, received their grades based on the following:

> *Appearance.* Neatness, clean and properly fit clothing, shave and haircut, military
>   bearing, standing and sitting erect, cleanliness.
> *Demeanor.* Calm, poise, confidence, and enthusiasm.
> *Courtesy and cooperation.*

Aggressiveness. Voluntary, constructive participation.
Grammar. Ability to phrase questions and answers clearly, concisely, and to the
   point.
Voice. Volume, clearness, and pronunciation.
Honesty. In answering questions. Bluffing will be penalized.[6]

Drill (also known euphemistically as "Leadership Lab") served as the quintessential example of military regimentation and discipline on the campus and was usually conducted in a visible public space such as a quadrangle or an open athletic field. ROTC cadets sometimes spent as much time drilling as they spent in the classroom. Duquesne University freshmen spent sixteen hours per week in drill; sophomores spent ten.[7] At the University of Texas and elsewhere all demerits a student had accumulated had to be worked off by the end of a semester in still more drill. Furthermore, all Texas cadets were required to join an ROTC cadet organization "to enhance the prestige of the Army ROTC and make it an organization in which every cadet can take pride." Several of these organizations were precision drill teams.[8]

The redesigned national Air Force ROTC (AFROTC) curriculum model of 1953 required 480 contact hours in the academic year and 232 hours of summer camp.[9] Like the Army and Navy curricula, it was squarely military and professional (see Table 1).

Educators largely agreed with the military's stated goals, but saw the problem differently. They argued that officer candidates needed a wider range of civilian courses to complement the military courses and to expose them to new fields in the humanities and social sciences as well as traditional civilian courses relevant to the military. In the two decades following World War II, American higher education adopted a more international focus. Universities developed several courses and concentrations in area studies, humanities, and social sciences designed to better inform American undergraduates about the world in which the United States was increasingly involved. Ironically, however, the tight military focus of ROTC programs meant that they remained largely apart from these changes.

A tension therefore existed between the steadfastly acknowledged national need for officer training programs on civilian campuses and the awareness that civilian and military curricula operated independently, to the detriment of both. If ROTC had a place in the curriculum (and few doubted that it did), then it was surely not the same place as English, philosophy, or chemistry. University of Texas officials noted that ROTC "is not to be viewed as a scholarship program" but as a specifically military program.[10] It is illustrative, then, that at many schools, including the University of Pittsburgh, ROTC had the same curricular standing as physical education. Students could complete four semesters of either one to fulfill a graduation requirement.[11]

By the late 1950s, when the bipolarity of the early years of the Cold War was replaced by a more complicated understanding that included the Third World, educators developed several ideas to more fully integrate the military and civilian instructional programs.[12] Furthermore, the 1957 Soviet launch of Sputnik reawakened an interest in officer education just as it reinvigorated interest in American education generally, especially in the secondary schools. Such an environment proved fertile for developing civilian suggestions for changes in the ROTC curriculum. By the late

1950s, as the specter of McCarthyism faded and as academics reached rough consensus on ideas for including civilian courses, civilian faculty and administrators

Table 1

Air Force General Curriculum, 1953

	Courses	Contact Hours
Air Science I: The Airplane and the Air Age (Freshmen)	Introduction to AFROTC	4
	Introduction to Aviation	16
	Fundamentals of Global Geography	10
	Security Organizations	14
	Military Instruments of National Security	15
	Leadership Lab (drill)	
Air Science II: Elements and Potentials of Air Power (Sophomores)	Careers in the USAF	6
	Moral Responsibility of Air Force Leaders	1
	Introduction to Aerial Warfare	3
	Targets	6
	Weapons	14
	Aircraft	10
	Bases	6
	Operations	14
	Leadership Lab (drill)	30
Air Science III: The Air Officer in the Air Age (Juniors)	Introduction to Advanced AFROTC	2
	Air Force Command and Staff	8
	Problem Solving	20
	Communication in the Air Force	25
	Instruction in the Air Force	10
	Military Justice System	5
	Weather	15
	Air Navigation	15
	Air Force Base Functions	15
	Preparation for Summer Training	5
	Leadership Lab	30
Air Science IV: Leadership and Air Power (Seniors)	Career Guidance	4
	Air Force Chaplain	1
	Leadership Seminar	40
	Military Aviation	15
	Wargaming	45
	PAS [Professor of Air Science] Time	5
	Commissioned Service	10
	Leadership Lab	30

Source: Air Force Historical Research Agency, Maxwell Air Force Base, Ala., *History of the Air Force Reserve Officers Training Corps* (Maxwell Air Force Base, Ala.: Air University, 1966), supporting document no. 4, Table 3.2.

lobbied the services to include civilian courses relevant to the military in the ROTC curriculum.

Substitution and the Use of Civilian Courses

Various ideas for incorporating civilian faculty into the ROTC curriculum developed independently at several institutions nationwide. One such plan, developed in 1957 by the Association of Naval ROTC Colleges and Universities (ANROTCCU), a group made up of presidents and provosts of those schools hosting Navy ROTC units, involved using civilian faculty as guest lecturers to broaden and to supplement the instruction provided by uniformed personnel.[13] The use of civilian lecturers, the ANROTCCU argued, could upgrade the quality of academic instruction while at the same time play upon one of the traditional strengths of ROTC. Part of the logic of training officers at civilian colleges was that they were exposed to civilian ideas and therefore entered the officer corps with a broad base of knowledge. It made military sense, then, to take advantage of civilian faculty with knowledge relevant to the general education of the junior military officer.

Some universities were already developing plans to use civilian faculty to teach whole classes, not just selected lectures. At the same time that the ANROTCCU developed its plan, the Ohio State University Air Force ROTC detachment experimented with civilian instructors teaching such courses as International Tensions and Security Organizations, Fundamentals of Global Geography, Military Aspects of World Political Geography, Communicating in the Air Force, and Problem Solving and Leadership Management.[14] In this pilot program, civilian Ohio State faculty taught courses that were military in their orientation, but more reflective and academic than those taught by uniformed officers.

The Air Force supported the Ohio State plan and also asked universities to develop specific courses to meet Air Force requirements that would be taught by civilian professors. These courses could be offered as a joint military-civilian enterprise and would be open to any student.[15] Ohio State, for its part, was anxious to have other schools try what was becoming known as "substitution," or the replacement of vocational, military courses by academic ones. Civilian faculties strongly resisted joint courses because they feared outside influence on course design, but viewed substitution more favorably because it involved use of existing academic offerings. By 1960, 48 of the Air Force's 176 schools used some variant of substitution.[16] To be sure, this meant that 128 schools still used strictly military curricula, but an alternative was emerging.

The localized nature of the ROTC curriculum meant that experiments with civilian faculty could be carried out at any school whose faculty and administration showed an interest. The University of Pittsburgh indicated interest early on and developed its own plan for substitution without significant input from Ohio State officials. A 1959 review of the Army and Air Force ROTC programs at Pitt by the Faculty Senate's Educational Policies Committee found that what was known as the "academic" portion of the ROTC curriculum did not merit the credit awarded, and that "the academic courses might profitably be replaced by courses already existing in our regular department offerings." The committee recommended that "The University should request the ROTC to consider a change in their curriculum which would recognize the suitability of regular departmental courses for credit toward military commissions. . . . These courses would be taught by regular non-military university faculty" in the departments of Political Science and Geography.[17]

The Air Force approved Pitt's suggestions quickly and enthusiastically. For the spring trimester of the academic year 1959–60, the Air Force authorized Pitt to substitute two courses, International Relations and Political Geography, for air science courses. University of Pittsburgh Faculty Senate Chairman Alan Rankin told Pitt's faculty that the changes in the Air Force curriculum and discussions with the Army for changes in its curriculum had improved the academic legitimacy of ROTC:

> In undertaking the study of the steps necessary to implement the Senate Committee's recommendations, we have found the Departments of the Air Force and the Army favorably disposed to change. . . . The Army is hoping to give purely academic courses a greater emphasis, to reduce on-campus instruction in purely military subjects and to allow ROTC credit for courses in psychology, mathematics, and physics.[18]

Reducing military subjects taught on campus would be necessary to allocate time to the new, substituted courses. Military subjects, under most proposals considered for national implementation, would be moved to the summer camps.

Substitution succeeded because it met the interests of military and civilian officials. Substitution met both the services' desires for more broadly educated students and higher education's desire to raise the academic standard of ROTC course instruction in the absence of military officers with advanced degrees and teaching experience. The University of Pittsburgh's Alan Rankin attended a 1959 Air Force ROTC conference and noted with pleasure the participants' warm reception of substitution:

> [T]he most interesting feature of the conference was the unmistakable sentiment on the part of both institutional representatives and the Air Force officials to liberalize the AFROTC program in exactly the same direction as the University of Pittsburgh is moving. Our regular academic courses taught by our regular faculty members is the coming order of the day, and I would anticipate little difficulty in getting approval from the Air Force headquarters for whatever modifications in our curriculum we might want to undertake.[19]

The following year, Pitt introduced eighteen more hours of substitution for Air Force ROTC. Freshmen took speech and one class from mathematics, natural sciences, social sciences, languages, or the humanities. Juniors took Social Psychology and either Technical or Expository Writing. Seniors took World Politics and Political Geography. Duquesne University officials established a list of twenty-nine classes in eleven departments that cadets could take to fill the 150 contact hours that the Air Force turned over to civilian departments.[20] Cadets at Pitt and Duquesne still took a core primarily composed of military subjects, as noted above, but these were now complemented by civilian courses.

Substitution also enabled students to fit ROTC more easily into their increasingly demanding schedules. The services had long recognized time constraints as an important detriment to Advanced ROTC enrollment. A 1950 study noted that students named "Too time consuming" as their most frequent complaint.[21] In 1961, the Army ROTC Advisory Panel noted in its report that "some educators are exerting pressure to further reduce on-campus military instruction. They consider that reduction is necessary in order to integrate student academic requirements and to reduce the impact of the pressure of time on the students."[22] A conference sponsored by Ohio State sold substitution by praising its ability to "reduce the overload on advanced students."[23]

Substitution thus had many virtues: It made the program more appealing to a large minority of military officials, primarily in the Army and Air Force, who wanted ROTC to move in the direction of education, not training; it pleased educators by injecting academic material from a wide variety of disciplines into the vocational curriculum; and it pleased undergraduates by allowing them to count their classes twice—once as credits for their diploma and once as requirements for their commission. This last advantage partly explains why the services so quickly adapted substitution into their curricula. The Army felt compelled to respond to Air Force plans that made it easier for students to enroll and complete ROTC training, especially on those campuses where both programs existed, lest they should lose potential cadets to the other service. Substitution was not widespread enough to allow the services to save money by assigning fewer officers, but it did bring all of the above advantages without adding any additional costs. Therefore, the new programs steadily spread, making at least a portion of the ROTC student's curriculum more academic and demonstrating the benefit of close working relationships between the military and its academic partners.

The ROTC Vitalization Act

In 1964, in response to declining enrollments and the increased demands that American higher education was beginning to place on its students, Congress passed the ROTC Vitalization Act to update the program and to prepare it for its future as the nation's primary source for active-duty officer procurement. The bill's passage represented a victory for traditionalists in the Congress and the Pentagon who viewed ROTC "more as a means of teaching Americanism than of turning out officers."[24] Ironically, the losers in the Vitalization Act debate were the ROTC staff officers and higher education officials who drafted the initial legislation. The Vitalization Act, as passed, ended substitution and left a legacy of suspicion among academics.

ROTC staff officers had hoped that the bill would remove drill and military training to summer camps to free up time on campus for academic subjects. To many ROTC officers, drill seemed antiquated; junior officers in the more managerial military environment of the 1960s did much less marching than had their predecessors and were subjected to much less formal discipline. As a 1964 AFROTC report declared, "Today's Air Force has little time or need for yesterday's emphasis on drill."[25] Military managers, in Janowitz's formulation, relied on "the technical proficiency of their team members" not the "formal authority structure" of years past.[26] As the new military was based much more around consensus than authority, devoting hours of valuable on-campus time to drill seemed wasteful. Moving drill to summer camps could leave campus ROTC instructors with more time to teach military management, decision-making, and critical thinking skills.

The final bill, however, legislated twice the number of total contact hours in the four-year program that the Air Force plan had called for (360 as opposed to 180). Congressional traditionalists such as House Armed Services Committee Chairman F. Edward Hébert (D–Louisiana) believed that the sight of young men participating in drill inspired patriotism and self-sacrifice on the nation's campuses. Hébert also supported expanding high school ROTC (also included in the bill) and continuing mandatory ROTC at land-grant schools. The ROTC staffs opposed both measures as wasteful and unmilitary. Hébert blocked the Vitalization Act, however, until Junior ROTC had been created and the ROTC program itself had been implicitly redefined as a citizenship program first and a military training program second.

As such, the law did not specify, as the Air Force ROTC planners had hoped, that the reduction in contact hours should come at the expense of technical military subjects like drill. Army headquarters recommended, consistent with congressional intent, eliminating the substitution of civilian-taught courses for those taught by uniformed personnel and reducing the time spent on academic subjects, which, of course, contravened the desires of educators.[27] The new curricula, then, affirmed the centrality of technical, military subjects at the expense of liberal arts courses. The high visibility and technical emphasis legislated by the Vitalization Act established a priority that educators, and even some military personnel, did not share. University hosts had hoped for less drill, both because of its visible contrast to other university programs and because even many ROTC instructors acknowledged that "Close order drill in many instances drives men away from the program."[28]

This lack of priority for academic subjects influenced the 1964 curricula introduced by both the Army and the Air Force. These new curricula deemphasized liberal arts courses and authorized the end of the substitution experiments of the late 1950s. University administrators, however, were "highly disposed toward this idea of substitution" and were reluctant to give it up.[29] Michigan and Princeton officials argued that the kinds of cognate courses substituted in the past were indeed military in that they were concerned with the political, cultural, technical, and social environment within which military officers operated. Substitution had the additional virtue of making ROTC "fit in with the 'social fabric' of the university campus." At Princeton, this fit meant that ROTC cadets took such courses as Economics and Non-Western Societies. Princeton President Robert Clifford protested the end of substitution to Secretary of the Army Stephen Ailes by arguing that "the Army must allow a degree of flexibility in the implementation of the new program so that it may be adapted to individual campus conditions. Each college must be encouraged to participate in the teaching of such courses in so far as its faculty talents permit."[30]

Michigan officials also protested the end of substitution, saying that the program had been a "terrific boon" to Michigan ROTC students. Michigan had been negotiating the substitution of two more courses into the AFROTC curriculum, Aeronautical Engineering and Business Administration, when university officials were informed that all substituted courses would be removed. Michigan administrative dean Robert Williams told the Air Force, "Some years ago the Air Force was the leader in this move toward the utilization of one course for two purposes, and it is the judgment of our group that the discontinuance of this practice can result only in harm to our combined efforts."[31]

The 1964 cancellation of substitution echoed the debates over its initial introduction in the late 1950s. The substitution debate touched on the larger issue of whether ROTC instruction was to be based on education, the preparation of cadets in subject areas with more general applications such as non-Western societies and business administration, or training, the narrower preparation of individuals to perform certain tasks such as drill, marksmanship, and map reading. American universities had been increasingly moving away from training in undergraduate instruction, even in specialized areas such as engineering and architecture, preferring instead to focus on more general problem-solving and management techniques.

Terminating substitution underscored ROTC's growing divergence from the general trend present in higher education. Educators attending a joint Army-civilian ROTC meeting at Ohio State University in 1965 warned the Army of the growing conflict: "ROTC on the campus is a 'paradox' to many students, since it gives heavy stress to 'training' (drill, rifle practice, map reading, etc.) at the same time instructors in other subjects are attempting to develop the student's conceptual powers through liberal education."[32] In the same year, Robert Williams warned the ROTC instructional staff in Ann Arbor that ROTC programs could not continue to exist at variance with changes in the general development of higher education:

> I am fully convinced that in the foreseeable future educational philosophy in all areas will demand more and more emphasis on 'principles' and less on 'how to do it' courses. If this be true, the competition for the time of the undergraduate who may wish to secure a commission through ROTC activities will require, in my judgment, (1) full recognition of ROTC courses as general electives within degree requirements, and (2) the critical review of the course content which justifies full utilization of the courses for degree programs.[33]

Williams's statement proved prophetic. As campus opposition to the military and to ROTC built in the late 1960s, the on-campus curriculum became a point of contention with academics, many of whom had been sorely disappointed with the Vitalization Act. The visibility that the Act mandated inadvertently made ROTC a much easier target for campus protest than it might otherwise have been. As a result, academic intrusion into the ROTC curriculum became almost inevitable.

The Vietnam Era

Radical student opposition to ROTC units on campus has been well documented; it symbolized the student movement at American colleges and universities. To date, however, no serious examination of the impact of the anti-ROTC movement exists. I contend that as newsworthy and visible as the student movement was, the military and most university administrators paid little formal attention to it, at least as far as it concerned ROTC.[34] Plenty of reliable evidence existed that the protesters constituted a small, if vocal, minority on all campuses. Two polls taken at the University of Pittsburgh in 1968 showed 70 percent of undergraduates were in favor of continuing ROTC with credit.[35] Even at Kent State, a poll taken in 1970 *after* the shootings in May showed three in four students in favor of retaining ROTC and half in favor of it remaining accredited.[36] Only at one of the ninety National Association of State Universities and Land-Grant Colleges (NASULGC) member schools, SUNY–Buffalo, did a student referendum show a majority of students opposed to ROTC's continuance.[37]

The 1969 National College Poll found that "Most students . . . are antiwar. They are not necessarily antimilitary. There is quite a distinction between the concepts."[38] Indeed, this distinction is crucial. The College Poll found that 60 percent of the nation's undergraduates believed that America was wrong in sending troops to Vietnam, but that 80 percent were at the same time in favor of voluntary ROTC programs and 59 percent were in favor of those programs receiving academic credit.[39] There was thus a considerable amount of evidence to demonstrate two important points: that the student radicals did not speak for the vast majority of the nation's undergraduates, and that anti-Vietnam sentiment did not necessarily correlate to anti-ROTC sentiment.

Of much greater concern to the services than student protests were the numerous formal investigations made by faculty committees into ROTC. According to one Air Force official, "While the threats and actual incidents of violence are by far the most dramatic manifestation of the anti-ROTC sentiment, the 67 institutional investigations of the AFROTC curriculum [in academic year 1968–69] are considered more significant by AFROTC."[40] By June 1970, forty-six of the ninety members of the NASULGC had formal studies underway to investigate ROTC and its place on campus.[41] Illinois, Kent State, Michigan, and Pittsburgh produced a total of nine such reports between April 1968 and September 1971.

These reports were written either by ad hoc faculty senate committees tasked to examine ROTC and its relationship to the university or by education and curriculum committees that had official supervision over university courses and faculty appointments, but had not customarily exercised that authority over ROTC in the past. These reports represented the faculties' chance to reassert that leverage. The reports revealed that the central concerns of the majority of the university communities were the same as those expressed in the 1950s. In short, the turmoil accompanying the protests against the war in Vietnam increased the attention faculties focused on ROTC, but did not fundamentally change the issues or even introduce new ones.

Rather, Vietnam gave the universities the motive and the opportunity to correct long-standing anomalies presented by the ROTC programs. Critics later charged that academics set conditions that were intended solely to make it impossible for ROTC to remain on campus; this was simply not the case. Instead, the faculties issued reports that challenged ROTC to reform itself to become more a part of the university along the same criteria applied to other programs.

The most common feature these reports shared was a recommendation that ROTC no longer be sanctioned by the university through the blanket awarding of academic credit in liberal arts colleges. These recommendations were largely related to the faculties' opinion of the quality and content of ROTC courses. As a common prelude to open discussion of ROTC, faculty subcommittees reviewed ROTC courses to determine how the quality of ROTC courses matched up to those in the regular university offerings. They unanimously criticized ROTC courses for being, in the words of Michigan's 1969 Gindin Committee, "shockingly bad . . . simply inappropriate to a liberal arts education . . . appalling . . . conjectural, non-analytical, cheaply moralistic, and often blatantly propagandistic."[42] Pitt's Student Affairs Committee further criticized ROTC as lacking "humanistic or political analyses."[43]

Many of the reports also recommended, in the words of the faculty of Tulane University, that "courses with substantive political or policy content . . . should be a part of the ROTC program; they should be offered by the appropriate academic departments, however, and should not be taught by military officers on active duty."[44] Such a recommendation recalled the substitution programs of the late 1950s that academics had lauded. Substituted courses, taught as they were "under the auspices of one of the degree granting colleges or schools," would, of course, receive full credit.[45]

Military history was one course often cited as suitable for substitution. Faculty members wanted military history taught by professional historians rather than by ROTC instructors, most of whom lacked any formal training in history. Military history, furthermore, was already an established course at many institutions and civilian faculty believed that ROTC cadets could benefit from these courses. In 1970, 112 Army ROTC hosts had civilian-taught military history courses already in place that were acceptable to the Army; the University of Michigan alone had seven such courses.[46]

These reports argued for substantial reform in ROTC. For all of the many changes recommended by these faculty committees, however, what is perhaps more significant is what the faculties did not recommend. Only one faculty report, from Dartmouth College, recommended that ROTC be removed from the campus. Elsewhere, faculty motions and petitions introduced to demand that the university sever all ties to ROTC were routinely defeated by margins of five to one or higher. Even recommendations that ROTC retain ties to the university, but be moved off-campus were soundly defeated. Some individual faculty members saw these moves as a first step toward removing ROTC from the campus, but the faculty committee reports did not argue that ROTC was an all-or-nothing venture. They could, and did, make recommendations designed to fit ROTC into the fabric of their campus community.

Indeed, despite the changes they recommended, the faculty committees remained firmly committed to their traditional support of ROTC. The NASULGC's Special Subcommittee on ROTC Policy rejected the idea that "the civilian academic community, public or private, is antithetical to association with national security affairs."[47] The Princeton and Pitt reports explicitly stated that there was value in ROTC because it prevented the formation of a military caste and it infused the military with civilian ideas. These reports, then, were not arguing for abolition of ROTC, but for reform. ROTC's traditional strengths outweighed the faults academics saw in the program. Therefore, they made a series of recommendations designed to bring ROTC in line with regular academic offerings and thereby produce a program worthy of remaining on the campuses on its own merit.

Military response to these reports was mixed. As noted, some military criticism of drill and the on-campus instruction of technical subjects had been building since 1960. Observers in the late 1960s continued to criticize drill as outdated, irrelevant, mindless, and embarrassing to the student. Survey evidence suggested that drill was even unpopular among people who supported all other aspects of the program.[48] Reducing the importance of drill promised to reduce a source of tension among both supporters and opponents. "Leadership laboratory," noted one Army officer, "may well be the program's worst enemy."[49] Or, as a Navy officer argued, "It's the weapons and the drilling that get up their dander, so, hell, let's throw them a crumb."[50] The officer's casual willingness to abandon drill indicates that its training value was minimal, a mere "crumb." The services had few problems accepting the faculty recommendations to remove academic credit for the drill portions of the ROTC curriculum and some officers were willing to consider removing drill from the campus altogether.

Knowing that the consensus of informed opinion within the military had become decidedly receptive to reducing drill, Michigan's Robert Williams began a meeting with the Navy Bureau of Personnel by saying, "let us begin by dismissing the drills, spit and polish, corps activity, or whatever you care to call it as a vital [and] necessary part of the ROTC program."[51] In a similar vein, University of Illinois Provost Dayton Pickett told a representative of the Army War College that the "portion of military leadership skills acquired through participation in drill is seen as a highly questionable component of a university curriculum."[52]

Because many uniformed and civilian military officials agreed that drill was overemphasized in the ROTC curriculum, they did not strenuously object to university desires to deny credit for drill or to reduce the number of hours spent on drill. Indeed, many military officials had opposed the changes of the Vitalization Act that made drill as much as one-third of the curriculum's total contact hours. The hours devoted to drill became the common target of military and educational officials who pushed to reintroduce substitution.

In the late 1960s, the services worked with universities to encourage various substitution suggestions on individual campuses. These ideas varied significantly in scale and scope nationwide. The Massachusetts Institute of Technology, for example, replaced drill with cadet presentations on aspects of engineering and physics relevant to the military.[53] Elsewhere, substitution ideas became more complex. In

1969, the Army authorized all Professors of Military Science (PMS) to introduce substitution, encouraging the substituted courses to be from the following: American military history, world military history, diplomatic history of the United States, political geography, American government, international relations, geopolitics, studies of developing countries, international trade and finance, psychology, sociology, group dynamics and human relations, calculus, chemistry, biology, physics, geology, foreign language, political philosophy, organization and management, computer science, and statistics.[54] "This is a forward move, in my judgment," noted Michigan's Robert Williams, "and it is my understanding that the departments concerned and the assistant deans in the college who work with the students and their programs are delighted about this movement."[55]

Most universities introduced some form of substitution, but nationwide the University of Michigan introduced the most elaborate series of curricular changes. Michigan introduced substitution for Army and Air Force drill periods for the Fall semester, 1969. Air Force freshmen took political science, juniors took two courses in aeronautical engineering, and seniors took international politics and either two industrial engineering courses or two business administration courses.[56] Army freshman cadets substituted political science and speech, and sophomores substituted geography and history. For the 1970–71 academic year, the Navy agreed to require its midshipmen at Michigan to take American military history, national security policy, calculus, physics or chemistry, and computer science.[57]

Because of Michigan's widespread use of and support for substitution, the Air Force selected the university as one of four to test a new "alternate curriculum" that made more elaborate use of substitution ideas than had previously occurred anywhere. The alternate curriculum was an Air Force initiative designed to give AFROTC cadets access to a wider array of subject areas. In place of Air Science 102 (United States Military Forces), freshmen cadets selected a course from the following: Humanities 101 (*Iliad, Oedipus, Canterbury Tales*); Humanities 102 (*Hamlet, Candide, Civil Disobedience*); Freshman Composition; Shakespeare; Creative Writing; and Great Books. Introduction to International Politics replaced Air Science 201 (Introduction to Defense Policy) for sophomores.[58]

The services generally supported a limited amount of substitution because they saw value in having junior officers with broader educations and a firmer grounding in areas such as engineering, administration, and even the humanities. They did, however, urge that a significant amount of the corps training remain the preserve of uniformed officers. As the University of Pittsburgh PMS told the administration: "although the Military Science Department could expand its use of the University courses provided they met the overall objectives, a course composed solely of University courses is not desirable" because the ROTC faculty needed to know their cadets well enough to evaluate their ability to succeed in the military. Instead, Pitt's ROTC instructional staff proposed a "team" approach—civilian instructors would teach the academic portion of the curriculum and military instructors the corps preparation aspects.[59]

The University of Illinois also supported the team-teaching concept and introduced the following team-taught courses: Military Map and Photo Analysis, the United States Defense Establishment, American Military History, Principles of Military Instruction, Military Law, and Principles of Military Leadership.[60] The services were generally more enthusiastic about team-teaching than they were about substitution because team-teaching allowed the ROTC staffs to maintain more contact hours with their cadets. It also meant less fundamental change to the program; team-teaching had already existed informally on some campuses in the form of guest lectures from civilian faculty. Team-teaching was less popular with academics who resisted sharing intellectual ownership of their curricula, but, even when employed on a limited basis, team-teaching combined with substitution helped the ROTC programs answer charges that their instruction was substandard.

ROTC courses had frequently been criticized as intellectually bankrupt and devoid of contextual analysis or practice in decision-making skills. While most of these criticisms came from academics, some military officers had themselves questioned the academic value of much of the ROTC curriculum. As a group, they understood that changing the curriculum would require the marginalization of some of the most martial of the ROTC courses. As one ROTC instructor noted, "Personally I do not believe that subjects we now teach such as small unit tactics, tactical communications, and other fundamental military subjects are college level courses. They are important and should be taught . . . but not [on campus] for academic credit."[61]

Much of the military and civilian motivation for improving the quality of the ROTC courses centered around making the nonmartial courses worthy of receiving academic credit on their own merit. Several military and civilian officials held to the position that without academic credit, ROTC programs would have a difficult time attracting quality students, especially if plans to create an All-Volunteer Force succeeded. Without quality students, they argued, the ROTC programs themselves were in jeopardy and the services thereby risked being constituted of less than the best men (and increasingly after 1969, women) available for the job.

To some civilian faculty, continued denial of credit for ROTC in liberal arts colleges risked isolating the military from sources of humanistic thought, undermining what for many was the most important justification for ROTC's existence: the university's ability to influence the military through the production of officers with a broader educational background than that of service academy-trained officers. These questions, then, recalled issues that dated back to ROTC's creation in 1916. Michigan President Robben Fleming wrote in 1969:

> As to the larger question of whether ROTC ought to be on the campuses, I tend to favor it despite my own connection with civil liberties causes. My reasons are that I fear a professional army in a democracy; that I think the infusion of officer talent from non-professional ranks promotes the concept of a civilian army; that in twenty-two years on three campuses which have ROTC programs, I have yet to find the faintest hint that somehow the military is dominating the campus; and I believe that officer standing is a legitimate outlet for those students who must serve some time in the military anyway.[62]

To many educators, not having ROTC was a far greater danger than having it.

Keeping ROTC On Campus

ROTC debates had traditionally centered around three interrelated concerns: the quality of the courses; the quality of the instructors; and the degree to which their professional, military orientation conflicted with the overall mission of civilian colleges, especially liberal arts colleges. Still, to many professors and administrators, military efforts in the Vietnam era to upgrade their courses had been successful. The dean of the University of Pittsburgh's College of Arts and Sciences (CAS), Jerome Schneewind, noted in 1972 that "The present ROTC program . . . is richer and more academically interesting" than it had been in 1968. This improvement, he believed, argued for a reexamination of the no-credit policy in place in CAS.[63]

As Schneewind's observation suggests, faculty and administrators who attended ROTC courses and examined the materials used were generally pleased with the new courses, except notably those courses they understood to have no application outside a single profession. In this vein, a 1974 faculty committee review of the ROTC courses at Michigan found that many courses could "stand on their own academic merit" and were "compatible with what we understand to be appropriate to the liberal arts curriculum." The committee recommended credit for all ROTC courses except: Introduction to Small Arms, Conduct of Military Operations, Ship Systems II, Naval Operations, and Amphibious Warfare.[64] "Armies must salute, of course," said the executive committee's chairman, "but we don't want anything to do with that."[65] A separate subcommittee report recommended a reappraisal of the no-credit policy at Michigan, noting that "[W]e found in the current course offerings none of those deficiencies that were found by the preceding ROTC subcommittee in the 1968–69 programs and which caused it to render the negative verdict leading to the withdrawal of all credit. The disparity between the earlier appraisal and the present one, we believe, is due not to different standards of evaluation or to a shift in academic values, but reflects a real change in the nature of ROTC courses."[66]

At many schools where the status of ROTC survived the Vietnam period intact, ROTC staffs redesigned curricula intentionally to put the ROTC courses up for review on the same standards as those obtaining for non-ROTC courses. The goal was "to provide a program compatible with the evolving environment of the academic community" and therefore fully capable of existing alongside other courses and making significant academic and intellectual contributions of their own.[67] At Duquesne University, which did not remove credit for ROTC courses, the Military Science department completely overhauled its curriculum in 1972 "to develop interesting, challenging, and stimulating courses worthy of academic credit for presentation to ROTC and non-ROTC students." Duquesne's new ROTC program included courses co-designed with three civilian departments.[68] Duquesne also introduced a Military Science minor, unanimously approved by the university's curriculum committee and open to any undergraduate. It included two courses taught through the Military Science department, one from the Sociology department, one from the Political Science department, and one from a list that included courses from five other civilian departments.[69]

In 1976, Illinois's College of Liberal Arts and Sciences, arguing that it was "educationally sound to enable our students to participate in [ROTC] programs" permitted students to count six ROTC credits toward graduation, far less than the fifteen credits of the 1950s and 1960s, but a clear change from the zero-credit policy passed in 1971. Illinois had, in effect, replaced the prewar blanket credit policy with a new policy that accredited only those courses, all above the freshman level, that the faculty believed merited consideration alongside civilian courses. Here as elsewhere, a reconsideration of credit resulted from a faculty belief in the genuine improvement in the quality of ROTC courses and the desire of the ROTC instructors to be a part of the campus community. As Illinois Provost J. W. Peltason informed the Army, "[A]s important as the actual granting of credit, has been the *way* in which credit was granted—via a rigorous evaluation process. ROTC work is thus being viewed as it should be: a serious, academically legitimate program of pre-professional education."[70] At Illinois, ROTC's reaccreditation symbolized an acknowledgment of ROTC's improved quality and its appropriateness to the university's mission.

Other symbols of university faith in the ROTC curricula existed as well. In 1975, all ten universities using the "alternate curriculum" introduced by the Air Force in 1972 returned to the standard AFROTC curriculum. The alternate curriculum had included courses from the Humanities, English, History, and Political Science departments, thus permitting greater use of civilian instructors. Michigan officials had, by 1975, been sufficiently impressed with changes to the standard curriculum to permit the Air Force to return to its regular course offerings.[71] The University of Nebraska did so as well, reflecting that faculty's views "as to [the standard curriculum's] quality."[72]

Conclusion

This process of negotiation and reform, then, was the agent of change to the ROTC program. Radical students and the occasional violence they directed against ROTC units seized media and popular attention, but they never succeeded in convincing faculty, administrators, or even their fellow students that ROTC had no place on American campuses. Rather, reform was the result of a convergence of interests, the military's in keeping open a channel for the recruitment of inexpensive, high-quality officer candidates, and higher education's in maintaining a role in officer production programs. This convergence did not always represent agreement. Many academics, such as Michigan President Robben Fleming, supported ROTC specifically because they did not trust the military. Still, ROTC served the practical and ideological interests of both institutions.

By 1980, as a result, the ROTC program had a firmer curricular standing than it had in 1950. Its educational programs had more systematic connections to the general offerings of American universities and colleges and the programs faced much less sustained opposition from faculty oversight committees. Each side had positions for which it was willing to fight, but, more importantly, both the military and the officials in charge of higher education believed that ROTC had to be saved and brought more fully into the American university.

Notes

1. I would like to thank John Modell of Carnegie Mellon University, Peter Karsten of the University of Pittsburgh, Daniel Holbrook of Marshall University, and James Toner of the USAF Air War College for their comments on drafts of this essay. Due to policies that close archival materials for as long as thirty-five years or more, I had to omit the Ivy League from this study.

2. Nathan Pusey, *American Higher Education* (Cambridge, Mass.: Harvard University Press, 1978), 45.

3. Morris Janowitz, *The Professional Soldier* (Glencoe, Ill.: Free Press, 1960), 45.

4. [W. H. S. Wright], Headquarters, United States Continental Command, Fort Monroe, Va., Senior Division Army ROTC Program, 27 Jan. 1964, box 1228, Secretary of the Army General Correspondence, Record Group (RG) 335, Suitland Federal Records Center, Suitland, Md.

5. Deichelmann to all PAS, n.d., app. 74, History of the Air University, August to December 1952, Air Force Historical Research Agency, Maxwell Air Force Base (AFB), Ala.

6. University of Texas, Department of Military Science, Army ROTC Cadet Regulations, 1962, T378.764 U ZMIL, University of Texas Center of American History, Austin, Tex. Emphasis in original.

7. Brooks to chairman, Council of Instruction, 22 May 1958, ROTC 1958–65 folder, ROTC box 1, Duquesne University Archives, Pittsburgh, Pa.

8. University of Texas, Department of Military Science, Army ROTC Cadet Regulations, 1962.

9. Supporting document no. 7, p. 13, chap. 5: "Transfer of Air Force ROTC Responsibilities," History of the Air University, January to June 1952, Air Force Historical Research Agency, Maxwell AFB.

10. Macdonald to Smith, 27 Oct. 1958, VF 30/A.b, Naval Science 1958–59 folder, President's Office Records, University of Texas Center of American History.

11. *The University of Pittsburgh College Bulletin, 1950–51,* 81.

12. These debates are closely tied to complex debates about the amount of academic credit that universities should award for ROTC classes and the acceptable level of professional qualifications for military officers. Due to space limitations, I have not dealt with these subjects here.

13. Durgin to Nutting, 29 Jan. 1957, ROTC-Naval Science folder, box 26, President David Henry Papers, University of Illinois Archives, Urbana, Ill.

14. Harding to Henry, 10 Apr. 1958, ROTC folder, ibid.

15. Rogers to Henry, 5 May 1958, ibid.

16. Ray Hawk, "A New Program for the AFROTC," *Journal of Higher Education* (February 1960).

17. Rosenberg to Litchfield, 30 Apr. 1959, folder 186, class. no. 55/1, University of Pittsburgh Archives, Hillman Library, Pittsburgh, Pa.

18. Alan Rankin, Report on Progress in Implementing the Recommendations of the Educational Policies Committee of the University Senate with Regard to the ROTC Programs, 13 Nov. 1959, ibid.

19. Rankin to Litchfield, 18 Dec. 1959, folder 190, class. no. 55/1, University of Pittsburgh Archives.

20. Walsh to deans, 29 Aug. 1960, ROTC 1958–65 folder, ROTC box 1, Duquesne University Archives. The departments were: Biology, Business, Chemistry, Economics, History, Mathematics, Modern Languages, Physics, Political Science, Psychology, and Sociology.

21. Grant Advertising, Research Studies for ROTC, [1950], box 1228, Secretary of the Army General Correspondence, RG 335, Suitland Federal Records Center.

22. Report of the Meeting of the Army Advisory Panel on ROTC Affairs, 25 Apr. 1961, folder 188, class. no. 55/1, University of Pittsburgh Archives.

23. "ROTC Proposals," *New York Times,* 17 July 1960, sec. 4.

24. "Rot-Cee Ranks," *Newsweek,* 24 Feb. 1964, 84.

25. The New Air Force Curriculum, 29 May 1964, folder 14, Army and Air Force ROTC Correspondence, 1962–1964, Robert White Papers, Kent State University Archives, Kent, Ohio.

26. Janowitz, 41.

27. [Wright], Senior Division Army ROTC Program, 27 Jan. 1964.

28. Worthy to Dangerfeld, 13 Aug. 1964, box 123, ROTC-Military Science, Henry Papers, University of Illinois Archives.

29. Williams to Committee on ROTC Affairs, 31 Mar. 1969, Advisory Committee on ROTC Affairs folder, University of Michigan Vice-President for Academic Affairs, box 39, University of Michigan Archives, Bentley Historical Library, Ann Arbor, Mich.

30. Clifford to Ailes, 27 Dec. 1964, box 1228, RG 335, Suitland Federal Records Center.

31. Williams to Stone, 8 Dec. 1964, ROTC-Air Force, 1964 folder, box 39, University of Michigan Vice-President for Academic Affairs, University of Michigan Archives.

32. National Association of State Universities and Land-Grant Colleges (NASULGC) Circular #38, 4 Nov. 1965, ROTC General, 1965 folder, ibid.

33. Williams to ROTC Staff, 19 Nov. 1965, ibid.

34. University of Michigan President Robben Fleming "never thought there were more than 20 or 25 really disruptive, violent people around campus." He also believed that at Michigan the student radicals "didn't play into [the decision-making on ROTC] very much." Robben Fleming, interview with the author, Ann Arbor, Mich., 12 Dec. 1995.

35. Paul Stoller, "Students Appeal for Vietnam Negotiations in Referendum," *Pitt News,* 25 Jan. 1968, 1; "SG Referendum," ibid., 20 Nov. 1968, 1. The January referendum attracted 2,422 voters. The November referendum attracted 3,869 voters.

36. Kent State University News Service, 14 Aug. 1970, folder 44, Charles Kegley May 4th Materials 34.13, box 102, Kent State University Archives.

37. NASULGC Circular #154, 30 June 1970, folder 42, ibid.

38. James Foley and Robert Foley, *College Scene: Students Tell It Like It Is* (New York: Cowles Books, 1969), 126.

39. Ibid., 31, 130–31.

40. Blake to AFROTC Advisory Panel, 13 Aug. 1969, box 235, ROTC-Air Science, Henry Papers, University of Illinois Archives.

41. NASULGC Circular #154, 30 June 1970.

42. James Gindin, Carl Cohen, John LaPrelle, and Locke Anderson, Report from the Curriculum Committee on the Issue of Accreditation for ROTC, 25 Mar. 1969, University of Michigan Navy ROTC Accreditation/Relations file 1969, North Hall, University of Michigan.

43. Student Affairs Committee of the University Senate Recommendation, 4 Apr. 1968, folder 239, class. no. 3/1/1, Provost Office Files 1965–1970, University of Pittsburgh Archives.

44. Report of the Tulane University Faculty, 30 July 1969, folder 236, ibid.

45. Cornell University, Report of the Special Faculty Committee on Military Training, 14 Nov. 1969, folder 233, ibid.

46. List of Universities and Colleges in the United States Offering Specialized Courses in Military History, [1970], HRC 326.6, United States Army Center of Military History, Washington, D.C.

47. NASULGC Special Subcommittee on ROTC Policy Report, June 1968, box 15, 10/3/57, NASULGC Papers, University of Illinois Archives.

48. In 1968, only 11 percent of AFROTC junior and senior cadets agreed with the statement that "Most [freshman and sophomore] cadets look forward to and enjoy participating in military drill periods." Only 40 percent of senior cadets agreed with the statement: "Cadets marching in the ranks learn discipline, respect for authority, and *esprit-de-corps.*" Five percent agreed that

"The campus respects and encourages drill." The author of the article concluded that these data argued for a "reconsideration [of AFROTC's] position on drill." See "Drill and Corps Training," *Air Force ROTC Bulletin,* December 1970, 23.

49. William F. Muhlenfeld, "Our Embattled ROTC," *Army,* February 1969, 21.

50. Quoted in David Rosenbaum, "Campus Attacks on ROTC Stir Pentagon," *New York Times,* 19 April 1969, 19.

51. Robert Williams, Meeting with the Bureau of Naval Personnel, 21 Aug. 1969, Advisory Committee on ROTC Affairs folder, box 39, University of Michigan Vice-President for Academic Affairs, University of Michigan Archives.

52. Pickett to Spitler, 18 Dec. 1970, Armed Forces folder, box 70, ser. no. 24/1/1, Chancellor's Office Papers, University of Illinois Archives.

53. Gerald Perselay and Raymond Grenier, "New Directions in Corps Training," *Air University Air Force ROTC Education Bulletin,* April 1971.

54. Senior Division Army ROTC Program of Instruction, 7 Aug. 1969, box 15, 10/3/57, NASULGC Papers, University of Illinois Archives.

55. Williams to Fleming, 23 May 1969, ROTC folder, box 10, University of Michigan Presidents Papers, University of Michigan Archives.

56. Blake to Fleming, 26 Mar. 1969, ibid.

57. Fleming to the Regents, 9 June 1970, ROTC folder, box 15, ibid.

58. Supplementary Memorandum of Understanding, 15 Nov. 1971, ROTC-Air Force 1971 folder, box 40, University of Michigan Vice-President for Academic Affairs, ibid.

59. University of Pittsburgh Department of Military Science (Army ROTC): An Overview, January 1969, class. no. 90/6-B, University of Pittsburgh Archives.

60. Subcommittee of the Military Education Council for the Review of the Army ROTC Curriculum, 29 Mar. 1971, box 1, ser. no. 27/3/1, University of Illinois Archives.

61. Irvin to Grosschmid, 9 Aug. 1972, ROTC 1972–1976 folder, ROTC box 1, Duquesne University Archives.

62. Fleming to Pemberton, 19 Mar. 1969, ROTC folder, box 10, University of Michigan Presidents Papers, University of Michigan Archives.

63. Quoted in Cindy Morgan, "FASC Ponders Re-opening ROTC Credit Issue," *Pitt News,* 28 Jan. 1972, 1.

64. Miller to Witke, 22 Feb. 1974, University of Michigan Navy ROTC Accreditation/Relations File 1973–1974, North Hall.

65. Quoted in Sara Rimer, "LSA Committee Approves Amended ROTC Plan," *Michigan Daily,* 15 Jan. 1975.

66. Report of the Subcommittee on Military Officer Education Programs to the LSA [Literature, Science, and the Arts] Committee on Curriculum, 14 January 1975, University of Michigan Navy ROTC Accreditation/Relations File 1973–1974, North Hall, University of Michigan.

67. [I. J. Irvin], Department of Military Science Annual Report, SY 72-73, 23 Aug. 1973, Annual Report 1973–74 folder, ROTC box 1, Duquesne University Archives.

68. Department of Military Science Annual Report, SY 73-74, 25 July 1974, Annual Report 1973–76 folder, ibid. The departments were: Political Science, Sociology, and History.

69. Bambery to McCulloch, 13 Feb. 1975, ROTC 1972–1976 folder, ibid. The departments were: History, Political Science, Psychology, Speech, and Business.

70. Peltason to Leslie, 27 Apr. 1976, box 86, ROTC-Military Science, President John Corbally Papers, University of Illinois Archives. Emphasis in original.

71. Fleming to Brickel, 15 Dec. 1975, USAF Correspondence folder, 1973–1976, box 139, University of Michigan Vice-President for Academic Affairs, University of Michigan Archives; Air University History, Academic Year 1975–1976, app. E, Air Force Reserve Officers Training Corps, vol. 10, pt. 1, series K239.01, 31, Air Force Historical Research Agency, Maxwell AFB.

72. Roskens to Corbally, 1 Feb. 1977, ROTC-Air Force, box 103, Corbally Papers, University of Illinois Archives.

Postwar U.S. Marine Corps Commissioning Sources and Junior Officer Training

Herbert M. Hart

In 1775, Congress commissioned thirty-two-year-old Samuel Nicholas as the first Marine officer.[1] Formal training for Marine officers, however, would not begin for another half-century. Archibald Henderson, Marine Corps commandant from 1820 to 1859, required that all new officers receive basic training under his personal supervision at the Marine Barracks in Washington, D.C.[2] From these beginnings, the Marine Corps came to draw upon a variety of sources for both its regular and reserve officers and developed training programs to prepare them for commissions and to shape them into effective junior officers. A substantial part of my thirty-seven-year career in the Marine Corps was in assignments involving officer education and training, particularly officer selection and junior officer preparation. Although this essay broadly surveys Marine Corps commissioning sources and junior officer training since 1945, its focus is on the 1950s through the early 1970s when I was directly connected with those programs.

During World War II, the Marine Corps had earned recognition on the battlefield and was America's favorite, but, along with the other services, experienced steep postwar personnel reductions—from a peak of 485,833 men and women in August 1945 to 92,000 in July 1947. Thereafter, the Marine Corps rebuilt to its assigned postwar level of 108,200, with two divisions, one on the West Coast at Camp Pendleton, California, and other on the East Coast at Camp Lejeune, North Carolina.[3]

Four principal commissioning sources, all in existence before the United States entered World War II, would provide new officers for the Marine Corps in the postwar decades. These sources were the United States Naval Academy, the Naval Reserve Officers Training Corps (NROTC), the Platoon Leaders' Class, and the Officer Candidates' Course. There would also be a number of new programs, including a Woman Officer Candidate Course, and programs to commission aviators, lawyers, and enlisted men.

After commissioning, all new Marine lieutenants (except women until 1973 and, for a time, aviators) attended The Basic School at Quantico, Virginia, a descendant of the Marine Corps School of Application established in 1891. The purpose of the school's basic officer course has always been to further junior officers' professional knowledge, enhance their *esprit de corps,* and develop the leadership traits and warfighting skills required of a Marine company-grade officer, especially those needed by a rifle platoon commander.

Although provision for Naval Academy graduates to be commissioned into the

Marine Corps had been on the books since the early 1880s, only after 1915 with some prompting of the Navy by Congress could the Corps count on a steady supply. The experience of World War II and the improved working relationship with the Navy further smoothed the flow. Since 1945, the number of Naval Academy graduates becoming Marine officers has steadily risen: forty-eight in 1945–54, sixty-one in 1955–64, and ninety-eight in 1965–74. Most recently (1991–96), 898 graduates received Marine commissions, roughly 16 percent of each graduating class.[4]

In 1946, the NROTC expanded significantly. It consisted of the six original colleges and universities with programs dating from 1926 and many new institutions for a total of fifty-two. At the original schools, such as Northwestern University, the NROTC restarted with former members of the Navy's V-12 program (a major source of college-trained Navy and Marine officers during the war) who desired to continue on campus as undergraduates and as officer candidates. In 1947, they were joined by civilians who had taken a national competitive test for appointment as midshipmen in the Naval Reserve in the NROTC units. Also included, for the first time, were several hundred enlisted men who had qualified on the test. They were sent to the Naval Training Center, Great Lakes, Illinois for an indoctrination course—supposedly to get them better prepared in English and mathematics but also to process them for their future roles as midshipman officer candidates and to select their colleges.

As midshipmen in the Naval Reserve, NROTC students were undergraduates with the freedom to major in most fields as long as they took mathematics through trigonometry and physics, primarily as preparation for naval science courses such as navigation. In addition, they took other naval science courses, including gunnery, seamanship, and naval justice. After their junior year, Marine officer candidates took special Marine Corps subjects. All participated in two-hour drills once a week, the only time during the academic year that officer-type naval uniforms were worn.

All NROTC midshipmen attended two, eight-week summer cruises—the freshman-sophomore cruise aboard a combatant ship; and the sophomore-junior cruise, a two-week period of training in amphibious warfare at Little Creek, Virginia, followed by six weeks of aviation indoctrination at the Naval Air Station, Pensacola, Florida. Between their junior and senior years, the Marine officer candidates also attended an eight-week indoctrination at The Basic School. By the time the first postwar class of NROTC Marine officer candidates (I among them) reached this stage of training, it was 1950 and the Korean War had started. We were very aware of the heightened tension at Quantico and the dramatic increase in north-south railroad freight traffic on the tracks that paralleled our barracks.

The Platoon Leaders' Class (PLC), which had begun in 1935 but was suspended in the summer of 1941, was destined to become the major source of Marine Corps reserve officers and, ultimately, many career regular officers after 1945. It revived on a major scale immediately after the war; in the summer of 1947 a total of 433 college students attended training at Quantico.[5] They were mostly freshmen and sophomores and came from colleges throughout the country. Prewar restrictions on what colleges were eligible for the program had been lifted and applications for participation were accepted from any accredited college. Students in the PLC attended six-week training sessions after either their freshman and junior, or sophomore and junior years; or in a continuous twelve-week course after their junior year. After successfully completing training and receiving their college degrees, the candidates were commissioned as reserve second lieutenants and ordered to The Basic School.

Both the PLC and the Officer Candidates' Course (OCC), the fourth of the prewar commissioning programs to be continued following the war, came under the direction of the Marine Corps Officer Candidates School at Quantico. Initiated in November 1940, the OCC was able to expand or contract rapidly according to the need for officers. Although aimed primarily at college graduates, many candidates had also been selected from the enlisted ranks during the war. The latter had to complete a Pre-Officer Candidate School and then the OCC itself.[6]

After World War II, the OCC no longer accepted enlisted officer candidates. In place of the OCC, the Marine Corps established new commissioning avenues for meritorious enlisted men. The Meritorious Noncommissioned Officer Program enabled Marines recommended by their commanding officers to be appointed commissioned officers; later the program was extended to meritorious former noncommissioned officers (NCOs) no longer on active duty and to reservists on extended tours. In the summer of 1949, the Officer Candidate Screening Course (OCSC) was established for enlisted men with college degrees and a minimum score of 120 on the General Classification Test (GCT). The candidates were given four weeks in which to demonstrate their ability; the successful were commissioned and ordered to The Basic School for further training. After the start of the Korean War, the OCSC expanded and was opened to reservists on active duty. In January 1951 it was extended to both regulars and enlisted reservists who held degrees, could attain a minimum score of 120 on the GCT, and could meet the age limits.[7]

After the Korean War and the stabilization of the Marine Corps with three divisions and aircraft wing teams (with the addition of the 3rd Marine Division and the 3rd Marine Aircraft Wing), the need for officers also stabilized. Permanent officer procurement offices were established at district headquarters and other locations, usually in major cities, around the country, and permanent relationships were established with college administrators. This organizational structure provided continuity of effort and points of contact between the colleges and the Corps. In the late 1950s, the title Officer Procurement Officer was changed to Officer Selection Officer (OSO) to make it more palatable on campuses, especially when it came to the programs for women. J. Walter Thompson, the New York public relations agency, assisted the selection officers in their efforts. The company had represented the Marine Corps as a public service since before World War II (some of its executives were Marines in World War I and had maintained ties with top officers in the Corps). Even in the post–Korean War period, the J. Walter Thompson team that supported the officer procurement effort charged only actual expenses; all other support was provided without cost to the Marine Corps.

Before World War II, aviators for the Marine Corps had come from the ranks of lieutenants who had served two years as ground officers and from the Marine Aviation Cadet Program (MARCAD). After the war, the Corps drew heavily for aviators on the Naval Aviation Cadet (NAVCAD) program (the MARCAD program had been suspended when the war started) which required only that participants have two years of college. When the NAVCAD program resumed after the war, there was considerable competition between the Navy and the Marine Corps for its graduates.

In 1956 I traveled to Pensacola as the publicity officer for the officer programs. The trip's purpose was research for an information booklet on Marine flight training. After I arrived, the senior Marine officers at the Naval Air Station pressured me about the need for a Marine flight program separate from the Navy's. That point was vigorously emphasized one evening in the kitchen of the quarters of Col. John L. Smith, World War II ace and Medal of Honor recipient, who was the senior Marine at the air station. He wanted to convince the Marine Corps to resume the MARCAD program, and he insisted that I get the word back to Marine Corps headquarters in Washington. My promise to deliver this message was the price for escape from the kitchen!

Whether or not there was any connection, the MARCAD program resumed in July 1959 when a dozen men reported to Pensacola for training. Eleven were enlisted Marines, the twelfth a civilian; all had two years of college. They underwent eighteen months of training that culminated in commissions and pilot wings.[8] The program continued through the mid-1960s with up to 225 accessions annually until Fiscal Year (FY) 1968 when the number dropped to 62 and the program ended. By this time, aviation officer candidate programs that had been initiated in 1955 in conjunction with the PLC and the OCC were providing sufficient college graduate accessions to meet the service's needs. An aviation cadet program was no longer necessary.[9]

Officer candidates in the PLC (Aviation) program attended the same two six-week summer sessions as did the other PLC participants, but upon commissioning were guaranteed assignment to flight training if they qualified. The program averaged slightly more than 100 newly-commissioned flight students annually from FY 1958 through FY 1965. The Aviation Officer Candidate program, in which all precommissioning training was conducted in one course after college graduation, averaged about 125 accessions per year during the same period. Participants in both programs went directly to flight training rather than to The Basic School after commissioning.[10]

Recognizing its need for lawyers, the Marine Corps initiated the PLC (Law) program in 1959. Here the initial target was law school students. Those recruited attended the two, six-week sessions at Quantico, were commissioned after completing the training, and continued in law school. Those who joined the program early enough often were able to be promoted while still in law school, sometimes coming on active duty as captains after passing the bar examination.

The program produced remarkable results. It was tested during the period that I ran the Marine Corps officer selection office in Chicago. A Marine lawyer, a captain, was detailed to the office and initially contacted one of my college classmates who had become the dean of Northwestern's law school. At the same time, we prepared a brochure describing the program and mailed it to law schools in the Illinois-Wisconsin region. The result of this intensive effort was that our office met the program's entire national quota the first year.

A significant advantage enjoyed by all of the PLC programs was that as soon as a full-time student was sworn in, he started earning longevity for pay. This meant that a PLC officer candidate who had been in the program for two years had already qualified for the first pay raise when commissioned; those enrolled in the program for four, or sometimes even five years, enjoyed a significant monthly pay advantage when they came on active duty over officers commissioned from other sources.

During World War II women had again come into the Marine Corps, resuming a tradition that had begun in World War I. As members of the Women's Reserve, they served in support functions such as air control, ordnance loading, administration, disbursing, and motor transport, thus "freeing a Marine to fight." At least the 18,460 women of Marines in World War II had a more palatable nickname, "WRs," than the "Marinettes" of the earlier war. By 1947, however, only a small nucleus had been retained in the Corps.[11]

Women were integrated into the regular Marine Corps establishment in 1948. Although women officers with World War II service were recalled to duty, the initiation in 1949 of the Woman Officer Candidate Course met the need for junior officers. The course at Quantico was heavy on grooming and Marine Corps indoctrination and orientation to prepare the women officer candidates for administrative jobs, and to lead enlisted women in that support specialty.

Recruiting college women into the program was a problem complicated by resistance both in the civilian world to women in the military and, in the Marine Corps, to women in the Corps. When I was publicity officer in the officer procurement branch at Marine Corps headquarters from 1954 to 1957, for example, the resistance was so strong that the slot for a woman to direct the program was left vacant and assigned to me. In 1955 Col. Julia E. Hamblet, the director of Women Marines, complained personally to Marine Corps commandant, Gen. Lemuel E. Shepherd, that nothing was being done for the program. This got the attention of the officer procurement branch. Advertising posters and a four-color information booklet were quickly produced using New York models and a photographer, and a woman was assigned to direct the program.

The women also wanted a motion picture to support officer recruiting but antiquated Department of the Navy procedures stood in the way. The by now responsive officer procurement branch used friends in the Pentagon to photograph a film, scripted and produced it in-house, and leaned on the Marine Band to provide musical background. The J. Walter Thompson agency arranged for a female television anchorwoman to narrate the film which was edited with the help of the Motion Picture School at the Pensacola Naval Air Station, and the East Coast Motion Picture Production Team at Quantico. The final result, *Leading Ladies,* was approved by the Department of the Navy and became the standard in the recruiting field for at least fifteen years, with the Navy's mild admonition that the no-cost production had been done "outside of channels."

The Vietnam War brought considerable turmoil to Marine Corps officer procurement. Expansion in the size of the Corps meant a requirement for many more lieutenants. Temporary programs were once again instituted for meritorious NCOs, and overall officer accessions jumped from 4,907 in FY 1966 to 5,542 in FY 1967.[12] When draft calls were high or when young married men lost their deferments beginning in late 1965, the OSOs spent more time sorting through applications than they did trying to generate them.[13]

Increasing public dissatisfaction with the war both on and off campus adversely affected Marine officer programs. Some colleges refused program representatives permission to visit their campuses, or situated established programs at inconvenient locations where few students encountered them. At Northwestern, the NROTC was moved off campus entirely and the university senate attempted to deny academic credit for naval science courses. Only through strong support from some faculty members and alumni was the credit issue settled; the problem of the off-campus site was not corrected for several years.

In FY 1969, the PLC realized only 77 percent of its quota of 2,200 and the OCC, only 76 percent of 2,679. To meet officer requirements, the Marine Corps resorted to commissioning men with only two years of college and promising them an opportunity to finish school after a period of satisfactory service. The NROTC program was similarly affected that year, achieving only 168 officer accessions out of the 195 programmed. The Corps hoped to improve these numbers by adding fourteen OSOs to the procurement effort.[14]

With the end of the Vietnam War, Marine officer programs stabilized and the procurement picture improved considerably. Nonetheless, continued command attention was needed to assure an adequate flow of qualified candidates.

In 1971 senior Marine Corps officers began to raise questions about the effectiveness of officer procurement programs and officer training. Lieutenant General William G. Thrash, commanding general of the Marine Corps Development and Education Command, convened a panel to study the programs for nine months and either validate or recommend changes to them. Major General Louis H. Wilson, commanding general of the Education Center and former commanding officer of The Basic School (and destined to be Marine Corps commandant, 1975–79), assumed overall direction of the study. Wilson took the study seriously, showing a personal interest in it.

Although the Vietnam War was in its final stages, there was no direct link between it and the study. In fact, training at Quantico had already started to change. When Wilson took over the Education Center in 1971 and visited The Basic School, he immediately ordered that the "Vietnam Village," a mock-up of conditions in Vietnam, be converted to reflect combat in a built-up, urban-type environment. "We are," I remember him saying, "no longer fighting the last war."

The study panel's senior member was Col. William F. Saunders, Jr., then in charge of the Marine Corps Extension School (responsible for administering the Corps' correspondence courses). Several panel members reviewed the senior and intermediate programs and I (then the operations officer at The Basic School) examined the junior officer and noncommissioned officer programs. The panel first consulted the twenty studies it could locate on officer training generally and Marine officer training specifically. Panel members then visited twenty-five schools, including those of the other services.

In May 1972 the study panel produced its report—more than 300 pages plus 25 annexes.[15] Several of the report's recommendations involved changes to the pre- and post-commissioning junior officer programs. The Marine Corps implemented some of the changes immediately, others after several years.

The study panel's work first affected the women officer programs. From 1949 to 1973, women's officer candidate and basic officer training were entirely separate from the programs for men. The courses for women were much different in content and significantly shorter; training was conducted at a separate location at Quantico; and the programs were run by the Woman Officer School which was not connected administratively to The Basic School.[16] Initially, the study panel had believed that there was nothing wrong with these programs and did not intend to consider them. But during a visit to Lackland Air Force Base in Texas, I noted that male and female officer candidates trained together and that both staff and trainees were enthusiastic about the integration. After returning to Quantico I briefed General Wilson, recommending that junior officer training be integrated and pointing out that the director of the Woman Officer School concurred. Wilson directed that changes be made to prepare for integration. Plans to construct new facilities on the main base for the women were canceled and they were quartered in The Basic School training area.[17]

The result was not true integration, however. The course syllabus for the women still differed from the men's and was less than half its length. Moreover, the women were in separate platoons, training with the men only when there was an obvious benefit. In 1976, General Wilson, now the Marine Corps commandant, approved The Basic School's recommendation that the male course be somewhat reduced in length and that female officers be more fully integrated into the course. In this version of integration, that began early in 1977, the women remained in separate platoons, but completed the same syllabus as the men. Moreover, during field and physical training several women were attached to each male platoon.[18] About three years later there was a backlash. At that time I was the director of public affairs for the Marine Corps and was present during a briefing when Gen. Robert H. Barrow, who succeeded Wilson as commandant, learned that a woman had graduated first in one of The Basic School classes. Concerned that the basic course had been softened, he commented: "We are going to have to walk back the cat."

Despite General Barrow's misgivings, today Marine Corps basic officer training is fully integrated. Both sexes are quartered in the same buildings, although in separate rooms. Women and men are assigned to the same platoons, and receive classroom instruction and participate in field and physical training together. In the OCC, women and men occupy the same barracks but with the living areas partitioned. Field and physical training are conducted separately.[19]

A second important study panel recommendation concerned infantry officer training. On the theory that all officers had to be trained as infantrymen, all had gone through the same infantry-oriented Basic School. Those not going into infantry assignments after graduation then went on to special schools such as supply, armor, artillery, aviation, and the like. The infantrymen went to the field with no further training. The study panel pointed out that this was to the disadvantage of the infantry officers who had to compete throughout their careers with non-infantry officers who would receive advanced training in their specialties. In its report, the panel recommended that an add-on course be established for infantry officers. Such a course would better prepare them for their future assignments and, at the same time, make integration of the male and female courses easier because the amount of infantry training received by all officers in The Basic School could be reduced, thus shortening the length of the basic officer course.[20]

The panel's recommendation was not immediately implemented, in part because of a misunderstanding by the staff of The Basic School. With a change in personnel at the school, the message got through that infantry officers in the basic course were being short-changed. In 1976, General Wilson approved the new course for infantry officers.[21]

Another of the 1972 report's recommendations related to the missions of the officer candidate and the basic officer courses. The study panel had noted that the purpose of the officer candidate courses was to graduate individuals with the "physical, mental and leadership qualities necessary for an effective Marine Corps officer." The panel took the position that if this were true, there was no need for the basic course. For this reason, its report recommended that the mission of the officer candidate courses be modified to: "indoctrinate, motivate and screen candidates for commissioning." The Basic School would then complete the task of producing effective Marine Corps officers. This change in wording was approved.[22]

Over the fifty-plus years since World War II, the Marine Corps has acquired its new officers from a number of sources but in a more formalized way than before the war. Those sources for both regular and reserve officers have remained fairly constant since the end of the Vietnam War. In FY 1997, the Marine Corps commissioned 1,374 lieutenants from the following sources:

Source	Number	%
PLC	528	38.15
OCC	388	28.03
USNA	161	11.63
NROTC	151	10.91
Enlisted	130	10.11
Other	16	1.16
Total	1,374	99.99

Women represented 7.59 percent and minorities 19.15 percent of this total.[23] Today, whatever their commissioning source, new lieutenants then enter that great leveler, The Basic School, where both men and women train together to acquire the professional skills essential for lieutenants of Marines.

Notes

1. J. Robert Moskin, *The U.S. Marine Corps Story,* rev. and updated (New York: McGraw-Hill, 1987), 25.

2. *Home of the Commandants* (Quantico, Va.: Marine Corps Association, 1995), 66.

3. Moskin, 420, 423.

4. John E. Greenwood, "The Corps' Old School Tie," U.S. Naval Institute *Proceedings* 101 (November 1975): 48–51; and interview with Lt. Col. Dennis Sabal, Officer Procurement Branch, Headquarters, U.S. Marine Corps, Washington, D.C., 8 and 11 Dec. 1997.

5. Bernard C. Nalty and Ralph F. Moody, *A Brief History of U.S. Marine Corps Officer Procurement, 1775–1969,* rev. ed. (Washington, D.C.: Headquarters, U.S. Marine Corps, 1970), 9, 16.

6. Ibid., 9–10, 13–14.

7. Ibid., 17–18.

8. U.S. Marine Corps Press Release 786-59, Washington, D.C., 11 July 1959.

9. Nalty and Moody, 21–22, 28.

10. Ibid., 21, 28.

11. Mary V. Stremlow, *A History of the Women Marines, 1946–1977* (Washington, D.C.: History and Museums Division, Headquarters, U.S. Marine Corps, 1986), 3–6.

12. Nalty and Moody, 20–23.

13. Ibid., 23.

14. Ibid., 24–25.

15. Report of U.S. Marine Corps Study Panel, *Education of the Corps, 1973–1993* (Quantico, Va., May 1972); copy in Breckenridge Library, Marine Corps Education Center, Quantico, Va.

16. Stremlow, 125–27.

17. *Education of the Corps,* pts. 2:31–44, 7:1–2; Lt. Col. Herbert M. Hart memorandum for the record, subj: *Education of the Corps, 1973–1993,* 31 May 1977; copy attached to report. I prepared this memorandum in 1977 to be included in the copy of the 1972 report that was to be permanently retained.

18. Stremlow, 129–36.

19. Interview with Lt. Col. Edmund F. Flores, executive officer, The Basic School, Marine Corps Eduction Center, 9 Mar. 1998.

20. *Education of the Corps,* pts. 2:12–13, 7:2.

21. Hart memorandum for the record, 31 May 1977; and U.S. Marine Corps News Release JWS-218-76, Washington, D.C., 8 Nov. 1976.

22. *Education of the Corps,* pts. 2:12–13, 7:1.

23. Statistics provided by the Officer Procurement Branch, Headquarters, U.S. Marine Corps.

Section VI
Civil-Military Relations and the Making of Officers in Developing Countries

Other than the conduct of war, perhaps no single issue has occupied scholars of military affairs more than armed forces' potential for or actual intervention in politics. Especially for military forces in postwar Africa, Latin America, the Middle East, and Southeast Asia, that focus has heavily dominated English-language studies. Not without good reason. By 1985, notes Alfred McCoy in his essay on the Philippine Military Academy, 58 of the 109 countries in what used to be called the Third World were subject to military rule. In searching for causes of the military's invariably disruptive and often deadly political involvement, scholars have investigated the roles played by officer social origins and education and training systems in considerable depth.[1] Each of the following essays—on the Philippines, Morocco, Mexico, and four nations in South America's Southern Cone—explores the relationship between the process of officer formation and the predilection for the military to enter— or stay out of—the political arena in the latter half of the twentieth century.

The Philippine Military Academy's Class of 1940 refrained throughout their careers from intruding into national politics; graduates of the Class of 1971 led six abortive coups d'état in the latter half of the 1980s. McCoy, a professor of Southeast Asian history at the University of Wisconsin, maintains the different behavior of the two groups of officers resulted both from the nature of their socialization at the academy and the conditions of their service as junior officers. Both classes formed strong internal bonds forged by the rigorous academy experience, particularly subjection to hazing during their initial, or "plebe," year. For the Class of 1940, the shared suffering of World War II intensified their ties to each other. Class members called upon those personal relationships to enforce the military's apolitical stance in the postwar Philippines—a subordination to civil supremacy reflective of the American concept of military professionalism that had been inculcated at the academy. For the Class of 1971, however, hazing in the postwar academy had turned into brutality, and the intrusion of radical nationalist sentiments and political activism during the late 1960s subverted the habit of subordination. When Philippine President Ferdinand Marcos declared martial law in 1972 and politicized the military, the young officers of the Class of 1971, accustomed to practicing brutal hazing at the academy, became torturers who terrorized the regime's political opponents. "Through their extraordinary duties in the service of dictatorship," writes McCoy, "they acquired special skills and a quality of arrogance that would later inspire their attack on the state."

The French masters of Morocco, according to Moshe Gershovich, intended that graduates of the Royal Military Academy established in their North African colony in 1919 would not only officer native units but, after their service, would join other elements of Morocco's indigenous elite as administrators in the Protectorate.

293

Gershovich, who teaches history at the Massachusetts Institute of Technology, shows, however, that the French plan to draw academy cadets exclusively from leading Moroccan families only partially succeeded; over time, entrants came from broader circles in Moroccan society, softening the academy's colonial image and helping to preserve the school's existence after Morocco achieved independence in 1956. Despite their second-class status in the French Army, academy-trained officers were largely unaffected by Moroccan nationalism and remained loyal to France throughout the colonial period because they possessed "a distinctive identity and the notion, ingrained at the academy, that they were destined to become the leaders of their society."

In Mexico, suggests Roderic Camp, the *Heroico Colegio Militar* (for officer candidates) and the *Escuela Superior de Guerra* (for lieutenants and captains) have distinctively shaped the Mexican officer corps since the 1920s and 1930s. The two institutions, argues the director of Tulane University's Mexican Policy Studies program, have imparted a nationalism suffused with suspicion of foreigners, promoted a caste-like mentality isolating the officer corps from society, emphasized subordination both to civil authority and to the highly centralized and inflexible Mexican military hierarchy, and stressed an exaggerated discipline designed to reinforce hierarchical rigidities that also compromises individual integrity. These values have centered officers' loyalty squarely on the armed forces' organization and have kept them out of politics.

Unlike the Mexican military, armed forces in South America have a long history in the twentieth century of intervening in politics, frequently through direct seizure of power—coups d'état that by the 1970s and 1980s, according to Frederick Nunn, assumed institutional form. With a focus on Argentina, Brazil, Chile, and Peru, Nunn, a professor of history and international relations at Portland State University, indicates that military education extending throughout officers' careers has become an integral part of professional militarism in South America. In his view, this education has expanded their capacity for critical thought and the desire to apply it to the solution of national problems, thereby increasing civilian political systems' vulnerability to military intervention. South American officers have always worried about threats to national sovereignty and the potential for "subversion of national values." Today, in a period of accelerated change, they blame those threats—including redemocratization, transnational capitalism, the internationalization of popular culture, and the United States' apparent intention to redefine Latin American military roles to suit its conception of the post–Cold War security environment—on outside influences. The thinking of South American officers, writes Nunn, has reflected remarkable "comparability and consistency" for decades, and seems likely to retain those characteristics.

Note

1. Concern about the civil-military nexus has been so pervasive that analysis of officer preparation's connection to military effectiveness has been almost completely neglected in English-language works on the armed forces of most nations in Africa, Latin America, the Middle East, and Southeast Asia.

Closer Than Brothers: Two Classes at the Philippine Military Academy

Alfred W. McCoy

In August 1967, the superintendent of the Philippine Military Academy (PMA), Gen. Reynaldo Mendoza, stood for the last time before the Corps of Cadets, over four hundred strong, in their dress-gray uniforms decorated with rows of brass buttons. In his valedictory, the general celebrated the academy's democratic mission and urged the cadets to honor its motto—"integrity, courage, loyalty." Nearly thirty years before, as a young cadet in the academy's first entering class, he had used those words as a refrain in lyrics that became the school song, "PMA, Oh Hail to Thee." As his wife wept and the cadets rose to applaud, General Mendoza stepped down from the podium, climbed into his blue Volkswagen "Beetle" sedan, and drove himself into retirement and into history. He went with pride knowing that he and his Class of 1940 had an unblemished record—no scandals, no corruption, and no coups.[1]

Swelling the ranks of the corps that day were the youngest cadets, the 148 "plebes" of the future Class of 1971. As colonels twenty years later, they would lead six abortive coups d'état before retreating into the guerrilla underground for a campaign of robbery, kidnapping, and terror bombings. Today those plebes are about to assume command of the military, and their class captain, Gregorio "Gringo" Honasan, has recently been elected to the Philippine Senate.

In retrospect, this seems a symbolic moment. Here we can glimpse a basic generational and cultural change in the Armed Forces of the Philippines (AFP). These two classes—1940 and 1971—span its entire history from founding in 1936 to the present. Through their eyes, we gain a unique vantage point on tumultuous events that have shaped the Philippine military over the past half-century. Marked differences in the character of these two classes—one opposing coups and the other plotting them—alert us to the military's changing role in Philippine political life.

Though separated by thirty years, both classes, in their turn, faced similar political decisions that make the comparison telling. In the 1953 presidential elections, the opposition candidate, convinced that the incumbent would win by fraud, pressed Class '40 to prepare a coup. They refused, adopting an apolitical stance that they would maintain until their retirement in the mid-1960s. As long as Class '40 was in command, the Philippine military, almost alone in Southeast Asia, did not engage in coup plotting. Then, in the 1986 presidential elections, the opposition, again certain that the incumbent would cheat, pressed the leaders of Class '71 to plan a coup. They agreed. When President Ferdinand Marcos did indeed win by fraud, the class led a bungled coup attempt that sparked the country's famous "people's power" uprising of February 1986. When the crowds installed their candidate Cory Aquino in the palace, the disgruntled colonels of Class '71 made five more coup attempts that greatly troubled their nation's transition to democracy.

The contrast is both striking and revealing. In the splendid isolation of their new academy, Class '40 imbibed American colonial values of military professionalism. Bound together by the harsh ritual of male initiation known as "hazing," the Class of 1940 forged a strong, collective identity during their cadet days. These bonds, stiffened by the shared suffering of war, produced a masculinity synonymous with a sense of military honor that made the Philippines the only major Southeast Asian nation free from coups in the two decades after World War II. Graduating on the eve of war, some fifty members of the class fought on Bataan, suffered the "Death March," and experienced months of starvation in Japanese prisoner-of-war camps. After leading anti-Japanese guerrillas across the archipelago, they rejoined the regular service after the war and led the battalions that defeated the Huk Communist guerrillas in the 1950s. As they rose up the chain of command, Class '40 helped build an armed forces that distanced itself from the partisan pressures that soon overwhelmed the postwar state. At the highest echelons a decade later, they were the senior officers whom President Marcos forced out when he began to politicize the military in preparation for his martial-law dictatorship.

Instead of defending the nation against invasion, the Class of 1971 was shaped by their service to dictatorship. Only months after their graduation, Marcos declared martial law with the near-unanimous support of his generals—a constitutional coup that made the military the foundation for his authoritarian regime. These young lieutenants soon found themselves fighting in a bloody civil war against Muslim insurgents and interrogating suspected subversives. Over the next decade, their service as torturer-interrogators emboldened the colonels of Class '71 in a bid for power, contributing to a climate that produced nine coup attempts in the late 1980s— perhaps the most of any nation in the world. Under the banner of the Reform the Armed Forces Movement (RAM), these officers sparked the turmoil that toppled Marcos in February 1986, led rebel soldiers to the gates of Malacañang Palace in August 1987, and seized much of Manila for a week in December 1989. In this last attempt, the coup plotters came remarkably close to capturing the state. Watching this succession of coups by their fellow PMA alumni in the late 1980s, surviving members of Class '40, long retired and well into their seventies, condemned the rebel officers, rued the military's politicization, and urged a wholesale purge of the cadet corps.

Reflecting changes in the wider society, the academy's institutional ethos had undergone a slow, subtle change that contributed to this contrast. While the cadets intensified the culture of group bonding, the institution's capacity to inculcate professional ideals seems to have declined. During the late 1960s, moreover, nationalism began to challenge military professionalism for the cadets' ultimate allegiance. But both of these factors pale by comparison with the intensive politicization of the military that came when Marcos made it the fist of his martial-law regime.

Despite these many contrasts, the two classes are, in certain significant respects, similar. Whether aggressively apolitical like Class '40 or passionately committed to political change like Class '71, both held values shaped, as cadets and lieutenants, by the dominant political order of their day, whether colonial Commonwealth or nationalist dictatorship. Both classes played upon the personal and institutional loyalties among academy alumni to realize their political aspirations. Just as Class '40 used their identity as a "batch" to discourage any political involvement, so military rebels later tugged at school ties to pull Class '71 in a diametrically opposite direction—into coup attempts and a movement for radical political change. For serving Filipino officers in the postwar era, trying to remain "apolitical" and resisting partisan pressures required, in a certain sense, as much active political engagement as did plotting a coup d'état.

Comparison of these two classes enables us to engage one of the most problematic aspects of the military in the Third World—the remarkable persistence of coups. Ironically, by forming an army for defense against foreign invasion, newly independent states, the Philippines included, have created the means of their own destruction. As Edmund Burke once noted, "an armed, disciplined body is, in its essence, dangerous to liberty."[2] Indeed, by the mid-1970s over a third of the member states in the United Nations had changed governments by coup d'état.[3] Among Latin America's twenty nations, all but two have experienced coups since 1945.[4] During the same period, the military seized power in seven of Southeast Asia's nine major nations. By 1985, the military ruled 58 of 109 countries that make up the so-called Third World.[5]

Although the dangers that a standing army poses for a democracy should be obvious, most specialists have overlooked this fundamental problem.[6] One notable exception, S. E. Finer, defined the key question in his classic *The Man on Horseback*: "Instead of asking why the military engage in politics, we ought surely ask why they ever do otherwise. . . . The military possess vastly superior organization. And they possess *arms*." We should not, he argues, ask why the military "rebels against its civilian masters, but why it ever obeys them."[7] The answer to this latter question should be obvious: the military obeys only when it wants to. What is it, then, that makes an army willing to subordinate itself to civil authority?

Study of the Philippines provides some answers. While postwar civilian regimes fell to coups in Burma, Thailand, and Indonesia, the Philippine military held back. How can we explain this unique restraint by Filipino officers, when their peers elsewhere in Southeast Asia were enmeshed in coups and counter-coups? Their resistance to political involvement springs, at base, from socialization into subordination through both academy training and later military service. At the PMA, future Filipino officers were influenced by two strong forces—a formal indoctrination into the ideal of civil supremacy and a less formal, but lasting, peer-bonding to classmates. In general, the PMA was reasonably effective in instilling the values of professionalism in a majority of the regular officers. Particularly among prewar classes, the academy's formal curriculum of drill and classwork imbued cadets with a respect for civil authority. After graduation, these prewar graduates found these lessons in civil supremacy challenged, but ultimately affirmed, by active duty in a democracy. The academy's

bonding to class and classmates represents a more variable influence. Early graduates, such as the Class of 1940, could draw upon their peer bonds to reinforce shared
professional values of apolitical service. When later PMA alumni, such as Class '71,
were politicized by service to dictatorship, these same class ties would pull classmates into compromising alliances or coup plots. Thus, to understand why Filipino
officers have decided to obey or disobey, we have to understand both the nature of
their socialization at the academy and the conditions of their service as junior
officers.

Class of 1940

On 15 June 1936, 120 new cadets boarded a train in Manila for the Philippine
Military Academy, recently established in the mountain city of Baguio. At the end of
the rails, the cadets transferred to Benguet Auto Line (BAL) buses for the five
thousand foot climb up the zig-zag road to Baguio City. As the buses pulled into the
campus, some forty upperclassmen, all transfers from the superseded Constabulary
Academy, were waiting to greet the new arrivals with a ritual called "hazing." Even
fifty years later, these cadets would recall their reception at the academy as traumatic:

> The drive up the . . . road to Baguio did not take long. Soon they began to feel
> the change in altitude as their ear drums began to ache. They smelled the scent of
> pine needles. When they got to the Teachers Camp grounds, a beautiful sight met
> their eyes: a formation of upper-class cadets, smart in their khaki uniforms of
> blouses, Sam Browne belts, short pants, and puttees. "Wow! O-h-h-h! That's how
> we will look like soon too. . . ."
> Then all hell broke loose! A command was given and the upperclassmen, like
> a pack of hungry wolves, ran to where the BAL buses were parked and shouted at
> the incoming cadets to . . . form a straight line in front of a table, pull their necks in,
> stop rolling their eyes, shoulders back, guts in, etc., etc., etc. The memory of those
> first few moments will forever be engraved in the minds of the group.[8]

For the next nine months, from 5:50 A.M. to 10:00 P.M., seven days a week, the
Class of 1940 shared an ordeal that not only trained them to become career officers
but bonded them together for life. As the first class to study under the new academy's
four-year curriculum, the arrival of these cadets that June marks the start of the
modern Philippine military. Indeed, at their graduation four years later, Gen. Vicente
Lim, the army's deputy chief of staff and the first Filipino graduate of West Point
(1914), would describe the seventy-nine remaining members of Class '40 as "the
beginning of the life of the Philippine Army. When one of these boys becomes the
Chief of Staff in the future then we will have a real Army in the Philippines."[9]

The new academy

The opening of the PMA was, as General Lim implied, part of a larger design to
create a new army in preparation for Philippine independence in 1946. Under Commonwealth Act No. 1, known as the National Defense Act of 1935, President Manuel
Quezon was establishing a Philippine army inspired, in large part, by Switzerland's

citizen army. With the advice of his aides, Gen. Douglas MacArthur and Maj. Dwight Eisenhower, Quezon planned that a small cadre of regular officers would train a mass of civilian reservists. By 1946, the Philippines would have a regular force of 430 officers and 10,000 soldiers backed by a 400,000-man reserve.[10] Although the twenty-year-old conscripts would get just six months of rudimentary training, the men who drilled them would be regular officers educated in military science and committed to a thirty-year career.[11]

Forming an officer corps was the most difficult part of this enterprise. As Quezon put it: "The heart of an army is its officers."[12] To train these officers, the Defense Act provided for the establishment of the Philippine Military Academy at Baguio; its curriculum would be modeled, quite explicitly, on the U.S. Military Academy at West Point.[13] In the words of the Commonwealth's vice-president, the PMA was "the foundation stone of the entire military establishment," providing "the leadership necessary to knit together a scattered and loosely connected citizen army into one whole, living, pulsating, homogenous machine that can fight with courage."[14]

Quezon and his American advisors scrutinized the technical details of their new military, but they did not grasp, in a fundamental sense, the long-term political risks entailed in its creation. Under this legislation, the Commonwealth was investing a self-selecting elite of regular officers with control over the nation's armed forces and its weapons of mass destruction.

If we study the inherent logic of the National Defense Act, the Philippine state's only defense against future coups d'état seemed to lie in the socialization of its officers. By implication, such restraints relied upon two mechanisms: the subordination of the military hierarchy to civil authority; and the indoctrination of regular officers, at a military academy, to accept their place in this chain of command. Quezon, apparently mindful of Latin America's frequent military coups, set these barriers against a possible politicization of his army. Most importantly, he tried to mute the inherent threat of a standing army by indoctrinating its regular officers into a respect for civil supremacy. In effect, the Commonwealth made the PMA the country's ultimate defense against a future coup attempt.[15] As Quezon himself put it in his 1936 report on national defense: "The surroundings of the Academy, the everyday life of the cadet, and the strict military training to which he is subjected, aim to produce a mental attitude which will make him place duty to country over and above any consideration."[16]

Preoccupied with immediate issues of armaments and mobilization, Quezon and his legislature left the new academy's design to the military itself. In Article IV, Section 30, the National Defense Act stipulated, in language devoid of detail, that: "There shall be established a military training school to be named the Philippine Military Academy, for the training of selected candidates for permanent commission in the Regular Force."[17] Translation of these broad outlines into a working academy became the task of senior Filipino officers and their American advisors. In particular, Quezon's chief advisor Douglas MacArthur, a graduate and former superintendent of the U.S. Military Academy, demonstrated a great faith in the capacity of its curriculum to inoculate Filipino officers against the influence of the country's pervasive politics. Not only did MacArthur make his alma mater the model for the new academy, he endorsed two of its Filipino graduates to be its first superintendent and commandant. Indeed, throughout the prewar period, the PMA's senior officers were "selected invariably from among the Filipino graduates of the United States Military Academy."[18]

In taking West Point as their model, Filipino leaders and their American advisors were adopting a particular kind of military academy. In a comparative study of St.-Cyr, Sandhurst, and West Point, Correlli Barnett found that "the essential and common factor . . . is the indoctrination with tradition: potent emotional conditioning in military myth, habits, and attitudes. . . ." Significantly, "At all three academies there are songs, slang, customs and ceremonies that link each annual class together for the rest of their army life; and to a slightly lesser extent link all old graduates." Indeed, "this indoctrination, together with drill and discipline . . . has been of greater importance than the changing detail . . . of academic curriculum." Unlike St.-Cyr and Sandhurst, however, West Point remains "more Prussian than the Prussians"—that is, more persistent in using discipline to mold its cadets during the first or so-called "plebe" year.[19] In 1906, echoing G. Stanley Hall's influential work *Adolescence,* a West Point instructor articulated the academy's theory of molding young males:

> At the period of adolescence, when character is plastic and impulse wayward . . . control and constraint are the essential forces for impressing permanent form upon young manhood. If the material can be removed from contaminating impurities . . . and kept in its mould until it has set, the best has been done that education can do . . . , provided that the mould is a noble one.[20]

At the broadest level, the PMA became a total institution that would, like West Point, eradicate any trace of individualism and leave its institutional imprint upon every cadet. In 1938, the PMA yearbook described the Tactical department and its drill instructors as "a veritable forging shop in which the raw and crude materials are . . . purified of their undesirable qualities and turned out as finished products bearing the trade mark of the Academy." The aim of their constant drill and discipline was "the moulding of a well organized corps of officers in the highest state of efficiency, composed of thoroughly-trained, full-grown men." In their song "P.M.A. Forever," prewar cadets sang of their academy's capacity to make men:

> Within the walls of old and glorious P.M.A.
> They're molded to the real men that they should be—
> Men who can face the bitter realities of life
> With courage even in the midst of bloody strife.[21]

Beneath this broad design, the PMA also borrowed key aspects of the American academy, particularly its practices of hazing and honor. By the 1920s, West Point's distinctive institutional culture was well developed—the ideal of the officer as an educated gentleman, a unique honor system, and a brutal hazing. Complementing a curriculum that mixed the sciences, social sciences, and the humanities, West Point disciplined its cadets with drill, toughened them with sport, tested them by hazing, and refined them with drama, debating, and dancing.[22]

When applied at the PMA, the Filipino cadets recast these West Point traditions in ways that created an even stronger culture of male bonding. While the PMA's staff adopted West Point's cadet culture verbatim, its Filipino cadets reshaped these imported values through their own culture of masculinity, making plebe year and its hazing the academy's central rite of passage—from civilian to soldier, from plebe to cadet. The same West Point regimen that inculcated loyalty to an impersonal institution, the state, in America, also inspired in the Philippine context, a more exaggerated personal identification with a group of male classmates.[23]

Plebe year

In my interviews with surviving members of Class '40, I found their memories of the academy usually began with the entrance examination. Following the West Point system, the first batch of cadets for the PMA was selected by a national examination given to over six thousand applicants.[24] In a society where patronage was pervasive and obligations last a lifetime, the exams proved an effective exercise in meritocracy and social mobility. Drawn from every social strata and region of the archipelago, the class proved a diverse group. Although there were two graduate civil engineers and several Manila socialites in the class, most cadets were elevated from a broad, lower middle stratum of Filipino society.

Among the thirty graduates interviewed, seven were from poor households, fifteen from a broadly defined middle class, and only seven from families with major assets or high positions. Significantly, only two were "army brats," or children of military officers, and only one could trace any familial connection to the old Spanish colonial army. While their mothers stayed at home raising six or seven children, their fathers held jobs that ranged from senator to mechanic, photographer, or small farmer.[25]

All of Class '40 had graduated from high school, most were honor students, and some twenty-five had been valedictorian or salutatorian.[26] But these academic achievements had already been devalued by the flood of high school graduates, and most found advancement blocked by lack of funds for higher education. At the time they took the exams, many of the future cadets were unemployed, some were working students, and a few were getting by as laborers or lower echelon employees—road laborer (*caminero*), dock worker, or soldier.[27] Among the thirty interviewed, twenty-three cited the need for a free college education as their main reason for applying to the PMA.

As at West Point, the cadets of Class '40 spent their first year as "plebes"— untested neophytes subjected to endless, arbitrary harassment by upperclassmen. As in Arnold van Gennep's formulation of the "rites of passage," the plebe-year initiation proceeded through three distinct phases: a "separation" from both family and corps during the first three weeks of "beast barracks" when hazing was most intense; a transitional, or "liminal," period of qualified admission into the corps during the academic year; and, finally, a year-end ritual of full acceptance, or "reintegration," by upperclassmen called "recognition."[28] Through a constant series of reflexive command and response rituals, the plebes were, moreover, socialized into a chain of command where officers obeyed superior orders without hesitation. In the view of the upperclassmen, "the seniors . . . are the sovereign—all powerful and almighty"; while the plebes were "the lowest type of mammalia existing on earth."[29] Although the upperclassmen of Class '38 were aggressive, the academy's early demographics muted any tendency toward excess. In the transition from the small Philippine Constabulary Academy to the larger PMA, Class '38 had only 27 members to discipline the 120 plebes of Class '40—an imbalance that reduced the pressures of plebe year.

Instead of the year-long hazing of the old Philippine Constabulary Academy, the PMA adopted the West Point tradition of "beast barracks"—an initial month of intense harassment and hazing that ended with the start of classes.[30] "Beast barracks was a veritable hell barracks," the class historian wrote at the end of their plebe year. "With painful limbs and aching backs, we hated every day that came, some of us even cursed their fate."[31]

Chastened by the expulsions of staff and cadets from the old Constabulary Academy after a hazing scandal, the upperclassmen were cautious and officers were protective in their treatment of the incoming plebes. Consequently, upperclassmen disciplined them without physical abuse and were required to ask a plebe's permission to touch his person.[32] With a few memorable exceptions, the cadets of Class '38 were tough but not "sadistic." When the upperclassman Eugenio Lara caught one plebe, Ramon Alcaraz, masquerading as a cadet officer, he ordered hours of extra rifle drills—a punishment that the plebe himself felt was fair and even helpful.[33] As Lara explained: "I required of them more rifle drill and more exercises . . . to learn, not to haze, not to humiliate."[34]

There were strict limits. When Renato Barretto ('38) hit plebe Ramon Nosce ('40) with his sword "unintentionally or accidentally" during a double-time drill, the upperclassman was "busted" without mercy—stripped of his rank and punished by endless hours of marching back and forth.[35] Similarly, upperclassman Flor Acosta ('37) was rebuked with a forty-four-hour punishment tour for just touching plebe Pedro Yap ('40) to "stomach in" during formation and was still marching the parade ground during his graduation.[36] In that gap between the rules and the cadet code of silence, Class '40's experience of hazing was harsh enough to induce a solidarity from shared suffering. But it was not so harsh that it moved beyond testing to brutality, leaving lasting scars.

After nine months of humiliation, Class '40 were "recognized" as full members of the corps during graduation week of 1937. At the close of commencement, the band struck up "Auld Lang Syne." Then, the graduating class of 1937 separated themselves from the corps and stood at attention while the cadets passed in review before them under the command of the new first class. At the command "stack arms," the upperclassmen of Class '38 approached, one by one, to shake hands with the plebes they had harassed, criticized, and punished for the past nine months. It was an emotional ritual with "much weeping."[37] As Class '38 explained in their yearbook, "recognition is the . . . touchstone that transforms the uncomplaining Ducrot into an 'Immaculate' and effaces that social stigma that has brought him down to the level of 'hell cats.'"[38] Instead of "beast," "mammal," or "ducrot," upperclassmen would now address Class '40 respectfully as "cavalier" or "mistah"—a passage from animal, to subhuman, and, finally, human male.

Numbering nearly three hundred graduates, Class '40 and its underclassmen formed a substantial, even dominant, block of regular officers that defined the values of the postwar Philippine military. Rising from company-grade to general officer after the war, Class '40 infused the armed forces with a belief in civilian supremacy and military professionalism that militated against coups. During the threatened coups of 1953 and 1958, for example, their presence in the chain of command served, in subtle but significant ways, to restrain the inclination to power. Viewed in a regional context, the class was stepping down from the apex of command to modest pensions at a time when peers in Bangkok or Rangoon were clinging to power by force of arms. Through thirty years of military service, the Class of 1940 played upon bonds of loyalty among classmates to preserve their standards of military professionalism.[39]

Class of 1971

On 28 August 1987, Col. Gringo Honasan stood at the threshold of power. In the confusion that had followed Marcos's fall from power, people were despairing of democracy and looking for salvation from their country's endless political crisis. Just past midnight, the colonel, at the head of a column of two thousand heavily armed rebel soldiers, breached Manila's defenses and drove for Malacañang Palace at the city's center. Flying the banner of the Reform the Armed Forces Movement, his rebels attacked the palace gates at dawn. After a see-saw battle, they were finally repulsed by troops of the Presidential Security Guard. As they retreated through the old city's narrow streets, the RAM rebels were jeered by onlookers. In rage and humiliation, they fired randomly into the crowd of students with automatic weapons, killing eleven civilians and wounding fifty-four. At that moment, Colonel Honasan's bid to enter history by force of arms had passed.

In the broadest sense, RAM represented a revolt against the essence of military culture—subordination of officers within a chain of command and the further subordination of that military hierarchy to the civil state.[40] Its leaders were, in every sense, the Philippines' military elite. Not only were they regular officers and academy graduates, most were the outstanding members of their classes—regimental commanders, top athletes, and leading scholars. They were exemplary officers and natural leaders, men who, in the normal course of events, would have risen to the highest echelons of command. How then can we explain their careers marked instead by torture, terror, and coup d'état?

The well-springs of this politicization seem to have risen from the military's evolving relations with the postwar state as well as changes within the academy itself. Indeed, from its reopening in 1947, the PMA was a troubled institution. A form of harsh, physical hazing emerged in the immediate postwar years that tarnished the honor code and introduced an element of brutality into cadet life. Prewar PMA graduates had learned to respect civil supremacy from their indoctrination at the academy and their experiences as training officers for conscripts in a citizen army. By contrast, most of the future RAM leaders attended the academy during the late 1960s, a period of intense social ferment, and were influenced by the radical nationalist currents sweeping universities throughout the country. These ideals dominating university life in Manila percolated into the PMA, introducing, for the first time, politics as an element in cadet life. In the words of an alumni history of the PMA:

> By the second half of the 1960's the students had already decided to have a word in national affairs. . . . The corps of cadets at that time also showed the same sentiments, and at times, antagonistic fervor. The literary pieces written during this period possessed a political tone; it even sideswiped the administration. The clamor of change . . . was also emanating from within the cadets themselves.[41]

In a society that is often organized around personal ties, we need to focus on key personalities to grasp the import of this broader change. In their last months at the academy, Class '71's experience was influenced by two extraordinary personalities—their new instructor in politics, Lt. Victor Corpus, and their class president, Cadet Gregorio Honasan II, known by his nickname "Gringo." The alliance of these two at the academy had a strong influence upon the class, just as their break fifteen years later had an impact upon the armed forces.

In the mid-1960s, the PMA moderated its traditional engineering program by introducing courses in the humanities and social sciences taught by "the discussion method."[42] Reflecting the nationalist passions of the 1960s, several of the new instructors appointed in these fields were activists who introduced, at least informally, a strain of radicalism into the academy curriculum that later influenced the Class of 1971. Among the first and most influential of these instructors was Maj. Dante Simbulan. Although a member of PMA's Class of 1952, Major Simbulan was an acutely political scholar. His 1965 doctoral dissertation at the Australian National University was a sharp critique of the Philippines' entrenched elite. Through a survey of the postwar period, Simbulan concluded that the country was ruled by a hereditary *principalia* class of only 169 families that monopolized both economic and political power.[43] In the heady atmosphere of open discussions that allowed cadets to speak freely for the first time, Simbulan developed a close relationship with one of his students, Victor Corpus, exposing the cadet to radical politics from a nationalist perspective. The two developed a close bond, and Corpus maintained contact with Simbulan after the major resigned his commission and became a professor at the Philippine College of Commerce.[44] As PMA boxing champion and winner of the Athletic Saber at graduation in 1967, Corpus himself had a quiet charisma that marked him for future command.[45] He was, as it turned out, something more than a conventional military careerist.

After graduation from the PMA, Lieutenant Corpus joined *Kabataan Makabayan*, the Communist party's youth group, and was strongly influenced by its leader, Jose Ma. Sison. In May 1969, Sison went underground to establish the New People's Army (NPA) and launch a Maoist rural revolution. In early 1970, after three years' service in the Constabulary, Lieutenant Corpus returned to the academy as an instructor in political science, bringing with him his nationalist concerns.[46] In the first semester of that academic year, Class '71 took Government 411 with Corpus and found him "a committed and influential teacher who encouraged free thinking."[47] Some members of Class '71 later recalled that he "junked the curriculum and substituted . . . questions on how to stage a *coup d'état*."[48] Indeed, his supervisor in the political science program confirmed that Corpus openly advocated radical change, debating fellow faculty over whether it should come through "constitutional process, revolution or *coup d'état*."[49]

Outside the classroom, the young lieutenant and his cadets shared a sense of moral outrage over signs of petty corruption at the academy. The leaders of Class '71 were angered by what they later called, in their yearbook, the "despicable attempt by some persons to take advantage of the good name . . . of the Philippine Military Academy by . . . smuggling stereos and other articles in the guise of donations."[50] After witnessing abuses during his service with the Constabulary, Corpus was now disturbed to find officers grafting from the PMA's mess hall budget—experiences that made him doubt the military's ability to live up to its ideals.[51] In a clear breach of regulations, Corpus began meeting outside of class with the leaders of Class '71, including First Captain Gringo Honasan.

"Lt. Victor Corpus was my instructor in political science," Honasan recalled fifteen years later. "He was, in fact, our class advisor. Most of the radical thinking of our class was his influence."[52] With their common family background, the two evidently formed a close bond. Corpus was the nephew of a career officer, Col. Edmundo Navarro, a member of the PMA's Class '40; while the cadet's father, Col. Romeo Honasan, was a member of Class '43. Both, moreover, had experienced the heady eruption of activism in Manila's universities in the late 1960s—Corpus as a young officer and Honasan as a student for two years at the country's most radical campus, the University of the Philippines. Their conversations grew into plans for a "mass walkout" by the entire Corps of Cadets which the two believed, incorrectly, would be "a first in Academy history."[53]

In December 1970, only three months before Class '71 was due to graduate, Lieutenant Corpus and another instructor, Lt. Crispin Tagamolila, led twenty-nine Communist comrades in a spectacular raid on the PMA armory. After seizing hundreds of infantry weapons, Corpus disappeared into the Communist underground where he was reportedly training the NPA's fledgling guerrilla army.[54] His defection brought strict security to the academy that blocked Honasan's protest demonstration and cast a cloud over Class '71's graduation. Despite the furor, the instructor's influence over the cadets remained. In 1985, the leaders of Class '71 would launch the RAM movement with their famous "We Belong" demonstration at the academy's alumni parade—a protest march strikingly similar to the one that Lieutenant Corpus had planned with Cadet Honasan fifteen years before. And, in his later coup plots, Honasan seems to have stripped his mentor's tactics of their left ideology and applied them to establishing a military dictatorship.[55]

Like Victor Corpus, Cadet Gregorio "Gringo" Honasan II was an individual who, by force of personality, would leave a lasting imprint upon the armed forces. Even as a cadet, Honasan showed a unique capacity to project himself as the warrior ideal, becoming an absolutely dominant leader among his classmates. In striking contrast to Class '40, which had an informal leadership echelon, Honasan soon emerged as the single, preeminent leader of Class '71.

As the son of a career officer, a onetime candidate for the Catholic priesthood, and fitness devotee, Honasan arrived at the PMA with an appealing mix of idealism and athleticism. Every year for four years, his classmates elected him president of the class organization and, in his senior year, the tactical officers selected him as Cadet First Captain. Honasan also exercised a formal moral authority as ex-officio chairman of the powerful Honor Committee. Indicative of his powers of persuasion, he twice captured the academy's title as "best debater." Not only did he win varsity letters in five sports ranging from boxing to basketball, he earned the grade of "sharpshooter" with the M-1 rifle and won a gold medal in the 1970 National Wrestling Tournament. His biography in the class yearbook, *The 1971 Sword,* celebrates these achievements in an almost reverential tone: "Gringo!—Comfortably warm and constantly dynamic; enough determination and perseverance in achieving his goals; coupled with characteristic humility and compassion for the less fortunate."[56]

The class yearbook provides some clues that five members of Class '71 who would later lead the RAM had, during their academy days, formed a recognized leadership clique around Honasan. Among the six officers the class elected in its senior year, three were future leaders of the coup group—class president Honasan, vice-president Oscar Legaspi, and class historian Victor Batac. Among the dozen snapshots that illustrate the title page for their "Class History" in the yearbook, one shows twenty-eight classmates posing informally on the gangplank of a navy ship in what seems a casual jumble of smiles, stares, and sunglasses. Yet even in this fleeting moment of spontaneous, high spirits, the class arrays itself around three cadets who stand at a central, dominant position—Honasan center, Victor Batac right, and Eduardo "Red" Kapunan left.

Hazing at the postwar PMA

The germ of campus politics had a long incubation, but hazing's effect on Class '71 was more immediate. Following the West Point tradition, the PMA had, from its inception, adopted the practice of socializing entering cadets by putting them through several months of double-time drill under the close, harsh supervision of upper-classmen.

When the academy reopened in May 1947, the armed forces paid considerable attention to its careful reconstruction. For reasons that would have been impossible to predict, the returning cadets were extraordinarily brutal in hazing the plebes of Class '51. "There were sadistic hazing," explained Class '40's Deogracias Caballero. "Class '44 and '45 . . . had only one year and a few months service in the Academy and they . . . had not tried being upperclassmen for long. All they could remember was the indignities that they suffered as plebes and they took it out on the plebes." Indeed, Caballero recalled that these acting upperclassmen tried to "electrocute cadets at the tongue," made plebes take cold showers "every hour," and used their bodies as "punching bags."[57]

As new underclassmen entered the PMA for the next three years, Class '51 evidently discharged their own degradation by brutally hazing incoming plebes— transforming the "fourth class custom" into simple brutality. During their last "beast barracks" in 1950, Class '51 supervised a hazing that was particularly harsh. In one of several egregious incidents, upperclassmen systematically beat plebe Jose Arias so badly that he required medical treatment for rib injuries and internal hemorrhaging. "I had been accustomed, by then, to the socking and slapping and beating," Arias recalled in a public statement six years later:

> Receiving thrusts in the abdominal area with the butts of rifles was also something we were getting used to. But then, there was a group of us plebes who were getting the starvation treatment during which we were not allowed to take our meals. . . . One day, I must have been really starved that I dared to grab a piece of bread when I thought no upperclassman was looking. I was wrong . . . I was again subjected to rifle thrusts in the abdomen. On my way out, I knew that I was going to faint so I rushed to the infirmary where I lost consciousness.

When Cadet Arias finally gave the name of the upperclassman who had beaten him, the superintendent dismissed the guilty cadet. Even though he had simply complied with an order, Class '51 retaliated against Arias and, using the honor code sanction intended to punish cheating, ordered the "silent treatment." For months nobody, not even his classmates would speak to him, not even a word. When Arias reacted to another beating by threatening an upperclassman with his rifle, the academy gave the plebe a medical discharge. Six years after his hazing, Arias, sick and jobless, was still suffering periodic nervous breakdowns.[58]

This was not an isolated incident. With each year, the brutality grew to become something of a cancer metastasizing within the corps. Just as Class '45 hazed Class '51, and Class '51 in turn vented its anger on Class '52, so each batch of brutalized plebes became yearlings who immediately tried to recover from their own, very recent humiliations by degrading the new class of incoming plebes. Between 1950 and 1961, the academy investigated forty-eight cases of brutal hazing and dismissed forty-eight cadets, an average of four cadets a year.[59] Indeed, these dismissals represented nearly eight percent of all graduates during this eleven-year period, and were equivalent to an entire graduating class. Outside authorities tried to intervene, but with at best temporary success.[60]

By the time the plebes of Class '71 entered the academy in March 1967, the *Manila Times* reported that hazing was "no longer practiced." Reflecting the concerns of the PMA's new superintendent, Gen. Reynaldo Mendoza, one of the last members of Class '40 still on active duty, the newspaper reported that: "Hazing is an ugly act. It is bestial. It is never tolerated by Academy authorities." Indeed, when he took command of the academy in 1966, Mendoza found that a mass dismissal of cadets for hazing the year before had produced a marked "lull" in the practice. Pressured to readmit these cadets, Mendoza refused and threatened to resign if they were, thereby sending a clear signal to the PMA's staff and students. Thus, the incoming cadets of Class '71 would face what the *Manila Times* called a "stern or serious reception by upperclassmen" who would discipline them through the "fourth class system" on the dictum that "no one can command who has not learned to obey." Through a ritual testing, without any physical brutality, the system was supposed to serve "as an instrument that weeds out from the individual habits and personal attributes that are out of place in the military profession, and places in their stead such qualities as willing obedience to authority, loyalty, self control."[61] When he stepped down from command in August 1967, Superintendent Mendoza gave the corps a valedictory celebrating the PMA's fourth class custom as an instance of "true Democratic opportunities" which insured that "the laborer's brother competes with the senator's heir, the corporal's son gives orders to a general's favorite boy."[62]

Although there is no evidence of physical hazing, the plebes of Class '71 still found their first six weeks at the academy, confined in "beast barracks," a harsh "rite of passage." In the heady days after the 1986 "people's power" uprising when the RAM were still heroes, their leader Eduardo "Red" Kapunan recalled his first traumatic day at the PMA:

> We got down from the bus. We lined up in front of the grandstand. No cadets in sight. Then all of a sudden, martial music. From behind us came cadets, all in dress-white, then march. We said, "the cadets look beautiful." Then, they form in front of us, as if they are going to eat people. A shout: "Charge!" Suddenly: about face everyone. There were those among us who wanted to run back to the bus or bury their heads in the ground. . . . "Mr. Kapunan, where are thou? Double time!" We did double time. "Raise those knees! Push up!" We raised our knees and did push ups. It was real shock treatment! They issue mattress, shoes, briefs. While we were in line, we were being asked to do so many things, all at the same time. Chin up, stomach in, bend down, push up, double time, while you are fixing the duffle bag. Passed by the barber shop, shave all of our heads at the barber shop. You are carrying everything and your duffle bag, running and running. The rest is really hazy for me. I was just following orders. I didn't know what I was doing.[63]

After the first six weeks in "beast barracks," the plebes of Class '71, like those before them, began to develop a sense of class identity and a feeling of incorporation into the corps. "The uppies were very 'helpful' too," the class wrote in their yearbook, "They had to shape us into worthy cadets within six weeks. We think they did not fail. Anyway, Incorporation Day inevitably had to come on the 12th of May. 'Let the fourth classmen join the Corps.' . . . How did they ever rig up such sweet words! Dress coats almost burst with pride." These feelings intensified at the end of plebe year when the class won full acceptance with the ritual of recognition. "Soon March Week was with us," their yearbook narrative continues. "March 24—Recognition Day—each handshake, each tear had a special meaning. What was left was gratitude and understanding for those who shaped us into what we were then . . . a full-blooded yearling."[64] Thus, like those before them, Class '71 went through a ritual initiation that broke down their individual identities and then rebuilt them with a shared identity as full members of a class and the corps.

But this lull in physical brutality proved short lived. When Class '71 reached their third year, hazing revived and the academy was again forced to adopt strict measures to reduce the abuse. As new plebes arrived in April 1969, the superintendent ordered close supervision to assure a "mild" reception and assigned each plebe to an upperclass "guardian."[65] Despite these strict rules, hazing soon reached an almost uncontrolled frenzy of physical abuse. On 3 April, plebe Miguel Arucan, an athlete who had arrived at the academy in good health, was admitted to the PMA hospital, weak and bleeding, after only three days of "beast barracks." Ten days later, he died.[66] Although the academy's initial investigation ruled out hazing and listed his cause of death as "mysterious," the cadet's father insisted that Arucan had been in perfect health.[67] Four days after his burial, the *Manila Times* published a letter from another hospitalized plebe to his mother complaining that hazing had become "more sadistic" since the death. The anonymous plebe insisted that "Arucan died of over-exhaustion due to hazing," and reported that many plebes had been hospitalized, most with "broken ribs and stomach trouble."[68] At the later court martial, plebes testified about the ritual beatings of 3 April that led to Arucan's death: at dawn they were wakened by blows; they left their rooms through a gauntlet of upperclassmen who pummeled their stomachs; and were later ordered to squat on their desks while upperclassmen beat them bloody about the face and torso. After blows to the head, Arucan began bleeding profusely from the nose and was taken to the hospital. There he later died.[69]

A year later, on the eve of Class '71's graduation, one member of the class, Cadet Ruben Cabagnot, was found guilty for his role in Arucan's death and was punished by a six-month suspension of his commission.[70] An American scholar resident at the PMA in 1970–71 found the culture of hazing to be so pervasive that even the academy's determined attempts at prohibition could not diminish cadet enthusiasm for the practice.[71] Hazing's prevalence throughout these troubles raises questions about Cadet First Captain Gringo Honasan's leadership.

More broadly, this persistent abuse meant that Class '71 had experienced brutality and brutalization that could serve as an emotional gateway to the practice of torture. When one AFP regular officer was investigated for human rights abuses during the Marcos era, he reportedly replied: "But sir, I did no more to this NPA suspect than had been done to me as a plebe."[72] Similarly, a student activist from the University of the Philippines who was arrested in 1978 found that his torturers were PMA cadets who had been suspended from the academy for hazing and were now using skills learned from brutalizing the plebes on him and his fellow prisoners. Their "most characteristic torture" was a direct copy of a notorious PMA hazing practice called "the bridge." While the prisoner was suspended, head and feet, between two beds, the PMA cadets, or AFP officers, beat hard upon the back and belly, producing remarkably intense suffering.[73]

Young lieutenants

Graduating only months before Marcos declared martial law in September 1972, the Classes of 1971 and 1972, the heart of the future RAM, were commissioned into a military with a mission of repression. As the catalyst in a series of coups d'état in the late 1980s, the PMA's Class of 1971 is an apt case study of a breakdown in military socialization. Indeed, about 15 percent of the Class of 1971 participated in the December 1989 coup that came close to seizing state power.[74]

Within eighteen months of their graduation, Class '71 became the instruments of Marcos's martial law regime. Over the next fifteen years, the class and its successors arrested and tortured civilians who were their social and intellectual superiors. As the defenders of dictatorship, they learned to regard their own society as the enemy, and to use espionage, surveillance, arrest, and torture against people who, in normal circumstances, would have been their political superiors. Rather than defending their country on the field of battle, these young officers violated life-long sanctions, familial and military, by torturing helpless victims drawn, in many cases, from their society's leading families.

The Mindanao campaign both hardened and bonded many of these young officers. In the months after the declaration of martial law, Muslims in western Mindanao launched a major secessionist rebellion, seeking to establish an independent Islamic state. Faced with a serious threat to the country's integrity, Marcos expanded the military and concentrated his forces in the south for a five-year campaign. In this brutal religious war, many recent PMA graduates served as combat commanders and were hardened by the experience. For three years after graduation in 1971, Lieutenant Honasan served in Mindanao where he earned a reputation as a fearless, sometimes ruthless, fighter, winning at least one Gold Cross medal for "heroism."[75] According to an officer who served on the military court, Honasan "participated in the revenge killing of a Muslim leader, who allegedly killed his classmate." For reasons that are not clear, Honasan was found innocent in the initial hearing. When later evidence established his guilt, Honasan "was able to evade an inquiry . . . because of his links to the defense secretary."[76] Shared service in this bloody war also strengthened bonds among the young officers. In 1973 when Honasan was wounded in heavy fighting, Lt. Eduardo "Red" Kapunan defied his superiors and helicoptered in under enemy fire to rescue his classmate, making the two "bosom friends."[77] Years later Honasan would say of Red: "We have developed a special bond that is only derived from taking risks together, sticking your neck out for each other, and our long association since Academy days. We are even closer probably than brothers."[78] After combat service in this dirty war, many officers rotated to Manila for staff assignments that often involved them in torture and interrogation.

Such work armed the RAM leaders for future political warfare. Through their extraordinary duties in the service of dictatorship, they acquired special skills and a quality of arrogance that would later inspire their attack on the state. Through intelligence work, some young officers learned media manipulation. Lt. Red Kapunan, for example, spent a year in the early 1970s posing successfully as a Kyodo News Agency correspondent to spy upon foreign and domestic reporters. As Defense Minister Juan Ponce Enrile's aide for disinformation during the 1978 parliamentary elections, Navy Capt. Rex Robles manufactured black propaganda against ex-Senator Benigno "Ninoy" Aquino, charging, for example, that he was responsible for the murder of a union organizer on the family's plantation. These exceptional experiences gave the RAM officers a sense of superiority that underlay their later attempts to topple the state. For example, among the 108 graduates of Class '71, 15 would later be involved in two or more coups, 6 were murderers, and 5 had practiced torture.[79]

Ethos of torture

After Marcos imposed his martial-law dictatorship in 1972, torture became an instrument of state power. In 1981, Amnesty International reported that "members of the Armed Forces of the Philippines had been responsible for acts of unusual brutality for which they were not held accountable."[80] Amnesty discovered that most victims were confined in a Philippine gulag of "'safe houses' and military barracks not designed for holding prisoners."[81] Almost all political prisoners who passed through this military gulag were tortured. In 1992, the Medical Action Group, an association of Filipino physicians, found that 102 of the 120 political prisoners held at the National Penitentiary since 1972 had been severely tortured—48 subjected to Russian roulette, 19 electrocuted, 17 "fed feces," 5 sexually abused, and 4 buried alive. Of these 102 victims, 91 were tortured in a military detention center before being imprisoned.[82] After Marcos fell from power in 1986, an organization of former detainees, SELDA, estimated that, of the 70,000 political prisoners incarcerated by his regime, "35,000 . . . suffered some form of torture."[83] If this latter estimate in any way approaches reality, then a significant element of the Philippine military was directly involved in torture.

Why such systematic and brutal torture? Lacking the resources for a comprehensive machinery of repression, Marcos's martial-law dictatorship used torture in a symbolic way to terrorize the Filipino people into submission. An individual was picked up mysteriously and later a mangled corpse was dumped, or left "salvaged," along a roadside—with a clear message of terror for the wider society.[84] Significantly, only 737 Filipinos "disappeared" between 1975 and 1985, but nearly five times that number, some 2,520, were victims of this public display called "salvaging."[85]

In Marcos's theater of terror, military officers were not cogs in a machinery of control, conducting torture as dispassionate technicians. They were actors who personified the violent capacities of the state. If Marcos wrote a hidden script of terror for his "new society," then these young officers, the future RAM leaders included, were his players. When the president ordered the military to enforce his authoritarian rule, it was its lowest echelons, lieutenants and captains, who became his instrument of actual implementation. In the transcripts of the torture victims, it is lieutenants, not colonels or generals, who appear time and again as the actual torturers. Instead of the dull routine of garrison duty, many members of Class '71 were, within a few years after graduation, assigned to arrest, interrogate, and, ultimately, torture civilian dissidents. As they rose through the ranks during the decade, they took command of the regime's "safe houses" and prison camps.

There is strong evidence that most of the key RAM leaders were involved in regular torture interrogations under the Marcos regime. Indeed, torture victims or their advocates have named almost all the key RAM leaders—Gringo Honasan, Red Kapunan, Vic Batac, Hernani Figueroa, Rudy Aguinaldo, and Tiburcio Fusilero. Reading over these statements, it seems that the future RAM leaders did not operate as impersonal technicians, attaching electrodes and scaling the voltage calmly upward to extract information. They generally engaged in perverse, often sexual, tortures that involved an effort to establish psychological dominance over their victims.

Reviewing the records of human rights violations from the Marcos era reveals, first, that the RAM leaders were among the most abusive; and second, that there was a particularly psychopathic quality to their tortures. As detailed by their victims, the RAM boys often used the torture chamber to play games of domination, engaging in sexual torture of both men and women. Some, like Lt. Rodolfo "Rudy" Aguinaldo, embraced the experience, reveling in the role of the all-powerful who holds suffering and death in his hands. Other officers, such as Lt. Col. Hernani Figueroa, played the omnipotent by sending detainees off for torture by minions when they resisted his will. Moreover, their tortures had an explicit theatrical quality consonant with larger aims of the regime.

Political lessons of torture

Since these torture sessions form a large part of the RAM leaders' early military service, such experience may have shaped their belief in the political efficacy of violence. As torturers who subjugated strong-willed political activists through limitless violence, the RAM officers may have emerged from the "safe houses" feeling like supermen capable of seizing state power. Judging from the conduct of their later coup d'état, the experience of torture also gave them a false, inflated sense of the efficacy of violence.

As lead actors in the theater of terror, the RAM leaders discarded the military habit of subordination. Instead of the self-effacing manner of the soldier, the RAM leaders developed flamboyant, individual styles designed to attract media attention outside the AFP and a loyal following within. Instead of standard-issue weapons, the Defense Ministry's Security Office took on the air of "Q's" laboratory in a James Bond film. When I visited Colonel Honasan's security office in July 1986, his officers were toying with crossbows, Israeli assault rifles, and machine pistols. Instead of the standard AFP dress uniform for headquarters duty, the RAM boys marched about the Defense Ministry in jungle camouflage outfits with quick-draw holsters holding exotic weapons. In place of short military haircuts, the RAM boys grew flamboyant manes, beards, and mustaches. It was as if they had erupted out of cadet uniforms with their statement of constrained, controlled power, into a costume of lethal masculinity, with a volatile capacity for destruction and untrammeled individual will.

As the RAM's style leader, Col. Gringo Honasan cultivated an image of threat. A tank of a half-dozen piranhas greeted visitors to his office. Plastered to his office door was his personal statement: "My Wife Yes, My Dog Maybe, But My Gun Never." He entertained special visitors with a jar of dried ears slashed from the corpses of Muslim rebels in Mindanao. He encouraged journalists to write about exploits that communicated a sense of recklessness and threat—his habit of sky diving with pet python "Tiffany" around his neck; quick-draw shooting practice; and black-belt competition in both karate and Filipino *arnis* cane fighting.[86] His hair was a leonine mane, his uniform custom-tailored jungle fatigues, and his military name-patch read "GRINGO." Through charm and charisma, Colonel Honasan's aura of violence seduced other officers, making this sexual aura of violent *machismo* the norm for RAM's members.

Instead of rejecting their experiences in the "safe houses," RAM invested violence with a romantic power. On the first day of the EDSA revolution, Lt. Col. Red Kapunan told journalist Shelia Coronel the story of "combing the country in search of a hired assassin reportedly out to kill the defense minister. Kapunan found the assassin and later had him killed."[87] In the euphoria that followed the February revolution, Manila's women journalists hosted a party for eight of the reformist officers and were surprised when "the entire RAM showed up with bodyguards." In this informal, intimate atmosphere, Lt. Col. Tiburcio Fusilero (PMA '71), speaking in a calm voice, tried to impress three of the women reporters with his exploits. "After a few drinks, Fusilero was not that drunk," recalled Coronel. "He started talking about what a good officer he was and said he was given a list of 40 people to kill and only two got away. I did not ask questions. It was too frightening—armed men surrounding my house."[88] Another journalist present, Jo-Ann Maglipon, published a slightly disguised version of that encounter. "In a private moment," she begins, "reformist T.F. admits typing out a two-page hit list of Marcos enemies upon the declaration of martial law in 1972. He noted, with neither pride nor regret, that in that two-page list, only one got away; his were professional and clean hit jobs." When Maglipon asked "T.F." if he would kill again, Fusilero replied coolly, "Yes, it is my duty to obey."[89]

Coup against Cory

During the late 1980s, the Philippines suffered more coup attempts than any other country in the world. For nearly five years, coup threats dominated its politics, blocking major social reforms and slowing its economic recovery. All of these serious coups were led by RAM officers. At first allied with President Marcos's former defense minister, Juan Ponce Enrile, these colonels were his fist in a grab for power. After he failed in two coup attempts and was forced from office, the RAM leaders built their own following within the military and gradually mobilized a disaffected minority for ongoing revolt against civil authority.

Through these many attempted coups, the Class of '71 became the main force in a military revolt against the Aquino administration. Not only did these coups define the class, but the class itself defined the role of the coup d'état in Philippine politics. Official statistics compiled by the Davide Commission indicate that Class '71 accounted for ten among the seventy-seven officers involved in two or more coups, by far the highest for any single class. By contrast, Class '81 with the second highest total had only three officers involved in two or more coups. This data reveals two significant trends: first, it was PMA regulars, not reservists or integrees, who commanded coups against President Aquino; and, of equal importance, within this larger group of regular officers a single class, 1971, was clearly dominant.[90]

Knowing that so many members of Class '71 led so many coups leads us to ask why? What happened in the collective experience of this single academy class that attracted so many of its members to rebel against military socialization and the social order? Although such conclusions must remain speculative, it appears that the future RAM leaders may have gained a sense of the malleability of their society from their extraordinary service to dictatorship. Speculating further, it seems that torture may have become a liberating experience, freeing these officers from PMA's indoctrination into subordination. Breaking society's leaders so easily and so thoroughly in torture may have somehow led these young officers to believe that they could, through force and with impunity, impose their will upon the wider society.

In the months that followed Marcos's downfall, many RAM officers complained of Enrile's eclipse within the Aquino cabinet. Although the coauthors of the revolution, they were still remote from power. In the early euphoria surrounding President Aquino, they did not dare launch an immediate coup and could only position themselves for an opening. Rather than simply wait for circumstances to change, Enrile challenged the legality of Aquino's mandate and began to agitate for new elections. Similarly, the RAM leaders used their media presence to mount a subtle political propaganda campaign, questioning Aquino's mandate and arguing that Enrile should have a share of power. "Let's make something clear," Honasan told *Veritas* magazine in September 1986. "The military will never be able to determine, now or in the future, if our commander-in-chief . . . won the election." Echoing Enrile's call for new elections, Honasan added: "If support for Cory is as strong as her advisers say it is, then she should be willing to call election, why, even this afternoon."[91] Since power in a revolutionary regime springs from a share of its creation, rather than any electoral mandate, the colonels were determined to project themselves as the real heroes of EDSA. Their message was simple: Honasan and RAM were the authors of Marcos's downfall, therefore their patron Enrile should have the lion's share of power.

Although civilians were slow in seeing RAM's self-interest, senior officers saw the ambition behind their protestations of idealism. Asked in a 1990 survey to explain the causes of these many coups, pro-government officers cited RAM's "obsession with the power which they thought they had won but handed over to Corazon C. Aquino in February 1986."[92]

Reflections

In 1986, on the 50th anniversary of their admission to the academy, Class '40 began a four-year cycle of reunions that coincided, quite precisely, with six attempted coups d'état led by a younger generation of fellow PMA alumni. At this time of celebration, Class '40 were confronted with a military that mocked their principles of apolitical service. This jarring juxtaposition of reunion and revolt forced the class to reflect.

Class '40 did not shirk self-congratulation at their reunion. "What may be considered a most significant achievement," they wrote in their *Golden Book,* "is the fact that after 30 years of service . . . no member of the batch was ever tainted with the breath of scandal; no charges of unexplained or ill-gotten wealth; no charges of human rights violations . . . were placed against them."[93] In effect, they told their alma mater, in the words of the PMA song, that the "honor you instill" still "doth guide our will." Within the more relative standards they had adopted under the pressures of duty, the class considered themselves men of honor.

The class was sharply critical as it turned to those "solid ranks of gray" that had followed them out of the academy. The Class of 1940 found the 1987 coup, particularly the PMA's participation, abhorrent. In speeches and letters, the class "vehemently denounced the mutinous acts of the coup plotters."[94] At a reunion in October 1987, the class heard reports that "no cadet officer has been busted . . . and no dismissals appear to be contemplated." Since these cadets had taken up arms against the state, the class urged, in a collective letter to Chief of Staff Fidel Ramos, that "sterner measures should be undertaken."[95] They suggested that all cadet classes tainted by the rebels' ideology be wholly or partially purged from the academy to cleanse this fundamental corruption of its core values.

At their 50th reunion in 1990, Commodore Ramon Alcaraz, a classmate active in the anti-Marcos opposition, delivered the homily before the Corps of Cadets in the PMA chapel. The speech celebrated his class, condemned the corruption of its values, and called upon the cadets for renewal. Giving a certain poignancy to the occasion, the entire Class of 1990, seated in the pews on the eve of their graduation, had taken up arms in Colonel Honasan's coup three years before.

> Let us return to the basics. The bottom line is that PMA is supposed to produce officers and gentlemen. . . . It is toward this end that all cadets, without exception, are pledged to abide by the honor code. . . .
>
> I consider this year's Class of 1990 still fortunate to be graduating tomorrow as scheduled. You were yearlings during that bloody August 1987 coup attempt led by Gregorio Honasan whose father, Romeo, was my underclassman; I wonder what this father might have said about his son if he were alive today? For your questionable behavior during that treasonous coup, all of you should have been summarily dismissed. . . .
>
> To our idealistic officers, we challenge you to regain the honor and glory of the Old Army. You can do it. Our customs and traditions are still with us. Only buried by the after effects of martial law. . . .[96]

Class '40 seemed to conclude that friendships and values from "kaydet days" had been the main bulwark of their service as honorable officers. Their assessment was sometimes uncritical, and they were often ready to forgive a classmate's breach of conduct, whether of omission or commission, to preserve their unity. They were proud to have two classmates who had become chief of staff, but seemed incapable of distinguishing between one's opposition to Marcos's illegal maneuvers and the other's compliance. Honored collectively and affirmed individually by Marcos's choice of their classmates as chief of staff, the group was willing to overlook Gen. Segundo Velasco's capitulation to Marcos's politicization of the AFP in the late 1960s.

Instead of criticizing Velasco for his failings as chief of staff, Class '40 celebrated him. On the final page of the *Golden Book,* the class wrote: "General 'Gunding' Velasco, after he retired as Chief of Staff, was always very proud to state publicly . . . that during his tour, not a single member of the Class . . . ever approached him to ask for any favor that could be misconstrued as 'taking undue advantage' of class relationship."[97] At the apex of the military hierarchy, however, choices were complex and ethics were nuanced in ways that eluded the absolutes of the honor code. In terms of the PMA's standard for cadet honor in examinations, General Velasco may not have cheated as chief of staff, but he did fail to report fellow officers who had. The class could not admit Velasco's responsibility for trends within the military that they abhorred. And even if they had admitted his dereliction, they still might have found a way to exculpate a beloved classmate. In some obvious and significant ways, Class '40 was thus held together by many of the same blind bonds of loyalty that led later classes, particularly Class '71 into Colonel Honasan's coup.

At a more subtle level, however, there are still important distinctions between the two. Class '40 had somehow integrated professional values into these personal ties in ways that Class '71 had not. Although the Class of 1940 had moved far beyond the academy's all-male formations into family life and individual careers, their sense of solidarity inspired integrity. As cadets, they had learned, through invention and indoctrination, a military masculinity that balanced subordination and self-assertion, group loyalty and individual integrity. The class itself may have dissolved into memory in the decades since graduation, but its ideals and loyalties had intertwined in ways that helped to restrain an entire generation of military leaders from corruption and coups. Despite many individual failings, Class '40, bound together by code and camaraderie, had maintained a reasonable standard of professionalism throughout their thirty years of active duty.

As it turned out, these slender threads of class, corps, and honor were the warp of the Philippine Republic's constitutional fabric. Wispy though they might have been, these ties and shared ideals made the Philippine military, for over a quarter century, one of the few in Southeast Asia that had not attempted a coup d'état. Once this fabric began to fray, regular officers could turn their guns on the people to become instruments of a martial-law dictatorship or commanders of a coup d'état.

Notes

1. Interview with Gen. Reynaldo Mendoza (Ret.), Camp Aguinaldo, Philippines, 13 Aug. 1996; Maj. Rogelio S. Lumabas, "Whither the MAP," *Cavalier* 7 (November–December 1967): 5–6.

2. Kenneth W. Kemp and Charles Hudlin, "Civil Supremacy over the Military: Its Nature and Limits," *Armed Forces & Society* 19 (Fall 1992): 8.

3. Claude E. Welch, Jr. and Arthur K. Smith, *Military Role and Rule: Perspectives on Civil-Military Relations* (North Scituate, Mass.: Duxbury Press, 1974), ix.

4. Eric A. Nordlinger, *Soldiers in Politics: Military Coups and Governments* (Englewood Cliffs, N.J.: Prentice-Hall, 1977), 6.

5. Republic of the Philippines, *The Final Report of the Fact-Finding Commission (pursuant to R.A. No. 6832)* (Manila: Bookmark, 1990), 97.

6. Although the literature on civil-military relations contains many insights, there is no discussion of this essential point. For some examples, see Claude E. Welch, Jr., "Civilian Control of the Military: Myth and Reality," in Claude E. Welch, Jr., ed., *Civilian Control of the Military: Theory and Cases from Developing Countries* (Albany: State University of New York Press, 1976); Stephen D. Wesbrook, "Sociopolitical Training in the Military: A Framework for Analysis," in Morris Janowitz and Stephen D. Wesbrook, eds., *The Political Education of Soldiers* (Beverly Hills, Calif.: Sage Publications, 1983); Col. Robert N. Ginsburgh, "The Challenge to Military Professionalism," *Foreign Affairs* 42 (January 1964); Samuel P. Huntington, *The Soldier and the State: The Theory and Politics of Civil-Military Relations* (Cambridge, Mass.: Harvard University Press, 1967); Morris Janowitz, *The Professional Soldier: A Social and Political Portrait* (Glencoe, Ill.: Free Press, 1960); and Sheldon W. Simon, ed., *The Military and Security in the Third World: Domestic and International Impacts* (Boulder, Colo.: Westview, 1978).

7. S. E. Finer, *The Man on Horseback: The Role of the Military in Politics* (London: Pall Mall Press, 1962), 5–6.

8. Jose M. Mendoza, ed., *Batch '36 Golden Book* (Manila: PMA Class '40 Association Inc., 1986), 16. Aside from extended interviews with thirty of the seventy-nine members of Class '40, the most important single source for this history of the Class of 1940 is the *Batch '36 Golden Book*, a privately printed volume of 358 oversized pages crowded with text, tables, and photographs.

9. Vicente Lim, *To Inspire and To Lead: The Letters of Vicente Lim, 1938–1942* (Manila: privately printed, 1980), 99–100.

10. Florencio F. Magsino and Rogelio S. Lumabas, *Men of PMA: Volume I* (Manila: Authors, 1978), 69–70.

11. Robert H. Ferrell, *The Eisenhower Diaries* (New York: Norton, 1981), 8–10. For contemporary reports of the plan, see *New York Times*, 20 Nov. 1934, 23; 25 Nov. 1935, 1; 30 May 1936, 1; and 20 June 1936, 6.

12. *National Defense and Philippine Democracy. Address Delivered by the Honorable Sergio Osmeña Vice President of the Philippines at the Commencement Exercises of the Philippine Military Academy, Baguio, March 15, 1940* (Manila: Bureau of Printing, 1941), 10, No. 28003, box 1276, E-5, Record Group 350, U.S. National Archives, Washington, D.C.

13. Indeed, the academy's semi-official history describes the PMA as "the brainchild of General Dwight D. Eisenhower." See Cadet Corps, *The Academy Scribe* (Manila: Academy Scribe Organization, 1988), 50.

14. *National Defense and Philippine Democracy*, 7–8.

15. One study of the PMA argues that "all countries attempt to control their national armed forces through indoctrination of their professional military officer corps. The national military academy has long been regarded as a key institution in this process." See Ronald G. Bauer, "Military Professional Socialization in a Developing Country" (Ph.D. diss., University of Michigan, 1973), 1–2.

16. Commonwealth of the Philippines, *Report of the President of the Philippines on the Activities of the Philippine Army for the Period, January 1 to December 31, 1936* (Manila: Bureau of Printing, 1937), quoted in David Y. Nanney, "Philippine Military Academy" (ms., 1938), 6, Julius Ruiz Papers, Filipino American Historical Society, Seattle, Wash.

17. Mendoza, *Batch '36 Golden Book*, i.

18. Lt. Conrado B. Rigor, "The Philippine Military Academy," *Coast Artillery Journal* 84 (July–August 1941): 326.

19. Correlli Barnett, "The Education of Military Elites," *Journal of Contemporary History* 2 (July 1967): 22–23.

20. Charles W. Larned, "West Point and Higher Education," *Army and Navy Life* 8 (June 1906): 18.

21. *The Sword of Nineteen Hundred and Thirty-Eight* (Baguio: Cadet Corps of the Army of the Philippines, Philippine Military Academy, 1938), 46–48, 104.

22. D. Clayton James, *The Years of MacArthur: Volume I, 1880–1941* (Boston: Houghton Mifflin, 1970), 276–77.

23. Much of the literature on military academies deals with the formal institutional aspects of "socialization," i.e., academic curriculum and military drill. See John P. Lovell, "The Professional Socialization of the West Point Cadet," in Morris Janowitz, ed., *The New Military: Changing Patterns of Organization* (New York: Russell Sage Foundation, 1964), 119–57; and Janowitz, *Professional Soldier*, 127–39. By contrast, there is a much smaller literature that examines the aspect of socialization central to this essay—the less formal cadet culture involving hazing, bonding, and hierarchy. For examples of this approach, see Sanford M. Dornbusch, "The Military Academy as an Assimilating Institution," *Social Forces* 33 (May 1955); and Gary M. Wamsley, "Contrasting Institutions of Air Force Socialization: Happenstance or Bellwether?" *American Journal of Sociology* 78 (September 1972).

24. Cadet Corps, *Academy Scribe*, 53.

25. Mendoza, *Batch '36 Golden Book*, 15, 180, 264.

26. Letter to the author from Col. Deogracias Caballero, 22 Dec. 1995; interview with Gen. Horacio Farolan, Villamor Air Base, Philippines, 14 Mar. 1995. Among the thirty classmates interviewed, seven were valedictorians and another nine were honor students.

27. In the words of Class '40's own memoir, "A few may be considered as coming from the upper crust of society. But most of them belonged to the middle and lower classes. Their parents were farmers, teachers, small merchants or traders and minor government officials. A few were sons of lawyers, engineers or doctors." See Mendoza, *Batch '36 Golden Book*, 15.

28. Arnold van Gennep, *The Rites of Passage* (Chicago: University of Chicago Press, 1960), 10–11, 106–07; Mendoza, *Batch '36 Golden Book*, 17–19.

29. *Sword of Nineteen Hundred and Thirty-Eight*, 193.

30. Interview with Capt. Eugenio Lara (Ret.), Anaheim Hills, Calif., 2 Jan. 1994.

31. *Sword of Nineteen Hundred and Thirty-Eight*, 114.

32. Mendoza, *Batch '36 Golden Book*, 17–19.

33. Interview with Commodore Ramon A. Alcaraz (Ret.), Orange, Calif., 30 Dec. 1993.

34. Interview with Lara, 2 Jan. 1994; Ernesto O. Rodriguez, *Commodore Alcaraz: First Victim of President Marcos* (New York: Vantage Press, 1986), 122.

35. Interview with Col. Jose Mendoza (Ret.), Quezon City, Philippines, 22 Oct. 1988.

36. Interview with Col. Manuel Acosta (Ret.), a member of Class '40 and brother of Flor Acosta, Quezon City, 18 Mar. 1995.

37. Mendoza, *Batch '36 Golden Book*, 37.

38. *Sword of Nineteen Hundred and Thirty-Eight*, 201.

39. Carolina G. Hernandez, "The Extent of Civilian Control of the Military in the Philippines: 1946–1976" (Ph.D. diss., State University of New York at Buffalo, 1979), 169–72.

40. Dornbusch, 316–21.

41. Cadet Corps, *Academy Scribe*, 142.

42. Benjamin N. Muego, "Civilian Rule in the Philippines," in Constantine P. Danopoulos, ed., *Civilian Rule in the Developing World: Democracy on the March?* (Boulder, Colo.: Westview, 1992), 215–17.

43. Dante Simbulan, "A Study of the Socio-Economic Elite in Philippine Politics, 1946–1963" (Ph.D. diss., Australian National University, 1965).

44. Interview with Jose "Pete" Lacaba, Canberra, Australia, 14 Feb. 1988.

45. Cadet Corps, *Academy Scribe*, 227, 242.

46. Rigoberto D. Tiglao, "Rebellion from the Barracks: The Military as Political Force," in Philippine Center for Investigative Journalism, *Kudeta: The Challenge to Philippine Democracy* (Manila: Philippine Center for Investigative Journalism, 1990), 15.

47. Viberto Selochan, "Professionalization and Politicization of the Armed Forces of the Philippines" (Ph.D. diss., Australian National University, 1990), 67.

48. Jo-Ann Q. Maglipon, *Primed: Selected Stories 1927–1992* (Manila: Anvil, 1993), 227.

49. Col. A. P. Aguirre, *A People's Revolution in Our Time* (Quezon City, Philippines: Col. Alexander P. Aguirre, 1986), v–vi.

50. Jose G. Ayap, *The 1971 Sword* (Baguio City, Philippines: Cadet Corps Armed Forces of the Philippines, 1971).

51. Interview with Lt. Col. Victor Corpus, Quezon City, Philippines, 1 and 5 Apr. 1987.

52. *Mr. & Mrs.*, 21–27 Mar. 1986, 17.

53. Maglipon, 227.

54. Tiglao, 15.

55. Interview with Corpus, 1 and 5 Apr. 1987; Tiglao, 15.

56. Ayap, *1971 Sword*; *Philippine Daily Inquirer*, 7 Oct. 1994.

57. Interviews with Col. Deogracias Caballero (Ret.), Quezon City, 12 Oct. 1988; Gen. Reynaldo Mendoza (Ret.), Camp Aguinaldo, Quezon City, 14 Mar. 1995; and Col. Francisco del Castillo (Ret.), Quezon City, 18 Mar. 1995.

58. *Manila Times*, 10 July 1956.

59. Ibid., 29 June 1961.

60. Ibid., 11 July 1956, 17 July 1957.

61. Ibid., 16 Mar. 1967.

62. Lumabas, 5.

63. Maglipon, 226–27. (The original was a mix of Tagalog and English.) EDSA refers literally to Epifanio de los Santos Avenue, where a million Filipinos massed to overthrow the Marcos regime in the "people's power" uprising of February 1986.

64. Ayap, *1971 Sword*.

65. *Baguio Midland Courier*, 6 Apr. 1969.

66. *Manila Times*, 20 Apr. 1969.

67. Ibid., 19 and 22 Apr. 1969; *Baguio Midland Courier*, 20 Apr. 1969.

68. *Manila Times*, 26 Apr. 1969.

69. *Baguio Midland Courier*, 1 Mar. 1970.

70. *Daily Mirror*, 5 Apr. 1971.

71. Bauer, 40–42.

72. Selochan, 64–65.

73. Interview with Jose Duran, Quezon City, 19 Mar. 1995.

74. Republic of the Philippines, *Final Report of the Fact-Finding Commission,* App. J.

75. "Tungkol Kay Gringo B. Honasan," *Para Sa Bansa Gringo Honasan Sa Senado* (Quezon City, Philippines: Benny J. Brizuela, 1995 [campaign leaflet]).

76. Selochan, 188.

77. Shelia Coronel, "RAM: From Reform to Revolution," in *Kudeta*, 60; *Mr. & Mrs.*, 27 February–5 March 1987, 6.

78. *PC Journal*, July 1986.

79. Republic of the Philippines, *Final Report of the Fact-Finding Commission*, App. J.

80. Amnesty International, *Report of an Amnesty International Mission to the Republic of the Philippines 11–28 November 1981*, 2.

81. Ibid., 8.

82. *Philippine Graphic*, 18 Jan. 1993.

83. *New York Times*, 10 Nov. 1986.

84. The Philippine military apparently coined this neologism to describe its torture operations in the mid-1970s and it was soon taken up by the country's human rights groups who monitored such abuses. See Task Force Detainees, *Political Detainees of the Philippines Book 3* (Manila: Association of Major Orders of Religious Superiors, March 1978), 41–43.

85. Richard J. Kessler, *Rebellion and Repression in the Philippines* (New Haven, Conn.: Yale University Press, 1989), 137. Kessler's incomplete statistics on the number of "extrajudicial killings," or salvagings, in 1984 are supplemented by data from Rev. La Verne D. Mercade and Sister Mariani Dimaranan, *Philippines: Testimonies on Human Rights Violations* (Manila: World Council of Churches, 1986), 89.

86. For one example of this adulation, see *Malaya*, 29 Aug. 1987.

87. Coronel, 60.

88. Interview with Shelia Coronel, the journalist who hosted this reception for the RAM leaders, Manila, 5 Jan. 1988. For a description of this party where Fusilero made these revelations, see Neni Sta. Romana-Cruz, "Reformists Night Out: In Uniform But Into Fun," *Mr. & Mrs.*, 21–27 March 1986, 19–20.

89. Maglipon, 228.

90. Republic of the Philippines, *Final Report of the Fact-Finding Commission*, App. J. Integrees were reserve officers incorporated or "integrated" into the list of regular officers.

91. *Veritas Special Edition*, October 1986, 4.

92. Republic of the Philippines, *Final Report of the Fact-Finding Commission*, 471.

93. Mendoza, *Batch '36 Golden Book*, 349–50.

94. Ibid., 350.

95. Ibid., 125.

96. Commodore Ramon A. Alcaraz (Ret.), "Homily." Address was given at the Alumni Memorial Mass, Philippine Military Academy Chapel, Ft. Del Pilar, Baguio City, 17 Feb. 1990.

97. Mendoza, *Batch '36 Golden Book*, 350.

Military Education and Colonialism: The Case of Morocco

Moshe Gershovich

The history of cadet and young officer training in Morocco began in 1919 with the inauguration of the Royal Military Academy for cadet-officers. Located at the sixteenth-century imperial palace of Dar el-Beida in Meknes, the school was controlled and administered by French military and civilian authorities. It remained under French supervision until 1962, six years after Morocco had ceased to be a French colonial possession. Currently it still serves as the central institution where officers of Morocco's Royal Armed Forces are educated. This essay explores the academy's development and the manner in which young Moroccans were educated and trained to serve as officers in the ranks of the French Army during Morocco's subjugation as a French protectorate (1912–56), and offers a linkage between the colonial era and the immediate postindependence years in the history of modern Morocco.

The Precolonial Moroccan State, Its Military, Its Officers

Before turning our attention to the French Protectorate, a brief overview of Morocco's precolonial military tradition is essential in order to place later policies in their proper perspective. A viable and distinct Moroccan political entity has existed for over a millennium. For the last three and a half centuries that entity has been ruled by a single dynasty, the 'Alawis. With few exceptions (notably that of Sultan Mawlay Isma'il who introduced a standing infantry corps of black slaves imported from sub-Saharan western Africa), the Moroccan state was characterized by a weak central government (termed *Makhzan*) which often could not extend its rule over the tribal populations of the Atlas and Rif mountain chains. To protect its own population against the "domain of anarchy," the *Makhzan* relied upon loyal tribes termed *Jaysh*. Members of these tribes were recruited to man military expeditions, to extract taxes from the countryside, and to defend the urban centers. In exchange for their services they were granted state lands and an exemption from taxation.

The *Jaysh* system provided the precolonial Moroccan state with a limited degree of internal stability, but proved highly ineffective as Morocco began to confront strong European neighbors to its east (French Algeria) and north (Spain). Humiliating defeats to both these powers forced the *Makhzan* to initiate a military reform program. New, European-styled troops were created and trained by European advisors. The experiment yielded scarce results since most foreign advisors sought to

321

serve the imperialist interests of their governments rather than strengthen the forces safeguarding Moroccan independence.[1]

A significant Achilles heel of the "New Army" system concerned the manner in which officers were recruited, trained, assigned to posts, and promoted. The proclaimed policy sought to recruit as many different elements of Morocco's society as possible. This principle applied to the command level as well, although the selection of officers relied primarily upon loyalty to the ruling dynasty and the financial abilities of the candidates to maintain the units they commanded. As for the training process, a military engineering school was created in 1844 in Fez and a preparatory school for students sent to European centers of military education was established in 1873 in Tangier. As of the mid-1870s, about one hundred young Moroccans had completed that school before being sent abroad to England, France, Belgium, Italy, and Germany.

As the French Army penetrated Morocco in the 1900s, it brought its tradition of reliance on North African populations as sources of manpower for its colonial garrisons. The nineteenth-century *Armée d'Afrique* was composed of a mixture of European and African contingents. The latter were usually commanded by Europeans at the higher levels and by indigenous subalterns who could not rise above the rank of lieutenant.[2] Most Algerian rank holders were veteran soldiers chosen to serve as noncommissioned officers (NCOs) and officers in recognition of their experience and loyalty. It was rare for a Muslim Algerian to be sent to a French military school. No institution training indigenous cadet-officers existed in French North Africa prior to the establishment of the Moroccan military school in Meknes.[3]

After taking direct control of the Moroccan Army, the French introduced some European and Algerian commanders into its units but did not remove the existing Moroccan commanders. After the April 1912 uprising in Fez and the consequent dissolution of the Moroccan Army, the French readmitted most of the soldiers and officers who had not been involved in the revolt to the newly created Moroccan auxiliary troops within the French Army.[4] In 1914 these Moroccan officers were dispatched with their men to France where they participated in some of the bloodiest battles of World War I. The French command was generally satisfied with their performance, but was also aware of the need for replacements within their ranks as the war progressed, which was partially achieved by promoting experienced NCOs. This need also convinced the French supreme command to approve the creation of the Royal Military Academy in Meknes.

Colonial Philosophy and the Establishment of the Military School

The initiative to create a Moroccan military academy, however, did not come from the battlefields of the Western Front but rather from Rabat, seat of France's first resident-general in Morocco, Marshal Louis Hubert Lyautey.[5] When Lyautey came up with the idea of the school is unknown; the available documentation allows us merely to establish that he submitted it for the Ministry of War's approval in early

1918.[6] His reason for proposing a military academy could be linked to the principal concept upon which he sought to establish the Protectorate.

Lyautey rejected the colonial approach of the French administrative system in Algeria (known as the *Bureaux Arabes*) which sought to establish direct control over indigenous populations.[7] Instead he called for collaboration between the Protectorate and what he perceived to be the natural, leading indigenous elites, generically termed the notables or *grandes tentes*. Securing the loyalty of these elements was a task with which Lyautey and his subordinates occupied themselves from 1912 onward. World War I provided assurance of that loyalty and convinced the French authorities that this was the correct policy to pursue. Their problem was how to guarantee the continuation of this collaboration; how to enable those loyal elements to maintain control over their communities. The goal, in other words, was to ensure that the sons of these notables would replace their fathers upon retirement and serve the French cause as loyally and effectively. The obvious answer was to create a special educational network under French control to attract sons of notables. Such education would expose them to French culture, values, language, and manners. The military academy in Meknes was to become one of a number of schools created for that purpose.

A monarchist at heart, Lyautey envisaged a careful selection of sons of the Moroccan rural and urban aristocracy who, through a carefully planned pedagogical system, would become junior officers in the French Army. Their military careers, however, were meant to be merely the corridor which would lead them to their true destination, the administrative system of the *Makhzan* and the Protectorate network of native control. In short Lyautey wanted to prepare a second generation of Moroccan notables polished by French culture and ready to follow in their fathers' footsteps.

The principle of aristocratic recruitment appears in virtually every letter written by Lyautey concerning the Meknes school. He was involved with the school's development throughout the latter part of his residency, as reflected in the extensive correspondence he conducted with Paris and with local authorities under his control. To ensure his mastery over the academy he insisted that it be removed from the Ministry of War's control and placed under that of the Protectorate through its Service of Intelligence and Native Affairs. He also took it upon himself to find suitable financial resources to continue the project.[8]

The Problem of Recruitment

Attracting the "right" elements to enroll in the school thus became an important factor in the success or failure of the project. At the same time, however, to justify the expense of the school, two other factors had to be considered. There had to be a sufficient number of cadets in each class physically and intellectually able to serve in a French unit, and an equivalent number of active-duty vacancies to employ the graduates. Not always could these prerequisites be accommodated. Comparing enrollment in the first two classes, for example, clearly shows the difficulty in finding candidates of "suitable" origins: Of the fifteen students in the Class of 1919, fourteen were not members of "great families." The Class of 1920 had eighteen candidates of whom only three had come from such families.[9]

What caused this disappointing result? Lyautey believed at first that the key element was the insufficient enthusiasm of France's agents in the countryside, and he argued that a greater effort to "sell" the school to its potential notable clients could reverse the trend. However, one can find evidence that local administrative authorities were motivated to find the "right" sort of candidates. Instead, the major cause for the insufficient enlistment of notable sons could be found in the attitude of the "great families" themselves. The French failed to change the local aristocracy's reservations toward the army and their reluctance to commit their sons to careers within its ranks.

Many young Moroccans of noble origins preferred a traditional scholastic education to the Meknes Academy. Such was the case with the son of a local governor (*qa'id*) in the Taza region who refused to leave the Moulay Yousuf College in Rabat where he was studying to go "make the *barud*" (namely, learn how to fight) in Meknes. Discussing this incident in his report on the candidates for the 1936 class, the school's director at that time, Lt. Col. P. Tarrit, argued that Moroccan parents "let their sons grow lazy and allow them their caprices. They show no interest in their moral education and instruct them to serve as nothing but tribal clergy (*faqihs*)."[10] The future prominence of the traditional rural aristocracy was thus endangered as children of inferior social classes, including the urban proletariat, were constantly improving their skills through exposure to French education and began looking down on the sons of the countryside nobility. In short, concluded Tarrit, the Moroccan indigenous chiefs were becoming totally dependent upon France to maintain their position.

Tarrit's report, while presenting the most thorough French discussion of the problem of recruitment, rehashed themes originally raised fourteen years earlier by the school's first director, a Major Quetin. The latter's report included some encouraging data: forty-two young Moroccans, all of excellent origin, submitted their candidacy. The sudden interest in the school expressed by the "great families" was related directly to the graduation of the first class of cadets whose eleven members were all assigned to respectable command posts. This apparent success notwithstanding, Quetin was fully aware of the difficulties of recruitment in the long run. The problem was not merely one of quantity but also involved the quality of the preferred candidates. "Most well-educated Moroccan youth come from among petite bourgeoisie and not from the nobility," he stated. Therefore, "if quality is to be maintained, the school must 'democratize,' namely it must accept students of lower origins." His preference was to maintain the emphasis on noble descent and solve the quality problem by introducing a preparatory course and by lowering the age of entry from eighteen to fifteen. In other words, the Meknes Academy was to change its status from a two-year military college to a long-term boarding school ("*Prytanée Militaire Marocain*") where students would study for four to five years before graduating as junior officers.[11]

Lyautey approved these recommendations in principle but very little if anything was done to implement them while he remained in office.[12] Only after he had left Morocco were some reforms introduced to the academy. Quetin's "democratization" prescription was adopted and the school opened its gates to applicants from lower social strata. Thus, of the twenty-nine candidates for the 1936 class (ten of whom were finally admitted) only sixteen were sons of current or former *Makhzan* functionaries. Three were sons of subaltern police agents and ten came from families with no direct connection to the state administration.[13] The effort to base the military school on a homogeneous social element was thus only partially successful, even though the majority of Moroccan officers who graduated from the academy did belong to the higher echelons of Moroccan society.

The Challenge to the School's Existence

This process of reform in the late 1920s was directly related to a mounting external challenge which threatened the existence of the school. As long as Lyautey was in control in Rabat, the Meknes school was virtually immune to external pressure. Things changed when he resigned and left Morocco in October 1925. No longer shielded by its creator, the school was now exposed to criticism from some high-ranking officers on the metropolitan General Staff who had always been either skeptical of the likelihood of its success or overtly hostile to the concept it represented.

The opponents of the Moroccan Military Academy found a potential ally in the personality of Lyautey's successor as resident-general, Théodore Steeg, a metropolitan politician and former governor-general of Algeria who showed little interest in military affairs. Soon after Steeg's arrival in Rabat the opponents of the Meknes school got their opportunity to be heard. In March 1926 the instructional section of the military operations bureau of the General Staff in Paris asked Marshal Franchet d'Esperey, the inspector-general of the North African troops, to inspect the Meknes school. A chief concern of the investigation was related to the potential frustration which might develop in the minds of the Moroccan graduates who "despite their intellectual aptitude, get lost for a long time in a subordinate rank and quickly acquire a mentality of bitterness and discontent."[14]

It is uncertain whether the Ministry of War sought a serious reform of the school or wanted in fact a report which would go beyond that and recommend aborting the project altogether. In any case, this initiative provided a golden opportunity for the opponents of the "Lyautey formula." The selection of Franchet d'Esperey was most convenient for them as well. A veteran colonial soldier born in Algeria, he passed a significant part of his pre-1914 career in North Africa, including a short period in 1912–13 during which he had served under Lyautey as the commander of the troops in western Morocco. Lyautey had no great admiration for his second-in-command and made no effort to reverse his call to France. Thirteen years later, Franchet d'Esperey was given an opportunity to repay this unkindness, and he did.

In his report, submitted in late March 1926, Franchet d'Esperey regarded the school's annual 550,000 French franc (FF) budget as wasted and its original goal of attracting sons of great families as having failed. No member of the preeminent Moroccan families had been sent to Meknes; most of the candidates had been sons of secondary officials whose prestige in Morocco was marginal. He launched a sharp attack against the pedagogical program:

> Too much time is devoted to history. It is not wise to teach these young Moroccans of high society in great detail the diverse vicissitudes through which their country passed to our hand. We risk the unfavorable conclusions to which they might arrive one day. It is imprudent to press too far their general instruction. As long as Morocco is not occupied by a vast French and European population, one should always anticipate surprises.
>
> For our own security we should not teach our subjects too early and too quickly the means with which they could fight us. As these Moroccans are likely to end sooner or later their careers as civil leaders, the military instruction given to them in Meknes should not go beyond the necessary, and they should be prepared through special studies primarily for their future functions within the protectorate.

Examination of the careers of former graduates provided little ground for dissatisfaction. However, Franchet d'Esperey's report dealt with the potential danger that the ambitious graduates of the "Moroccan Saint-Cyr" might find it difficult in the future to be content with the mediocre careers awaiting them:

> Our Moroccan troops . . . have no need for officers like those who graduate from the military school of Meknes. For the last fourteen years since they were organized, during which they fought practically everywhere, it has been possible to discover within their ranks enough NCOs who, through their proven loyalty, their manner of service, and their merits, have proven to be worthy of being promoted to officers. It is mostly for them that vacancies in the regiments should be preserved.
>
> In short . . . I consider it necessary to install full direction of civilian authority over that institution and to replace its military personnel with civilians as quickly as possible.

According to Franchet d'Esperey this opinion was shared by Steeg, who preferred to leave the school in its current status for political reasons, but envisaged its future transformation into an agricultural institute.[15]

Amidst such pressure, and deprived of support from the residency, the school's future seemed doubtful. Its survival depended in part on the willingness of the academy's management to compromise temporarily on some of its principles and partially accept certain reforms. Such was the case in 1928 with the accommodation within the walls of the Dar el-Beida palace of a course for French and Moroccan NCOs serving in Goum formations.[16] Still, the strongest argument, ultimately saving the school, was the service record of its graduates within the ranks of the French Army. The general image of these Moroccan officers was highly positive, as frequent reports from their regimental commanders revealed. Even Franchet d'Esperey had to admit in his report that of the twenty-two officers who had graduated between 1921 and 1925 only two failed to satisfy requirements and were discharged from the army. The key to this achievement was to be found in the pedagogical formula employed in Meknes.

Structure and Curriculum

Originally constructed as a two-year military college, the Meknes Academy changed its format, as we have seen, to include a preparatory course along with the cadet course within its new framework of an extended boarding school. The two programs were interrelated as most cadets would begin their studies at the preparatory stage. The passage from one level to the other was not automatic, however. It involved a physical test as well as a written final examination. Students who failed either of these tests returned to their families and joined those who had elected civilian careers. Since the size of the cadet class corresponded to projected vacancies in Moroccan units, the preparatory course admitted that number of students plus a small surplus to allow for the choice of the best candidates and to make up for occasional accidents.

The annual admission ratio changed frequently during the early years of the school's existence before the exact needs of the rapidly expanding Moroccan units could be determined.[17] This issue became a constant bone of contention between Rabat and Paris as the Protectorate authorities pressed to augment the number of students while the Ministry of War sought to reduce it. The controversy derived from the conflicting views regarding the loyalty and potential hostility of young and ambitious Moroccan officers toward France. The compromise reforms forced upon the school in the late 1920s included a quota of five new students per year designed to gradually arrest the academy's growth.[18]

In addition to his origins, the candidate's admission to the preparatory course depended on his elementary education and fluency in French. The minimum entrance age was fifteen and studies lasted for four years. At age nineteen the student could sign up for military service, become a cadet, and start another four-year period of study. At the end of this period he would graduate as an *"aspirant de 1° classe,"* or reserve second lieutenant. He could then be promoted to second lieutenant on the active list after two years of probation (*stage*) in a regular Moroccan unit. A Moroccan officer thus worked ten years studying and training to accomplish by age twenty-five what a French counterpart would normally achieve three to four years earlier.

Most preparatory students paid an annual tuition fee of slightly over 2,000 FF. Each year one or two particularly gifted students of modest social origins were admitted free of charge with a stipend to sustain them throughout their studies. The date the school started admitting such nonpaying students is unknown, but the practice was probably related to the trend toward "democratization" and perhaps derived from the reforms of the late 1920s.

Documentation dealing with the curriculum at the academy is scarce and relates by and large to nonmilitary subjects in both the preparatory and cadet courses. Studies were conducted in French and devoted largely to French literature and history. It is important to note the total disregard of the study of Arabic or any of the three Berber dialects prevailing in Morocco. The language question emerged critically only in 1946, when several candidates complained that their command of Arabic had deteriorated so much that it was inferior to that of the men they were supposed to command. The school's director at the time, a Lt. Col. de Jenlis, looked into the issue and found no record that Arabic had ever been taught in the school. He rapidly appointed an Arabic instructor to teach fourteen hours per week.[19]

The Francocentric approach could be seen clearly from the teaching of history in the preparatory course. A report prepared in 1926 by its instructor, Jean Allais, indicated the abnormality of the curriculum offered to Moroccan youth under French auspices. The 1925 history course was divided into two parts. The first unit—thirty-four lessons—was entitled The Origins of Contemporary France. The second unit—twenty-six lessons—dealt with modern France. Only one of these sixty lessons taught the students something about their own heritage: a lecture on the Arab conquests of the seventh and eighth centuries. Nothing related to Morocco's history was offered in this course, but the history of feudal France and the absolute monarchy were thoroughly studied. When dealing with modern history, the instructor chose to diverge from the manual and ignore altogether events such as the Franco-Prussian War of 1870–71 "in order to prevent the presentation of an unfavorable vision of France."[20]

Once Out of School: Military Careers of the Academy's Graduates

How did the graduates of the Moroccan Military Academy fare once they had successfully passed their final examinations and began their active military service? To what extent did the patterns of their careers match those of French and other officers serving under the tricolor?

In terms of their status one can detect two conflicting tendencies. On the one hand, the interwar metropolitan General Staff sought to establish uniformity among France's colonial troops, including indigenous officers, regardless of their racial or national origins. On the other hand, the Protectorate leadership insisted upon the unique title of "Moroccan officer" in its effort to preserve some symbolic pretense of Moroccan sovereignty. On a less formal level Moroccan officers often received special attention from their regimental and battalion commanders, many of whom had spent a significant portion of their military careers in Morocco and cherished the "heroic days of pacification" as the most significant period of their lives. Thus, while the practical aspects of their military service were governed by regulations applicable to all non-European officers, Moroccan officers carried with them a distinctive identity and the notion, ingrained at the academy, that they were destined to become the leaders of their society.

Contrary to this faltering image, however, the promotion venues offered to these men were very limited given the contemporary dogmas prevailing within the French military establishment. During the interwar period the highest rank to which all non-European officers could aspire was captain, but very few actually reached that rank by the outbreak of the war in 1939, and none was a Moroccan. They were also very limited in the roles offered to them since their administrative talents were usually deemed inferior to those of European officers, and were typically confined to subaltern command positions.[21] No Moroccan officers were permitted to command French officers of equal rank, even when the latter were inferior in seniority.

Graduating from his military school a year or two older than his French counterpart, the Moroccan officer was then subjected to the two-year probationary period to confirm his second lieutenant rank. His subsequent promotion was also much slower; some Moroccans spent nine years as second lieutenants and received their second stripes only a year or two before their retirement. The slower pace of promotion had an obvious financial consequence: the salaries of Moroccan officers were substantially lower than those of French officers of equal or lower rank and lesser seniority. In some cases French NCOs earned more than Moroccan officers serving in the same unit. This state of affairs was well known to the French supreme command and measures were often discussed during the late 1930s to decrease the gaps between French and colonial officers. However, very little actual improvement was made by the outbreak of World War II.

Given these circumstances, it is interesting to note that the disillusionment of these Moroccan officers did not translate into overt disloyalty. Embittered as they may have been about their deprivations, they still appreciated the advantages of their status within their own society. Their salary, which in 1931 fluctuated between 1,270 and 1,700 FF per month, ranked well above the average for *Makhzan* functionaries. Upon retirement they could count on a comfortable pension added to the income from the administrative post they were almost automatically assured.[22] The military also provided a respectable career promising an excellent prospect for social prominence, especially for those Moroccans of relatively modest social origins who managed to enter the Meknes school. As long as France prevailed in Morocco, it was practically unthinkable for a Moroccan officer to challenge the status quo with an expression of disobedience.

Real reforms in the status of Moroccan officers began only during World War II. A decree of 7 March 1940 made it theoretically possible for non-European officers to reach the rank of major and its equivalents. Another reform reduced the probationary period from twenty-four to eighteen months, and shortened the period between promotions. This allowed some Moroccans to become captains by 1942. However, they were still denied direct command over troops and had to remain seconds to French officers. These last traces of inequality were reduced to a minimum only after the war.

The Royal Military Academy and the Moroccan Officer Corps in the Twilight of the Protectorate and in the Postcolonial Era

More than fifty Moroccan officers were serving in the French Army at the outbreak of World War II and their number almost doubled during the war. Five were killed during the Battle of France and five others later in the war. Thirteen officers were taken prisoner by the Germans in 1940; six escaped captivity and the other seven were repatriated before the war was over. Seven officers were injured in the 1940 campaign and six others from 1943–45. Moroccan officers were the recipients of 242 citations, decorations, and medals.[23]

During the war the academy trained and graduated sixty-two cadets—almost as many as the figure for the 1925–39 period. The accelerated pace was achieved by reducing the duration of studies from four years to twenty-six months. The selection criteria were also lowered to allow more candidates to be admitted. To make room, the preparatory course was abolished in 1939 and resumed only in 1946. All these factors, plus the inability to expose the cadets to actual field experience prior to their graduation, resulted in a considerable deterioration of the standards set during the 1930s, a shortcoming which the school command sought to correct once the war was over.

Along with the resumption of the prewar program came a program drawn by the metropolitan General Staff to reduce considerably the number and size of active-duty Moroccan units. Thus, in early 1946 news was suddenly spread that the army would not need any new Moroccan officers for 1948 and 1949. Students preparing to graduate in those years found their future careers jeopardized as rumors circulated about a plan to terminate their contracts by failing them on the final examination. The shame involved in their sons returning home without having graduated caused the parents to act. Several influential tribal chiefs appealed to Resident-General Eric Labonne and demanded a change of policy. Faced with this pressure, the school offered to allow all current students to graduate and receive their ranks. Most were then expected to be sent back to their homes and wait for vacancies.[24] The outbreak of the Indo-Chinese War in 1947 solved this problem, as Moroccan units were hastily reformed and sent to Vietnam.

Another problem affecting the school during the immediate postwar period was the impact of Moroccan nationalism. As of the early 1930s, following the issuing of the "Berber Dahir" which increased the autonomy of the customary tribal judiciary system, Morocco's urban centers underwent several waves of unrest and anticolonial agitation.[25] The French sought to suppress this activity and prevent its spread among their Moroccan troops, which, unlike other areas of French policy in Morocco, did not exercise a policy of ethnolinguistic segregation. That standard had been in effect for the Moroccan Military Academy and officer corps where Arabs and Berbers studied and served together. However, in October 1948 the French command devised a plan to eliminate the Arabs' access to military education with the inherent consequence of building the future Moroccan officer corps out of Berbers alone.

This objective was to be achieved through the attachment to the Meknes Academy of the Berber school at Azrou, a high school which had been inaugurated in 1927 and reorganized in 1942 as a semimilitary program commanded by a French colonel. Used as a political device to combat the "Arabization" of the Berber mountain tribes, the Azrou school catered to families of former Berber soldiers and trained their sons to serve in local administrative posts such as the Berber judicial system and the postal, telephone, and telecommunications networks. The October 1948 reform sought to enhance the level of the Azrou school by turning it into a paramilitary boarding school in the Meknes Academy style. As such, Azrou was to become the major supplier of candidates for the cadet course in Meknes, which was to become uniquely Berber.[26] However, the plan failed to yield the envisaged fruit. In fact it was likely never attempted. Later records of the school's history show that its student population remained ethnically mixed.[27] As for the potential impact of the nationalist ideology, the available documentation indicates that the school and its students remained by and large unaffected.[28]

While seeking alternative plans for the education of Moroccan officers, the French modified the structure of the Meknes Academy and reduced the emphasis on the military training it provided. An instruction issued on 15 March 1949 added to the cadet course two new options which students completing their preparatory stage could choose to continue their educations. These included the newly created Moroccan School of Administration and some metropolitan *grandes écoles* which were called upon to open their gates to Moroccans. No longer restricted by expected vacancies in Moroccan units, the school augmented its ranks considerably, from 55 students in 1949 to 102 students in 1951, and 155 in 1955. The socioeconomic preferences were also modified as the percentage of students of modest social origins increased.[29]

Lyautey's ideal of a future Moroccan leadership based on the "natural" aristocratic elite, and formed by French education, had been transformed almost beyond recognition by the end of the Protectorate. Even though they expressed allegiance to the old master's maxims, Lyautey's successors had to adjust to the new post–World War II currents and reshape the school in a manner which led it to a different destination from that which he had intended. By enhancing the "democratization" of the school, however, they may have contributed to its disassociation from its colonial image, which allowed it to be accepted as respectable institution in independent Morocco.

The termination of the Protectorate in March 1956 did not diminish the importance of the Meknes Military Academy, nor did it end French involvement in its administration. In the treaty with the sultan reestablishing Moroccan independence, France committed itself to "lend its assistance to Morocco towards the constitution of a national army."[30] Acting toward that end, in early May 1956 the French military began transferring several Moroccan infantry and cavalry battalions to the newly created Royal Armed Forces (FAR). A total of twenty thousand men were placed at the disposal of the Moroccan government. This contingent which included 120 officers and 1,500 NCOs formed the core of the FAR's initial order of battle. They were reinforced by a team of several hundred French officers and soldiers who had volunteered to serve as specialists and train Moroccans to assume diverse technical tasks.[31] Other than manpower, the French Army also contributed vehicles and other equipment and arms at a total value of 144 billion FF.

The automatic transfer of officers from the French (and Spanish) Armies was insufficient to ensure the smooth functioning of the FAR.[32] The need for new officers caused the Moroccan government to ask for and receive French assistance. A group of 167 hastily recruited high school graduates was sent in late 1956 to France where they completed a special infantry officer course and then returned to Morocco. Meanwhile the French were asked to continue their management of the Meknes Military Academy. The length of the course of studies at the cadet level was considerably shortened to allow for the quick graduation of many new officers. Infantry officers were sent directly to their units after a two-year course. Cadets who chose other branches of service were sent to French military schools for an additional year of training.

In 1960, four years after the end of the Protectorate, there were still more than one hundred French military men in Morocco. Most served in various administrative and technical positions which the Moroccans lacked the trained manpower to fill. A particularly strong French presence was found in the Meknes Academy where the cadre of thirty-five instructors included only seven Moroccans. When French technical aid to the Moroccan Army ended in late 1960, a sufficient number of Moroccans had become available to replace the French instructors.

While admitting new cadets and constantly expanding its ranks during the immediate postindependence era, the Moroccan officer corps were required (due to an agreement reached in July 1956 between Crown Prince Hassan and the leadership of the anti-French "Liberation Army") to integrate within their ranks elements from the disbanded armed resistance movement.[33] The uneasy coexistence between those who had served the colonial order and those who fought it did not last for long. Most former resistance leaders who had been admitted to the FAR in 1956 were soon isolated within its officer corps and eventually forced into early retirement. Among them were a few graduates of the Meknes Academy who had joined the armed wing of the nationalist movement during the last months of French rule over Morocco. Such was the case of Capt. Mahjoubi Ahardan who commanded a unit of the "Liberation Army" after being dismissed from his rural administrative post due to his nationalist activity.[34]

Their lack of overt support for Moroccan independence was not held against Moroccan officers coming from the French and Spanish Armies. Upon his triumphant return from exile as the king of independent Morocco, Muhammad V sought to establish special ties with them. Since they could not count on any sympathy from the nationalist leadership, the officers had no alternative to ties with the Palace. Satisfying their material needs and promoting their pride became a prime concern of the king and was provided for by frequent promotions, pay raises, and similar means. It was the erosion of this system of monarchic patronage and the consequent cumulative dissatisfaction within the high command of the late 1960s which posed the most serious threat to the throne, by then held by King Hassan II. The coup attempt of July 1971 was the first episode in a rapid chain of events which forced the king to reevaluate his policy toward the army and began a new era in the relationship between the Moroccan state and its military.[35]

Notes

An earlier and somewhat different version of this article was published in *Middle Eastern Studies* 28 (April 1992).

1. The most detailed study of the nineteenth-century military reform movement in Morocco is Wilfrid Rollman's "The New Order in a Pre-Colonial Muslim Society: Military Reforms in Morocco 1844–1904" (Ph.D. diss., University of Michigan, 1983).

2. On the eve of World War I the first Algerian officer reached the rank of captain: Khalid b. al-Hashimi al-Jazairi, also known as Emir Khaled. A grandson of 'Abd al-Kadir, the leader of the initial resistance to France's conquest of Algeria during the early 1840s, Khalid was an exceptional case from his admission to Saint-Cyr in 1897, throughout his military career, and finally with his political activity within the ranks of the young Algerian national movement. See Ahmed Koulakssis and Gilbert Meynier, *L'Emir Khaled, premier Az'im? Identité algérienne et colonialisme française* (Paris: L'Harmattan, 1987).

3. In 1942 the Vichy government opened a school for Algerian and Tunisian officers near Algiers and a training center for platoon commanders at Cherchell. See Anthony Clayton, *France, Soldiers and Africa* (London: Brassey's Defence Publishers, 1988), 256.

4. The uprising involved several units posted at Fez whose soldiers revolted and killed their French instructors and other European residents of the city. It occurred a few days after the signing of the Protectorate treaty but was also an outcome of French efforts to force too rapid a modernization on the Moroccan Army. See Edmund Burke III, *Prelude to Protectorate in Morocco: Precolonial Protest and Resistance 1860–1912* (Chicago and London: University of Chicago Press, 1977), 180–87.

5. On Lyautey's tenure in Morocco and his personal impact on the shaping of French policy there, see William A. Hoisington, Jr., *Lyautey and the French Conquest of Morocco* (New York: St. Martin's Press, 1995); and Daniel Rivet, *Lyautey et l'institution du protectorat française au Maroc, 1912–1925*, 3 vols. (Paris: L'Harmattan, 1988).

6. See Lyautey's letter #844/TM, 27 Jan. 1918. The project was approved several months later by Premier and War Minister Georges Clemenceau, in his letter to Lyautey #6235/9-11, 3 Sept. 1918. Both letters are in the Diplomatic Archives of the French Foreign Ministry at Nantes [hereafter MEA/Nantes], serié Maroc/Direction de l'Intérieur [DI], file 79.

7. The *Bureaux Arabes* were created in the 1840s by Marshal Thomas Bugeaud, in an effort to solidify French rule over the Algerian hinterland. The methods adopted by the *bureaux* resulted in the destruction of local elites and the imposition of direct French rule over the indigenous population. On the history of France's native policy in North Africa, see Kenneth J. Perkins, *Qaids, Captains and Colons: French Military Administration in the Colonial Maghreb 1844–1934* (New York: African Publishing Co., 1981).

8. The change of authority was approved in late 1923, as mentioned in Lyautey's instruction #188 IM, 1 Dec. 1923, MEA/Nantes/Maroc/Direction des Affaires Indigènes [DAI], file 357.

9. The details are drawn from an undated letter sent in late 1921 by the school's director Major Quetin to Lyautey, MEA/Nantes/Maroc/DAI 357.

10. Report #700/EM, 28 May 1935, by the school director, Lieutenant Colonel Tarrit, found in the personal papers of Marshal Franchet d'Esperey at the Military Archives at Vincennes, *Service Historique de l'armée de terre* (hereafter SHAT), serié K, file 162.

11. Quetin to Lyautey, n.d., 1921, MEA/Nantes/Maroc/DAI 357.

12. Lyautey to Quetin, 11 Dec. 1921, ibid.

13. Tarrit's 1935 report #700/EM cited in note 10 above.

14. Letter #3070 3/11-2, 9 Mar. 1926, from the instruction section of the military operations bureau to Marshal Franchet d'Esperey, SHAT, serié 3H, MAROC file 604.

15. An undated report prepared in late March 1926 by Franchet d'Esperey, ibid., files 604 and 608.

16. On the controversy over reforms see then Major Tarrit's notes of previous correspondence, 20 Mar. 1929, MEA/Nantes/Maroc/DI/79. The Goums were small semiregular infantry

and cavalry units scattered throughout the Moroccan countryside. Their 150–170 man units included some 10–15 French officers and NCOs aided by some 5–10 Moroccan NCOs. The course for these subalterns was aimed at enhancing the cohesiveness of these units and rewarding the men whose service was usually tougher than that of regular NCOs. Another reform considered in 1927 was to move another NCO course from Taza to Meknes. While in power, Lyautey had opposed such a move, fearing the infusion of mediocre social elements into the academy.

17. Whereas in 1919 there were twelve Moroccan infantry battalions combined in two regiments and five cavalry squadrons combined in one regiment, by the turn of the next decade there were already eight infantry and four cavalry regiments. On the development of the Moroccan troops within the French Army during the interwar period see Clayton, 262–81.

18. The metropolitan General Staff pressed to limit the annual number of admissions to two students and to increase at the same time the number of NCOs promoted to officer ranks through the Taza course. See letter #11089/II, 20 June 1928, quoted in Tarrit's notes of 20 Mar. 1929, and the Ministry of War's final word on the subject, 17 Nov. 1928, #2964-9/11, MEA/Nantes/Maroc/DI/79. The five-student limitation remained in effect throughout the 1930s, as reflected in a request forwarded on 2 Aug. 1939 by the supreme commander of the troops in Morocco, General François, to augment the number to eight.

19. See de Jenlis's report #491/EMC to the political section in Rabat, 30 July 1946, MEA/Nantes/Maroc/DI/79. As of the early 1950s, the school introduced English as a third language taught in the cadet course. According to the recent informal testimony of a former instructor in the Royal Military Academy, the school still conducts most of its curriculum in French rather than Arabic.

20. Report, 31 Oct. 1926, ibid. The teaching of military history at the cadet course was also French and European centered, as is evidenced in an undated curriculum from the mid-1940s. The rich Muslim tradition of warfare was totally absent from the course, which instead presented various examples from ancient Greek military history.

21. Some exceptions should be noted, however, such as the case of Lt. Si Kittani b. Hamou who served in 1939 in the military cabinet of Resident-General Charles Noguès. Kittani became the first Moroccan general to serve in the French Army. Even though he remained loyal to the French to the end of the Protectorate, Kittani refused to take part in the 1953 French scheme to depose Sultan Muhammad and replace him with a member of the 'Alawi family who would be more favorable to the French interest. His reserved conduct during that affair worked in his favor after the sultan's return from exile. As king, Muhammad V appointed Kittani to head the Moroccan Royal Armed Forces.

22. These details were drawn from an article on the academy, published in the Protectorate's official Arabic-language journal, *El-Sa'ada*, #3661, 7 Mar. 1931.

23. Note by the school commander, Lieutenant Colonel Denis, on the wartime history of the Royal Military Academy, 10 May 1944, MEA/Nantes/Maroc/DI/79. A 1946 report elevated the number of officers killed to fourteen and the number of wounded to twenty-seven.

24. On the scandal and the reversed decision, see letter #258 CMK/2 by General des Essarts, military commander of Meknes, 14 Mar. 1946, ibid.

25. On the rise of Moroccan nationalist and anticolonialist sentiments, see John Halstead, *Rebirth of a Nation: The Origins and Rise of Moroccan Nationalism 1912–1944* (Cambridge, Mass.: Harvard University Press, 1967); William A. Hoisington, Jr., "Cities in Revolt: The Berber Dahir (1930) and French Urban Strategy in Morocco," *Journal of Contemporary History* 13 (July 1978).

26. The entire plan, including the residency instruction of 13 Oct. 1948, can be found in a dossier in MEA/Nantes/Maroc/DI/79.

27. Information regarding the ethnic consistency of the school was provided in a note of 20 Feb. 1953 by Colonel d'Arcimoles. Only 30 percent of the 112 students were labeled as "sons of notables" and the majority came from rural backgrounds. These figures refer to all the students in the school and not merely to the cadets. It is clear that during the last years of the Protector-

ate the French promoted more Berbers than Arabs to command posts, a practice adopted by the royal family during the first decade of Moroccan independence.

28. The only indication of such an influence appeared in a letter dated 26 Dec. 1955 by the school's director, Lieutenant Colonel Delhumeau, MEA/Nantes/Maroc/DI/79. Referring to his students' state of mind, he noted a sentiment that Morocco should get closer to the Arab world and affirm its Muslim character. An indication of this sentiment was the expression of hatred toward the Jews, who were accused of disloyalty. Another sign was the expressed wish of several students to watch Egyptian movies in school. All this did not jeopardize the basic everyday discipline of the school, nor did it affect the relationship between the students and their French instructors. All students expressed the hope that they would be chosen to represent the school in the July 14th parade in Paris.

29. For example, the Class of 1955, the last one to be enrolled before independence, included 38 students chosen from among 112 candidates. Only 11 were sons of notables and functionaries and 6 others were sons of veteran combatants. Twelve came from merchant families, 6 were sons of farmers, and 3 came from working-class families. Only 4 students paid full tuition while the others received complete or partial scholarships.

30. This pledge was stipulated in the protocol annexed to the Franco-Moroccan Declaration of 2 Mar. 1956.

31. In July 1957 there were 150 French officers, 550 NCOs, and 130 soldiers serving in FAR units. They served in every position except for the infantry, and carried out technical and supervisory tasks. Serving on an individual, voluntary basis, French personnel wore Moroccan uniforms and received their salaries from the Moroccan treasury with additional compensation from the French Army. Useful information on the assistance provided by French military men in the formation of the FAR and their status within its ranks was provided by one of those volunteers, Captain Callery, in a research paper he submitted to the *Centre des Hautes Études Administratives sur l'Afrique et l'Asie Modernes* (C.H.E.A.M.) in January 1960 (#3235), entitled *Les Forces Armées Royales Marocaines.* A wider perspective of the events of that initial period in the history of the FAR can be found in Alain Bibillan's article, "Le rôle des Forces Armées Royales dans l'état marocain," which was submitted in 1963 to C.H.E.A.M. (#3871) and published anonymously three years later in Leo Hamon, ed., *Le rôle extra-militaire de l'armée dans le Tiers-Monde* (Paris: Presses Universitaires de France, 1966).

32. The Spanish, who controlled northern Morocco and two smaller enclaves along the Atlantic coastline, recruited Moroccan soldiers to their armed forces. Moroccan soldiers formed a significant portion of the troops supporting Gen. Francisco Franco's revolt against the Republican government in Madrid in 1936. After he assumed control over Spain, Franco continued to rely on Moroccan troops and even promoted a Moroccan officer, Mezziane, as the first non-Spanish general in his army. Mezziane returned to Morocco in 1956 and was appointed inspector-general of the Moroccan Army.

33. On the history of that movement, see Zaki M'barek, *Résistance et Armée de Libération: portée politique, liquidation, 1953–1958* (Tangier: Éditions du Nord, 1987).

34. Ahardan was later one of the founders of the *Movement Populaire* and held various administrative and ministerial posts, including minister of defense. See William Spencer, *Historical Dictionary of Morocco* (Metuchen, N.J. and London: Scarecrow Press, 1980), 22.

35. On the 1971 coup attempt, see John Waterbury, "The Coup Manqué," in Ernest Gellner and Charles Micaud, eds., *Arabs and Berbers: From Tribe to Nation in North Africa* (London: Duckworth, 1973). On the role of the military in Morocco's postcolonial national life, see J. Régnier and J. C. Santucci, "Armée, pouvoir et légitimite au Maroc," in M. Teitler, ed., *Élites, Pouvoir et Légalité au Maghreb* (Aix-en-Provence, France: C.N.R.S., 1973).

The Education and Training of the Mexican Officer Corps

Roderic Ai Camp

Among professional groups providing leadership in the contemporary world, perhaps the single, most important formal credential that both establishes their access to decision-making circles and has increasingly become a de facto requirement for joining these inner circles, is education generally, and higher education specifically. Education has become a determinant of who becomes a leader, as well as a fundamental, formal characteristic of leadership. The officer corps, as one of the more important groups in most nation-states, both in performing its own professional mission and in carrying out other, broader political tasks, is not immune to this universal leadership quality.

Except possibly for similar social origins, higher levels of education most characterize military and civilian (whether religious, cultural, or economic) leaders. This pattern has emerged largely in response to the almost exponential growth of knowledge workers and knowledge industries in this century.[1]

Mexico is not an exception to these general trends, and the Mexican military shares in the larger societal patterns. With respect to education, however, the officer corps can be distinguished from other Mexican leadership groups in several ways. First, compared to all other groups it is the least educated in terms of level or quality of education, lagging far behind civil professionals in college and postgraduate education (see Table 1). Second, like the Catholic Church in Mexico, it has provided its own educational institutions for developing leaders, rather than relying on those staffed and attended by nonmembers of the profession. Third, it has concentrated its military education in a small number of schools; indeed, the bulk of the officer corps (largely made up of army officers) has been educated in just two institutions since the Mexican Revolution of 1910.

Describing the breadth of functions performed by education may help explain why it has become such a significant variable in the background of Mexican military officers. Typically, we think of education as performing a cognitive function, imparting knowledge to both students and teachers. Within the officer corps, that cognitive function has been directed at creating a "professional" soldier—an individual trained in the credo, skills, and mission of his profession. In Mexico, as in many other countries, military academies were established specifically to provide this type of preparation, but they also served to eliminate self-trained or self-designated officers who had emerged on the battlefield from 1910 to 1920 as leaders of Mexico's popular revolutionary army. This victorious institution became a state ally after 1920.

Mexican leadership, political and military, believed formal education would accomplish professionalization in the political sense as well. It is crucial to recall that the civilian political leadership making these decisions from 1920 through 1946 had themselves been top revolutionary army officers.[2] Military politicians concluded that the best way to safeguard the new political model they were crafting, and to provide continuity to its political leadership, was to weed out military political opponents from the officer corps through stricter, formal, professional credentials. They were convinced that such controls would steer military officers away from political careers.

Mexican leaders saw military educational institutions as the means to these ends. In 1920, President Alvaro Obregón ordered the prerevolutionary *Colegio Militar* reopened. It closed briefly in 1925 for repairs, reopening again the following year, staffed by officers trained in the United States and Europe, or in Mexico prior to 1911. It became the *Heroico Colegio Militar* in 1949. A little more than a decade after the *Colegio Militar* reopened, the president in 1932 instructed the secretary of defense and the director of military education to establish the *Escuela Superior de Guerra,* considered to be the capstone educational experience for junior officers thought to be general officer material.[3]

These two institutions also performed a complementary task, molding a new, junior officer corps in postrevolutionary values. This socializing function, designed to establish a coherent set of military values, involved developing a sense of loyalty to the armed forces' organization, and to the institutions' position vis-à-vis civil institutions in society. An examination of the military schools' curricula over time, and the behavior of their students in the classroom, suggests some important general features of junior officer education, features that have remained in many respects unchanged from the 1930s through the 1970s. During the 1970s, the *Escuela Superior de Guerra* added new courses, invited outside lecturers from the intellectual, political, and business worlds, and increased its curriculum's sociopolitical content.[4] It is helpful to highlight some of these values to understand education's impact on junior officer mentality and roles, as well as civil-military relations broadly conceived.

Table 1

Historic Patterns of Education among Mexican Generals and Civilian Politicians, 1884–1935

Educational Level Achieved	Generals (%)	Politicians (%)
Primary	29	5
Secondary	7	4
Normal	3	3
Preparatory	31	12
University	29	73
Total	99	97

Source: Mexican Political Biography Project.
Note: Percentages do not add up to 100 due to rounding.

Nationalism

Mexican officers were given a strong dose of nationalism; specifically, they viewed the United States suspiciously, particularly given its subversive and overt intervention in Mexican affairs, and its violation of Mexican sovereignty in 1914 in the invasion of the port of Veracruz, and again in 1916 as the U.S. Army chased Pancho Villa throughout northern Mexico after his attack on Columbus, New Mexico. Under a traditional concept of professionalism in which the army's mission is to defend the nation against external enemies, Mexico considered the United States as its primary threat and developed an appropriate defense strategy. Realizing it could not prevent the United States from invading its national territory, Mexico planned to resist through a popular war strategy similar to that used by the masses during the 1910 Revolution, and by applying the guerrilla concepts popular in the Third World after the Communist victory in China in 1949.

Ironically, the Mexican Army also has viewed the United States military as a professional force worth emulating, especially following the Axis defeat in World War II. Mexico ultimately became a U.S. ally during the war, sending a small expeditionary force from its air force to the Philippines.[5] The military has long used American military manuals in many of its courses, although typically the manuals tend to be out-of-date. Generally, U.S. Military Academy officers visiting the *Colegio Heroico Militar* and the *Escuela Superior de Guerra* have described the professional quality of the education as mediocre at best.

The historical antagonism between Mexico and the United States affects not only how nationalism is construed within the armed forces, but also the degree to which the Mexican military collaborates with the United States, and the level of training Mexican officers receive from the United States. The extensive U.S. Military Assistance Programs common to the Latin American region have had little influence in Mexico; basically, Mexicans have resisted them, particularly U.S. officers training Mexican officers or troops on national territory. It is not always clear whether military officers would have preferred to increase their collaboration with the United States, or whether they were being guided by civilian superiors' decisions.

As is so often the case within the Mexican military, contradictions abound. Since World War II, the American Cold War mentality definitely influenced Mexican military officers' views. In fact, potential officer candidates often expressed anti-Communist views as a reason to join the army, and to enroll in the *Heroico Colegio Militar*. This mentality is not dead among Mexican officers. As recently as 1995, a division general, one of fewer than twenty on active duty, indicated to the author in discussing the *Ejército Zapatista de Liberación Nacional* uprising in Chiapas in January 1994 that its primary cause could be attributed to outside Communist infiltrators, a view no serious analyst has offered.[6]

Despite this ideological affinity, any Mexican military officer viewed as pro-American will have difficulty being promoted, and will be warned by superiors if perceived in that light. The most prominent indicator of anti-American paranoia, and underlying a deeper institutional value, is the armed forces' prohibition against offic- ers speaking to foreigners generally, and especially to Americans, without official permission from their superiors. General officers are under the same prohibition, but consider it demeaning. Nevertheless, the restriction suggests the degree to which the Mexican armed forces wishes to separate itself from its American counterpart, and more generally, reflects its closed, suspicious attitude toward outsiders. U.S. military attachés repeatedly have attested to the difficulties they encounter in estab- lishing a trusting relationship with their Mexican peers, and in obtaining the most basic information from them.

It is therefore doubly ironic that Mexican Army, and even more so Air Force and Navy officers, receive advanced training at bases and academies in the United States. Indeed, nearly all the Mexican military's officer foreign training takes place there (even though the *Escuela Superior de Guerra* was originally patterned after the French model).[7] Most of this training involves specialty instruction in tactical and technical skills; for example, in armored vehicle employment for army personnel, or in submarine warfare for navy officers. More recently, however, Mexican officers have pursued advanced studies at U.S. command schools. In fact, Mexico's two leading officers, the secretary of defense and the secretary of navy, Gen. Enrique Cervantes Aguirre and Adm. José R. Lorenzo Franco, have spent considerable time in the United States, Cervantes Aguirre as an assistant military attaché and military attaché in Washington, D.C., and Lorenzo Franco as a student at the Naval War College, and as an assistant naval attaché in Washington, D.C.

Caste Mentality

A second value emphasized within Mexican officers' educational experiences is a high level of secrecy and protection from the outside world, especially from prying civilian eyes. The degree of ignorance in Mexico about the armed forces, about the defense budget, even about the role the military should play, stems from the atti- tudes engendered in the military academies and the civilian authorities' tolerant posture. This inclination toward a protected institutional environment can be blamed on both the civilian and military leadership. Civilian leaders reached a modus vivendi with military leaders including an agreement not to interfere in military affairs. While this understanding may have been essential to achieve civilian supremacy over the military, it has led to an artificial and detrimental relationship between the military and society, one in which the military always receives praise from politicians, and a pro- tected status in the media and from intellectuals regarding open discussion of the military's merits or attributes.

The military reinforces this caste-like mentality early in the officer education process. The *Heroico Colegio Militar* is essentially a preparatory school. Most cadets enroll at age seventeen, complete four years, and graduate as second lieuten- ants in the traditional military occupational specialties. At the time he enrolls, how- ever, the typical cadet is required to have only a primary and secondary education.

Of those officers reaching general rank and holding top defense posts in 1990 (likely to be the most well-prepared educationally), none had obtained a preparatory education before enrolling in the *Colegio Militar*. Among naval officers of equal rank and position, only 10 percent had completed a preparatory program. Those junior officers thought to have general officer potential and selected for training at the *Escuela Superior de Guerra*, typically are first lieutenants and captains, although for a period of time captains and majors were more common.

The junior officer destined to become a general officer will have attended both academies. In recent years officers have begun listing their military education from the *Escuela Superior de Guerra* as a *licenciatura* (equivalent to a college or professional degree). That Mexico's most successful officers spend six to seven years in training, early in their careers, when they are most susceptible to socializing, helps develop a protective layer around their perceptions and experiences.

Military officers spend most of their time living and eating within the confines of the academies (both located in Mexico City, the political capital) where they are supervised and largely taught by other officers. A few civilians teach languages, and the *Escuela Superior de Guerra* often boasts a single U.S. officer who teaches English. Essentially, however, junior officers are not exposed to civilian instructors' viewpoints. When civilian instructors are used, as in the case of the *Colegio de Defensa Nacional*, which educates only lieutenant colonels and colonels who are on track for promotion to general, they typically are known personally to someone in the military and, in many cases, have kinship ties to career military families.

The level of protective layering sought and achieved by the Mexican military reinforces its isolation from other leadership groups and from society generally. It has also led to the military's reluctance to share even the most basic information about itself with other sources. For example, the military will not reveal the names of the generals presently in command of Mexico's thirty-six military zones, although this is essentially public information in each respective zone. More importantly since 1989, in an attempt to shut itself off from inquiring eyes, Mexican and foreign, the military stopped distributing its official journal, *Revista de Ejército y Fuerza Area*, to the Library of Congress, the University of Texas, Harvard University, and the *Colegio de México* (in Mexico City). Without the journal, no outsider can learn much of anything about changing army perceptions, behavior, and policies.

Hierarchy

Perhaps the most important value ingrained in the minds of Mexican cadets and young officers, especially at the *Escuela Superior de Guerra,* is subordination to superiors. Obviously, obeying orders and taking direction from superiors are part and parcel of every military organization, formal and informal. In Mexico, however, the hierarchical command structure is among the most inflexible and centralized in the world. As Frederick Nunn has argued:

Leadership and discipline of an army, of a collectivity, were based on hierarchy and subordination, on obedience. These in turn were predicated on willingness to submit. Without esteem and admiration for the leader, discipline might become a matter of fear and intimidation. And these had been disdained in print regularly since the early years of the century. In Mexico, just as in other countries, "men were not automatons, but men subject to emotions, feelings, complexes, and passions."[8]

In Mexico the hierarchical command structure has three features. First and foremost, the secretary of defense is the most important decision-maker in the armed forces because the army constitutes about 130,000 troops and officers, compared to some 50,000 in the navy and air force combined. The secretary of defense commands the officer structure with an iron hand, and it is no accident that he is often referred to in private as a "little god." Biases existing in the promotion structure and in policy-making are increasingly apparent in the decisions of each secretary of defense (or navy), an officer who has led his service in each presidential administration for a full presidential term. In fact, the secretary of defense has been the most stable position in the Mexican cabinet since the 1950s.[9]

The secretary of defense rules with such an authoritarian hand that he controls many of the most insignificant personnel decisions, such as approving individual leave requests. This centralized control has led to abuses and to many private criticisms by senior officers, who are given prestigious assignments with little decision-making authority. Thus officers learn very early in their careers that they exercise little autonomy, must obey their superiors, and will obtain few command skills. In a recent internal document, dated June 1995, officers complained that the promotion process at the higher levels should follow guidelines regarding time in grade, and even more significantly, that the entire command structure should be decentralized to give greater authority to regional commanders in the field.[10]

Second, the curricula in the military academies instill junior officers with a strong respect for their commander in chief, the president of Mexico. Military subordination to civil authority is also deeply ingrained in the minds of the officer corps throughout the training process. On public occasions, the defense and navy secretaries frequently profess their loyalty to the president or civilian administration. In turn, presidents typically praise the armed forces and the officer corps for their service to country and their institutional loyalty.

Of equal importance is that through the public educational system, all Mexican citizens grow up learning that the military has been subordinate to civil authority since the 1930s, and that it should not be directly involved in political matters. Officers, just as any other Mexicans, are also products of a familial upbringing and of a largely public grammar school education; these additional experiences reinforce what they learn in their own professional schools.

Third, and finally, the secretaries of defense and navy are appointed by the president. In recent years, however, these individuals have not been known personally to the incoming president.[11] Consequently, presidents rely on the incumbent secretaries of defense and navy to recommend the most appropriate candidates for these offices. Given the extraordinary power wielded by the secretaries, officers hoping to succeed in the upper echelons realize they must subordinate themselves to those interests rather than presidential preferences. This pattern suggests that officers may establish, at least informally, greater loyalty to the secretary of defense or navy than to any individual president. The oath taken by *Escuela Superior de Guerra* students in 1987, stating that an officer's loyalty redounds to the benefit of the armed forces first, and country second, illustrates this point.[12]

Discipline

Complementing the rigid, hierarchical structure in the armed forces is the discipline designed to reinforce the hierarchical rigidities that is imposed on younger officers in the military academies. The second-tier school, the *Escuela Superior de Guerra,* performs the key task of weeding out those first lieutenants and captains who have too much imagination and self-esteem to tolerate such treatment. It can be argued that given the mediocre, military and nonmilitary content of the *Escuela Superior de Guerra's* curriculum, its disciplinary efforts are its foremost contributions.

The *Escuela Superior de Guerra's* disciplinary influence can be illustrated in two ways. First, the school boasts an extraordinarily high washout rate. Those officers who fail to complete the program, and therefore fail to obtain the coveted *diplomado* insignia, have a lesser chance of going beyond lieutenant colonel. The failed officers rarely wash out because they lack academic ability, but because they will not tolerate the abuse directed toward them by their instructors, most of whom are captains and majors who have graduated previously.

Second, this exaggerated level of discipline, matching a similar level of hierarchical rigidity, can be seen in the common practice among students who are forced, as a group, to cheat in order to complete their assignments. This perversion of the military education process not only compromises a basic value, the importance of integrity, which might set the military apart from some other leadership groups, but also reinforces an alternative set of values among budding officers. In other words, junior officers learn that they must do what they are told (hierarchy reigns supreme), that they must obey regardless of the justice or integrity of the command (discipline above all else), and that honesty is not valued within the officer corps (institutional and personal self-interest above integrity), at least that modeled by instructors directly within the educational process.[13]

In contrast, civilian instructors at the *Colegio de Defensa Nacional* suggest that they detected a change among that generation of officers who reached their classrooms in the late 1980s and who were cadets during or after the 1968 student massacre in Tlatelolco. They describe these post-1968 officers as more willing to raise questions in class, to debate issues among their peers and with the instructor, and to deviate from the formal strictures of discipline imposed on them from earlier, military educational experiences.

Recruitment

Military academies in Mexico function beyond value formation and professional education by also performing recruitment. This function is by no means minor. Indeed, it can be argued that military educational institutions are the most important locus of career advancement within the Mexican military. It is not just a question of each officer obtaining the "correct," formal educational credentials. It is true, as previously suggested, that seven out of ten officers hoping to achieve general rank will have graduated with a *diplomado de estado mayor* from the *Escuela Superior de Guerra*. But all officers will have passed through the *Heroico Colegio Militar,* unless they are from the medical officer corps. There is essentially no other route available for army officers than through a military academy. The navy does have a direct commission program, so occasionally civilian administrators with military rank, who have not attended naval academies, can be found in important posts.

The academies possess considerable prestige within the Mexican military. One way to measure their level of importance is to examine the careers of Mexico's most successful officers. Interestingly, nearly every flag officer has taught at one or more of these academies, particularly at the *Escuela Superior de Guerra*. For example, among all officers reaching brigadier general in the army or air force, or rear admiral in the navy, 83 percent taught at the *Heroico Colegio Militar*, the *Heroico Escuela Naval Militar*, or the *Escuela Superior de Guerra*. The remaining 17 percent also had taught at applied military schools. A teaching assignment, therefore, is a coveted position for an ambitious junior officer.

Individuals who are selected to direct their alma maters typically pursue successful careers beyond that post leading to top positions in the defense or naval secretariat, or as commanders of major military and naval zones. Between 1940 and 1988, of the fifteen directors of the *Heroico Colegio Militar,* all reached top positions in the secretariat of defense, and three, Marcelino García Barragán, Gilberto Limón Márquez, and Enrique Cervantes Aguirre, became defense secretaries.

Attending or teaching at the military academies, given the small size of each cadet or officer class, brings these officers in close contact with one another, especially those enrolled in the same military occupational specialty at the *Heroico Colegio Militar,* or in each class at the *Escuela Superior de Guerra*. The *Heroico Colegio Militar* has recruited approximately 250 students yearly in the early 1990s, smaller than the average high school class in the United States. Fewer than thirty officers graduate in the third-year class at the *Escuela Superior de Guerra*.

These shared experiences, whether as students or teachers, at the two academies training junior officers lead to life-long friendships and contacts crucial to career success. An analysis of the thirty-one general officers in the Mexican secretariat of national defense from 1988 to 1994 demonstrates their extent and importance. For example, the secretary of defense, Antonio Riviello Bazán, taught two of his appointees, Gen. Tomás Salgado and Gen. Antonio Clemente Fernández Peniche, at the *Escuela Superior de Guerra*, and the first of these officers joined the 43rd

Infantry Battalion under Riviello's command when he graduated. (Salgado recently was appointed as police chief of the Federal District.) Riviello's assistant secretary, Gen. Alfredo Ochoa Toledo, taught four additional appointees as an instructor at the *Escuela Superior de Guerra*. He attended the *Heroico Colegio Militar* with eight of his subordinates in the defense secretariat, was a classmate of another collaborator at the *Escuela Superior de Guerra*, and overlapped his training with five other officers at the latter academy.

Conclusions

These educational contributions to junior officer values and careers have been significant in the Mexican armed forces for many decades. It is interesting, therefore, that in the 1995 internal document, army and air force officers pointed to a number of concerns, implicitly or explicitly impacting the quality and direction of their educational experiences. What are these concerns and how do they affect the patterns described above?

In the first place, the officers requested that entrance standards be raised at the *Heroico Colegio Militar,* specifically that preparatory education be required for all cadets. It is difficult to know precisely why such a request was made, but it can be surmised that officers now believe they should be better educated, both to improve their own professionalism, but also to equal more closely their peers' educational preparation, particularly that of politicians, with whom they have felt a sense of inferiority in the past.

The consequences of such an increased educational requirement would be significant. Because preparatory education is pursued by those Mexicans following an academic rather than a vocational track (in the United States they would be considered college bound), it would produce a better caliber of entrant. Moreover, it would delay by several years, and therefore age cadets entering the *Heroico Colegio Militar,* decreasing somewhat their susceptibility as teenagers to military values. Furthermore, a slightly older group of recruits with the educational experience of the typical preparatory program would possess broader individual backgrounds and sophistication.

Second, and noteworthy among the officers' concerns, are demands that the armed forces examine the issue of recruiting civilians to the military and, equally important, that the number of civilians be increased among the training staff. The officers asked to have civilian professionals and specialists in the military, not only in teaching functions, but assigned to other appropriate tasks, and remunerated accordingly. With respect to teaching, they called for "creating attractive salaries" for professors. Increasing the number of civilians within the military, especially as teachers, has interesting implications for maintaining the military's caste-like quality. A greater civilian presence in the academies and the armed forces would expand contacts between civilians and the military, and introduce the possibility for broader influences on junior officers during their formative period.

The third request in the 1995 report having direct implications for education is the officers' desire to rely more heavily on professional training abroad, specifically involving special forces, military police, infantry, armor, engineering, and medical units. Although large percentages of flag officers in all three services have trained in the United States, approximately only one out of twenty officers of all ranks has received foreign training. Despite nationalistic sentiments among older officers, the officers who authored this report obviously believe their training is inadequate, and that foreign training is essential. Such training naturally exposes officers to the values and technical expertise of their instructors, adding additional influences to what has been a fairly tightly controlled, homogeneous military education.

Fourth, the report called on the military university system to offer "open careers." Although the officers did not specify precisely what they meant, they suggested that they be given the opportunity to obtain degrees in fields other than narrow, military sciences. In other words, an officer might be trained within the social sciences, including economics. The request for "open careers" has two implications. Such typical civilian academic offerings would expose junior officers to a much broader curricular content, give them more in common with their civilian peers, and make their degrees more compatible with and useful in the civilian world. In addition, although not expressed in these terms, such a change might enable civilians to enroll in military university system programs.

Fifth, and finally, the officers implied criticism of the academies' content, arguing that the curricula at the *Escuela Superior de Guerra* and the *Colegio de Defensa Nacional* did not correspond either to the national reality or to the realities faced by the armed forces. The implication is that they want more courses focusing on sociopolitical problems and national security issues rather than merely tactical and strategic issues pertaining to military sciences. Considering their emphasis elsewhere in the document on antiguerrilla training, and interestingly, on the armed forces' carrying out "real" social work in rural villages, the officers believe that they require more knowledge about social conditions in Mexico, that they need to help alleviate these conditions before they generate guerrilla-initiated violence, and that their future focus, short and long term, should be on domestic threats to national security.

Notes

1. James Davison Hunter, "Religious Elites in Advanced Industrial Society," *Comparative Studies in Society and History* 29 (April 1987): 361.

2. These were Gens. Alvaro Obregón, 1920–24; Plutarco Elías Calles, 1924–28; Pascual Ortiz Rubio, 1929–32; Abelardo Rodríguez, 1932–34; Lázaro Cárdenas, 1934–40; and Manuel Avila Camacho, 1940–46.

3. Lyle McAlister, *The Military in Latin American Socio-Political Evolution: Four Case Studies* (Washington, D.C.: Center for Research in Social Systems, 1970), 206.

4. Michael J. Dziedzic, "Civil-Military Relations in Mexico The Politics of Cooptation" (Unpublished paper, University of Texas, Austin, 1983), 25; and William S. Ackroyd, "Descendants of the Revolution: Civil-Military Relations in Mexico" (Ph.D. diss., University of Arizona, 1988), 123.

5. It is interesting to note that the majority of these junior officers provided the leadership of the small, Mexican Air Force for years to come.

6. The best appraisal of this in English is by Stephen J. Wager and Donald E. Schulz, *The Awakening: The Zapatista Revolt and Its Implications for Civil-Military Relations and the Future of Mexico* (Carlisle Barracks, Pa.: Strategic Studies Institute, U.S. Army War College, 1994).

7. According to a leaked U.S. Defense Intelligence Agency report, Mexican officers during 1994–95 obtained training in Britain, Spain, and Israel, as well as in various Latin American countries. The British instructed them in laying land mines, the Chileans and Argentineans in antiguerrilla tactics, and the Guatemalans put them through the brutal antiguerrilla Kaibil course, which only two out of six soldiers passed. *El Financiero*, 16 Sept. 1996, 4.

8. Frederick M. Nunn, *The Time of the Generals: Latin American Professional Militarism in World Perspective* (Lincoln: University of Nebraska Press, 1992), 88.

9. President Carlos Salinas was forced to fire his initial Secretary of Navy Adm. Mauricio Schleske Sánchez in June 1990, when it was discovered that the officer possessed "inexplicable wealth." According to an investigation published in *Proceso,* he and others at naval headquarters were running a drug-smuggling operation. Neither the president nor the secretariat of the navy ever mentioned reasons for the dismissal. See Carlos Marín, "Inexplicablemente rico, Schleske omitió declarar sus residencias en Houston," *Proceso*, 3 Aug. 1990, 8–13; and *Proceso*, 23 July 1990, 8.

10. "Programa de desarrollo del ejército y fuerza aerea mexicanos," Unpublished manuscript, June 1995, 1–14.

11. Author's interviews with ex-presidents José López Portillo (1976–82), Mexico City, 1991; and Miguel de la Madrid (1982–88), Mexico City, 1990.

12. Article 145, Secretariat of National Defense, *Reglamento de la Escuela Superior de Guerra*, 31.

13. As Michael Dziedzic suggests, the presence of group cheating at the *Escuela Superior de Guerra* is more compatible with turning out officers willing to break the rules in return for personal advancement than to sacrifice for their country. "The Essence of Decision in a Hegemonic Regime: The Case of Mexico's Acquisition of a Supersonic Fighter" (Ph.D. diss., University of Texas, Austin, 1986), 28.

Beyond Classrooms and Parade Grounds: Education, Professionalism, and Militarism in South America

Frederick M. Nunn

The professional officer of the late twentieth century has an educational background that contrasts more than it compares with that of his or her counterpart a century earlier.[1] Education in the military, just as in many other professions, has become a process of "life-long learning" with professional development programs originating and expanding to meet work place or other professional needs. This is particularly characteristic of Latin America, and is specifically demonstrable in southern South America, the Southern Cone, where the professionalization process began in the late nineteenth century. It is that process, especially its educational component, that more than any other cause, led to the effect we call the institutional *golpe de estado*, the coup d'état writ large.

In the second half of the twentieth century the majority of South Americans spent decades under military rule. As time wore on, military rule assumed an institutional guise and ceased to represent cliques and coteries of officers led by personalistic *caudillos*. Military rule in the 1970s and 1980s was institutionalized and purposeful, not designed to bridge a gap between civilian administrations. Ambitious though they might have been, leaders of military movements in Argentina, Brazil, Chile, and Peru knew they led armed forces that were, to one degree or another, professional.

In the professionalized armies of the Southern Cone the educational process envisioned a century ago developed slowly over the early decades of this century. By the outbreak of World War II, each of the countries just named had specialty and staff schools where military academy graduates studied in order to advance in the career of arms. Junior officers knew they had to do well in their coursework if they wanted to become senior officers. Indeed, cadets knew from the onset of their studies that the *escuela, colegio,* or *escola* was not the end of their education. A military career meant study as much as drill, and later research as well as maneuvers.

"Knowledge is power," said Francis Bacon four hundred years ago. "In a time of troublesome change, it is more true than ever that knowledge is power," said John F. Kennedy nearly four centuries later. Both Bacon and Kennedy lived in interesting times, *times of change*. In such times knowledge was brought to bear on situations at hand: whether on political struggles in early modern Europe or on those of the Cold War. Knowledge is power still in the early stages of the New World Order.

347

Knowledge gained by cadets in Southern Cone armies in this century has been brought to bear often on situations at hand. As young officers they have increasingly followed middle ranking and senior officers whose continuing education (and whose knowledge, one trusts) qualifies them to lead, to plan, and to execute policy. From the end of World War II senior officers have followed other seniors whose education has gone beyond the level of command and staff school to that of advanced courses at national defense and security schools at home, and specialized schools abroad. From Brazil's *Escola Superior de Guerra* or Peru's *Centro de Altos Estudios Militares* it has been a brief quick-march to seizure of power. Attempts to link training at the School of the Americas or other U.S. institutes to political action are less successful by scholarly standards than those making a connection between advanced coursework in Latin American schools and participation in institutional *golpes* at home.

What I am asserting here (and what I have shown elsewhere) is that professional education in South America is an integral part of the development of professional militarism. Professional militarism and its political vehicle, the institutional *golpe,* simply would not have existed without officer education after the classroom and beyond the parade ground. The education of junior officers can achieve only so much. Essays by Roderic Camp, Moshe Gershovich, and Alfred McCoy included in these pages make this pretty clear. For this reason, I propose to concentrate on one of the ways in which ongoing schooling of academy (used generically) graduates can be measured and assessed as to content.

There is a correlative relationship between rank, educational level, and ability to think critically about national problems. A similar relationship also exists between the acceleration of cultural, economic, political, and social change and the application of knowledge gained in the military educational process to the solution of those problems. Ideation exists among cadets, to be sure, and rank brings privileges and authority, but continuing education increases the capacity for critical thinking. When opportunities come along for officers to ideate, wield authority, and think critically about national problems, civilian political systems may be vulnerable.

George Orwell reminded us of what Euripides and Thucydides wrote in the fifth century B.C., and of what George Santayana would say about remembering the past in order to understand the future. Orwell had one of his characters in *1984* insist that control of the past meant control of the future and that control of the present meant control of the past. Today in southern South America the past is up for grabs, the future more than uncertain. Officers of all ranks and educational levels know this. They think and write about it.

We know they know this because of what they write. They care a lot about the past, "the good old days." They are uncomfortable with the present, always have been. And they worry about future uncertainties, always will. They are no longer useful counterpoises to communism or populist nationalism. Their educational systems are in a state of flux unlike anything seen since the professionalization process began under French and German tutelage a century ago.

What they write—they being professional officers and select civilians—in their professional journals is both the result of the educational process that produces junior officers and the continuing or advanced education that produces commanders and staff officers. What they write is based on the knowledge attained as cadets, the ideation expected of leaders, and the critical thinking required of those who grapple with defense, security, and professional development at the dawn of a new international paradigm that has already provided challenges to governments of the Southern Cone.

Their writing expresses the thought and self-perception of professional officers of all ranks. As in most other professional journals, degree of sophistication usually can be linked to rank and educational level; so can degree of intensity. Intellectuality and ideological content also increase with time and professional status. Those who write are both students and teachers; those who read are cadets, advanced students, and commanders. The education of junior and senior officers, this is to say, is a symbiotic process. What appears in the professional journals of the armies of Argentina, Brazil, Chile, and Peru evinces the intrinsic qualities of military education from beginning to advanced levels.

There is much consistency and much to compare in the thought and self-perception of Latin American professional army officers.[2] Over the decades officers writing in their professional journals have expressed remarkably similar thoughts about democracy and politics, military-civilian (and, of course, *civil-military*) relations, the importance of spiritual values, the military as a unifying social and developmental force, ideological struggle, and the differences between military and civilian education.

Military men (and lately women) think much alike and do so consistently, national frontiers notwithstanding. Decade after decade the same issues concern them all—many of the same issues that concern civilians, too, by the way. Officers from South American countries have always worried, for example, that substantive shifts in national domestic and foreign policies put defense capabilities at risk and that accelerated change on the international scene may do the same. They have found fault for decades with the way "things seem to be going." This holds for most of Latin America as much as it does for the rest of the hemisphere and for Eurasia.

Recently the pace of change has made comparability and consistency more significant than in the past. The uncertainties and the perceptions that they create frustrate military leaders in more ways than they did, say, at the beginning of the Cold War or on the eve of either of the two world wars. When I say "over the decades" I mean "for the past century."

In three recent essays I analyzed, compared, and contrasted Latin American officer corps' thought and self-perception in different contexts. In an essay of 1995 I discussed consistent and comparable thought on (re)democratization expressed in the published works of Argentines, Brazilians, and Peruvians for the 1950s and 1960s, juxtaposing it with that of Chileans from the 1990s.[3] My conclusion was that much of what was being thought and said by the latter had been said already by their Southern Cone neighbors decades before. Neither the phenomenon nor the reaction to it was all that new, I thought. I still think so. For good measure I reviewed what Spanish officers had been writing about since the end of the Franco regime in 1975 and what Paraguayans had written since the end of the *Stronato* in 1989. In both form and content, professional military thought and self-perception were comparable and evinced consistency across national borders and decades.

This comparability and consistency in form and content is mirrored by that of the (re)democratization process itself. In assessing perceived roles of Latin American armed forces and in analyzing the ideological content and intellectual quality of their thought, it is a good thing to remember that they "have been here before." Equally important, "(re)democratization" is not a term very widely used by the Latin American military. Needless to say I understand that history does not repeat itself, so "having been here" is a relative concept.

(Re)democratization has occurred on more than one occasion in the recent pasts of Argentina, Brazil, and Peru. The process that has caused a flurry of speculation worldwide recently is one that—from a Eurocentric viewpoint—appears to have broken out with regime changes in southern Europe in the mid-1970s: Greece, Portugal, Spain. From a Western Hemispheric viewpoint Brazilians endured this experience beginning in the 1940s with the end of the *Estado Novo*. Argentina began the process in the mid-1950s after Juan Perón was driven from power, and Peruvians have undergone it in fits and starts from about the same period. That each had to begin all over again at various junctures in the 1960s, 1970s, and 1980s substantiates the assertion that the process can be a long and difficult one for some countries. What happened in Iberia and Greece is, by European standards only, worthy of note as a beginning. What has happened over and over again in the Southern Cone cannot but have affected the officer educational process at all levels.

Latin American army officers have had plenty of experience with the process throughout their careers. From the time they "just nibbled at the outer crust of learning," to use Michel de Montaigne's felicitous phrase, officers have been well aware of the assets and liabilities that attend liberal democracy and international capitalism. Their intellectual and ideological focus has been consistently on democracy's weaknesses, its vulnerabilities. For this they have often been accused of authoritarian designs, of being *golpistas* and worse. Allowing for a degree of accuracy in negative appraisals of their political acumen and behavior (it is hard to credit Alfredo Stroessner, say, with foresight regarding what democracy needed in order to take root in the South American heartland), it is reasonable to say that officer corps' thought and self-perception in Argentina, Brazil, Chile, and Peru reflects support for functional (and theoretical) democracy—as long as it works to the advantage of state, nation, and society, and, needless to say, the military profession.

In two other contexts I juxtaposed Southern Cone officer corps' thought and self-perception both at the beginning and near the end of the Cold War.[4] Comparability and consistency reigned. Uncertainty, wishful thinking, a certain nostalgia for the known, worries about the unknown, fear (and occasional loathing) of civilian rule characterized essays by officers that dealt with change. For good measure I looked at literature dating from the early 1990s from countries beyond the region—Bulgaria and Canada—to test a simple hypothesis: the degree of comparability and consistency in all professional military thought and self-perception (that is, not just Latin American) increases with the internationalization of military culture regardless of national or international context. In other words what I had argued in *Yesterday's Soldiers* (1983) and *The Time of the Generals* (1992) was still valid. Educational systems I described in those works cannot have escaped the effects of national and international events. If the preceding continues to hold, both comparability and consistency should characterize military thought and self-perception set down in writing during the early stages of the New World Order.

Two phenomena condition thought and self-perception deriving from academy, institute, and school experiences in this contemporary context. First, is the devolution process, which begins with the transition to civilian rule, becomes adjustment to another wave of democracy, and ends for all practical purposes with another adjustment to the internationalization of the second phenomenon: life in the 1990s and beyond. This second phenomenon is the result of trends, both political and economic, that began in the 1980s well before the advent of the New World Order. Latin America's "lost decade" was a bitter learning experience for the military, whether in or out of power, in most of the region. The end of the Cold War was an added shock.

Peruvian officers had turned over control of government and a collapsed economy to civilians in 1980. The "revolutionary government" that arose out of their *golpe* of 1968 was both an ideological and economic disaster. Their Argentine counterparts had left government following the Falklands debacle. Control of government devolved upon Brazilian civilians in 1985 with the blessing of the armed forces. In both Argentina and Peru internal conditions continued to be unstable as insurgents, restless officers, and terrorists threatened new civilian regimes on more than one occasion into the 1990s.

In all of these countries economies revived slowly. Guerrillas, narcotraficking, and crime plagued the Peruvian military into the 1990s. Brazilians struggled to put the economy in order, withstanding political scandals and instability all the while. Argentines had to face up to the *guerra sucia* and a lost international conflict. The Chileans finally entered an institutionalized transition framework in 1988 and left power in 1990, by which time the Berlin Wall had been definitively breached, then destroyed. To assert that the education of cadets and junior officers escaped these effects would be folly, or worse.

The rapidity of events triggered by—and here one searches for cause and effect compatibility—the debt crises of the 1980s, *glasnost* and *perestroika,* hardline U.S. policies toward the USSR, the declaration of European peace in 1990, the break up of the Eastern Bloc and Warsaw Pact, the realignment of Europe, the proclamation of the New World Order, the Gulf War, the triumph of neoliberal economic doctrines, and the Columbian quincentennial all accelerated the pace of change in Latin America to such a degree that one might think the things that contribute to comparability and consistency no longer obtained. This may be the case but I do not think so. Even should it be, this does not mean that new factors cannot appear to create new conditions for comparability and consistency. Whether owing to new or old causes, the effects of change do obtain, and comparability and consistency do too.

Insofar as I can tell, Latin American cadets and officers are still taught to see time's passage and perceive change as occurring in two principal ways. They think (hence write) in terms of *changing times* and *times of change*. These terms are more than conceits. Change occurs constantly, the pace of change varies. At certain points *the times* influence how change occurs. In military writing this latter type of change is associated with the international scene. The former, the inexorable kind of change, is associated more with national progress and development.

The late 1980s and early 1990s are viewed as *times of change*. International influences are probably more important in Latin America today than they have been at any time in this century, including the onset and tenure of the Cold War, and their impact on the educational process, consequently, has increased. We will soon be at a historical juncture where it will be possible to take a sophisticated look at officer corps' thought and self-perception during both the (re)democratization and New World Order era(s), and do more than speculate about the linkages between officer corps' education and ideation and political roles. Indeed, some scholars and policymakers have already begun to do so.[5]

The propitious time to assess anew the intellectual and ideological directions of the Latin American military is drawing near. In 1999 Chileans will go to the polls to elect their second civilian president thoroughly detached from military-civilian transition issues since the devolution of power in 1990 (August Pinochet Ugarte, commander-in-chief of the army, 1973–1998; junta leader and president, 1973–1990, retires in 1998). Argentines, Brazilians, and Peruvians already have either elected or re-elected a president for two terms since the transition from military rule. The two-term span will be sufficient time over which to assess just what military education and career development have produced in the way of officers who express institutional thought and self-perception in print for their colleagues to read. I suspect this assessment will indicate that comparability and consistency continue to mark both. This will indicate in turn both long-term professional stress on key issues and short-term policy interests. Early in the next century will be a good time to again place Latin America in a world context, for that context to have more definition.

Now, though, we can explore the subject a bit, the way a seafarer probes the shores of a new continent. Using the concept *"times of change,"* I want to offer some preliminary conclusions on what I think are the principal subjects of thought and self-perception, results of educational trends within the military, at a point in history where accelerated change is viewed as a product of international influences. It needs to be borne in mind that traditional roles of armies are essentially international. At least that is what South Americans claim.

Several subjects have figured to one degree or another in the writing of Argentine, Brazilian, Chilean, and Peruvian officers for years, which certainly supports a continuing case for historical comparability and consistency. All are sources of consternation today, indicating commonality among the major military powers of the Southern Cone. All can be found imbedded in postgraduate curricula. All are linked to one of the military's primordial missions.

A review of the most recent military literature from these countries shows that traditional themes are still important, adjusted as always, owing to the passage of time, in terms of language and specificity of application. Leadership, the "gift" of command (*don [dom] de mando*), obligatory military service, civic action, developmental roles, relations with limitrophe states, the spiritual qualities of the military profession, and nostalgia for the past still grace the pages of military journals. These journals, of course, are officially vetted. They often publish essays prepared during the postgraduate educational process; specialty, staff, command, and other advanced course writing assignments find their way into print, thus extending the educational process in more ways than one, just as they did before and during the Cold War. Whether or not junior officers and cadets read faithfully the literature of their profession we cannot know for certain, but if they do read, they read a controlled dose of institutional lore.

The end of the Cold War paradigm has not yet diminished the significance of another object of military intellectual attention, one that has roots in the pre–Cold War era. Growing to extraordinary lengths between 1945 and 1990 was the corpus of literature dealing with "subversion of national values."

If the world Marxist-Leninist movement is no longer a threat to Argentine, Brazilian, Chilean, and Peruvian military organizations, other forms of communism and continued subversion of "national values" are. Coupled with continual acts of terrorism and linked to narcotraffic, subversion, and terrorism, both provide military leaders with a lot to worry about. Acts of terrorism do still occur—witness the episode at the Japanese embassy in Lima early in 1997—but their overall political significance has diminished. Random acts in Argentina, Brazil, and Chile probably pose no serious threat to a government or to democracy. Peru's long struggle with *Sendero Luminoso* and *Movimiento Revolucionario Tupac Amaru* has concluded—one trusts. But officer-authors still worry that internal disorder might again rear an ugly head.

The New World Order is very disorderly. Narcotraffic-related terrorism remains a problem; so does ideological agitation. Both are associated with what really troubles officers—or so their writings would lead us to believe. What has them worried at this juncture is the subversion (and here I use the term in the dictionary, not the ideologically charged sense) of "national values." These are in essence patriotism, reverence for national heroes, Roman Catholicism, the sanctity of life and family, and collections of cultural "icons" peculiar to each country.

Accelerated change, the internationalization of popular culture, the sexual revolution, neoliberal economic schemes, world economic-political multipolarity, consumerism, and materialism all put at risk those traditions and values that professional

Latin American military organizations hold dearest. That most of these traditions and values are associated with an idealized past goes without saying. That they are integral to the educational process is a given.

Antonio Gramsci's ideas on creating revolutionary scenarios by creating revolutionary culture have made their biggest comeback in Argentina and Chile. Cultural change in terms of new moral and value systems puts at risk religion and family, according to military writers. Protestant fundamentalism, indeed Protestantism itself is frequently seen as a disruptive force, the Reformation as cataclysmic. Hilaire Belloc's, anti-Protestant, anti-Semitic diatribes still find their way into Argentine military literature.[6] Homosexuality, abortion, divorce, feminism, rock music, individualism, alternate styles of garb, modern art, television, advertising, Madonna, and *The Simpsons* have all been singled out as detrimental to "national values." *Times of change* can be most uncomfortable. They challenge the very essence of the profession.

There has always existed a tendency to associate accelerated change with outside forces. Those responsible for what transpires on the international scene are somehow responsible for the degradation of national values. Never mind that armed forces oversaw the dismantling of nationalized economies in Argentina, Brazil, and Chile, or that they looked outside Peru for solutions to its economic woes, foreign enemies exist—still. This professional xenophobia probably dates from the post–World War I recession, the Great Depression, the economic instability associated with World War II, and the ensuing Cold War. But these are no longer seen as immediate causes of the erosion of "national values." Military leaders evince a lack of comfort with the way things are going with youth, women, religion, families. They lament that citizens no longer spend spare time the way they used to. They attribute the assault on tradition and "national values" to outsiders. They learn this as cadets and learn more about it in advanced courses, in which, again to cite Montaigne, they are made to "consider what [they have] just learnt from a hundred points of view and apply it to as many different subjects. . . ."

In their latest guises liberal democracy and free-enterprise capitalism are hypnotically compatible, linked together as never before. Civilian leaders embrace them, encouraged by international financial agencies and political leaders from Europe, the United States, and Japan. The international aspect of current economic orthodoxy raises questions (and will raise more) about democracy's ability to assure sovereignty. This is a complex relationship at the core, regardless of how hypnotically compatible it may appear.

National sovereignty was a major factor in each of the institutional *golpes* of the second half of this century. If it was not the Communists or other leftists who threatened national sovereignty, it was the capitalists. This latter threat distinguished, in both form and content, what happened in Peru from what happened in Argentina, Brazil, and Chile. But it did not distinguish Peru from those countries with respect to the imposition of professional militarism: willingness and propensity to provide solutions to national problems based on a military ethos. Threats to sovereignty elicit military reactions, and always have.

If together democracy and transnational capitalism pose threats to national sovereignty as professionally defined and popularly conceived, are they immune from sanction? If national resources fall into the hands of outsiders, if control over unexploited or unexplored hinterlands is turned over to environmentalists or entrepreneurs from other countries, if mining booms are cut short by international causes supporting the rights of indigenes, or if joint mining ventures create permeable frontiers, is not national security threatened? If "national values" are subverted (again a nonideological usage), are not both security and sovereignty at the mercy of both national democratic and international capitalist interests? If so, what should be done about it and who should do it? These are questions asked by instructors in postgraduate military schools.

I am by no means suggesting that *golpes de estado* are imminent in the Southern Cone. Indeed, extant evidence indicates otherwise. It is apparent, though, from a reading of Argentine, Brazilian, Chilean, and Peruvian military sources dating from the 1980s and reaching into the 1990s, that military writers have proportionately as much latitude in expressing their *inquietudes* and *incertidumbres* as civilians have always had. So what I am suggesting is that traditional attitudes prevail, no matter how much praise democracy receives from *uniformados*. Civilians have not yet proven themselves willing or capable of agreeing with the military view on national security, defense, and sovereignty, although some significant steps have been taken in most countries toward that goal. Variation in degree of agreement and compatibility varies greatly from country to country still.

For the nonce, armies of the Southern Cone countries are content to register traditional complaints, objections, and concerns in print but not by political action. The former method of protest has always been the case, the latter often the product of change. What has changed again is the context in which education, ideation, and critical thought take place. Denied much of their ability to apply pressure, or granted it under circumscribed conditions, and well aware of the consequences of political action, military intellectuals (and others who just write a lot), lament the likes of "sex without love," teenage rebellion, and the invasion of their weekends by satellite television. They worry that armed terrorism might pose problems nullifying efforts at pacification and national reconciliation. They criticize international nongovernmental organizations that, they say, seek to limit nationally directed penetration and exploitation of hinterlands. Communism may not be dead; subversion of "national values" assumes more than one guise.[7] Diminution of the extent of national sovereignty signifies diminution of the military's constitutionally mandated role of defense, and this threatens those who proudly sport the uniform.

Another subject of military thinking is the historical and post–Cold War behavior of the United States. From the beginning of the Cold War to its end the United States was undisputed leader of the "Free World" and military "hegemon" of the Western Hemisphere. Tensions between East and West subordinated North-South tensions in the Americas, at least until the 1980s. Advisory groups, missions, courses in the United States, the doctrine of reciprocal assistance—these all linked most

Latin American armed forces to those of the United States. But relations were never comfortable. At no time was the U.S. military presence in Latin America or the educational opportunities it afforded Latin Americans favorably compared in military literature to those of Germany or France prior to World War I or World War II.

Beginning in the early 1990s the United States made it clear to erstwhile hemispheric allies in the worldwide struggle against Marxism-Leninism that their services were no longer necessary. U.S. policymakers began emphasizing international peacekeeping activities, regional defense cooperation, and the war on the narcotics trade. There were strong international pressures to reduce arms expenditures, to disarm, to abolish obligatory service, and to turn professional armies in Latin America into police-like organizations. The leadership role heretofore played by the United States was just not applicable to the new order of things thought some Latin Americans. This led more than a few to think they had been courted by the United States only as long as a hemispheric world power perceived there were threats to its own interests within that hemisphere. In the past six or seven years a spate of military literature has appeared that if not overtly anti-U.S., is implicitly critical of either the new role or the "nonrole" currently played by the former military titan of the hemisphere, and of the policies it advocates for Latin American armies, navies, and air forces. Forces of an international nature were behind accelerated change again.

To military leaders of the Southern Cone the idea of becoming glorified police is less than attractive. They do not want to transform *escuelas, academias, escolas,* and *colegios* into police training centers. Their reactions to the U.S. wishes for post–Cold War military role redefinition vary according to local circumstances. The Argentine military, haunted by the Falklands fiasco, is resigned to being a reliable source of (all-volunteer) talent for United Nations peacekeeping missions. And they have done a pretty good job at this. Although the Argentine defense budget has shrunk and obligatory service (for decades a mainstay of the South American professional military ethos) has been eliminated, Argentines see their prestige restored by participating in peacekeeping. More power to them, as it were.

To the north, Brazilian officers do the same in terms of peacekeeping—in quality if not quantity—but they resent a perceived U.S. role in encouraging environmentalist and ecological programs in the hinterland, especially in the Amazon basin. They complain about budget reductions and they fear for the future of obligatory service. They worry that Argentina will supplant Brazil as the United States' prime South American military partner.

Over on the Pacific side of the continent, Chilean officers show little or no sympathy for current U.S. policies toward the armed forces of its hemispheric partners. Having left power intact and having bequeathed (as they see it) a thriving economy to fellow Chileans, *uniformados* from *el rincón más apartado del mundo* (the farthest corner of the world) resist becoming lackeys of a country that shows neither the willingness nor the ability to lead the way toward protection of traditional military values, much less their "national values." Indeed, like their Argentine and Brazilian counterparts, Chileans see themselves as integral parts of national development: securing internal frontiers of progress and integration, working on infrastructural improvement, exploring oceanic frontiers, and leading in scientific and technological innovation.

The Peruvian armed forces remain wary of internal conditions that so recently brought the country to the brink of collapse. Fifteen years of internal war helped them reestablish professional prestige, which diminished so markedly when their regime collapsed in the late 1970s. Revolutionary terrorism has allowed the armed forces, especially the army and air force, to improve its public image.

Each of these countries has had some external frontier problems, and there still exists something of a nineteenth century European power-balance relationship in South America. Each, therefore, can make some claim to the need for armed forces able to counter a foreign threat. Each of these armed forces can also make some claim to legitimacy of an internal "tutelary" role: Peru and Chile most convincingly—but for different reasons. All have been involved in peacekeeping—Argentina and Brazil most prominently. In their professional military literature all demonstrate a conviction that conventional international war is still a possibility. All are less than enthusiastic about national volunteer armies, policing, fighting the narcotics wars, and regionally integrated defense forces. They see the former roles as forced abdication of their national developmental mission and a "diminution of national values." They see the latter as an internationalization of these values no less insidious than the brand of internationalization they opposed for the better part of this century. Sovereignty is still important in an age of accelerated transnationalization of cultures, economies, polities, and societies.[8] Thus anyone who advocates too much change can be viewed as threatening their primordial role.

Defense of the homeland against all threats to state, nation, and society, be such threats domestic or foreign, socioeconomic, political, or cultural, remains the primordial role of the South American armed forces, especially of the continent's armies. They differ strikingly from Caribbean counterparts, and always have, for obvious reasons.[9] South Americans know full well what the impact of the U.S. presence or professionalism can be if not tempered by "national values" and national professional traditions. They are not about to give up on their past in order to be part of someone else's future. The consistency and comparability of their thought and self-perception make this clear. So does the education of their officers.

Only time will tell if consistency and comparability will diminish or remain as they have since the professionalization process began a century ago. As they have moved from cadet classrooms and cadet parade grounds to advanced coursework and political action they have become "life-long learners," teachers of others, and often political activists.

Notes

1. For examples of pioneering late nineteenth and early twentieth century military education reform, see Frederick M. Nunn, *Chilean Politics, 1920–1931: The Honorable Mission of the*

Armed Forces (Albuquerque: University of New Mexico Press, 1970); and idem, *The Military in Chilean History: Essays on Civil-Military Relations, 1810–1973* (Albuquerque: University of New Mexico Press, 1976).

2. See as examples, Frederick M. Nunn, *Yesterday's Soldiers: European Military Professionalism in South America, 1890–1940* (Lincoln: University of Nebraska Press, 1983); and idem, *The Time of the Generals: Latin American Professional Militarism in World Perspective* (Lincoln and London: University of Nebraska Press, 1992).

3. See Frederick M. Nunn, "The South American Military and (Re)Democratization: Professional Thought and Self-Perception," *Journal of Interamerican Studies and World Affairs* 37 (Summer 1995).

4. See Frederick M. Nunn, "The Roles of Civilian Experts in the International Strategic Community and Military Professionalism in the New World Order," in Ernest Gilman and Detlef E. Herold, eds., *Democratic and Civil Control Over Military Forces: Case Studies and Perspectives*, NATO Defense College Monograph Series, No. 3 (Rome: Nato Defense College 1995); and idem, "Latin American Military-Civilian Relations Before and After the Cold War: Some Thoughts, Questions, and Hypotheses," UNISA *Latin American Report* 13 (University of South Africa, 1997).

5. Following is a select list of meetings held in the 1990s in which military professionals, academicians, and policymakers have addressed the subject at hand from a variety of standpoints, sponsorships, and perspectives: Woodrow Wilson Center/Florida International University conference, "Between Public Security and National Security: The Police and Civil-Military Relations in Latin America," Washington, D.C., October 1993; NATO Defense College, International Research Seminar on Euro-Atlantic Security, "Democratic and Civil Control over Military Forces," Rome, December 1994; U.S. Southern Command/U.S. National Defense University, Institute for National Strategic Studies symposium, "Partners in Regional Peace and Security," Miami, Fla., May 1995; U.S. Army Center of Military History, Conference of Army Historians, "The Early Years of the Cold War, 1945–1958," Washington, D.C., June 1996; U.S. Air Force Academy Military History Symposium, "Educating and Training Junior Officers in the 20th Century," U.S. Air Force Academy, Colo., November 1996; Ford Foundation/Universidad Torcuato Di Tella conference," "La cuestión cívico-militar en las nuevas democracias de América Latina," Buenos Aires, May 1997; U.S. National Defense University, Center for Hemispheric Defense Studies, 'Hemispheric Conference on Defense Education for Civilians," Washington, D.C., September 1997; Leiden University, Centre for Non-Western Studies, workshop, "The Soldier and the State in the ABC Countries," Leiden, The Netherlands, October 1997.

6. See Nunn, "Latin American Military-Civilian Relations Before and After the Cold War."

7. For examples of consistency in argumentation over the years, compare the following examples from Argentina, Brazil, and Peru with examples cited in notes, supra: Colaboración Curso Básico de Comando 1980, "El eurocomunismo," *Revista de la Escuela Superior de Guerra* (Argentina) (January–February 1981); Gen. Ramón G. Díaz Bessone, "Guerra revolucionaria en la Argentina, 1959–1978," *Revista Militar* (Argentina) (January–June 1978); Gen.(Ret.) Héctor Rodriguez Espada, "Las fuerzas armadas y la subversión terrorista," *Revista Militar* (October–December 1995); Col. Nelson Abreu do O'de Almeida, "Forças Armadas: Apenas Segurança Externa?" *A Defesa Nacional* (March–April 1989): 23–34; Lt. Col. Osmar José de Barros Ribeiro, "O Tráfico de Drogas no Mundo e no Brazil," *A Defesa Nacional* (October–December 1995); Dr. Jesús Lazo Acosta, "Análisis sicosocial del terrorismo," *Revista Militar del Perú* (September–December 1985); and Lt. Col. Roberto Vizcardo Benavides, "Ejército del Perú: Seguridad, paz, y desarrollo," *Revista del Ejército del Perú* (30 November 1995): 5–8. For Chilean sources see Nunn, "The South American Military and (Re)Democratization." I cite only month or date of issue for these and following journal articles owing to discrepancies in time, number, and dates caused by occasional delays and cancellations of publication.

8. What follows is a sampling, no more, of implicit and explicit criticism by South Americans of the U.S. hegemonic position vis-à-vis hemispheric defense and security priorities following the end of the Cold War: Maj. (Ret.) Sergio Toyos, "Fuerzas armadas y guardias nacionales," *Revista Militar* (May–June 1995); Gen. Ricardo Etchevary Boneo, "El Tercer Mundo dentro del pensamiento estratégico de los EE.UU.," *Revista Militar* (January–June 1996): 46–52; Col. (Ret.) Roberto Miscow Filho, "A Função e O Papel das Forças Armadas," *A Defesa Nacional* (July–September 1993): 89–95; Col. Valmir Fonseca Azevedo Pereira, "O Relacionamento Militar Brasil x Estados Unidos," *Revista do Exército Brasileiro* (4th trimester 1996); Lt. Col. Cristián Le Dantec G., "El NAFTA, los tratados de libre comercio y su posible repercusión en la defensa nacional," *Política y Estrategia* (Chile) (May–August 1995): 33–45; Col. Juan R. Galecio Araya, "La situación mundial actual: Algunas reflexiones," *Memorial del Ejército de Chile* (2nd Trimester 1996); Lt. Col. Otto Guibovich Arteaga, "La doctrina militar propia: Necesidad o imperativo," *Revista del Ejército del Perú* (7 June 1995); and "Chavín de Huántar," *Actualidad Militar* (Peru) (May–June 1997), a paean to the 22 April 1997 operation that freed, "without either advice or assistance of any foreign power with experience" hostages of the Tupac Amaru revolutionary movement held in the Japanese embassy.

9. See Frederick M. Nunn, "Foreign Influences on the Armies of the Southern Cone: Historical and Regional Perspectives," paper presented at the conference on "The Soldier and the State in ABC Countries," Centre for Non-Western Studies, Leiden University, Leiden, The Netherlands, 24–25 October 1997.

Section VII
Educating and Training Officers for the Twenty-first Century: Contemporary Perspectives

Probably no consensus is possible on the universally applicable values and attitudes or on the knowledge and skills desirable for future officers. All might agree that officers must accept death or permanent injury as inherent characteristics of their profession, and be willing to subordinate the individual self to group needs and objectives.[1] Beyond these, opinions begin to diverge. Obviously, what an officer must know and be able to do changes over time and place, and responds to a variety of influences internal and external to the military profession.

Although much less evident, the personal attributes officers are expected to possess vary in similar ways. Some of the essays in this volume have made clear, for example, that "character" (and the qualities it usually embodies—integrity, moral and physical courage, loyalty, duty, self-discipline) have been defined as much by social context as by any catholic standard of officership. Thus, the Germans and the British equated character with social class well into the twentieth century; the socialization process at the Philippine Military Academy generated among its cadets a distorted notion of loyalty to fellow classmates; and early in the 1990s, West Point, in response to the increasing diversity in American society and the Corps of Cadets, put "consideration of others" on a par with "honor" as the academy's two "bedrock values."[2]

Producing officers equipped for the next century will mean correctly anticipating the knowledge and skills demanded by the changing nature of both war and society. Just as challenging, however, will be the task of balancing the various components of officer preparation—value formation, the level and content of general education, and the amount and timing of instruction in professional theory and practice. As several essays in this volume have demonstrated, the cost of disequilibrium has often been high.

The authors of the four essays in this section, all active or retired officers, offer personal perspectives intended to influence the future course of officer education and training. Three of the essays have been adapted from remarks given at the U.S. Air Force Academy's 17th Military History Symposium in November 1996. Although all four concentrate on United States cadet and junior officer preparation, the authors' views have relevance for commissioning programs elsewhere.

John Flanagan, a retired brigadier general in the U.S. Air Force Reserve, graduated from the Air Force Academy in 1962. Within four years and while still a lieutenant, he was flying a light observation aircraft as a forward air controller supporting ground forces in combat in Vietnam. Flanagan describes how each aspect of his academy experience—military training, physical conditioning, mental stress, academic coursework, and value development—had a direct and practical application to his performance and survival in combat. His recollections underline the point that whatever role initial military schooling may have in preparing officer candidates to assume larger responsibilities in their careers, it must first get them ready to answer an early, "come as you are" call to war.

In the 1990s, argues David Smith, the U.S. Naval Academy is failing to prepare its midshipmen to serve effectively in the Navy's high-tech fleet. Smith, a 1957 Naval Academy graduate and retired Air Force Reserve officer, holds at fault the academy's "liberalized" curriculum and military environment, and its neglect of the traditional mission to produce graduates educated in the naval profession who can perform immediately as capable junior officers. To remedy the deficiencies he perceives, Smith proposes numerous reforms. Among them are that midshipmen major only in mathematics, science, and engineering disciplines, and that the academy be "remilitarized" by granting fewer privileges to midshipmen and by toughening the first-year indoctrination program to emphasize that learning to follow must precede learning to lead.[3]

No matter how pressing the need for future officers to be scientifically and technically competent, the history of officer education and training, as many of the essays in this volume have shown, warns loudly against neglecting the social and behavioral sciences and the humanities. Although some of its disciplines are often the "odd man out" in officer education, study of the humanities—of history, literature, philosophy, and the fine arts—stimulates critical thought and provides a sense of value and ultimate purpose in human affairs. Major General Josiah Bunting, superintendent of the Virginia Military Institute and a novelist and combat veteran, reinforces the humanities' importance in cadet education, especially biography's power to inspire emulation. Part of the mission of national military academies, writes Bunting, must be "to kindle a lifelong habit of intellection, of unquenchable intellectual curiosity, to lay deeply the habits of inspiration from books, from art, from music." For Bunting, men and women of action, particularly, must be men and women of deep reflection.[4]

Before retiring from the Royal Air Force, Air Vice-Marshal Tony Mason served in several officer education and training posts including air secretary with the responsibility to identify and prepare officers for high command, and as an exchange officer in the U.S. Air Force Academy's History department. He is also a graduate of service colleges in both nations, and an accomplished historian of air power. Mason outlines the security environment air force officers are likely to face early in the twenty-first century, and suggests that it will not require sweeping change in officer education but "reinforcement and expansion." Air education should reinforce officers' ethical values and their professional awareness (by emphasizing military studies related to air power). It should also expand officers' capacity for judgment and their sensitivity to cultural diversity. In seeking to mirror its parent society, however, the military must not incorporate "incompatible civilian values and priorities" into officer preparation.

Notes

1. For insightful analyses in the context of officer education of the tensions inherent in the military's subordination of the individual to the collectivity, see Pat C. Hoy II, "Soldiers and Scholars," *Harvard Magazine* (May–June 1996); and Monte D. Wright, "In Defense of the Terrazzo Gap," *Air Force Magazine* 58 (April 1975).

2. Dave R. Palmer, former superintendent of the U.S. Military Academy, discusses "consideration of others" and character development at West Point in *Shaping Junior Officer Values in the Twentieth Century: A Foundation for a Comparative Perspective,* Harmon Memorial Lectures in Military History No. 39 (USAF Academy, Colo.: U.S. Air Force Academy, 1996).

3. In June 1997, the report of the Special Committee to the Board of Visitors of the U.S. Naval Academy proposed many changes at the academy, but found no "systemic flaws." Among the Special Committee's recommendations were that the academy place more emphasis on social and behavioral sciences in the curriculum, monitor the balance between engineering/technical and social sciences/humanities studies to meet the needs of the service, and allow midshipmen more freedom in choosing academic majors. See Report of the Special Committee to the Board of Visitors, United States Naval Academy, *The Higher Standard: Assessing the United States Naval Academy* (U.S. Naval Academy, June 1997). M. D. Van Orden, a retired Navy admiral, and Stanton S. Coerr, an active-duty Marine Corps captain, have sharply criticized the Special Committee report in "Reverse Engineer the Academy: Toward Restoring Service Integrity," *Strategic Review* 25 (Fall 1997). The authors second much of what Smith argues, adding that the academy should no longer admit women. A study of over 1,500 Naval Academy graduates from the Classes of 1976–80 found no relationship between academic major and the performance and retention of junior officers through their initial tour of duty. See William R. Bowman, "Do Engineers Make Better Officers?" *Armed Forces and Society* 16 (Winter 1990).

4. For other explanations of the part played by the humanities in officer education, see Josiah Bunting, "The Humanities in the Education of the Military Professional," in Lawrence J. Korb, ed., *The System for Educating Military Officers in the U.S.* (Pittsburgh, Pa.: International Studies Association, 1976); and Jesse C. Gatlin, "The Role of the Humanities in Educating the Professional Officer," *Air University Review* 20 (November–December 1968).

From Classes to Combat: Lessons for the Junior Officer*

John F. Flanagan, Jr.

I am alive today because of the knowledge and experience I acquired at the United States Air Force Academy. This afternoon, I am going to talk about the lessons of training, of education, and of value development I learned there; not from the perspective of a lecture hall or the athletic fields or simulators or field exercises, but from the perspective of the jungle, the rice paddies, and the skies over Vietnam where I served as an Air Force forward air controller, or FAC, in 1966.

A FAC was unique in that period. Our armed forces were geared to responding to Soviet and Warsaw Pact tanks thundering through the Fulda Gap where we would fight while outnumbered to stop the attack. Suddenly, in Vietnam we were thrust into a jungle environment, a guerrilla war, something that was alien and that we did not understand. We could not find the targets in the jungle with our supersonic fighter-bombers, so we re-created the FAC, a specialty employed in World War II and the Korean War—part infantryman, part pilot—whose job was to work with the ground troops, and from the air or sometimes from the ground, to identify the targets and to direct devastating air power from the fighter-bombers on those targets.

My view of Vietnam in 1966 was that the war was in transition from being a Vietnamese war fought with American assistance to becoming an American war fought in Vietnam. While in Vietnam I watched this change take place. During my first two and one-half months, I served with units from the Republic of Korea (very few people know that the South Koreans, "the ROKs," had committed two army divisions and a marine brigade to Vietnam). They were the fiercest fighters in the theater, and feared by the Vietcong because they understood the culture. I also had the opportunity to serve with the U.S. Army's elite 101st Airborne Division, and then, finally, the highlight of my career (where probably I learned more about camaraderie and values, and, tragically losses) was the eight months that I spent with the Army's Project Delta (the Special Forces or Green Berets) known today as Delta Force, the counterterrorist unit.

Let me first address the lessons of military training that I learned at the Air Force Academy: the assault course and physical conditioning, combat skills, rifles and bayonets, the things that we went through as basic cadets, and that later on, as upperclassmen, we taught to the underclassmen.

*Adapted from Brigadier General Flanagan's remarks delivered at a luncheon on 20 November 1996 during the United States Air Force Academy's 17th Military History Symposium.

I remember that as a cadet I cleaned and cared for my M-1 rifle; crawled, bayoneted, and qualified with it for four years. That rifle was mine. Three-and-a-half years later, by candlelight in a tent at the height of the monsoons in the central highlands of Vietnam, I was cleaning my M-16, not to avoid demerits, but to stay alive.

I also recall field training at the academy when we lived and operated from two-man pup tents. It was called FASE (Forward Airstrip Encampment) and we learned about camouflage, cover, concealment, and small-unit patrols. Today the cadets go through the lessons of Combat Survival Training. In Vietnam, I remember going out on one of my first patrols with the Koreans, making sure I kept a safe interval so that a single round or a mortar shell would not take me out and the person in front of me. I also remember watching where the person in front of me walked so that I stepped in the very same place to avoid punji stakes and trip mines.

On Saturday mornings at the academy, we had SAMIs, Saturday A.M. inspections. Then, and everyday when we left for class, our rooms had to be in absolutely perfect condition, everything in its place. In March 1966, in Vietnam I was on Highway 1 near a village called Phu Cat, out in the field serving with a Korean battalion. We lived in tents in bunkers. The tents were below ground because periodically we were getting shot at, sometimes with automatic weapons fire, and, of course, by the ever-present rocket-propelled grenades and mortars. As at the academy, I kept the equipment in my tent in order.

One night we found that our perimeter was being probed; tracer fire started whipping across the tops of the tents and we extinguished all of our lights. As I huddled there in the pitch dark, I reached for my equipment. Instead of the socks and jocks and underwear I had as a cadet, now I reached for smoke grenades, hand grenades, magazines with tracers, and magazines without tracers. Even in the pitch dark I could find the respective handsets connected to the radios—the HF, UHF, VHF, and FM radios that were our links to the outside world. Underneath the orderly stacks of equipment were things that I hoped I would never use, my bandage pack and morphine syringe. But the lessons of that SAMI—of putting everything in its place and keeping it there—are the ones that I lived by on that fateful night, as we successfully drove off the attack.

As cadets we had an exercise called "clothing formation," during which we had an absurd combination of uniforms that we were required to change into in a minimum of time. The combination included everything from ponchos and helmet liners, to sneakers and rifles. The upperclassmen would be on us as we would race back to our rooms to put on the proper uniform, then out onto the terrazzo (the marble-embedded mortar quadrangle surrounded by the buildings of the academy's central cadet area), standing at attention, being inspected, and then racing back again if we made a mistake. Some people might think of that as harassment, but it proved to be vital training for me.

In my tour with the 101st Airborne at Tuy Hoa, I was serving in the command post and an urgent message came over the radio: "We have had a patrol ambushed. They are shot up badly. We can't get the medevac (medical evacuation) helicopter in.

We need a FAC." I raced from the command post, grabbing my maps and my signal operating instructions, racing by the tent to pick up my survival vest, my survival radio, my flak vest, my headset, my rifle, my ammunition, my signal flares, and my signal mirror, buckling the equipment on as I raced to the airplane, not forgetting any piece. Here was a lesson that those upperclassmen at the academy taught me—how to get dressed, and get dressed with the right equipment in a big hurry. I launched in my O-1 (small, single-engine observation aircraft) and fortunately we got some A-1E Skyraiders (piston fighter-bombers) in there and were able to save the wounded soldiers.

Physical conditioning throughout the four years at the academy—whether it was intercollegiate or intramural sports, or simply physical fitness and testing—also proved vitally important in Vietnam. I recall the words of my classmate, Russ Goodenough, in 1966. When I saw Russ at Cam Ranh Bay, he was flying F-4 jet fighter-bombers. He had been shot down over Laos. Russ, who was a fine athlete in good condition, spent one hour on the ground running for his life, armed only with a .38 revolver. He said, "John, I was never so tired in all my life." But the lessons of conditioning and athletics carried through, and were probably among the reasons that he lasted long enough to be rescued.

The mental conditioning—stress training, mental discipline, and certain physical exercises (sometimes called hazing) that I underwent at the academy paid off for me in Vietnam. At the academy, the upperclassmen would gang up on an underclassman, usually a "doolie" (freshman) and put him through stress—try to break him—by forcing him to recite "knowledge" (basic facts all "doolies" were expected to have memorized) or to do pushups. I can recall that some—for some reason it is always the little cadets—liked to pick on me. Three of them would gather around me, and they would look up at me and say, "I'll bet you would love to smash us, wouldn't you, Mister?" "Yes, sir." "You mean you want to hit me?" "No, sir." And on and on it went.

What they were trying to do was to develop my mental toughness—my ability to handle stress. I can tell you about mental toughness and stress. It is when you are over the tops of the jungle and you have three radios in the airplane (replacing the three upperclassmen), and there is a jungle patrol down on the ground that has been ambushed and you are trying to get helicopters in there to get them out. In the meantime, you have aircraft circling overhead that are running out of fuel with ordnance that needs to be expended. All of a sudden anti-aircraft artillery rounds start exploding all around, and you are wondering where it is coming from. All this time you are trying to keep your cool and maintain control of the battle. Where did you learn to handle that stress? For me it was on the terrazzo and in the hallways of the Air Force Academy.

My training, however, was not complete. I had never gone through jump school. When I ended up with Project Delta, they very quickly recognized the shortcoming in my training. I would insist, "I'm a pilot. Pilots never jump out of functioning airplanes. You don't understand." They replied, "Don't make no difference, Lieutenant. You gotta jump." "Okay, Sarge," I replied. So for two days they put me through an abbreviated and grueling jump training program, and I made my first jump. I said, "What about the reserve chute?" They said, "Don't worry, you won't get a chance to use it." And I went out on a static line out of a "Gooney Bird" (C-47 transport aircraft) at 600 feet with the Vietnamese Airborne Rangers. I became jump qualified. They had the last laugh. The drop zone was a graveyard.

In addition to military training, I want to discuss the lessons I learned from the academy's academic courses. In chemistry class we studied the chemical composition of explosives, of C4, and learned that it would not explode unless it had a detonator. In Vietnam we carried C4 in our pockets in little slices because it made a great heat tablet for heating our coffee.

In aerodynamics courses, we spent hours working with models in wind tunnels. We studied the drag and lift equations and understood the principles of flight. In Vietnam, we applied this knowledge in a bizarre manner. During the monsoons, the winds were up to sixty and seventy knots, with gusts even higher, and our airplanes were being blown right out of their chocks in the tie-downs. We put sandbags on the wings and the tail to create turbulent flow and to spill the lift, insuring that our little O-1s did not get blown over.

Mathematics, physics, and astronautics courses were also directly relevant to what I was doing in Vietnam. Think of the apparently simple problem of dropping a flare at night, which really turns out to be a vertical sine wave. It is an aircraft in a shifting elliptical orbit with a free-falling flare governed by Newton's universal law of gravity. The flares that fall are driven horizontally by an unknown variable, a wind vector. On the nights of 24 and 25 March 1966, we dropped 250 flares using five flare ships (aircraft dispensing flares) controlled by two FACs—this time on the ground— one of them myself, the other offset in the other direction as we gave north and south and east and west bearings, keeping those flares over our forces. Why did we need the flares? The Vietcong were attacking out of the Phu Cat mountains and they were using civilian noncombatants as shields as they assaulted the Korean lines. Only because we kept those flares precisely centered over that assault force were the Koreans able to flank the Vietcong. Their sharpshooters picked off the enemy soldiers, and the civilians lived to come back another day.

I recalled a lesson from my history courses the very next day as we put together a coalition force: Vietnamese Air Force fighter-bombers, Korean infantry, and a U.S. Air Force FAC. We went back into the Phu Cat mountains and absolutely leveled that base camp. In this coalition force of Vietnamese, Koreans, and Americans, I do not think the highest rank involved was above a lieutenant.

In political science courses, we studied the various forms of government and military structure in many different countries. In spite of these studies, nothing prepared us for the convoluted military and government structures of the provinces of South Vietnam. We learned that the province chief was both a civilian head and a military head serving as a political appointee of the Saigon central government. We quickly discovered that sometimes the air strikes that we were asked to put in were not directed at an enemy camp, but at someone who had not paid their taxes. This was the political environment that we were fighting in, and we were supposed to win the hearts and minds of the people!

From the academy classroom came also the lessons of economics, of market theory, and of marginal costing. In Vietnam, we kept the supply lines open. We used our helicopters to take rice, rubber, and coffee to the ports and to the provinces so that people could stay employed and the Vietcong would not overrun their lands. I learned something about private enterprise and business when I later discovered why the rubber plantations were never attacked or bombed—in Vietnam the rubber industry paid taxes to both the Vietcong and the Saigon governments.

At the academy we studied foreign languages, the languages of our adversary, the Warsaw Pact led by the Soviets, and those of our NATO allies. In Vietnam we did not have very many Vietnamese linguists. Nor could we find anybody who spoke Korean. But fortunately, Indochina had been the sphere of colonial France, and my academic training in French stood me in good stead. I found out that virtually all the officers in Vietnam and all the bureaucrats spoke French, and I was able to communicate with them.

In geography and cartography classes at the academy, we poured over maps, learned the nuances of contour lines, how to determine the height of a waterfall, vegetation patterns, and the changes in colors in the vegetation. A professor at the academy taught me those things and insisted that I learn them thoroughly. Because of this schooling in map reading, I was able to identify precisely the location of a friendly force that had been mistaken for an enemy unit and was about to be attacked by "friendly fire."

In addition to military training and academic courses, the academy molded many of our values. In Vietnam, two of those values came to have special meaning for me. First, the tough academy environment that we all shared forged strong bonds of loyalty among us. We were committed to each other. Second, we lived by an honor code—not to lie, cheat, or steal, or to tolerate among us anyone who did—that strengthened our sense of integrity, our moral courage.

On 17 August 1966, a Special Forces team commanded by a Vietnamese lieutenant, Vu Man Thong, and U.S. Army Sgt. Norm Doney was inserted into the jungle near the Cambodian border. Early one morning, they heard sounds in the jungle, and as one of the Vietnamese sergeants crept forward, he discovered a Vietcong unit doing physical training at sunrise. Sergeant Doney called for an air strike, and a FAC, Lt. Carlton S. "Skinner" Simpson, responded.

When Skinner arrived over the area, he contacted the team, and the team signaled him with a mirror. Together, they directed an air strike with devastating results: the bombs came in while the Vietcong were still counting cadence. The Vietcong realized that they had been compromised and that there had to be some alien party (our intrepid Special Forces team) in the area. They began searching, eventually calling in an entire regiment to look for the team. Norm Doney and his compatriots were surrounded—six people surrounded by a regiment of Vietcong.

Back at headquarters, we started asking for air strikes, going through the emergency channels, while Skinner Simpson hung over the target, directing strike after strike. Doney finally came on the radio and said, "I don't know if we are going to make it. They're closing in on us."

At that time a flight of F-100 jet fighter-bombers arrived in the area. The fighters checked in, and Skinner Simpson started to brief them on the situation. One of the fighter pilots said, through his oxygen mask: "Skinner, is that you? This is Rob." And Skinner clicked his mike twice in acknowledgment and fired a smoke rocket to mark the location of the enemy forces.

Those F-100 fighter pilots pressed the attack with cluster bomb units and napalm so closely that the number two man returned to Bien Hoa with a three-foot section of tree in the leading edge of his wing. Because of their skill and effort in the face of heavy ground fire, we were able to rescue the Special Forces team. They made that extra effort because Rob Pollack, the lead fighter pilot, had been Skinner Simpson's roommate at the academy.

Physical bravery is demanded of many in a combat zone. Fewer, however, are called upon to demonstrate moral courage. Moral courage is the courage to stand for what you believe and to do what you think is right. In 1966 a team of two Americans and four Vietnamese were inadvertently inserted by helicopter into Laos, creating a political embarrassment for the United States. Attempts were made to rescue the team. A helicopter was subsequently shot down, and the team was given orders to escape and evade, to get out on their own.

Finally, the monsoons broke sufficiently so that we could launch a search. A U.S. Army lieutenant colonel called in a FAC, a first lieutenant, and said, "Lieutenant, what are your intentions?" The lieutenant replied, "Well, sir, I know the escape and evasion plan of those two Americans, and I intend to overfly and search their intended route in Laos." The lieutenant colonel looked at the lieutenant and said, "Lieutenant, you're not to search for the team."

The lieutenant in turn said, "Sir, I'm an Air Force officer. My rules of engagement are such that I can fly over Laos and I can put in air strikes with the approval of a Laotian official." The lieutenant colonel summoned a major into the tent and reiterated his directive, saying, "I was telling the lieutenant that he is not to fly to search for the team, and that if he goes down, we're not going to search for him or rescue him."

The lieutenant walked out of the tent and wondered what to do now. He had made a personal commitment to his two comrades, Sgts. Russ Bott and Willy Stark, that he would search for them. Every Special Forces reconnaissance sergeant in that camp stood there waiting to get in the back seat of that airplane to go and search for their comrades. What should the lieutenant do? Should he stand in front of them and say, "No, it can't be done?" But how could he take anyone with him knowing that if they did get shot down, they would not be rescued?

You have probably guessed that I was that first lieutenant. We searched for Russ and Willy. We did not find them, but we found two of their Vietnamese compatriots who confirmed to us that Russ and Willy had been taken prisoner and led away by the North Vietnamese. Sometimes the courage expected in combat is not just physical courage, but the more difficult moral courage.

I learned many lessons from my training and education as a cadet, and tried to apply them to the battlefield, some successfully, and some not so successfully. Let me briefly discuss how those lessons and that experience relate to the current and future environment for junior officers.

As we proceed through the era of high technology in warfare, simulations are producing a revolution in training. We began with simulators and computer models for strategic force deployment, and now have simulations for tactical options and assaults. Today, we have the virtual battlefield, the virtual Apache helicopter, and rangeless pilot training for the Air Force. One danger that lies in this technology is that it can create a false sense of proficiency and competence in the junior officer. In combat there is no such thing as virtual fatigue or virtual leeches or virtual heat exhaustion or hypothermia. There is no virtual fear or virtual blood. They are real.

We are moving toward a digitized battlefield in the twenty-first century. We are looking to equip everybody with a personal communication device that would fit in their uniform pockets and link them to anyone automatically through internetted voice-data imagery. We are clambering for more technically trained officers. The danger is that technology is separating the leader from his troops, both in training and in war-fighting.

This occurred in Vietnam as we employed the helicopter in combat. It gave us mobility and firepower, and an increased ability to observe and communicate. Eventually, we created the "command and control" helicopter. Now the colonels became squad leaders as they flew over the battlefield, all-seeing and surfing the net as they outfitted their helicopters with five and six FM radios so that they could get into every battalion, company, and platoon net. They interjected themselves into every battle, bypassing the chain of command, destroying and eroding trust, respect, and integrity—that intangible bond that exists between a leader and his combat soldiers.

Is there a danger that in the next war the lieutenant is going to call up his screen and it is going to say "Attack." And he will turn to his men and say, "We have to attack." And the men will say, "Why?" And the lieutenant will say because "The computer told me so." Are we once again on the verge of destroying the trust between the combat leader and his troops?

Social change can be as important as technological change for the military. Along Highway 1 in Vietnam, the Koreans captured a Vietcong nurse (or she said she was a nurse) who carried the same AK-47 as the other prisoners, and her magazine bandoleers were no lighter or smaller than anybody else's. I know of no bullet, no shell, no missile that is smaller or slower or stamped as made for women only. We should remember this as we totally assimilate women into our military forces.

In the 1960s and the 1970s, the Reserve Officers Training Corps (ROTC), a critical source of leadership for our military, was challenged and driven from some of our best campuses, denying the services the leadership that this country needed. As our leadership degraded, we suffered the consequences. The atrocities of My Lai were the product of leaders who would never have been commissioned in another era or another time. Today, that same scenario is unfolding, but in a different guise.

In July 1996, the assistant secretary of the army for manpower and reserve affairs, in an address to the Reserve Officers Association, said: "We are contemplating in the Army, of taking 2,121 billets of captains out of ROTC, and assigning the mission to the reserve components." I am not saying that the reserve components are not capable, but I think the training of cadets and newly commissioned junior officers is one of the most important tasks that the services can undertake, and deserves the most competent instructors and mentors.

Those are our lessons. Those are our challenges. I look at them from the perspective of one who has served, who has learned and gained. We have faced adversity, hardship, death, and in some cases ridicule, but we survived with our values intact, and can face eternity alone with them. We must impart our experience and our values to our successors, for we are not obligated to pass the warrior sword to those unable to hold it as high as we did.

The U.S. Naval Academy: How It Came to Be What It Is Today and How to Fix It

David A. Smith

Over the years, the public's principal interest in the three U.S. military academies has been the annual football competitions, ending each fall with the Army-Navy game—a hundred-year-old rivalry. But during the past decade, public attention has been drawn increasingly to a series of scandals including, astonishingly, cheating, sexual harassment, grand theft of automobiles, child molestation, drinking, and drug use. Although none of the academies has completely escaped the problems, the United States Naval Academy appears to have experienced the greatest difficulty, especially in recent years.

In 1996, James Barry, a former naval officer who was then a civilian professor of seven years' standing at the academy, published a very critical article in the *Washington Post* claiming that the Naval Academy was "adrift" and that the Navy's recent behavioral problems stemmed from the environment and problems at the academy.[1] This charge focused the public's attention on more fundamental questions regarding the academy's health.

What has *not* been brought to the public's attention is another phenomenon—the fact that all of the military academies have changed significantly since the end of World War II. The changes came in response to pressures and initiatives dating from the 1960s that have resulted in the academies—and especially the Naval Academy—becoming mini–liberal arts universities instead of the military-technical institutions of yore.

At Annapolis, at least, the changes fall into three general categories: a liberalized curriculum, decreased militarization, and the failure to stay true to the Naval Academy's real mission. These will be discussed in turn, followed by a plan for change that I believe absolutely necessary to save the Naval Academy for future generations.

The Liberalization of the Academies' Curricula

The early curriculum at both the U.S. Military Academy at West Point and the U.S. Naval Academy at Annapolis was engineering-based, and it remained so until after 1957. During these years, academy programs focused on the operation of weapons systems, supported by mathematics, science, and general engineering. The focus reflected the belief that officers must understand mathematics and the sciences to be able to follow the rapid changes in weaponry, transportation, and logistics.

Although social and behavioral sciences were added to the West Point and Annapolis curricula after the turn of the century, both programs retained their engineering orientation throughout the first part of the twentieth century.

In 1957, coincident with Sputnik and reforms in many other U.S. colleges and universities, significant changes were made in the curricula at West Point and Annapolis, offering additional engineering and science courses and increasing their technological rigor. The Air Force Academy at Colorado Springs graduated its first class in 1959, also with a curriculum that had a solid science, mathematics, and engineering content.

From the 1960s to the mid-1980s, however, the academies liberalized their curricula. They expanded their offerings to include majors for all cadets and midshipmen; a steadily increasing number were in the liberal arts, thereby diluting the overall technical content of the earlier curricula. The liberal arts programs have attracted so many students that today at both West Point and Colorado Springs, over 40 percent of all cadets major in fields other than science or engineering. The number of Annapolis midshipmen electing liberal arts majors has also increased significantly, from 19.8 percent in 1987 to 34 percent in 1996, and it is reported that the Class of 1999 will approach 40 percent. At the same time, less than 2 percent of 1996 graduates majored in each of electrical engineering, marine engineering, and naval architecture.[2]

What began in the 1950s as increased rigor and broadening of the science and engineering curricula in response to Sputnik and the realities of technology resulted in a weakening of technological content and in a proliferation of liberal arts majors. With student bodies of just over four thousand each, the Naval Academy now offers nineteen academic majors, the Military Academy twenty-two, and the Air Force Academy twenty-six.[3] The available majors cover virtually all the liberal arts, science, and engineering disciplines found in large, selective colleges and universities. Since the academies are relatively small, purely undergraduate institutions not associated with a major university, questions have to be asked about the quality of the many programs and about the cost-effective expenditure of resources to support them.

Looking to the future needs of the armed forces, one finds compelling evidence that advances in technology will continue to present challenges to U.S. security. Proliferation of technological advances, particularly information technology, will make traditional warfighting ever more complex and difficult.[4] Defense experts argue that future U.S. military preeminence depends on using technology's competitive edge. "By applying our advantage in this area," argues Adm. Paul David Miller, "we will better ensure that existing capabilities continue to provide security for the nation, even as traditional force levels grow smaller."[5] It is widely understood that a redesigned force structure and methodology cannot be accomplished without new military technology and that this technology is the key to future national security capabilities.

In a recent interview, Adm. Jay Johnson, the chief of naval operations, outlined the key technological upgrades now in progress throughout the Navy. "Tomorrow starts today," he stated, and "it is all about the magic of new technology and bringing ourselves into the next century with new applications and new systems."[6] In this respect, one of his priorities is to "ensure that naval education and training is brought up to date." He does not describe how the nearly 40 percent of Naval Academy graduates studying the social sciences and the humanities can be expected to operate these advanced, technology-based weapons and support systems.

There should be no disagreement that the armed forces have entered a new era of technology-based warfare. This fact dictates increased educational and training requirements for both officers and enlisted personnel. There are few low-skilled jobs left in the armed forces today. The question, then, is what should be the role of the liberal arts in an academy preparing officers for future high-tech warfare? Should cadets and midshipmen receive a broad, and integrated but ancillary underpinning in the social sciences and the humanities, as is now required at most science and engineering schools, or should many of them continue to be allowed to concentrate on liberal arts?

Offering majors in the liberal arts creates two contradictory problems. First, it drives up the cost of operating the academies by requiring additional faculty of a specialized nature, including extra overhead for staffing additional academic departments. Second, within a fixed budget, it reduces the resources available to each of the other disciplines being taught, making it difficult to offer programs that are competitive with those available in many public and private colleges and universities.

Although most would agree that military officers should have a sound basis in the social sciences and the humanities, the current trend in the numbers of cadets and midshipmen now majoring in nontechnical fields gives the impression that the academies are, in reality, liberal arts institutions. American taxpayers are probably largely unaware that they are paying much more for liberal arts graduates from the military academies than would be the cost of obtaining them from other schools. To the extent that they are aware, they are likely to disapprove.

The Demilitarization of the Naval Academy

Together with the liberalizing of the curriculum at Annapolis, there was a corresponding liberalization of the military environment in which the midshipmen lived and were trained. This trend included more time away from the academy during weekends and holidays; more access to automobiles and civilian clothes; lifting the restriction against drinking within the city of Annapolis; reducing the rigor of plebe (first-year midshipmen) indoctrination; liberalizing the rules for residing in Bancroft Hall, including reveille and lights-out; and easing of dress codes and inspections.

Some observers see changes in the military life at a "military academy" as causal factors in some of the scandals reported over recent years. Often cited are reminders that the young men and women at the service academies reflect the society from which they are drawn, and that it may be expecting too much to try to hold them to the same honor standards demanded of those who attended the academies during earlier, more benign times.

Other observers, however, believe that problems with honor and ethics at the Naval Academy actually started with poor indoctrination of new midshipmen during plebe summer and plebe year, and that these problems have been perpetuated throughout the program as a result of the demilitarized and relaxed environment.

Plebe indoctrination is the beginning of the militarization process at the academy; it sets the tone for all midshipmen for all four years. The importance of the transition from civilian youngster to midshipman and then to naval officer cannot be overestimated. It is vitally important to the quality of future officers, just as boot camp is vitally important to the quality of the Navy's enlisted people, who all become part of the same combat and combat support team during wartime and other military operations.

For much of this century at Annapolis it was believed that "he who would learn to lead must first learn to follow," and plebes found their lot to "be largely that of a follower."[7] Midshipmen were told that "Discipline means a prompt, willing responsiveness to commands. The best discipline is self-discipline. . . . The nature of military organization requires that every individual and unit be responsive immediately to the direction provided at the top."[8]

One historic difference between officers graduating from the Naval Academy and those coming from other sources such as the Naval Reserve Officers Training Corps or Officer Candidate School has been the ability of academy graduates to accept responsibility early and to carry out professional duties more quickly and with less on-the-job training. In the past, Naval Academy graduates were clearly recognized as being better prepared than new officers from other sources. The plebe system provided the opportunity for midshipmen to learn to handle responsibility earlier than did most other new officers: "The characteristics of reliability, imagination, and perseverance on an everyday basis are important tools of an officer's trade . . . these same attributes are the foundation of responsibility, and soon become a way of life for a properly indoctrinated plebe."[9]

Beyond plebe year, this process continued, although the academy applied its regulations and its system of accountability with decreasing severity until, during first-class (senior) year, midshipmen had a great deal of freedom. Along with that increased freedom, however, came increased responsibility.[10] This responsibility traditionally included oversight of the brigade, enforcement of the honor system, and supervision of the plebes.

In recent years, however, things have changed. The plebe indoctrination program presents a significant contrast to that of previous years. Unlike the 1975–76 Naval Academy catalog, the 1993–94 catalog says nothing about "testing and developing" or about "requiring midshipmen to stand on their own feet," to "respond promptly and intelligently to orders," and—most important—"to measure up to the highest standards of character, honor, and morality."[11]

The current indoctrination system has been described by midshipmen in *Shipmate* (the academy's alumni association magazine) as the three upper classes working together to teach and to instruct, rather than bantering with and badgering the plebes. Today plebes ask questions about and exchange views on current events with members of the upper classes; there are no more one-sided barrages of questions from upperclassmen. King Hall, the midshipmen's dining hall, has been transformed into a "true wardroom." No longer do plebes want to climb the Herndon Monument for "carry-on" (their formal recognition as members of the brigade of midshipmen). Today plebes are "champing at the bit" to pick up the standard and cry, "follow me."[12] What has happened to the need to learn to follow first?

Plebe year should provide the opportunity to weed out the people who cannot make it through the program. It should be a period of testing as well as tempering. But, competition, achievement, and excellence are giving way to diversity, sensitivity, double standards, and "goals." The toughness of plebe year is being curtailed. As one 1982 graduate laments: "King Hall is not supposed to be a 'true wardroom.' . . . It is a training ground. The midshipmen in King Hall are not officers yet. They are undergoing training to become officers."[13]

The 1993 Naval Academy Strategic Plan established the new plebe indoctrination system described above to "develop leaders—not just survivors—by imbuing midshipmen with the qualities sought in leaders in the fleet."[14] The Strategic Plan, however, neither described the qualities "sought in leaders in the fleet" nor told just how the new process would provide these qualities.

No one wants a "boot camp" environment for plebe indoctrination. Upperclassmen should not emulate boot camp drill instructors by yelling in the face of plebes, nor should they physically abuse them. Hazing of all types is forbidden, and rightly so. Nevertheless, plebe year should not be an easy passage. An essential ingredient is stress—physical, mental, and most important, emotional. Plebes have to be overloaded to force them to decide what is important and to teach them to use every scrap of time in every day. The wardroom idea is fictitious, since there can be no true collegiality involving those who are untested. Plebe indoctrination is a long, slow, caring task, one that should be a principal responsibility of the most mature and experienced midshipmen (not all classes). It is not mean or sadistic and, in fact, properly executed can be full of humor and great experiences.

Completion of plebe year should signify the midshipman's successful accomplishment of a long, tough task without having fallen by the wayside. It further provides a basis for the behavior and conduct of upperclass midshipmen throughout the four-year program. Easing this process to suit the fashion of the times is an error that will be regretted.

The Naval Academy's Changing Mission

The essence of the Naval Academy—its life and mission—is set forth in two short paragraphs—the first can be found in the official academy mission statement, the second in John Paul Jones's "Qualifications of the Naval Officer." In 1957, the Naval Academy mission statement, as printed in that year's *Lucky Bag* (the class yearbook), was:

Through study and practical instruction to provide the midshipmen with *a basic education and knowledge of the naval profession*; to develop them morally, mentally, and physically, and by precept and example to indoctrinate them with the highest ideals of duty, honor, and loyalty, in order that the Naval Service may be provided with graduates who are *capable junior officers in whom has been developed the capacity and foundations* for future development in mind and character, leading toward a readiness to assume the highest responsibilities of citizenship and Government. (emphasis added)

The current Naval Academy Strategic Plan contains a different, shorter mission statement:

To develop midshipmen morally, mentally, and physically and to imbue them with the highest ideals of duty, honor, and loyalty in order to provide graduates who are dedicated to a career of naval service and have potential for future development in mind and character to assume the highest responsibilities of command, citizenship, and government.[15]

An examination of the two mission statements reveals that the italicized portion of the 1957 statement is absent from the later version. What has been removed is a dedication to a specific type of education—a knowledge of the naval profession—and the objective of providing graduates who are capable junior officers. The current statement is very general and could apply to many officer commissioning programs, but not to a military academy charged with the unique mission to provide officers with a certain type of education for the Navy and Marine Corps who can start performing immediately as capable junior officers.

The other passage reflecting the essence of the Naval Academy is attributed to John Paul Jones, widely recognized as the father of the United States Navy. Part of his famous statement outlining the qualifications desired in a naval officer is well known by every graduate of the Naval Academy and still rings clear across the centuries:

It is by no means enough that an officer of the Navy should be a capable mariner. He must be that, of course, but also a great deal more. He should be as well a gentleman of liberal education, refined manners, punctilious courtesy, and the nicest sense of personal honor.[16]

The academy's mission statement and John Paul Jones's words, if followed, should provide all the guidance necessary for naval leaders to structure and to carry out a program of education and training for midshipmen at the Naval Academy. However, the revisions to the mission statement and the recent troubles at the academy argue that the academy is no longer operating to properly fulfill its real mission or to inculcate the desired qualifications of the naval officer.

As previously discussed, the Naval Academy has been doing things with respect to curriculum and discipline that are outside the strict interpretation of these fundamental, guiding documents. Furthermore, at the academy, the task of preparing officers for operational assignments and combat has taken a back seat to other objectives. For instance, sixteen Naval Academy graduates of the Class of 1996 were sent off to medical school at government expense. That means in addition to the

$250,000 spent on each during four years at Annapolis, the Navy is committing vast additional sums to put them through six years or more of very expensive medical schooling—perhaps as much as $1 million each for basic medical school, residency, and specialty training. Where in the mission statement is it suggested that the Naval Academy exists to train pre-med students?

Athletics is another area where a gap exists between what the academy is doing and what it should be doing. The varsity athletic programs are still designed to attempt to regain glories of old, such as the 1956 Sugar Bowl triumph under quarterback George Welsh. Now, after four different coaches, Navy has had only one winning season (1996) in the past fifteen years. Times have changed, and Navy can no longer compete successfully with nationally ranked schools in varsity sports. To even hope to be competitive today requires accepting football players with less-than-desired academic qualifications and then providing them remedial education at the Naval Academy Preparatory School (NAPS).[17]

As a result, the athletic program as now operated is counter-productive to the academy's mission and should be sized according to capability. Even West Point has seen the light, having recently joined a new football league (called C-USA) and leaving the National Collegiate Athletic Association's Division 1A ranks. The Naval Academy should abandon expectations of national rankings in major sports and instead encourage those martial sports that are a part of the military ethos such as rifle, water polo, lacrosse, soccer, track, and sailing. Moreover, the academy should reverse past decisions to phase out certain varsity sports, including fencing, pistol, and boxing. These sports will provide both the athletic and leadership experiences to midshipmen that they rightly should have.

How to Fix the Naval Academy for the Future

The Naval Academy needs to undertake an intensive review of all of its programs and to conduct a rigorous restructuring for the twenty-first century to overcome its problems. This restructuring should include a new, modern curriculum.

The place to start is with a more modern and visionary mission statement:

> To provide midshipmen with the necessary education and training through academic, moral, and physical development such that graduates will have the capability to successfully serve as officers in the highly technical naval service of the twenty-first century and will have the potential to develop the qualities of mind and character necessary to assume the highest leadership responsibilities of command, citizenship, and government, with the highest ideals of duty, honor, and loyalty.

The Naval Academy, to remain a viable officer commissioning source, has to affirm clearly that to carry out its mission, it must commit to operating as a technological institution—a technological military academy, focused on and prepared for the rapidly shifting technology of modern warfare.

Action Plan for Change

The following specific areas of change should be examined and evaluated, and implementing action should be taken. In a time of seriously decreasing budgets, attention to costs becomes essential. The Naval Academy should be operated to carry out its mission in the most cost-effective way.

Academics

First, reduce the number of majors to permit faculty excellence in a few fields in lieu of mediocrity in many, to correspond more closely with the academy's mission and the Navy's needs, and to improve the academy's overall program.

Second, terminate the liberal arts majors' programs. Permit current midshipmen to graduate under current rules. All future midshipmen should be required to qualify for graduation in mathematics, science, or engineering fields.

Third, establish minimum academic standards for all entrants. There should be no special, lower standards—however they may be justified—for any individual candidate, whether that individual be a racial or ethnic minority group member, a female, or an athlete. Today, there is a growing supply of academically capable minorities and females graduating from high schools who can be recruited into the academy. It is only necessary to identify the sources and to compete for the candidates.

Fourth, reorganize faculty and academic departments to ensure delivery of an integrated, multidisciplinary technological curriculum. Computer science and operations research (quantitative studies) must be integrated with other technical courses. The degree program should emphasize basic technical knowledge, with a focus on those scientific and technical fields that are of primary interest to the military and are relevant to developing and operating future weapons and support systems.

Fifth, require all midshipmen to take a foreign language and to demonstrate proficiency in it. Two years of coursework represent the minimum exposure to a foreign language that should be required of all graduates. Even those studying a foreign language in high school should take at least one year in that language at the academy.

Honor and ethics

First, continue renewed emphasis on the honor code. Concentration should be on individual integrity. Individual honor is the basis of all behavior, whether on sports teams, in combat operations, or in leadership positions. The honor code should rest on the solid foundation of individual integrity, high moral standards, and a willingness by midshipmen to stand fully behind their oral and written statements. An honor system operated by midshipmen is probably preferable to one enforced by staff.

Second, administer the Naval Academy under the same honor code that the midshipmen live by.

Military life

First, remilitarize the academy. The environment should be fully military—that of a military academy devoted to the profession of arms, not that of a civilian university. Increase restrictions on midshipmen regarding absence from the Yard, dress, behavior in public, alcohol use, and access to and use of automobiles.

Second, end athletic scholarships through the Naval Academy Preparatory School. Restrict attendance at NAPS to those for whom it was established—meritorious enlisted personnel capable of attending the Naval Academy.

Third, require athletes to take the same coursework as other midshipmen. Abolish all perquisites for athletes such as special coursework, assignments, tutoring, lodging, and eating arrangements. Athletes should be treated the same as other midshipmen and have the same academic and military requirements. (Some observers believe that the creation of a separate athletic culture at the academy is one cause of today's problems.)

Fourth, reestablish varsity sports in those martial sports recently reduced to sports clubs—boxing, pistol, fencing. Emphasize participation in these types of sports.

Fifth, replace any lost income resulting from lack of sports played against schools such as Notre Dame on nationwide television with alumni contributions to the Athletic Association.

Conclusion

The Naval Academy's future depends on adopting a properly focused curriculum; higher standards of honor, loyalty, duty, and character; more rational intercollegiate sports programs; and increased militarization of academy life. This redirection will have to be accomplished with diminished resources. The program must be cost effective. It must, at lower cost, produce graduates who have undergone a unique education and training experience—one not available elsewhere at any cost, and one that provides the Navy and Marine Corps with incomparably educated officers who can be expected to remain in the service for a full career and who can compete favorably with officers from other commissioning sources.

Those officers who will lead and operate the new, highly technological armed forces cannot do so with an education deficient in the foundations of technology. History is replete with examples of new weapons and new technology being ignored by a country's military services only to see another country develop and use this technology against them.

Military service is a vocation, a calling different from any civilian occupation, and the military academies play a vital role in upholding the military traditions of the United States. They set the standard for measuring the performance of all military officers. The opportunity now exists to redirect the Naval Academy and the other service academies away from the pseudo-university track they have been on and back toward being the highly select military-technical schools they once were.[18]

Notes

1. James F. Barry, "Adrift in Annapolis," *Washington Post*, 31 Mar. 1996, C1. See also his piece in the paper's "Outlook" section a year later, 6 Apr. 1997.

2. Data provided by the service academies.

3. Ibid.

4. U.S. Joint Chiefs of Staff, *A Strategic Vision for Professional Military Education of Officers in the Twenty-first Century: Report of the Panel on Joint Professional Military Education* (Washington, D.C.: Joint Chiefs of Staff, 1995).

5. Adm. Paul David Miller, "The Military After Next," U.S. Naval Institute *Proceedings* 120 (February 1994): 41.

6. Interview of Adm. Jay Johnson, *Jane's Defence Weekly*, 19 Mar. 1997, 32.

7. *Reef Points, 1953–1954* [Handbook for the Brigade of Midshipmen] (U.S. Naval Academy, Annapolis, Md.), 18.

8. Ibid., 23.

9. LCdr Frederick N. Howe, Jr., "The Plebe Indoctrination System—A View From Seaward," *Shipmate* 33 (June 1970): 8.

10. Ibid., 10.

11. *The United States Naval Academy Catalog, 1975–1976*, 28.

12. Midshipman Brian Scott Knowles, "Brigade Activities," *Shipmate* 57 (November 1994): 42.

13. Steve Gatanis, "The Mail Boat," ibid. 58 (April 1995): 15–16.

14. "USNA Strategic Plan—First Update" (Annapolis, Md.: U.S. Naval Academy, n.d., ca. 1993), 9–10.

15. Ibid., 1.

16. *Reef Points*, 54.

17. Of 234 students at NAPS in the 1995–96 academic year, 71 were recruited athletes who had not been accepted into the Naval Academy for academic reasons. See Bill Brubaker, "An Unstated Mission," *Washington Post*, 8 Apr. 1996, A1.

18. The revised Naval Academy program can provide a standard for subsequent changes at the Military Academy at West Point and the Air Force Academy at Colorado Springs.

Liberal Education, the Study of History, and Generation X*

Josiah Bunting III

I am most honored to be here. The Virginia Military Institute (VMI) is two thousand miles east of the United States Air Force Academy and in many significant ways, about two hundred years behind you in our habits and comforts of living, our lifestyles. We are surely a vital legacy of the nineteenth century as this school seems to be a legacy and child of the twentieth and twenty-first. Everybody in Virginia admires the Air Force Academy in particular. We admire your double-digit football victories over huge land-grant institutions all over the west. We wonder to ourselves, how do you do that? And then we see that Navy and Army are able to do the same thing and we wonder how they do that too, but they do it. Here it must be the bracing air of the Rockies. About thirty years ago, I studied in England at Oxford. It seemed to me every third or fourth person I passed on the street was a Rhodes scholar from the Air Force Academy. I do not know if that tradition still continues, but I certainly hope it does.

VMI has its being in the Shenandoah, the beautiful valley of Virginia, and we live, so to speak, in American history. In the old town of Lexington, there are two well-known colleges. The other institution is named for the greatest American soldier of the eighteenth century, George Washington, and a man whom many argue was the greatest soldier of the nineteenth, Robert E. Lee. VMI is the alma mater of George C. Marshall, a great soldier, who is remembered, however, as much for his pursuits in peace as in war. General Marshall, like General Lee and General Washington, and, of course, "Stonewall" Jackson, was a great man somewhat in danger of becoming memorialized or marbled almost to the point where it is difficult for us to realize what these people must have been like as friends and colleagues.

These visible reminders of their legacies, the clichés of inspiration, lie very close at hand. But how and even whether the influence of these men can play some useful part in our own lives—can somehow kindle in our hearts a desire to emulate their virtues is another issue altogether. In a way, this question will be at the heart of my few comments this evening. I will be much shorter than the funeral oration of Pericles and no less brief than John F. Kennedy's wonderful inaugural address. I was thinking about that speech today in the context of current American politics and wondering what a distracted public would make of such an oration in 1996, either of those, but particularly the Kennedy inaugural.

*Adapted from General Bunting's remarks delivered at a banquet on 21 November 1996 during the United States Air Force Academy's 17th Military History Symposium.

We seem no longer to have the self-assurance and confidence in audiences of American citizens to speak in what was once called an "elevated style." But, of course, as a contemporary journalist reminds us, what we elect nowadays more frequently seems to be what we wish, namely, the mayor of the United States, not the president. That is to say candidates who will fix things that we want fixed. We want a certain kind of a CEO, not so much a head of state. We want, and this is a nonpartisan comment, a compound sentence connected with a "but."

Virtually all in this audience are military professionals or people who work closely with the military. You ponder and debate the great public issues. You cast your ballot. You unhesitatingly lead others and obey the elected officers of government. Everywhere we hear decried the absence of truly remarkable and capable candidates for public office. We shrug at the usual responses, "Well, why don't you run." Surely, then, we get the mayors, the governors, the senators, the representatives we deserve. They are us.

If this is true for American politics, it is true and directly relevant to the study of history in 1996—the study of history and all the liberating arts in our colleges, perhaps especially for the young people now at the academies, who in a way are like aberrations, refugees from "Generation X," if you will. Academy undergraduates represent a perfect national cross-section of talented members of a generation born in the mid- and late 1970s. But for most of its members, those not at schools as eminent as this one, the serious study of history seems a preposterous waste of time.

You may be refugees from Generation X, refugees from a postmodernist culture that mistrusts reason, and according to the *X Files* should trust no one. But you are a national cohort of refugees, a dwindling number who are the cadets and the younger officers in undergraduate and graduate programs in history. If one can appropriate the term, the younger cohort of this "X" Generation sees a culture in such profound change that it doubts anything history might provide would be relevant to its needs and interests. Nor, of course, is this generation used to the long periods of solitary study and reflection over the liberal arts disciplines generally.

On this subject, most historians and indeed most academics are notoriously silent. How and in what ways? By what alchemy of intellection and inspiration does the serious study of history do undergraduates any good, undergraduates who are going to be laypersons and not historians? You may think the questions appalling, but I think the failure of academic historians to answer them in a way that provides a compelling answer for an educated lay public, particularly its young members, is perhaps also appalling. That failure is symptomatic indeed of a profession that is becoming, in many respects, a self-perpetuating guild of specialists laying up for themselves granaries of recondite facts that are not knowledge and that may not lead to wisdom.

Gore Vidal laments, "Nowadays I read too many of those magazines that seem to have taken the place of books for those still able to read." We know what he means: laying waste to our powers. When we read at all, it is to pick up something near to hand: *Time, USA Today, Sports Illustrated.* How about *Vanity Fair,* the latest issue with yet another condensed version of a new life of Pamela Harriman or perhaps another take on the design of jurors at the second O. J. Simpson trial?

We are fingering several separate issues here, all of them directly relevant to our curricula, to our study of the liberal disciplines in our military academies. First, members of a generation generally mistrustful of objective knowledge and with a span of attention of grasshoppers. Second, a generation of professors, most of them less attracted to undergraduate teaching than to certain forms of research, that rarely considers the larger purposes to which their historical researches are to be put. Third, the unexamined question of how such study may seriously influence the behavior of women and men in the world of politics, government, war, and business in the 1990s. And fourth, a culture in which the leisure to read and to think is nonexistent, and even when it is available is almost discouraged. We must all of us wrestle with these issues.

Overwhelmingly, young people in this audience will become serving officers or professionals in the world of affairs, not full-time writers or philosophers or perhaps even academics. How, then, will you organize your days and your years always to assure that your brain continues to be nourished and refreshed by history, by literature, by philosophy? Or will you join a large cohort of recent graduates for whom a four-year college education is a steamer trunk to be stuffed into an attic to molder while your brain prepares for a lifetime of moldering?

Alexander Hamilton, a colonel at Yorktown, secretary of the treasury in the first Washington cabinet, wrote political classics on the backs of old envelopes. The greatest Hamiltonian of them all, Teddy Roosevelt, almost the most famous law school dropout of them all, I might add, spent six weeks in law school and said it was a complete waste of time. What a brilliant man! He published thirty books in his lifetime, some on botany and biology, others on history and at least two while he was president. Harry Truman's reading was ceaseless and voluminous. He probably knew more history than most of us in this room. Harold MacMillan, who went through the Suez crisis, read Jane Austen lying on a sofa giving directions. Not necessarily a role model, but you see my point.

The point is indeed simple. Men and women of action in particular need what a frenetic culture does seem to deny them, books, leisure, the habit of thinking deeply as the authors' partners. Two weeks ago a voice on television commanded me to "Sculpt the 'abs' of your dreams." Presumably a large number of those whom, in your profession, you will be called upon to defend have such dreams as these. Perhaps that form of education we call "liberal" will immunize you against them. I hope so.

The mission of our national military academies must, in part, be that of the great liberal arts colleges. It must be at least in part to kindle a lifelong habit of intellection, of unquenchable intellectual curiosity, to lay deeply the habits of inspiration from books, from art, from music. Twenty years ago we would hear the phrase "positive addiction," meaning aerobic workouts to the point where you feel ill if you do not have them. I mean a positive addiction here of the mind, one that not only survives, but positively dominates a life of active service in civilian or military pursuits. I am not talking so much of graduate degrees, of directed research, of credentialling, but simply about the working mind, about Emerson's thinking, particularly his capacity for wonder, for imagination.

Somehow, there has grown up in our American culture over the last 200 years the ignoble prejudice that men and women of action cannot, even must not be men and women of reflection—that men and women of habitual reflection must somehow be disqualified from timely, resolute action: "The native hue of resolution is sicklied o'er with the pale cast of thought." The time spent reading a work of European history or contemporary literature must be given over to making a visible bustle in the office. In not so different a context, both Zbigniew Brzezinski and Henry Kissinger are reported to have told new and younger associates in Washington, D.C., one of them Adm. John Poindexter, that they must henceforward live only on their intellectual capital. And that during their assignments in Washington, their official labors alone would totally suck dry whatever reservoirs of energy, idealism, determination to read and keep learning beyond the limits of whatever duties they would be assigned. This is surely not wise advice, nor wise policy, nor a wise way to transact business, especially the nation's business.

Something interesting has happened and in no more than a generation and a half. There are men and women in this room old enough to remember in the late 1950s our easy expectation that our lives at the end of the twentieth century would be dominated by leisure. Machines would accomplish everything. Automation would free us to do virtually any task we wished without interruption. Only in military academies would there exist feverish regimens in which people worked at all times. But in these projections of leisure for the rest of the culture, all were shown in tennis clothes playing cards while riding in automobiles that were guided by hidden beams. Life was perpetual summer. The world lay before us like a land of dreams.

The opposite has happened. There is no leisure, or if there is, it is a grim by-product of a service economy that lays off unexpectedly in order to compete successfully or unsuccessfully in other markets. More to our point, most of us, for whatever reason, literally resist making use of leisure for ourselves out of some misbegotten idea that we must be visibly and actively participating in unceasing labor in whatever profession or enterprise claims our allegiance.

Here is a good 1970s term for you, one unique to our culture: "the bottom line." The bottom line is this. All of our military colleges and all of our liberal arts colleges are failing their constituencies if they do not graduate officers who, because of cultivated intellectual appetites begun in places like this one, resist full immersion in the American cult of dutiful hyperactivity—the kind of mechanical existence portrayed in Dave Matthews's hit, "Ants Marching." Somebody said to me the other day, VMI cadets should not be allowed to listen to "Ants Marching." They have too well-developed a sense of irony. How many people in the room know the song? It is not one an academy glee club would sing.

The study of history (provided it is lifelong and unceasing) and particularly the study of historical biography must be a staple of every young military officer's diet.

The great historians, particularly the great British historians, have always written for an audience, not of specialists, not of historians, or graduate students, but of an educated lay public, an educated public of civilians and military professionals, not themselves experts, but who should be reading what experts produce. These historians have found fulfillment, not only in a disciplined attempt to reconstruct the past, fastening on those elements often without conscious thought they think are useful, but also simply for the joy of writing what an earlier age called narrative history. Their obligations to literature and rhetoric were as heavy as their obligations to research.

I mention very briefly the study of biography, perhaps I should say an addiction to biography and especially to biographies of people who were once called "men of action." Educated eighteenth-century Americans gloried in such works, many by Plutarch. Their heroes were the heroes of republican Rome. In these lives, the Americans saw a satisfying corroboration for their own instincts about government, about culture, about citizenship, about how they should behave. They saw individual lives whose challenges and dangers seemed not so different from their own. They believed, and I hope many of you yet believe, that learning about such men should and could not only instruct, but also inspire the emulation that might follow.

The results of such study, that is to say, might not only entertain us. It might help make us learned. It might even make us wise. Such wisdom, possibly, could penetrate the boundary, the memory between the intellect and action, between mind and character. In their view, learn and grow wise might be translated to learn and grow better. It is the greatest single question and each generation must reanswer it in all of education. In what way does how we learn and how we continue to learn reliably influence the way we act in the world of affairs? Incidentally, the virtues that characterize these noble Romans that our ancestors loved to read about comprised austerity, forbearance, selflessness, resolution—virtually everything our culture nowadays does not celebrate.

Biography and autobiography present the great as they were at twenty or thirty or forty, not as aged icons or marble statues or irascible old people. No young cadet who studies in any of our military academies can fail to have learned that every great general was once a lieutenant or a major, who like a Jesuit novice, invariably had his own dark nights of the soul, his own failures, his own dark disappointments. There is no better time to judge a man or a woman's character than during the years after he or she has been denied something or failed at something in which he or she believed deeply or wanted badly. History and biography in this regard are unexcelled lifelong teachers.

Joseph Conrad wrote that, "A great work of imaginative literature carries its justification in every line." The study of history by men and women of action carries its justification precisely in its capacity to enlarge our own meager ambits of direct experience, and in its plain demonstration of the general consistency of human nature through history in the realms of politics, government, international relations, and war. Thucydides hoped, as he said modestly, but he meant believed, that his great history of the war in the Peloponnese would be a history for all time: "For men who wish to see clearly the events that have taken place and in accordance with human nature will happen again and again in the future in the same or similar way."

In the widening and enriching of our own humanity, in your humanity as military women and men, and I speak particularly to the young cadets in the audience, the Duke of Burgundy in *Henry V* is given a searing line: "They grow as soldiers will that nothing do but meditate on blood." We may adopt it metaphorically to another century whose horrors reduce those of Agincourt to a tiny scale. We must, as has been said, learn and know the value of what we may be called upon as soldiers to destroy. For surely history instructs us we will go to war again, and almost certainly in this generation to fight for country and cause against other men as certain of the justice of their cause as we are of ours.

It is plainly time to subside. Let me do so with a particular challenge to every cadet, undergraduate, young officer, and young instructor present. Before you leave your undergraduate training, your undergraduate studies, make a list on a piece of paper or in your journal if you keep one (and you should) of historical topics that for one reason or another continue to claim your attention, things that you continue to read about. You are developing a propensity for that topic. Resolve at some point in your life to sit down and write a book about it and, moreover, to do it while you are also serving in some other duty. It is not necessary to be a professional historian and to write great works of history to devote your time, full time, to the study of history.

I think when men and women of action who are in other professions continue to do their own researches and to follow their own hearts and instincts about topics that are useful, great history is often the result. The roster of men and women who have written history, and often great history, but only as determined amateurs is long and distinguished. Let me put at the head of this list that of perhaps the greatest historian who ever wrote, Thucydides of Olorus, the historian of the Peloponnesian War, and the last, the greatest man of action of this century who received the Nobel Prize for literature, Winston Churchill. The French statesman Georges Clemenceau famously noted that war is too serious a matter always to be left to generals. Perhaps history is too serious a matter to be left only to professional historians.

The Challenge of the Twenty-first Century: Balancing General Education, Military Training, and Professional Studies*

R. A. Mason

Superintendent and distinguished guests, colleagues, it is a very great privilege and a pleasure for me to be invited to contribute, for the fourth time, to the Military History Symposium, even as a Redcoat in the last plot of the graveyard. I served my apprenticeship as a Department of History bag carrier for a previous generation of visiting prima donnas, so I doubly appreciate the courtesy and hospitality which has greeted us again this year.

In the early summer of 1970, Brig. Gen. Robin Olds, commandant of cadets, fired nine air officers commanding (AOCs) in one day. I was invited by Vince Hart, the Group AOC and a battle-hardened Marine colonel, to move across from the Department of History as a temporary replacement. When Colonel Hart sought the approval of General Olds, Robin said, "Hey, wait a minute, wait a minute. A Brit has no authority under Congress and as sure as hell the writ of the Queen's commission doesn't extend as far as Colorado." Whereupon Colonel Hart said, "Sir, you know that, I know that, but the goddamn cadets don't know that." Fortunately, no cadet asked and I did not tell, and so I probably avoided all kinds of violations.

In that same year, a heavily decorated British paratroop general and scholar, Sir John Hackett, gave the United States Air Force Academy's Harmon Memorial Lecture in which he defined the British military ethic, a collection of moral principles associated with the profession of arms in the United Kingdom:

> The human qualities which the military life demands include fortitude, integrity, self-restraint, personal loyalty to other persons, and the surrender of the advantage of the individual to the common good. The military profession is unique in one very important respect. It depends upon these qualities, not only for its attractiveness, but for its very efficiency.

In 1995, the British Army was very explicit officially and publicly about the reasons for an ethic. It is the soldier's obligation to follow orders in the face of an enemy and to do his duty despite the risk of death or injury. It is that operational liability with the possibility of self-sacrifice accepted by every soldier on enlistment

*Adapted from Air Vice-Marshal Mason's remarks delivered on 22 November 1996 at the concluding session of the United States Air Force Academy's 17th Military History Symposium.

that marks the armed services as being essentially different from the rest of society. According to the Army:

> Officers are expected to set the example—to show moral courage. An officer's personal conduct must be beyond reproach. Integrity is essential to leadership, to moral courage and consistency of approach. If an officer's conduct calls into question his or her integrity, or brings the Army into disrepute, the trust or respect of those he or she is privileged to command is placed in jeopardy and the right to hold a Queen's commission may be forfeited.

Such concepts are deeply rooted. Plato argued that efficiency in war demanded professionalism requiring exclusive attention, the greatest skill and practice, the brave, the fit, and the spirited, gentle to their fellows and fierce to their enemies. Otherwise, he pragmatically observed, they will prevent the enemy from destroying the city by doing it first themselves.

Thomas More's ideal soldier had similar characteristics. If the Utopians could not avoid war, command of the army should be given to one of tried virtue and prowess. Their soldiers would rather die than give up an inch, sustained by a knowledge of chivalry and feat of arms and by the wholesome and virtuous opinions absorbed in their childhood. The fiercest battle would see a band of chosen young men who would be sworn to live and die together.

This early identification and recognition throughout history of a need for a military ethic is hardly surprising, considering the nature of war, which at the moment perhaps, seems rather remote from us here in Colorado Springs in 1996. War remains the province of destruction, despair, danger, desolation, demoralization, violence, and brutality. In Carl von Clausewitz's philosophical abstract, war is an act of force and there is no logical limit to the application of that force.

There is an inherent conflict of values here, in the use of violence to achieve an objective by democratic societies professing adherence to the rule of law. In war, the soldier's license to kill and destroy must be constrained. In peacetime, he must, in Socrates's expression, be denied the inclination, the motivation, and the opportunity to destroy the city, and the military ethic buttresses that constraint.

There, I think, is the answer to the question why we need an ethic. I suggest that it is still the inculcation of this ethic which sets a military academy apart from other education and training establishments.

As I understand it, in the United States both the ROTC (Reserve Officers Training Corps) and OTS (Officer Training School) add it to a broader general education. Only in a military academy is it the *raison d'être*. In preparation for a military profession, as Gen. Dave Palmer, former West Point superintendent, so felicitously put it, there is the need for a sustained reinforcement of character on an everyday basis, not a specific program or a specific orientation. It is the pursuit of an ideal which lies much deeper in Western democratic culture than just medieval romantic folklore. But it is, I would argue, the pursuit of an ideal. It is not the imposition of moral absolutes.

In this context, I would address three questions. What will be expected of young officers in the next decades? What will be similar and different? And how should their responsibilities influence the content of their preparation? Let me deal with the easy one first.

I would identify ten features likely to influence the security environment faced by the Air Force Academy's Class of 1997 and its immediate successors.

First, there will be no direct threat to the national integrity of the United States; no immediate, obvious need for its defense. Second, there will, however, be a need for constant vigilance against the emergence of any future threat. Third, there will be continued concern for regional stability and U.S. interests worldwide. Fourth, commitments will frequently be unpredictable. Fifth, responsibilities will include "peace-keeping"—a term used very loosely—and humanitarian operations. Sixth, new dangers will arise from terrorism and from international crime, particularly narcotics. Seventh, there will, most probably at the same time, be an extended period of dwindling resources allocated to defense. And eighth, most military operations will be exposed to international media, with instant impact on public opinion and government policy.

Those eight features are also awaiting the 1997 classes of Annapolis and West Point. The ninth is awaiting all air force graduates worldwide. It is the impact of technology, not just as a general military revolution—because I would argue that the military revolution which took place at the beginning of this century was probably greater than the one that is taking place now—but the specific impact of technology: on the interaction of space, the manned aircraft, and the unpiloted air-breathing vehicle.

The tenth feature is one which is unique to the Air Force Academy's Class of 1997. It is the twenty-first-century inheritance of the U.S. Air Force—an unchallenged aerospace world supremacy, an awesome responsibility, and a potential for good or ill unparalleled in the history of warfare. That is a pretty sweeping statement when you think about it. The tenth is one which I, as a Brit, take more than just a passing interest.

The implications of those ten features for junior officers can be, I believe, equally easily summarized. There will not always be a clearly defined enemy and the use of force may be subject to tight, unpredictable, and uncongenial political control. The range of possibilities, duties, and operational environments will broaden. Junior officers will be exposed suddenly and frequently to widely differing cultures and regions. They may carry much greater personal responsibility; to the point of individually enhancing or damaging national credibility because of the constant interaction between operations, media coverage, and public and international opinion. The domination of combat by the manned aircraft will gradually be shared with space, with unpiloted vehicles, and with information warriors—with far-reaching implications for traditional air force culture.

So what needs to change in the content of the preparation of junior officers? I look to reinforcement and expansion rather than sweeping innovation. I would reinforce the inculcation and fostering of personal qualities for the reasons I have already outlined: integrity, loyalty, commitment, but above all, responsibility—responsibility for one's self and one's own actions, responsibility for colleagues—

Again, I go back to the point raised as to the cohesion of the British Army during World War I, which was sustained not so much by willingness of men to die for King and Country, but by the willingness of men to die for their buddies. Responsibility goes four ways. It starts with one's self. It extends to one's contemporaries and one's close friends. It goes to one's subordinates. It goes to one's seniors. This is still one of the first lessons a junior officer in all three British services is taught.

I would seek to expand the development of judgment. Usually in military education, there is a tension because of superficially incompatible objectives: discipline and individuality, conformity and questioning, responding and innovating, determination and flexibility, imagination and objectivity, fire and dispassion.

More than ever before, junior officers must be prepared to recognize these apparent incompatibilities as integral and permanent in their chosen profession. They must develop the intellect and the strength of character to identify the time for thought and the time for action, the time for conformity and the time for initiative, the time for consolidation and the time for innovation. More than at any previous time in the short history of aerospace power, mental flexibility built on strength of character and good judgment will be indispensable to the continued success of the U.S. Air Force, and, indeed any other air force for that matter.

It may not make for an easy life for instructors. My own privilege, apart from working in a superb department with great friends, was to work alongside officers from other departments including Majs. Mike Dugan, Bob Oaks, and Erv Rokke, among many others.* As deputy head for military history I had, among my junior officers, a promising but rather barbary young captain named Ron Fogleman.** Lessons of leadership, for example, far outlast those learned in the classroom.

In the classrooms we need to reinforce professional awareness by emphasizing those subjects which distinguish the military studies here from those anywhere else, including the evolution, potential, and constraints of aerospace power: the military exploitation of the third dimension *by* man and woman, not necessarily *with* man and woman. These should be studied in the context of the history of warfare, of the fluctuating impact of technology, of the interaction of war and politics, especially the often uncomfortable subordination of the military method to the political objective. We need to expand still further the junior officer's awareness of, and sensitivity to, cultural diversity.

But we also need to be cautious. We need to be cautious in seeking to follow social trends because they will constantly change and we shall be playing catch-up, perpetually out-of-step—especially when those trends may be easier to adopt than to reject. I hate having to present the Royal Air Force (RAF) as a negative model for anything, but Cathy Downes*** very comprehensively and accurately described

---

*Gen. Michael J. Dugan, USAF, Ret., former USAF chief of staff; Gen. Robert C. Oaks, USAF, Ret., former commander U.S. Air Forces in Europe; and Lt. Gen. Ervin J. Rokke, USAF, Ret., former president U.S. National Defense University.

**Gen. Ronald R. Fogleman, USAF, Ret., former chief of staff.

***See her contribution in this volume.

the self-destructive process in the 1960s, which finished in the loss of Cranwell College Academy Course. I know that my RAF colleagues are trying hard to rebuild from there and are doing a tremendous job but now under very difficult circumstances.

In seeking to mirror our parent society, let us be careful about cliché. Reflect ethnic and gender diversity, yes, but do not incorporate incompatible civilian values and priorities. Should cultural diversity be subordinated to the pursuit of a unifying military ethic? I have no doubt about the answer: Yes.

Let us be cautious also in harnessing technology to educational methods. Are we going to enhance the relationship between instructor and his, or her, cadet or are we going to replace it by impersonal communication?

Moreover, are we in danger of seeking to quantify unquantifiables? Recall Professor I. B. Holley's superb sentiments that in training, we can quantify proficiency, but we educate to stimulate, to expand perspectives, to enhance judgment, to allow the flexibility, to encourage the flexibility to deal with the unexpected and indeed to shape the unexpected. Education is rather like a diet. You can easily measure the efficiency of a digestive system, but the longer term impact of food on a body may be influenced by many other factors. Whenever a position is defended, whatever it is, wherever it is, on exclusively quantifiable grounds, the intangible factors are discounted. Yet in education, they are frequently the most important.

The success of preparation for the military profession and the extent of the inculcation of the military ethic can only be measured by their impact on the parent service in peace and in combat. Without nurturing by the parent service, young ideals will wither and die. The environment and atmosphere of the receiving service are likely to be more important in the longer term than the uncertainty and the rate of change of the cultural environment from which the cadet came in the first place.

I would conclude as I began with a quote from Gen. Sir John Hackett's Harmon Memorial Lecture of 1970. He said—and General John wrote this, of course, when there were no female cadets:

> A man can be selfish, cowardly, disloyal, false, fleeting, perjured, and morally corrupt in a wide variety of other ways and still be outstandingly good in pursuits in which other imperatives bear on those upon the fighting man. He can be a superb creative artist, for example, or a scientist in the very top flight and still be a very bad man. What the bad man cannot be is a good sailor or soldier or airman. Military institutions thus form a repository of moral resource which should always be a source of strength within the state.

In determining the balance of continuity and change in the preparation of our junior officers, that sentiment would seem to be still a guiding light.

Select Bibliography of English-Language Books Covering Officer Social Origins, Selection, Education, and Training

General/Comparative

Abrahamsson, Bengt. *Military Professionalism and Political Power.* Beverly Hills, Calif. and London: Sage, 1972.

Childs, John. *Armies and Warfare in Europe, 1648–1789.* New York: Holmes & Meier, 1982.

Corvisier, André. *Armies and Societies in Europe, 1494–1789.* Trans. Abigail T. Siddall. Bloomington and London: Indiana University Press, 1979.

Dixon, Norman. *On the Psychology of Military Incompetence.* New York: Basic Books, 1976.

Duffy, Christopher. *The Military Experience in the Age of Reason.* New York: Atheneum, 1988.

Fidel, Kenneth, ed. *Militarism in Developing Countries.* New Brunswick, N.J.: Transaction Books, 1975.

Finer, S. E. *The Man on Horseback: The Role of the Military in Politics.* 2d ed., enlarged, rev., and updated. Boulder, Colo.: Westview, 1988.

Gutteridge, William F. *Military Institutions and Power in the New States.* New York: Praeger, 1965.

Hackett, John W. *The Profession of Arms.* The 1962 Lees Knowles Lectures. London: Times, 1963.

Hale, J. R. *Renaissance War Studies.* London: Hambledon, 1983.

―――. *War and Society in Renaissance Europe, 1450–1620.* Baltimore, Md.: Johns Hopkins University Press, 1985.

Huntington, Samuel P. *The Soldier and the State.* Cambridge, Mass.: Belknap Press, 1957.

Janowitz, Morris. *The Military in the Political Development of New Nations: An Essay in Comparative Analysis.* Chicago and London: University of Chicago Press, 1964.

―――. *The Professional Soldier: A Social and Political Portrait,* 2d ed. New York: Free Press, 1971.

―――, and Jacques Van Doorn, eds. *On Military Ideology.* Rotterdam: Rotterdam University Press, 1971.

―――, and Stephen D. Wesbrook, eds. *The Political Education of Soldiers.* Beverly Hills, Calif., London, and New Delhi: Sage, 1983.

Kennett, Lee. *The First Air War, 1914–1918.* New York: Free Press, 1991.

Kourvetaris, George A., and Betty A. Dobratz. *Social Origins and Political Orientations of Officer Corps in a World Perspective.* Denver, Colo.: University of Denver, 1973.

―――, eds. *World Perspectives in the Sociology of the Military.* New Brunswick, N.J.: Transaction Books, 1977.

Martin, Michel Louis, and Ellen Stern McCrate, eds. *The Military, Militarism, and the Polity: Essays in Honor of Morris Janowitz.* New York: Free Press, 1984.

McFarland, Stephen L., and Wesley Phillips Newton. *To Command the Sky: The Battle for Air Superiority over Germany, 1942–1944.* Washington, D.C. and London: Smithsonian Institution Press, 1991.

Moskos, Charles C., Jr., ed. *Public Opinion and the Military Establishment.* Beverly Hills, Calif.: Sage, 1971.

———, and Frank R. Wood, eds. *The Military: More Than Just a Job?* Washington, D.C., New York, and London: Pergamon-Brassey's, 1988.

Ralston, David B. *Importing the European Army.* Chicago: University of Chicago Press, 1991.

Stephens, Michael D., ed. *The Educating of Armies.* New York: St. Martin's, 1989.

Teitler, G. *The Genesis of Professional Officers' Corps.* Trans. C. N. Ter Heide-Lopy. Beverly Hills, Calif.: Sage, 1977.

Vagts, Alfred. *A History of Militarism.* Rev. ed. New York: Meridian Books, 1959.

Van Creveld, Martin. *The Training of Officers: From Military Professionalism to Irrelevance.* New York: Free Press, 1990.

Van Doorn, Jacques. *The Soldier and Social Change.* Beverly Hills, Calif. and London: Sage, 1975.

———, ed. *Armed Forces and Society: Sociological Essays.* The Hague and Paris: Mouton, 1968.

———, ed. *Military Profession and Military Regimes: Commitments and Conflicts.* The Hague and Paris: Mouton, 1969.

Van Gils, M. R., ed. *The Perceived Role of the Military.* Rotterdam: Rotterdam University Press, 1971.

Welch, Claude E., ed. *Civilian Control of the Military: Theory and Cases from Developing Countries.* Albany: State University of New York Press, 1976.

Wells, Mark K. *Courage and Air Warfare: The Allied Aircrew Experience in the Second World War.* London: Frank Cass, 1995.

Africa (General)

Gutteridge, William F. *Armed Forces in New States.* London: Oxford University Press, 1962.

Argentina

Corbett, Charles D. *The Latin American Military as a Socio-Political Force: Case Studies of Bolivia and Argentina.* Miami, Fla.: Center for Advanced International Studies, University of Miami, 1972.

Goldwert, Marvin. *Democracy, Militarism and Nationalism in Argentina, 1930–1966: An Interpretation.* Austin: University of Texas Press, 1972.

Potash, Robert A. *The Army and Politics in Argentina, 1928–1945: Yrigoyen to Perón.* Stanford, Calif.: Stanford University Press, 1969.

———. *The Army and Politics in Argentina, 1945–1962: Perón to Frondizi.* Stanford, Calif.: Stanford University Press, 1980.

———. *The Army and Politics in Argentina, 1962–1973: From Frondizi's Fall to the Peronist Restoration.* Stanford, Calif.: Stanford University Press, 1996.

Australia

Coulthard-Clark, C. D. *Duntroon: The Royal Military College of Australia, 1911–1986.* Sydney: Allen & Unwin, 1986.

Cunningham, I. J. *Work Hard, Play Hard: The Royal Australian Naval College, 1913–1988.* Canberra: AGPS Press, 1988.

Encel, S. *Equality and Authority: A Study of Class, Status, and Power in Australia.* London: Tavistock, 1970.

Frost, R. E. *RAAF College & Academy, 1947–86.* n.p.: Royal Australian Air Force, 1991.

Mortensen, K. G. *An Australian Army Cadet Unit, 1945–1977: Dismissal and Reveille.* Parkville, Victoria, Australia: Gerald Griffin, 1978.

Smith, Hugh. *Preparing Future Leaders: Officer Education and Training for the Twenty-first Century.* Canberra: Australian Defence Studies Centre, Australian Defence Force Academy, 1997.

———, ed. *Officer Education: Problems and Prospects.* Duntroon, Australia: University of New South Wales at the Royal Military College, 1980.

Wilson, L. G. *Skilled Hands at Sea: The Story of Her Majesty's Australian Ship Nirimba and Naval Training in New South Wales, 1855–1983.* Rev. ed. Quakers Hill, New South Wales, Australia: H.M.A.S. Nirimba Heritage Trust, 1984.

Austria

Barker, Thomas M. *Army, Aristocracy, Monarchy: Essays on War, Society, and Government in Austria, 1618–1780.* New York: Columbia University Press, 1982.

Deak, Istvan. *Beyond Nationalism: A Social and Political History of the Habsburg Officer Corps, 1848–1918.* New York: Oxford University Press, 1990.

Duffy, Christopher. *The Army of Maria Theresa: The Armed Forces of Imperial Austria, 1740–1780.* New York: Hippocrene Books, 1977.

Rothenberg, Gunther E. *The Army of Francis Joseph.* West Lafayette, Ind.: Purdue University Press, 1976.

Sondhaus, Lawrence. *The Habsburg Empire and the Sea: Austrian Naval Policy, 1797–1866.* West Lafayette, Ind.: Purdue University Press, 1989.

———. *The Naval Policy of Austria-Hungary, 1867–1918: Navalism, Industrial Development, and the Politics of Dualism.* West Lafayette, Ind.: Purdue University Press, 1994.

Bolivia

Corbett, Charles D. *The Latin American Military as a Socio-Political Force: Case Studies of Bolivia and Argentina.* Miami, Fla.: Center for Advanced International Studies, University of Miami, 1972.

Brazil

Einaudi, Luigi, and Alfred Stepan. *Latin American Institutional Development: Changing Military Perspectives in Peru and Brazil.* Santa Monica, Calif.: RAND, R-586-DOS, April 1971.

Schneider, Ronald M. *The Political System of Brazil: Emergence of a "Modernizing" Authoritarian Regime, 1964–1970.* New York and London: Columbia University Press, 1971.

Stepan, Alfred. *The Military in Politics: Changing Patterns in Brazil.* Princeton, N.J.: Princeton University Press, 1971.

Canada

Douglas, W. A. B. *The Official History of the Royal Canadian Air Force.* Vol. 2, *The Creation of a National Air Force.* Toronto: University of Toronto Press, 1986.

Eayrs, James. *In Defence of Canada.* Vol. 1, *From the Great War to the Great Depression.* Toronto: University of Toronto Press, 1964.

English, Allan D. *The Cream of the Crop: Canadian Aircrew, 1939–1945.* Montreal and Kingston, Canada: McGill-Queen's University Press, 1996.

Goetze, Bernd A. *Military Professionalism: The Canadian Officer Corps.* Kingston, Ontario, Canada: Centre for International Relations, Queen's University, 1975.

Greenhous, Brereton, et al. *The Official History of the Royal Canadian Air Force.* Vol. 3, *The Crucible of War, 1939–1945.* Toronto: University of Toronto Press, 1994.

Harris, Stephen J. *Canadian Brass: The Making of a Professional Army, 1860–1939.* Toronto: University of Toronto Press, 1988.

Hatch, F. J. *Aerodrome of Democracy: Canada and the British Commonwealth Air Training Plan, 1939–1945.* Ottawa: Directorate of History, Department of National Defence, 1983.

Preston, Richard A. *Canada's RMC: A History of the Royal Military College.* Toronto: University of Toronto Press, 1969.

————. *In the Service of Canada: History of the Royal Military College Since the Second World War.* Ottawa: University of Ottawa Press, 1992.

Wise, S. F. *The Official History of the Royal Canadian Air Force.* Vol. 1, *Canadian Airmen and the First World War.* Toronto: University of Toronto Press, 1980.

Ziegler, Mary. *We Serve That Men May Fly: The Story of the Women's Division, Royal Canadian Air Force.* Hamilton, Ontario, Canada: R.C.A.F. (W.D.) Association, 1973.

Chile

Nunn, Frederick M. *Chilean Politics, 1920–1931: The Honorable Mission of the Armed Forces.* Albuquerque: University of New Mexico Press, 1970.

————. *The Military in Chilean History: Essays on Civil-Military Relations, 1810–1973.* Albuquerque: University of New Mexico Press, 1976.

China

Biggerstaff, Knight. *The Earliest Modern Government Schools in China.* Ithaca, N.Y.: Cornell University Press, 1961.

Fung, Edmund S. K. *The Military Dimension of the Chinese Revolution: The New Army and Its Role in the Revolution of 1911.* Vancouver: University of British Columbia Press, 1980.

Gittings, John. *The Role of the Chinese Army.* London, New York, and Toronto: Oxford University Press, 1967.

Jencks, Harlan W. *From Muskets to Missiles: Politics and Professionalism in the Chinese Army, 1945–1981.* Boulder, Colo.: Westview, 1982.

Joffe, Ellis. *The Chinese Army after Mao.* Cambridge, Mass.: Harvard University Press, 1987.

Li, Lincoln. *Student Nationalism in China, 1924–1929.* Albany: State University of New York Press, 1994.

Liu, F. F. *A Military History of Modern China, 1924–1949.* Princeton, N.J.: Princeton University Press, 1956.

McCord, Edward A. *The Power of the Gun: The Emergence of Modern Chinese Warlordism.* Berkeley, Los Angeles, and London: University of California Press, 1993.

Nelsen, Harvey W. *The Chinese Military System: An Organizational Study of the Chinese People's Liberation Army.* 2d rev. ed. Boulder, Colo.: Westview, 1981.

Powell, Ralph L. *The Rise of Chinese Military Power, 1895–1912.* Princeton, N.J.: Princeton University Press, 1955.

Price, Jane L. *Cadres, Commanders, and Commissars: The Training of the Chinese Communist Leadership, 1920–1945.* Boulder, Colo.: Westview, 1976.

Rawlinson, John L. *China's Struggle for Naval Development, 1839–1895.* Cambridge, Mass.: Harvard University Press, 1967.

Sheridan, James E. *Chinese Warlord: The Career of Feng Yü-hsiang.* Stanford, Calif.: Stanford University Press, 1966.

Sutton, Donald S. *Provincial Militarism and the Chinese Republic: The Yunnan Army.* Ann Arbor: University of Michigan Press, 1980.

Swanson, Bruce. *Eighth Voyage of the Dragon: A History of China's Quest for Seapower.* Annapolis, Md.: Naval Institute Press, 1982.

Whitson, William W. *The Chinese High Command: A History of Communist Military Politics, 1927–71.* New York, Washington, D.C., and London: Praeger, 1973.

Cuba

Dominguez, Jorge. *Cuba: Order and Revolution.* Cambridge, Mass.: Harvard University Press, 1978.

Czechoslovakia

Rice, Condoleeza. *The Soviet Union and the Czechoslovak Army, 1948–1983: Uncertain Allegiance.* Princeton, N.J.: Princeton University Press, 1984.

Eastern Europe (General)

Király, Béla K., and Walter Scott Dillard, eds. *The East Central European Officer Corps 1740–1920s: Social Origins, Selection, Education, and Training.* Highland Lakes, N.J.: Atlantic Research and Publications, 1988.

Ecuador

Fitch, John Samuel. *The Military Coup d'Etat as a Political Process: Ecuador, 1948–1966.* Baltimore, Md.: Johns Hopkins University Press, 1977.

Egypt

Heyworth-Dunne, J. *An Introduction to the History of Education in Modern Egypt.* London: Luzac, 1939.

Vatikiotis, P. J. *The Egyptian Army in Politics: Pattern for New Nations?* Bloomington: Indiana University Press, 1961.

France

Adriance, Thomas J. *The Last Gaiter Button: A Study of the Mobilization and Concentration of the French Army in the War of 1870.* New York, Westport, Conn., and London: Greenwood, 1987.

Artz, Frederick B. *The Development of Technical Education in France, 1500–1850.* Cambridge, Mass.: Society for the History of Technology and the M.I.T. Press, 1966.

Elting, John R. *Swords Around a Throne: Napoleon's Grande Armée.* New York: Free Press, 1988.

Griffith, Paddy. *Military Thought in the French Army, 1815–1851.* New York and Manchester, England: Manchester University Press, 1989.

Holmes, Richard. *The Road to Sedan: The French Army, 1866–1870.* London: Royal Historical Society, 1984.

Hood, Ronald Chalmers, III. *Royal Republicans: The French Naval Dynasties Between the World Wars.* Baton Rouge: Louisiana State University Press, 1985.

Kennett, Lee. *The French Armies in the Seven Years' War: A Study in Military Organization and Administration.* Durham, N.C.: Duke University Press, 1967.

Kiesling, Eugenia C. *Arming Against Hitler: France and the Limits of Military Planning.* Lawrence: University Press of Kansas, 1996.

Langins, Janis. *The Ecole Polytechnique (1794–1804): From Encyclopaedic School to Military Institution.* New York: Oxford University Press, 1981.

Martin, Michel. *Warriors to Managers: The French Military Establishment Since 1945.* Chapel Hill: University of North Carolina Press, 1981.

Mitchell, Allan. *Victors and Vanquished: The German Influence on Army and Church in France after 1870.* Chapel Hill: University of North Carolina Press, 1984.

Paxton, Robert O. *Parades and Politics at Vichy: The French Officer Corps under Marshal Petain.* Princeton, N.J.: Princeton University Press, 1966.

Porch, Douglas. *The March to the Marne: The French Army, 1871–1914.* Cambridge, England: Cambridge University Press, 1981.

Pritchard, James. *Louis XV's Navy, 1748–1762: A Study of Organization and Administration.* Kingston, Ontario, Canada: McGill-Queen's University Press, 1987.

Ralston, David B. *The Army of the Republic: The Place of the Military in the Political Evolution of France, 1871–1914.* Cambridge, Mass.: M.I.T. Press, 1967.

Scott, Samuel F. *The Response of the Royal Army to the French Revolution: The Role and Development of the Line Army, 1787–1793.* Oxford, England: Clarendon, 1978.

Walser, Ray. *France's Search for a Battle Fleet: Naval Policy and Naval Power, 1898–1914.* New York and London: Garland, 1992.

Germany

Abenheim, Donald. *Reforging the Iron Cross: The Search for Tradition in the West German Armed Forces.* Princeton, N.J.: Princeton University Press, 1988.

Bucholz, Arden. *Hans Delbrück and the German Military Establishment: War Images in Conflict.* Iowa City: University of Iowa Press, 1985.

———. *Moltke, Schlieffen, and Prussian War Planning.* New York and Oxford, England: Berg, 1991.

Clemente, Steven E. *For King and Kaiser! The Making of the Prussian Army Officer, 1860–1914.* Westport, Conn.: Greenwood, 1992.

Corum, James S. *The Luftwaffe: Creating the Operational Air War, 1918–1940.* Lawrence: University Press of Kansas, 1997.

———. *The Roots of Blitzkrieg: Hans von Seeckt and German Military Reform.* Lawrence: University Press of Kansas, 1992.

Craig, Gordon A. *The Politics of the Prussian Army, 1640–1945.* Oxford, England: Clarendon, 1955.

Demeter, Karl. *The German Officer-Corps in Society and State, 1630–1945.* Trans. Angus Malcolm. New York: Praeger, 1965.

Duffy, Christopher. *The Army of Frederick the Great.* New York: Hippocrene Books, 1974.

Forster, Thomas M. *The East German Army: A Pattern of a Communist Military Establishment.* Trans. Anthony Buzek. London: Allen & Unwin, 1967.

Gordon, Harold J., Jr. *The Reichswehr and the German Republic, 1919–1926.* Princeton, N.J.: Princeton University Press, 1957.

Herspring, Dale R. *East German Civil-Military Relations: The Impact of Technology, 1949–72.* New York: Praeger, 1973.

Herwig, Holger H. *The German Naval Officer Corps: A Social and Political History, 1890–1918.* Oxford, England: Clarendon, 1973.

Hughes, Daniel J. *The King's Finest: A Social and Bureaucratic Profile of Prussia's General Officers, 1871–1914.* New York: Praeger, 1987.

Kitchen, Martin. *The German Officer Corps, 1890–1914.* Oxford, England: Clarendon, 1968.

Moncure, John. *Forging the King's Sword: Military Education Between Tradition and Modernization—The Case of the Royal Prussian Cadet Corps, 1871–1918.* New York: Peter Lang, 1993.

Paret, Peter. *Yorck and the Era of Prussian Reform, 1807–1815.* Princeton, N.J.: Princeton University Press, 1966.

Rust, Eric C. *Naval Officers Under Hitler: The Story of Crew 34.* New York: Praeger, 1991.

Shanahan, William O. *Prussian Military Reforms, 1786–1813.* New York: Columbia University Press, 1945; reprinted New York: AMI, 1966.

Spires, David N. *Image and Reality: The Making of the German Officer, 1921–1933.* Westport, Conn.: Greenwood, 1984.

Thomas, Charles S. *The German Navy in the Nazi Era.* Annapolis, Md.: Naval Institute Press, 1990.

Wegner, Bernd. *The Waffen-SS: Organization, Ideology and Function.* Trans. Ronald Webster. Oxford, England: Basil Blackwell, 1990.

White, Charles Edward. *The Enlightened Soldier: Scharnhorst and the Militärische Gesellschaft in Berlin, 1801–1805.* New York: Praeger, 1989.

Ghana

Baynham, Simon. *The Military and Politics in Nkrumah's Ghana.* Boulder, Colo. and London: Westview, 1988.

Great Britain

Barnett, Correlli. *Britain and Her Army, 1509–1970: A Military, Political, and Social Survey.* New York: Morrow, 1970.

Beckett, Ian F. W., and Keith Simpson, eds. *A Nation in Arms: A Social Study of the British Army in the First World War.* Manchester, England: Manchester University Press, 1985.

Bidwell, Shelford. *The Women's Royal Army Corps.* London: Leo Cooper, 1977.

Bond, Brian. *British Military Policy between the Two World Wars.* Oxford, England: Clarendon, 1980.

———. *The Victorian Army and the Staff College, 1854–1914.* London: Eyre Methuen, 1972.

Bruce, Anthony. *The Purchase System in the British Army, 1660–1871.* London: Royal Historical Society, 1980.

Davies, Evan L., and Eric J. Groves. Dartmouth: *Seventy Five Years in Pictures.* Portsmouth, England: Gieves & Hawkes, 1980.

Downes, Cathy. *Special Trust and Confidence: The Making of an Officer.* London: Frank Cass, 1991.

Escott, Beryl E. *Women in Air Force Blue: The Story of Women in the Royal Air Force from 1918 to the Present Day.* Wellingborough, Northamptonshire, England: Patrick Stephens, 1989.

Fletcher, M. H. *The WRNS—A History of the Women's Royal Naval Service.* London: Batsford, 1989.

Godwin-Austen, A. R. *The Staff and the Staff College.* London: Constable, 1927.

Gray, T. I. G. *The Imperial Defence College and the Royal College of Defence Studies, 1927–1977.* Edinburgh: H.M.S.O., 1977.

Guggisberg, F. G. *"The Shop": The Story of the Royal Military Academy.* 2d ed. London: Cassell, 1902.

Harries-Jenkins, Gwyn. *The Army in Victorian Society.* London: Routledge and Kegan Paul, 1977.

Haslam, E. B. *The History of Royal Air Force Cranwell.* London: H.M.S.O., 1982.

Houlding, J. A. *Fit for Service: The Training of the British Army, 1715–1795.* Oxford, England: Clarendon, 1981.

Hughes, Edward A. *The Royal Naval College Dartmouth.* London: Winchester, 1950.

Lewis, Michael. *England's Sea Officers: The Story of the Naval Profession.* London: Allen & Unwin, 1939.

———. *The Navy of Britain: A Historical Portrait.* London: Allen & Unwin, 1948.

———. *The Navy in Transition, 1814–1864.* London: Hodder & Stoughton, 1965.

———. *A Social History of the Navy, 1793–1815.* London: Allen & Unwin, 1960.

Luvaas, Jay. *The Education of an Army: British Military Thought, 1815–1940.* Chicago: University of Chicago Press, 1964.

Marcus, G. J. *Heart of Oak: A Survey of British Seapower in the Georgian Era.* London: Oxford University Press, 1975.

Maurice-Jones, K. W. *The Shop Story, 1900–1939.* Woolwich, England: The Royal Artillery Institution, 1954.

Mockler-Ferryman, A. F. *Annals of Sandhurst: A Chronicle of the Royal Military College.* London: Heinemann, 1900.

Pack, Stanley, W. C. *"Britannia" at Dartmouth.* London: Alvin Redman, 1966.

Parker, Peter. *The Old Lie: The Great War and the Public School Ethos.* London: Constable, 1987.

Penn, Geoffrey. *HMS Thunderer: The Story of the Royal Naval Engineering College, Keyham and Manadon.* Emsworth, England: Kenneth Mason, 1984.

———. *Snotty: The Story of the Midshipman.* London: Hollis & Carter, 1957.

———. *"Up Funnel, Down Screw!": The Story of the Naval Engineer.* London: Hollis & Carter, 1955.

Reader, W. J. *Professional Men: The Rise of the Professional Classes in Nineteenth-Century England.* London: Weidenfeld & Nicolson, 1966.

Rodger, N. A. M. *The Wooden World: An Anatomy of the Georgian Navy.* Annapolis, Md.: Naval Institute Press, 1986.

Schurman, Donald M. *The Education of a Navy: The Development of British Naval Strategic Thought, 1867–1914.* Chicago: University of Chicago Press, 1965.

Shepperd, Alan. *Sandhurst: The Royal Military Academy and Its Predecessors.* London: Country Life Books, 1980.

Simkins, Peter. *Kitchener's Army: The Raising of the New Armies, 1914–1916.* Manchester, England: Manchester University Press, 1988.

Smyth, John G. *Sandhurst: The History of the Royal Military Academy, Woolwich, the Royal Military College, Sandhurst, and the Royal Military Academy, Sandhurst, 1741–1961,* London: Weidenfeld & Nicolson, 1961.

Spiers, Edward M. *The Army and Society, 1815–1914.* London and New York: Longman, 1980.

———. *The Late Victorian Army, 1868–1902.* Manchester, England: Manchester University Press, 1992.

Strachan, Hew. *Wellington's Legacy: The Reform of the British Army, 1830–54.* Manchester, England: Manchester University Press, 1984.

Taylor, John W. R. *C.F.S.: Birthplace of Airpower.* London: Putnam, 1958.

Thomas, Hugh. *The Story of Sandhurst.* London: Hutchinson, 1961.

Thoumine, R. H. *Scientific Soldier: A Life of General Le Marchant, 1766–1812.* London: Oxford University Press, 1968.

Turner, E. S. *Gallant Gentlemen: A Portrait of the British Officer, 1600–1956.* London: Michael Joseph, 1956.

Vernon, Philip E., and John B. Parry. *Personnel Selection in the British Forces.* London: University of London Press, 1949.

Vibart, H. M. *Addiscombe: Its Heroes and Men of Note.* Westminster, England: Constable, 1894.

Wells, John. *The Royal Navy: An Illustrated Social History, 1870–1982.* Phoenix Mill-Far Thrupp-Stroud-Gloucestershire, England: Alan Sutton in association with the Royal Naval Museum, 1994.

Western, J. R. *The English Militia in the Eighteenth Century: The Story of a Political Issue, 1660–1802.* London: Routledge & Kegan Paul, 1965.

Wilkinson, Rupert. *The Prefects: British Leadership and the Public School Tradition.* London: Oxford University Press, 1964.

Yardley, Michael. *Sandhurst: A Documentary.* London: Harrap, 1987.

Young, F. W. *The Story of the Staff College, 1858–1958.* Camberley, England: The Staff College, Camberley, 1958.

Zugbach, R. G. L. von. *Power and Prestige in the British Army.* Aldershot, England: Avebury, 1988.

Guatemala

Adams, Richard N. *Crucifixion by Power: Essays on Guatemalan Social Structure, 1944–1966.* Austin: University of Texas Press, 1970.

India

Chaturvedi, M. S. *History of the Indian Air Force.* New Delhi: Vikas, 1978.

Cohen, Stephen P. *The Indian Army: Its Contribution to the Development of a Nation.* Rev. Indian ed. Delhi: Oxford University Press, 1990.

Hastings, D. J. *The Royal Indian Navy, 1612–1950.* Jefferson, N.C. and London: McFarland, 1988.

Heathcote, T. A. *The Indian Army: The Garrison of British Imperial India, 1822–1922.* New York: Hippocrene Books, 1974.

————. *The Military in British India: The Development of British Land Forces in South Asia, 1600–1947.* Manchester, England and New York: Manchester University Press, 1995.

Mason, Philip. *A Matter of Honour: An Account of the Indian Army, Its Officers and Men.* New York: Holt, Rinehart & Winston, 1974.

Rosen, Stephen Peter. *Societies and Military Power: India and Its Armies.* Ithaca, N.Y. and London: Cornell University Press, 1996.

Sinha, B. P. N., and Sunil Chandra. *Valour and Wisdom: Genesis and Growth of the Indian Military Academy.* New Delhi: Oxford and IBH Publishing, 1992.

Indonesia

Sundhaussen, Ulf. *The Road to Power: Indonesian Military Politics, 1945–1967.* Kuala Lumpur: Oxford University Press, 1982.

Iran

Zabih, Sepehr. *The Iranian Military in Revolution and War.* London and New York: Routledge, 1988.

Israel

Cohen, Eliezer. *Israel's Best Defense: The First Full Story of the Israeli Air Force.* Trans. Jonathan Cordis. New York: Orion Books, 1993.

Gal, Reuven. *A Portrait of the Israeli Soldier.* Westport, Conn.: Greenwood, 1986.

Luttwak, Edward N., and Daniel Horowitz. *The Israeli Army, 1948–1973.* Cambridge, Mass.: Abt Books, 1983.

Perlmutter, Amos. *Military and Politics in Israel: Nation-Building and Role Expansion.* London: Frank Cass, 1969.

Rolbant, Samuel. *The Israeli Solider: Profile of an Army.* London: Thomas Yoseloff, 1970.

Rothenberg, Gunther E. *The Anatomy of the Israeli Army: The Israel Defence Force, 1948–1978.* New York: Hippocrene Books, 1979.

Williams, Louis. *The Israel Defense Forces: A People's Army.* Tel Aviv: Ministry of Defense Publishing House, 1989.

Italy

Gooch, John. *Army, State and Society in Italy, 1870–1915.* New York: St. Martin's, 1989.

Japan

Auer, James E. *The Postwar Rearmament of Japanese Maritime Forces, 1945–1971.* New York: Praeger, 1973.

Dore, R. P. *Education in Tokugawa Japan.* Berkeley and Los Angeles: University of California Press, 1965.

Evans, David C., and Mark R. Peattie. *Kaigun: Strategy, Tactics, and Technology in the Imperial Japanese Navy, 1887–1941.* Annapolis, Md.: Naval Institute Press, 1997.

Harries, Meirion, and Susie Harries. *Soldiers of the Sun: The Rise and Fall of the Imperial Japanese Army.* New York: Random House, 1991.

Humphreys, Leonard A. *The Way of the Heavenly Sword: The Japanese Army in the 1920s.* Stanford, Calif.: Stanford University Press, 1995.

Maeda, Tetsuo. *The Hidden Army: The Untold Story of Japan's Military Forces.* Trans. Steven Karpa. Ed. David J. Kenney. Chicago: edition q, 1995.

Peattie, Mark R. *Ishiwara Kanji and Japan's Confrontation with the West.* Princeton, N.J.: Princeton University Press, 1975.

Presseisen, Ernst L. *Before Aggression: Europeans Prepare the Japanese Army.* Tucson: University of Arizona Press, 1965.

Shillony, Ben-Ami. *Revolt in Japan: The Young Officers and the February 26, 1936 Incident.* Princeton, N.J.: Princeton University Press, 1973.

Korea

Kim, Jai-Hyup. *The Garrison State in Pre-War Japan and Post-War Korea: A Comparative Analysis of Military Politics.* Washington, D.C.: University Press of America, 1978.

Sawyer, Robert K. *Military Advisors in Korea: KMAG in Peace and War.* Washington, D.C.: Office of the Chief of Military History, U.S. Department of the Army, 1962.

Latin America (General)

Johnson, John J. *The Military and Society in Latin America.* Stanford, Calif.: Stanford University Press, 1964.

Lieuwen, Edwin. *Arms and Politics in Latin America.* Rev. ed. New York: Praeger, 1961.

Loveman, Brian, and Thomas M. Davies, Jr., eds. *The Politics of Antipolitics.* 3d ed., rev. and updated. Wilmington, Del.: Scholarly Resources, 1997.

Lowenthal, Abraham F., and J. Samuel Fitch, eds. *Armies and Politics in Latin America.* Rev. ed. New York and London: Holmes & Meier, 1986.

Nunn, Frederick M. *The Time of the Generals: Latin American Professional Militarism in World Perspective.* Lincoln and London: University of Nebraska Press, 1992.

————. *Yesterday's Soldiers: European Military Professionalism in South America, 1890–1940.* Lincoln: University of Nebraska Press, 1983.

Philip, George. *The Military in South American Politics.* London: Croom Helm, 1985.

Rouquié, Alain. *The Military and the State in Latin America.* Trans. Paul E. Sigmund. Berkeley: University of California Press, 1987.

Vargas, Augusto, ed. *Democracy Under Siege: New Military Power in Latin America.* New York: Greenwood, 1989.

Wesson, Robert, ed. *The Latin American Military Institution.* New York: Praeger, 1986.

Mexico

Camp, Roderic Ai. *Generals in the Palacio: The Military in Modern Mexico.* New York and Oxford, England: Oxford University Press, 1992.

DePalo, William A., Jr. *The Mexican National Army, 1822–1852.* College Station: Texas A & M University Press, 1997.

Ronfeldt, David, ed. *The Modern Mexican Military: A Reassessment.* San Diego: Center for U.S.-Mexican Studies, University of California, San Diego, 1984.

Middle East (General)

Be'eri, Eliezer. *Army Officers in Arab Politics and Society.* Trans. Dov Ben-Abba. New York: Praeger, 1970.

Fisher, Sydney Nettleton, ed. *The Military in the Middle East: Problems in Society and Government.* Columbus: Ohio State University Press, 1963.

Netherlands

Bruijn, Jaap R. *The Dutch Navy of the Seventeenth and Eighteenth Centuries.* Columbia: University of South Carolina Press, 1993.

Nicaragua

Millett, Richard. *Guardians of the Dynasty: A History of the U.S. Created Guardia Nacional de Nicaragua and the Somoza Family.* Maryknoll, N.Y.: Orbis Books, 1977.

Nigeria

Luckham, Robin. *The Nigerian Military: A Sociological Analysis of Authority and Revolt, 1960–67.* Cambridge, England: Cambridge University Press, 1971.

Peters, Jimi. *The Nigerian Military and the State.* London and New York: I. B. Tauris, 1997.

Pakistan

Cohen, Stephen P. *The Pakistan Army.* New Delhi: Himalayan Books, 1984.

Peru

Einaudi, Luigi. *The Peruvian Military: A Summary Political Analysis.* Santa Monica, Calif.: RAND, RM-6048-RC, May 1969.

———, and Alfred Stepan. *Latin American Institutional Development: Changing Military Perspectives in Peru and Brazil.* Santa Monica, Calif.: RAND, R-586-DOS, April 1971.

Poland

Malchar, George C. *Poland's Politicized Army: Communists in Uniform.* New York: Praeger, 1984.

Russia (Soviet Union)

Bayer, Philip A. *The Evolution of the Soviet General Staff, 1917–1941.* New York and London: Garland, 1987.

Beskrovny, L. G. *The Russian Army and Fleet in the Nineteenth Century: Handbook of Armaments, Personnel, and Policy.* Ed. and trans. Gordon E. Smith. Gulf Breeze, Fla.: Academic International Press, 1996.

Curtiss, John Shelton. *The Russian Army Under Nicholas I, 1825–1855.* Durham, N.C.: Duke University Press, 1965.

Donnelly, Christopher. *Red Banner: The Soviet Military System in Peace and War.* Coulston, Surrey, United Kingdom: Jane's Information Group, 1988.

Duffy, Christopher. *Russia's Military Way to the West: Origins and Nature of Russian Military Power, 1700–1800.* London and Boston: Routledge and Kegan Paul, 1981.

Erickson, John. *The Russian Imperial/Soviet General Staff.* College Station: Center for Strategic Technology, Texas Engineering Experiment Station, Texas A & M University System, 1981.

———. *The Soviet High Command: A Military Political History, 1918–1941.* New York: St. Martin's, 1962.

Fuller, William C., Jr. *Civil-Military Conflict in Imperial Russia, 1881–1914.* Princeton, N.J.: Princeton University Press, 1985.

Glantz, David M. *Stumbling Colossus: The Red Army on the Eve of World War.* Lawrence: University Press of Kansas, 1998.

Hagen, Mark von. *Soldiers in the Proletarian Dictatorship: The Red Army and the Soviet Socialist State, 1917–1930.* Ithaca, N.Y. and London: Cornell University Press, 1990.

Jacobs, Walter Darnell. *Frunze: The Soviet Clausewitz, 1885–1925.* The Hague: Martinus Nijhoff, 1969.

Jones, Ellen. *Red Army and Society.* Boston: Allen & Unwin, 1985.

Keep, John L. H. *Soldiers of the Tsar: Army and Society in Russia, 1462–1874.* Oxford, England: Clarendon, 1985.

Kenez, Peter, *Civil War in South Russia, 1918: The First Year of the Volunteer Army.* Berkeley, Los Angeles, and London: University of California Press, 1971.

Kilmarx, Robert A. *A History of Soviet Air Power.* New York: Praeger, 1962.

Lambeth, Benjamin S. *Russia's Air Power at the Crossroads.* Santa Monica, Calif.: RAND, 1996.

Menning, Bruce W. *Bayonets Before Bullets: The Imperial Russian Army, 1861–1914.* Bloomington and Indianapolis: Indiana University Press, 1992.

Miller, Forrestt A. *Dmitri Miliutin and the Reform Era in Russia.* Nashville, Tenn.: Vanderbilt University Press, 1968.

Reese, Roger R. *Stalin's Reluctant Soldiers: A Social History of the Red Army, 1925–1941.* Lawrence: University Press of Kansas, 1996.

Scott, Harriet Fast, and William F. Scott. *The Armed Forces of the USSR.* 3rd ed., rev. and updated. Boulder, Colo.: Westview, 1984.

Screen, J. E. O. *The Helsinki Yunker School, 1846–1879: A Case Study of Officer Training in the Russian Army.* Helsinki: Suomen Hist. Seura, 1986.

Van Dyke, Carl. *Russian Imperial Military Doctrine and Education, 1832–1914.* Westport, Conn.: Greenwood, 1990.

White, D. Fedotoff. *The Growth of the Red Army.* Princeton, N.J.: Princeton University Press, 1944.

South Africa

Frankel, Philip H. *Pretoria's Praetorians: Civil-Military Relations in South Africa.* Cambridge, England: Cambridge University Press, 1984.

Spain

Martinez, Rafael Banon, and Thomas M. Barker, eds. *Armed Forces and Society in Spain: Past and Present.* Boulder, Colo. and New York: Social Science Monographs; distributed by Columbia University Press, 1988.

Payne, Stanley G. *Politics and the Military in Modern Spain.* Stanford, Calif.: Stanford University Press, 1967.

Turkey

Shaw, Stanford J. *Between Old and New: The Ottoman Empire under Sultan Selim III, 1789–1807.* Cambridge, Mass.: Harvard University Press, 1971.

United States

Abrahamson, James L. *America Arms for a New Century: The Making of a Great Military Power.* New York: Free Press, 1981.

Ambrose, Stephen E. *Duty, Honor, Country: A History of West Point.* Baltimore, Md.: Johns Hopkins University Press, 1966.

———. *Upton and the Army.* Baton Rouge: Louisiana State University Press, 1964.

Armed Forces Staff College. *To Labor as One, The First 35 Years* [*Command History: The Armed Forces Staff College, 1946–1981*]. Norfolk, Va.: Armed Forces Staff College, 1982.

Ball, Harry P. *Of Responsible Command: A History of the U.S. Army War College.* Carlisle Barracks, Pa.: Alumni Association of the U.S. Army War College, 1984.

Bauer, Theodore W. *History of the Industrial College of the Armed Forces, 1924–1983.* Washington, D.C.: Alumni Association of the ICAF, 1983.

Benjamin, Park. *The United States Naval Academy.* New York: Putnam's, 1900.

Bittner, Donald F. *Curriculum Evolution: Marine Corps Command and Staff College, 1920–1988.* Washington, D.C.: History and Museums Division, Headquarters, U.S. Marine Corps, 1988.

Boynton, Edward C. *History of West Point.* New York: Van Nostrand, 1863.

Brown, Richard C. *Social Attitudes of American Generals, 1898–1940.* New York: Arno, 1979.

Cardozier, V. R. *Colleges and Universities in World War II.* Westport, Conn. and London: Praeger, 1993.

Centennial of the United States Military Academy at West Point, New York, 1802–1902. 2 Vols. Washington, D.C.: Government Printing Office, 1904.

Clifford, John Gary. *The Citizen Soldiers: The Plattsburg Training Camp Movement, 1913–1920.* Lexington: University Press of Kentucky, 1972.

Coffman, Edward M. *The Old Army: A Portrait of the American Army in Peacetime, 1784–1898.* New York and Oxford, England: Oxford University Press, 1986.

Coumbe, Arthur T. *U.S. Army Cadet Command: The 10 Year History.* Fort Monroe, Va.: Office of the Command Historian, U.S. Army Cadet Command, 1996.

Couper, William. *One Hundred Years at V.M.I.* 3 Vols. Richmond, Va.: Garrett & Massie, 1939.

Crackel, Theodore J. *Mr. Jefferson's Army: Political and Social Reform of the Military Establishment, 1801–1809.* New York: New York University Press, 1987.

Craven, Wesley Frank, and James Lea Cate, eds. *The Army Air Forces in World War II.* Vol. 6, *Men and Planes.* Chicago: University of Chicago Press, 1955.

Cullum, George W., comp. *Biographical Register of the Officers and Graduates of the U.S. Military Academy at West Point, N.Y., From Its Establishment, in 1802, to 1890; with the Early History of the United States Military Academy.* Vol. 3. 3d ed., rev. and extended. Boston: Houghton Mifflin, 1891.

Cunliffe, Marcus. *Soldiers and Civilians: The Martial Spirit in America, 1775–1865.* Boston: Little, Brown, 1968.

Dastrup, Boyd L. *U.S. Army Command and General Staff College: A Centennial History.* Manhattan, Kans.: Sunflower University Press, 1982.

Dederer, John Morgan. *War in America to 1775: Before Yankee Doodle.* New York and London: New York University Press, 1990.

Dupuy, R. Ernest. *Men of West Point: The First 150 Years of the United States Military Academy.* New York: William Sloane Associates, 1951.

———. *Sylvanus Thayer: Father of Technology in the United States.* West Point, N.Y.: United States Military Academy, 1958.

Earle, Ralph. *Life at the U.S. Naval Academy: The Making of the American Naval Officer.* New York and London: Putnam's, 1917.

Ebbert, Jean, and Marie-Beth Hall. *Crossed Currents: Navy Women from WWI to Tailhook.* Washington, D.C.: Brassey's, 1993.

Ekirch, Arthur A., Jr. *The Civilian and the Military.* New York: Oxford University Press, 1956.

Eliot, George Fielding. *Sylvanus Thayer of West Point.* New York: Julian Messner, 1959.

Ellis, Joseph, and Robert Moore. *School for Soldiers: West Point and the Profession of Arms.* New York: Oxford University Press, 1974.

Fagan, George V. *The Air Force Academy: An Illustrated History.* Boulder, Colo.: Johnson Books, 1988.

Finney, Robert T. *History of the Air Corps Tactical School, 1920–1940.* Maxwell Air Force Base, Ala.: Research Studies Institute, USAF Historical Division, Air University, 1955; new imprint, Washington, D.C.: Center for Air Force History, U.S. Air Force, 1992.

Fleming, Thomas J. *West Point: The Men and Times of the United States Military Academy.* New York: Morrow, 1969.

Forman, Sidney. *West Point: A History of the United States Military Academy.* New York: Columbia University Press, 1950.

Galloway, K. Bruce, and Robert Bowie Johnson, Jr. *West Point: America's Power Fraternity.* New York: Simon & Schuster, 1973.

Greenwood, John E. *Officer Education and Training in the United States Marine Corps.* Washington, D.C.: National War College, 1972.

Hattendorf, John B., B. Mitchell Simpson III, and John R. Wadleigh. *Sailors and Scholars: The Centennial History of the U.S. Naval War College.* Newport, R.I.: Naval War College Press, 1984.

Heise, J. Arthur. *The Brass Factories: A Frank Appraisal of West Point, Annapolis, and the Air Force Academy.* Washington, D.C.: Public Affairs Press, 1969.

Holm, Jeanne. *Women in the Military: An Unfinished Revolution.* Rev. ed. Novato, Calif.: Presidio, 1992.

Hudson, James J. *Hostile Skies: A Combat History of the American Air Service in World War I.* Syracuse, N.Y.: Syracuse University Press, 1968.

Jakeman, Robert J. *The Divided Skies: Establishing Segregated Flight Training at Tuskegee, Alabama, 1932–1942.* Tuscaloosa and London: University of Alabama Press, 1992.

Karsten, Peter. *The Naval Aristocracy: The Golden Age of Annapolis and the Emergence of Modern American Navalism.* New York: Free Press, 1972.

Keefer, Louis E. *Scholars in Foxholes: The Story of the Army Specialized Training Program in World War II.* Jefferson, N.C.: McFarland, 1988.

Kennedy, Gerald John. *United States Naval War College, 1919–1941: An Institutional Response to Naval Preparedness.* Newport, R.I.: Center for Advanced Research, U.S. Naval War College, 1975.

King, Irving H. *The Coast Guard Expands, 1865–1915: New Roles, New Frontiers.* Annapolis, Md.: Naval Institute Press, 1996.

Korb, Lawrence J., ed. *The System for Educating Military Officers in the U.S.* Pittsburgh, Pa.: International Studies Association, 1976.

Lee, Ulysses. *United States Army in World War II, Special Studies, The Employment of Negro Troops.* Washington, D.C.: Office of the Chief of Military History, U.S. Army, 1966.

Lovell, John P. *Neither Athens Nor Sparta?: The American Service Academies in Transition.* Bloomington and London: Indiana University Press, 1979.

Lyons, Gene M., and John W. Masland. *Education and Military Leadership: A Study of the R.O.T.C.* Princeton, N.J.: Princeton University Press, 1959.

Mahon, John K. *History of the Militia and the National Guard.* New York: Macmillan, 1983.

Manning, Thomas A., et al. *History of Air Training Command, 1943–1993.* Randolph Air Force Base, Tex.: Office of History and Research, Headquarters Air Education and Training Command, 1993.

Margiotta, Franklin D., ed. *The Changing World of the American Military.* Boulder, Colo.: Westview, 1978.

Marshall, Edward Chauncey. *History of the Naval Academy.* New York: Van Nostrand, 1862.

Masland, John W., and Laurence I. Radway. *Soldiers and Scholars: Military Education and National Policy.* Princeton, N.J.: Princeton University Press, 1957.

Maurer, Maurer. *Aviation in the U.S. Army, 1919–1939.* Washington, D.C.: Office of Air Force History, U.S. Air Force, 1987.

McKee, Christopher. *A Gentlemanly and Honorable Profession: The Creation of the U.S. Naval Officer Corps, 1794–1815.* Annapolis, Md.: Naval Institute Press, 1991.

Millett, Allan R. *The General: Robert L. Bullard and Officership in the United States Army, 1881–1925.* Westport, Conn. and London: Greenwood, 1975.

———, and Peter Maslowski. *For the Common Defense: A Military History of the United States,* rev. ed. New York: Free Press, 1994.

Mitchell, Vance O. *Air Force Officers: Personnel Policy Development, 1944–1974.* Washington, D. C.: Air Force History and Museums Program, U.S. Air Force, 1996.

Morden, Bettie J. *The Women's Army Corps, 1945–1978.* Washington, D.C.: Center of Military History, U.S. Army, 1990.

Morrison, James L., Jr. *"The Best School in the World": West Point, the Pre-Civil War Years, 1833–1866.* Kent, Ohio: Kent State University Press, 1986.

Nenninger, Timothy K. *The Leavenworth Schools and the Old Army: Education, Professionalism, and the Officer Corps of the United States Army, 1881–1918.* Westport, Conn.: Greenwood, 1978.

Newton, Wesley Phillips, and Robert R. Rea, eds. *An Account of Naval Aviation Training in World War II, The Correspondence of Aviation Cadet/Ensign Robert R. Rea.* Tuscaloosa and London: University of Alabama Press, 1987.

Palmer, Robert R., Bell I. Wiley, and William R. Keast. *United States Army in World War II, The Army Ground Forces, The Procurement and Training of Ground Combat Troops.* Washington, D.C.: Historical Division, Department of the Army, 1948.

Pappas, George S. *Prudens Futuri: The U.S. Army War College, 1901–1967.* Carlisle Barracks, Pa.: Alumni Association of the U.S. Army War College, 1967.

———. *To the Point: The United States Military Academy, 1802–1902.* Westport, Conn.: Praeger, 1993.

Reardon, Carol. *Soldiers and Scholars: The U.S. Army and the Uses of Military History, 1865–1920.* Lawrence: University Press of Kansas, 1990.

Reeves, Ira L. *Military Education in the United States.* Burlington, Vt.: Free Press Printing, 1914.

Sandler, Stanley. *Segregated Skies: All-Black Combat Squadrons of WWII.* Washington and London: Smithsonian Institution Press, 1992.

Sarkesian, Sam C., John Allen Williams, and Fred B. Bryant. *Soldiers, Society, and National Security.* Boulder, Colo. and London: Lynne Rienner, 1995.

Schneider, James G. *The Navy V-12 Program: Leadership for a Lifetime.* Boston: Houghton Mifflin, 1987.

Seager, Robert II. *Alfred Thayer Mahan: The Man and His Letters.* Annapolis, Md.: Naval Institute Press, 1977.

Sherwood, John Darrell. *Officers in Flight Suits: The Story of American Air Force Fighter Pilots in the Korean War.* New York and London: New York University Press, 1996.

Shindler, Henry, and E. E. Booth. *History of Army Service Schools.* Fort Leavenworth, Kans.: Staff College Press, 1908.

Shulimson, Jack. *The Marine Corps' Search for a Mission, 1880–1898.* Lawrence: University Press of Kansas, 1993.

Simons, William E. *Liberal Education in the Service Academies.* New York: Teachers College, Columbia University, 1965.

Skelton, William B. *An American Profession of Arms: The Army Officer Corps, 1784–1861.* Lawrence: University Press of Kansas, 1992.

Soley, James R. *Historical Sketch of the United States Naval Academy.* Washington, D.C.: Government Printing Office, 1876.

Spector, Ronald H. *Professors of War: The Naval War College and the Development of the Naval Profession.* Newport, R.I.: Naval War College Press, 1977.

Stiehm, Judith Hicks. *Bring Me Men and Women: Mandated Change at the U.S. Air Force Academy.* Berkeley: University of California Press, 1981.

Stillwell, Paul, ed. *The Golden Thirteen: Recollections of the First Black Naval Officers.* Annapolis, Md.: Naval Institute Press, 1993.

Stremlow, Mary V. *A History of the Women Marines, 1946–1977.* Washington, D.C.: History and Museums Division, Headquarters, U.S. Marine Corps, 1986.

Sweetman, Jack. Rev. Thomas J. Cutler. *The U.S. Naval Academy: An Illustrated History.* 2d ed. Annapolis, Md.: Naval Institute Press, 1995.

Todorich, Charles. *The Spirited Years: A History of the Antebellum Naval Academy.* Annapolis, Md.: Naval Institute Press, 1984.

Treadwell, Mattie E. *United States Army in World War II, Special Studies, The Women's Army Corps.* Washington, D.C.: Office of the Chief of Military History, Department of the Army, 1954.

U.S. Army Command and General Staff College. *A Military History of the U.S. Army Command and General Staff College, Fort Leavenworth, Kansas, 1881–1963.* Fort Leavenworth, Kans.: U.S. Army Command and General Staff College, 1964.

Vlahos, Michael, *The Blue Sword: The Naval War College and the American Mission, 1919–1941.* Newport, R.I.: Naval War College Press, 1980.

Weigley, Russell F. *History of the United States Army.* Enlarged ed. Bloomington: Indiana University Press, 1984.

Wise, Henry A. *Drawing Out the Man: The VMI Story.* Charlottesville: University Press of Virginia, 1978.

Venezuela

Burggraff, Winfield J. *The Venezuelan Armed Forces in Politics, 1935–1959.* Columbia: University of Missouri Press, 1972.

Vietnam

Pike, Douglas. *PAVN: People's Army of Vietnam.* Novato, Calif.: Presidio, 1986.

Contributors

Horst Boog was born in central Germany, and received a B.A. in 1950 from Middlebury College in Vermont and a Ph.D. in 1965 from the University of Heidelberg. He has been a German Air Force reserve officer, military analyst for NATO, and for over twenty-five years a historian at the Militärgeschichtliches Forschungsamt (German Armed Forces Office of Military History) in Freiburg. He retired in 1993 having been the office's Senior Director of Research and Head of the Research Department for the Second World War. He is the author of *Die Deutsche Luftwaffenführung, 1935–1945* [German Air Force command and leadership] (1982), and numerous articles on twentieth-century military history, especially air warfare, military operations and technology, and the German officer corps. He is co-author of *Verteidigun im Bündnis* [A history of the Federal German Armed Forces from its beginnings to 1972] (1975), *Der Angriff auf die Sowjetunion* [German preparations and campaign against Russia 1940–41] (1983), and *Der Globale Krieg* [Global war, 1941–1943] (1990). He has also edited *The Conduct of the Air War in the Second World War: An International Comparison* (1992), and contributed an introductory essay to *Sir Arthur Harris, Despatch on War Operations* (1995). After retiring, Dr. Boog has continued to work on the Miltärgeschichtliches Forschungsamt's ten-volume history *Das Deutsche Reich und der Zweite Weltkrieg* [Germany and World War II].

Josiah Bunting III, Major General, U.S. Army, is Superintendent of the Virginia Military Institute. He is a VMI graduate, class of 1963. After receiving a B.A. and M.A. from Oxford University, he entered the U.S. Army in 1966. During six years of service, he reached the rank of major with duty stations at Fort Bragg, North Carolina; Vietnam (Ninth Infantry Division); and West Point, New York, where he was an assistant professor of history and social sciences at the U.S. Military Academy. Among General Bunting's publications are *The Lionheads,* selected one of the "Ten Best Novels of 1973" by *Time* magazine; *Small Units in the Control of Civil Disorder* (1968); *The Advent of Frederick Giles* (1974); and *An Education for Our Time* (1998).

Philip D. Caine, Brigadier General, U.S. Air Force, retired, was commissioned as a Distinguished Military Graduate from the Air Force ROTC at the University of Denver. He graduated at the top of his pilot training class and, during his thirty-seven-year Air Force career, flew some 5,000 hours in various military aircraft. General Caine earned both his M.A. and Ph.D. in history from Stanford University, and served in the U.S. Air Force Academy's Department of History as an instructor, tenure associate professor, tenure professor, and acting department head. Among his other Air Force assignments were acting head of Project CHECO (Contemporary Historical Examination of Current Operations) in Vietnam, Professor of Strategic Studies at the U.S. National War College, and Senior Research Fellow at the U.S. National Defense University. In 1980 he became the first permanent professor under the Commandant of Cadets at the Air Force Academy and was also named Deputy Commandant of Cadets for Military Instruction. General Caine retired from the Air Force in 1992 and is the author of four books: *Eagles of the RAF* (1990), *American Pilots in the RAF* (1992), *Spitfires, Thunderbolts and Warm Beer* (1995), and *Aircraft Down* (1997).

Roderic Ai Camp joined the Tulane University faculty in 1991. He has also been a visiting professor at the *Colegio de México* and the Foreign Service Institute, and was a research fellow

413

at the Woodrow Wilson Center for International Scholars, Smithsonian Institution. He has received a Fulbright Fellowship on three occasions, as well as a Howard Heinz Foundation Fellowship for research on Mexico. He is a contributing editor to the Library of Congress' *Handbook of Latin American Studies* and to the *World Book Encyclopedia,* and serves on the editorial board of the journal *Mexican Studies.* His special interests include Mexican politics, comparative elites, political recruitment, church-state relations, and civil-military affairs. The author of numerous books and articles on Mexico, Dr. Camp's most recent publications include: *Politics in Mexico* (1996); *Political Recruitment Across Two Centuries, Mexico, 1884–1991* (1995); *The Successor* (1993); *Generals in the Palacio: The Military in Modern Mexico* (1992); and *Entrepreneurs and Politics in Twentieth Century Mexico* (1989). Presently, he directs the Tinker Mexican Policy Studies Program at Tulane.

D'Ann Campbell is currently Professor of History and Vice President for Academic Affairs at the Sage Colleges in Troy, New York. She received her B.A. from the Colorado College and, in 1979, her Ph.D. from the University of North Carolina. Her teaching career has included two years as a visiting professor of military history at the U.S. Military Academy. Her book, *Women at War with America: Private Lives in a Patriotic Era,* was published in 1984. She has also published numerous articles and reviews.

Elliott V. Converse III, Colonel, U.S. Air Force, retired, earned a B.A., M.A., and Ph.D. in history from Montana State University (1967), the University of Wisconsin–Madison (1972), and Princeton University (1984), respectively. During twenty-five yers in the Air Force, his assignments included the faculties of the U.S. Air Force Academy and the Air War College, and as commander of the Air Force Historical Research Agency. Since retiring in 1992, he has been an associate professor at Reinhardt College and a visiting professor at the Air Force Academy. He is the principal author of *The Exclusion of Black Soldiers from the Medal of Honor in World War II* (1997).

Roger Dingman received his B.A. from Stanford University in 1960, and then served as a U.S. Navy officer until 1962. He received his M.A. and Ph.D. from Harvard University in 1963 and 1969, respectively. He is currently an associate professor of history at the University of Southern California. His publications include: *Kindai Nihon no taigai taido* [Modern Japan and the outside world] (1974); *Power in the Pacific: The Origins of Naval Arms Limitation, 1914–1922* (1976); *Ghost of War: The Sinking of the Awa maru and Japanese-American Relations, 1945–1995* (1997); and numerous articles on American–East Asian relations. He has served as a visiting professor at the U.S. Air Force Academy and the U.S. Naval War College, and was the 1997 Distinguished Visiting Lecturer at the National Institute of Defense Studies, Japan Defense Agency.

Cathy Downes is Senior Research Officer of the New Zealand Defence Force. She holds an M.A. (Strategic Studies with Distinction) and a Ph.D. from the University of Lancaster, United Kingdom. She has held appointments at Harvard University's Center for International Affairs, the University of Melbourne, and Australian National University's Strategic and Defence Studies Centre. Her published works include: *Special Trust and Confidence: The Making of an Officer* (1991); *Security and Defence: Global and Pacific Perspectives* (co-editor, 1990); *High Personnel Turnover: The Australian Defence Force is not a Limited Liability Company* (1988); and *Senior Officer Professional Development: Constant Study to Prepare* (1989). Dr. Downes's current research concerns defense planning in an indeterminate threat environment, strategic policy and military doctrine formulation, military performance measurement, and the international uses of multinational military forces. She is a fellow of the 21st Century Trust, United Kingdom, and of the Inter-University Seminar on Armed Forces and Society; an associate of the Centre for Defence and International Security Studies, Lancaster, United Kingdom; and a graduate of the Australian Defence Industrial Mobilisation Course and of the U.S. Commander in Chief Pacific Symposium on East Asian Security.

John F. Flanagan, Jr., Brigadier General, U.S. Air Force Reserve, retired, graduated from the U.S. Air Force Academy in 1962. He spent the next five-and-a-half years on active duty as a pilot, flying 300 combat missions in Vietnam in 1966 as a forward air controller in the single-engine O1-E. After Vietnam, he continued in the U.S. Air Force Reserve, becoming deputy commander of the New York Air National Guard and retiring in 1995. He received an M.B.A. from Boston College in 1966. His business career includes spearheading the introduction of the Boeing 747 for American Airlines, serving as chief financial officer in Europe for a Hertz trucking subsidiary, and financing ships as treasurer of Holland America Lines. In 1970–71, he served in the White House as a Presidential Exchange Executive assigned to the U.S. Department of Transportation. General Flanagan is the author of *Vietnam Above the Treetops* (1992), a Military Book Club main selection, and has published articles in several journals and magazines. Currently, he maintains an international consulting practice, is chairman of the Aviation Management Department at St. Francis College in New York, and is president of the New York–Connecticut chapter of the National Defense Industrial Association.

Moshe Gershovich was born in Israel and served in the Israel Defense Force. In 1982 he earned a B.A. in Jewish and Middle Eastern history from Tel Aviv University, and in 1995 a Ph.D. from Harvard University's joint program in History and Middle Eastern Studies. His dissertation is entitled "French Military Policy in Morocco and the Origins of an Arab Army." Dr. Gershovich has published numerous articles concerning French colonialism and the military and political history of the Middle East and North Africa. Among them are "The Ait Ya'qub Incident and the Crisis of French Military Policy in Morocco," *Journal of Military History* (1998); "The Sharifian Star Over the Rhine: Moroccan Soldiers in French Uniforms in Germany, 1919–1925," *Al-Maghrib: Review of North African Studies* (1998); "French Control of the Moroccan Countryside: The Transformation of the Goums, 1934–1942," *The Maghreb Review* (1997); "Lyautey, Mangin and the Shaping of French Military Strategy in Morocco, 1912–1914," in *Proceedings of the Nineteenth Meeting of the French Colonial Historical Society* (1994); and "A Moroccan St.-Cyr," *Middle Eastern Studies* (1992). He is currently teaching at the Massachusetts Institute of Technology.

Herbert M. Hart, Colonel, U.S. Marine Corps, retired, was an infantry officer, who during thirty-seven years in the Marine Corps also served in several assignments related to officer selection and training. These included: publicity officer for officer programs, 1954–57; officer-in-charge of the Officer Selection Office, Chicago, Illinois, 1959–62; operations officer, The Basic School, Quantico, Virginia, 1971; member, Marine Corps Education Study Panel, 1971–72; academic director, Marine Corps Command and Staff College, 1972–73; and head of the Professional Education Branch, Director of Naval Education and Training, Department of the Navy, 1977–78. During 1973–77, Colonel Hart headed the Marine Corps Historical Branch, and when he retired in 1981 was Director of Public Affairs for the Marine Corps. After retiring from the service, Colonel Hart was Director of Public Affairs of the Reserve Officers Association and editor of its monthly *National Security Report.* Presently, he is Executive Director of the Council on America's Military Past, edits its monthly newspaper, and is writing a five-volume series on historic forts throughout the United States.

William R. Heaton is Academic Coordinator in the U.S. Central Intelligence Agency's Center for the Study of Intelligence, and is an estimates officer with the U.S. National Intelligence Council. He is also a colonel in the U.S. Air Force Reserve and serves as Reserve attaché to the People's Republic of China. Additionally, he is an adjunct associate professor of politics, Catholic University, Washington, D.C. Dr. Heaton received his B.A. and M.A. from Brigham Young University and his Ph.D. from the University of California, Berkeley. Formerly he was Professor of National Security Affairs at the U.S. National War College, 1981–84, and an associate professor of political science, U.S. Air Force Academy, 1973–78. He is the co-author of two books, *The Politics of East Asia* (1978) and *Insurgency in the Modern World* (1980), and has also published more than fifty articles on Chinese and East Asian political and military affairs in scholarly journals and magazines, including a chapter on Chinese defense policy for the Air Force Academy-sponsored *Defense Policies of Nations* (3d ed., 1994).

Holger H. Herwig received his B.A. from the University of British Columbia in 1965, and his M.A. and Ph.D. from the State University of New York at Stony Brook in 1967 and 1971, respectively. He is currently Professor of History at the University of Calgary. He has published numerous books and articles on military history including: *The German Naval Officer Corps A Social and Political History, 1890–1918* (1973); *Politics of Frustration: The United States in German Strategic Planning, 1889–1941* (1976); *"Luxury" Fleet: The Imperial German Navy, 1888–1918* (1980); *Biographical Dictionary of World War I* (co-author, 1982); *Germany's Vision of Empire in Venezuela, 1871–1914* (1986); *Wolfgang Wegener: The Naval Strategy of the World War* (1989); and *The First World War: Germany and Austria-Hungary, 1914–1918* (1997).

I. B. Holley, Jr., Major General, U.S. Air Force Reserve, retired, is Emeritus Professor of History at Duke University. He received his B.A. from Amherst College in 1940 and his M.A. in 1942 and Ph.D. in 1947 from Yale University. During World War II he served in the Army Air Forces as an aerial gunner. After the war, he continued in the U.S. Air Force Reserve, retiring in February 1981 as a major general. Professor Holley served as chairman of the Secretary of the Air Force's Advisory Committee on History for ten years. He now serves on the advisory boards for *Air Power History, Air Power Journal,* and the *Air Force Journal of Logistics.* His publications include: *Ideas and Weapons* (1953); *Buying Aircraft: Air Materiel Procurement for the Army Air Forces* (1964); and *General John M. Palmer, Citizen Soldiers and the Army of a Democracy* (1982). In 1974, Professor Holley delivered the Harmon Memorial Lecture at the U.S. Air Force Academy entitled "An Enduring Challenge: The Problem of Air Force Doctrine."

David R. Jones was born in Baltimore, Maryland, and raised in Halifax, Nova Scotia, Canada. Educated at Dalhousie, Duke, and Oxford Universities, he holds a doctorate in modern Russian history. From 1979 to 1991 he directed the Russian Micro-Project at Dalhousie University, and from 1988 to 1991 also held the chair of Military and Strategic Studies at Acadia University. During 1991–92 he was a Secretary of the Navy Fellow and professor at the U.S. Naval War College. The past editor of *The Soviet Armed Forces Review Annual* (vols. 1–11), at present he is editor of the *Military-Naval Encyclopedia of Russia and Eurasia* and an adjunct professor with Dalhousie University, Troy State University, and the Mountbatten Centre of the University of Southhampton, United Kingdom.

Eugenia C. (Jennie) Kiesling, a specialist in twentieth-century French military history and the history of military strategy and doctrine, is an associate professor of military history at the U.S. Military Academy. She received a B.A. in history (and rowing) from Yale University and an M.A. in ancient history (and rowing) from the Oxford University. Her Ph.D. in modern European military history (and bicycle racing) is from Stanford University. After a year as a Ford Post-doctoral Research Fellow at Harvard University's Center for International Affairs, she taught for six years at the University of Alabama, including one year as a Leverhulme Commonwealth Research Fellow at the University of Southampton, United Kingdom. She is the author of *Arming Against Hitler: France and the Limits of Military Planning* (1996), and an abridged translation from the French of Adm. Raoul Castex's six-volume *Strategic Theories* (1993).

Charles E. Kirkpatrick is a native of North Carolina. He earned a B.A. cum laude with honors in history from Wake Forest in 1969, an M.A. in German history the following year, and a Ph.D. in modern history in 1987 from Emory University where he was a Ford Fellow. He also holds diplomas from the U.S. Army Air Defense Artillery School, the U.S. Naval War College, the U.S. Army Command and General Staff College, and the U.S. Defense Language Institute. A Distinguished Military Graduate of ROTC, he was a career air defense artilleryman who commanded units in Germany and in Florida, was an assistant professor of history at the U.S. Military Academy, taught tactics and combined arms at the Air Defense Artillery School, and was a historian at the U.S. Army Center of Military History. He is presently historian for V Corps in Heidelberg, Germany, where in 1994 he also served as chief historian for the U.S. Army's World War II commemorative activities in Europe. He is the author of *Archie in the A.E.F.: The Creation of the Antiaircraft Service of the United States Army, 1917–1918* (1984); *An Unknown Future and a Doubtful Present: Writing the Victory Plan of 1941* (1991); and a number of articles and chapters in anthologies.

R. A. (Tony) Mason, Air Vice-Marshal, Royal Air Force, retired, is Professor of Aerospace Security in the Department of Political Science and International Studies, University of Birmingham. He holds degrees from St. Andrews, London, and Birmingham Universities. He is also a graduate of the U.S. Air Force Air War College and of the Royal Air Force Staff College. From 1969 to 1972, he served on exchange duty in the Department of History at the U.S. Air Force Academy. His last appointment before retiring from the Royal Air Force in 1989 was Air Secretary, with responsibilities that included the identification and preparation of officers for high command. For twenty years he has published and lectured worldwide on air power and international security. His ninth book, *Air Power: A Centennial Appraisal,* was published in 1994. His next, *Air Power Roles and the Technology Revolution,* will be published in 1998.

Alfred W. McCoy is Professor of Southeast Asian History at the University of Wisconsin–Madison. Educated at Columbia (B.A.), the University of California–Berkeley (M.A.), and Yale (Ph.D.), he has spent the past quarter-century writing about modern Southeast Asian history, with a focus on two topics—the political history of the modern Philippines and the politics of opium in the Golden Triangle. Professor McCoy is the author of several books on Philippine history, two of which have won that country's National Book Award—*Philippine Cartoons* (1985) and *Anarchy of Families* (1994). Recently he has completed a book entitled *Closer Than Brothers: Manhood at the Philippine Military Academy* to be published by Yale University Press in 1999. His *Politics of Heroin in Southeast Asia* (1972) has remained in print for a quarter-century and has been translated into eight languages. His study of Australia's heroin problem, *Drug Traffic: Narcotics and Organized Crime in Australia* (1980), became a national bestseller. In addition to his publications, Professor McCoy serves as a correspondent for the *Observatoire Geopolitique des Drogues* (Paris) and has given papers on the international drug traffic at conferences in Paris (1992), Palermo (1993), and Hobart (1996). He has also served as expert witness and consultant to the Australian Royal Commission of Inquiry into Drugs; the South Australian Parliament; the Board of Inquiry into Casinos, Victoria State Parliament; and the U.S. Deputy Assistant Secretary of Defense for Drug Enforcement Policy and Support.

Vance O. Mitchell is Senior Historian at EAST, Inc. of Chantilly, Virginia. A retired U.S. Air Force officer whose military career was primarily in reconnaissance and strategic airlift, he flew 107 combat missions in Southeast Asia, primarily in the AC-119K, was an instructor of navigation, and served as a historian at Air Force Headquarters. Dr. Mitchell is the author of *Air Force Officers: Personnel Policy Development, 1944–1974* (1996), and *The History of Air Force Intelligence, 1946–1963,* currently being declassified for publication. He is presently writing a second volume that will carry the Air Force air intelligence story through the Gulf War. He has given papers at symposiums sponsored by the U.S. Strategic Air Command, the Society for History in the Federal Government, and by the U.S. Air Force commemorating its 50th anniversary. Dr. Mitchell holds a B.A. from Texas Christian University, an M.A. from California State University, Sacramento, and a Ph.D. in American history from the University of California, Riverside.

Michael S. Neiberg is an assistant professor of history at the U.S. Air Force Academy. He received his B.A. from the University of Michigan and his M.A. and Ph.D. from Carnegie Mellon University. Dr. Neiberg teaches world and military history with a special interest in the interaction of military systems with the societies they serve. He is the co-author of two book-length studies on the history of women in the American military, *The Unwelcome Decline of Molly Marine* (1994) and *A History of Sexual Harassment in the U.S. Military* (1996). His most recent book, *Between Gun and Gown: ROTC and the Ideology of American Military Service, 1916–1980,* will be published by Harvard University Press in 1999.

Frederick M. Nunn is Professor of History and International Studies, and Vice Provost for International Affairs at Portland State University, Oregon. His graduate degrees (M.A., Portuguese; Ph.D., Latin American history and literature) are from the University of New Mexico. Professor Nunn is the author of numerous books, articles, and chapters on modern Chilean history and politics, military-civilian relations in Latin America, and fiction and history. His most recent book, *The Time of the Generals: Latin American Professional Militarism in World Perspective* (1992), received the 1994 McGann Prize of the Rocky Mountain Council for Latin American Studies. He has also been the recipient of American Philosophical Society, American Council of Learned Societies/Social Science Research Council, Guggenheim, and Fulbright grants and fellowships for research in Argentina, Brazil, Chile, Colombia, Mexico, Peru, Germany, and England. His current research deals with military-civilian relations in Latin America and Europe in the post–Cold War era.

Carl W. Reddel, Colonel, U.S. Air Force, has been Permanent Professor and Head of the Department of History, U.S. Air Force Academy since 1982. A specialist in Russian area studies and historiography, Colonel Reddel was educated at Drake University (B.S. Ed.), Syracuse University (M.A.), and Indiana University (Ph.D.). He has also studied at the Institute for the Study of the U.S.S.R. (Munich), undertaken research at the U.S. Army Russian Institute (Garmisch, Germany) and Moscow State University (Moscow), and was a post-doctoral fellow under Professor John Erickson at the University of Edinburgh. His publications include: "Russia Today: Leaving the Communist Period of Russian History," in *The New World Order in Historical Perspective* (1993); "The Competition: The Nature and Purpose of Soviet Power," in *American Defense Policy* (6th ed., 1990); "The Soviet View of Human Resources in

War," in *The Soviet Union: What Lies Ahead?* (1985); "The Future of the Soviet Empire: A Historical Perspective," *Air University Review* (1982); and "Transition from Peacetime to Wartime," in *The Soviet Military District in Peace and War* (1979). He also edited *Transformation in Russian and Soviet Military History* (1990), and contributed the introductory essay to vol. 1 of S. M. Soloviev's *History of Russia* (forthcoming).

Gunther E. Rothenberg is currently Professor of Military History, Purdue University, Indiana. He served at various times with the British Army, the pre-state Israeli forces, the U.S. Army, and the U.S. Air Force. After studying at the University of Chicago, he earned his Ph.D. at the University of Illinois in 1958 and has taught at Southern Illinois University, the University of New Mexico, and since 1973, at Purdue University. Currently he divides his time between Purdue and Monash University, Melbourne, Australia. In 1985 he taught at the Australian Royal Military College, Duntroon, and has lectured at advanced U.S. Army, Marine Corps, and Air Force schools. His publications include: *The Austrian Military Border in Croatia, 1522–1740* (1960); *The Military Border in Croatia, 1740–1881* (1966); *The Army of Francis Joseph* (1976); *The Art of Warfare in the Age of Napoleon* (1978); *The Anatomy of the Israeli Army* (1979); and *Napoleon's Great Adversaries* (1982). Also, he contributed two essays to *Makers of Modern Strategy* (1986) and has published articles in a number of historical journals.

David A. Smith, Colonel, U.S. Air Force Reserve, retired, is a research fellow at the Logistics Management Institute, McLean, Virginia. He received a B.S. in engineering with distinction from the U.S. Naval Academy in 1957, and an M.S. in management from Rensselaer Polytechnic Institute, Troy, New York, in 1966. Additional education includes the Industrial College of the Armed Forces, the Federal Executive Institute, and the Defense Systems Management School at the U.S. Navy Post Graduate School, Monterey, California. He has worked in the areas of manpower, personnel, and training policy analysis since 1966 in various capacities, including as an active-duty Air Force officer, as a civilian senior executive, and since 1988, as a private consultant. Selected assignments include the Office of the Secretary of Defense, the Executive Office of the President, and the U.S. Army Science Board. Colonel Smith has written and published a wide range of studies and papers on manpower, personnel, training, and service academy education issues including an article, "U.S. Naval Academy—Where to in the 21st Century?" U.S. Naval Institute *Proceedings* (1994).

Brian R. Sullivan received his B.A., M.A., and Ph.D. from Columbia University in 1967, 1971, and 1984, respectively. He was an active-duty U.S. Marine Corps officer from 1967 to 1970, serving in Vietnam in 1968–69. He taught history at Yale University, 1984–88; strategy at the U.S. Naval War College, 1988–91; and was a senior research professor in the Institute of National Strategic Studies, U.S. National Defense University, 1991–96. During the Gulf War, he served as an advisor on psychological operations and deception for the U.S. Assistant Secretary of Defense for Special Operations and Low Intensity Conflict. Dr. Sullivan has published numerous articles on Italian military and naval history, and on the history of Italian Fascism. He is the co-author of *Il Duce's Other Woman* (1993), and also editor of a forthcoming book from the Naval Institute Press: Romeo Bernotti, *Fundamentals of Naval Tactics and Fundamentals of Naval Strategy*. He is currently writing a book for the U.S. Space Command on strategic theory for war in space.

Tim H. E. Travers received his B.A. and M.A. from McGill University in 1963 and 1967, respectively, and his Ph.D. from Yale University in 1970. He is currently Professor of History at the University of Calgary. He has published widely on the effects of technology on warfare including: *How the War Was Won, Command and Technology in the British Army on the Western*

Front, 1917–1918 (1992); *The Killing Ground: The British Army, the Western Front and the Emergence of Modern Warfare, 1900–1918* (1987); and *Men at War: Politics, Technology and Innovation in the 20th Century* (co-editor, 1982). He has also recently contributed "Command and Leadership Styles in the British Army: The 1915 Gallipoli Model," *Journal of Contemporary History* (1994); "The Army and the Challenge of War 1914–1918" in *The Oxford History of the British Army* (1994); and "When Technology and Tactics Fail: Gallipoli 1915" in *Tooling for War* (1996).

Mark von Hagen succeeded Professor Richard E. Ericson as Director of the Harriman Institute, Columbia University in July 1995. After graduating from the Georgetown University School of Foreign Service in 1976 (B.S. in international relations), he received an M.A. in Slavic languages and literature in 1978, an M.A. in history from Stanford University in 1980, and a Ph.D. in history and humanities from Stanford in 1985. He is the author of *Soldiers in the Proletarian Dictatorship: The Red Army and the Soviet Socialist State, 1917–1930* (1990), and has published various articles on the Soviet military as well as on the cultural and intellectual history of Russia. His research interests include modern Russian and Soviet history, especially social and cultural history, the Red Army and non-Russian nationalities and nationality policy. He currently holds grants from the National Endowment for the Humanities, the Alexander von Humboldt Foundation, and the Ford Foundation. His latest book is *After Empire: Multiethnic Societies and Nation-Building: The Soviet Union and the Russian, Ottoman, and Habsburg Empires* (co-editor, 1997).

Malham M. Wakin, Brigadier General, U.S. Air Force, retired, currently serves as the Maj. Gen. William Lyon Professor of Professional Ethics at the U.S. Air Force Academy. He received his B.A. from Notre Dame in 1952, M.A. in 1953 from the State University of New York in Albany, and Ph.D. in philosophy from the University of Southern California in 1959. He was commissioned in 1954 and completed navigation school in 1955. His first flying assignment was with the 63rd Air Rescue Squadron at Norton Air Force Base, California. At the Air Force Academy, General Wakin served as an instructor, assistant professor, associate professor, and professor. He was appointed a permanent professor in 1964. General Wakin has authored numerous articles on ethics, leadership, and the military profession which have appeared in a number of professional and scholarly journals and as chapters in several books. He has also authored, co-authored, or edited four books: *War, Morality, and the Military Profession* (1986), *The Viet Cong Political Infrastructure* (1968), *Introduction to Symbolic Logic* (1969), and *The Teaching of Ethics in the Military* (1982).

Index